HUNTINGTON'S DISEASE

Volume 22 in the Series

Major Problems in Neurology
LORD WALTON OF DETCHANT, TD, MD, DSc, FRCP
PROFESSOR CP WARLOW, BA, MB, BChir, MRCP, MD, FRCP
Consulting Editors

OTHER MONOGRAPHS IN THE SERIES

Barnett, Foster and Hudgson: **Syringomyelia**
Dubowitz and Brooke: **Muscle Biopsy: A Modern Approach**
Pallis and Lewis: **The Neurology of Gastrointestinal Disease**
Hutchinson and Acheson: **Strokes**
Gubbay: **The Clumsy Child**
Hankinson and Banna: **Pituitary and Parapituitary Tumours**
Behan and Currie: **Clinical Neuroimmunology**
Cartlidge and Shaw: **Head Injury**
Lisak and Barchi: **Myasthenia Gravis**
Johnson, Lambie and Spalding: **Neurocardiology: The Interrelationships between Dysfunction in the Nervous and Cardiovascular Systems**
Parkes: **Sleep and its Disorders**
Hopkins: **Headache Problems in Diagnosis and Management**
Wood and Anderson: **Neurological Infections**
Critchley: **Neurological Emergencies**
Ross: **Syncope**
Donaldson: **Neurology of Pregnancy** SECOND EDITION
Porter: **Epilepsy: 100 Elementary Principles** SECOND EDITION
Harper: **Myotonic Dystrophy** SECOND EDITION
Harper: **Huntington's Disease**

HUNTINGTON'S DISEASE

Edited by

PETER S. HARPER, MA, DM, FRCP
Professor of Medical Genetics
University of Wales College of Medicine
Consultant Physician and Consultant in Medical Genetics
University Hospital of Wales
Cardiff, UK

W. B. Saunders Company Ltd   London   ·   Philadelphia
Toronto   ·   Sydney   ·   Tokyo

W. B. Saunders Company Ltd 24–28 Oval Road
London NW1 7DX, England

The Curtis Center
Independence Square West
Philadelphia, PA 19106–3399, USA

55 Horner Avenue
Toronto, Ontario M8Z 4X6, Canada

Harcourt Brace Jovanovich Group
(Australia) Pty Ltd
30–52 Smidmore Street
Marrickville, NSW 2204, Australia

Harcourt Brace Jovanovich Japan Inc.
Ichibancho Central Building, 22–1 Icibancho
Chiyoda-ku, Tokyo 102, Japan

© 1991 W. B. Saunders Company Ltd

All rights reserved. No part of this publication may be reproduced, stored in a retrieval system or transmitted, in any form or by any means, electronic, mechanical, photocopying or otherwise, without the prior permission of W. B. Saunders Company Ltd, 24–28 Oval Road, London NW1 7DX, UK.

British Library Cataloguing in Publication Data is available

ISBN 0–7020–1538–5

This book is printed on acid-free paper

Typeset by Paston Press, Loddon, Norfolk
Printed in Great Britain by Mackays of Chatham PLC, Chatham, Kent

George Huntington (1850–1916). Reproduced by courtesy of the Wellcome Institute for the History of Medicine, from Watson LA. (ed) (1896) 'Physicians and Surgeons of America'. Concord: Republican Press Association.

This book is dedicated to patients with Huntington's disease in Wales and elsewhere, and to their families.

Contents

Authors		ix
Series editor's foreword		xi
Preface		xiii
1	Introduction: a historical background Peter Harper and Michael Morris.	1
2	The clinical neurology of Huntington's disease Oliver Quarrell and Peter Harper.	37
3	Psychiatric aspects of Huntington's disease Michael Morris.	81
4	The natural history of Huntington's disease Peter Harper.	127
5	The neurobiology of Huntington's disease Oliver Quarrell.	141
6	Social and psychological aspects of Huntington's disease Audrey Tyler.	179
7	Management and therapy Michael Morris and Audrey Tyler.	205
8	The Epidemiology of Huntington's disease Peter Harper.	251
9	Genetic aspects of Huntington's disease Peter Harper.	281
10	Molecular genetic approaches to Huntington's disease Duncan Shaw and Sandra Youngman.	317
11	Genetic counselling in Huntington's disease Peter Harper and Audrey Tyler.	337
12	Predictive tests in Huntington's disease Peter Harper, Michael Morris and Audrey Tyler	373

Appendices
1 Huntington's disease research in Wales. A chronological list of publications — 415
2 Criteria for quantified staging of functional capacities — 418
3 Voluntary societies concerned with Huntington's disease — 420
4 Drugs used to treat chorea — 422
5 Cardiff presymptomatic testing protocol for HD — 426

Index — **429**

Authors

Peter S. Harper MA, DM, FRCP
Professor of Medical Genetics,
Consultant Physician and Consultant in Medical Genetics,
Institute of Medical Genetics,
University of Wales College of Medicine and University Hospital of Wales, Cardiff, UK.

Michael R. Morris MB, MSc, MRCPsych
Clinical Research Fellow, Honorary Senior Registrar in Psychiatry,
Institute of Medical Genetics,
University of Wales College of Medicine and University Hospital of Wales, Cardiff, UK.

Oliver W. J. Quarrell BSc, MD, MRCP
Clinical Research Fellow, Senior Registrar in Medical Genetics,
Institute of Medical Genetics,
University of Wales College of Medicine and University Hospital of Wales, Cardiff, UK.

Presently Consultant Clinical Geneticist, Centre for Human Genetics, University of Sheffield, Sheffield, UK.

Duncan J. Shaw PhD
Senior Lecturer in Molecular Genetics,
Institute of Medical Genetics,
University of Wales College of Medicine and University Hospital of Wales, Cardiff, UK.

Audrey Tyler BA, MSc
Research Social Worker in Medical Genetics,
Institute of Medical Genetics,
University of Wales College of Medicine and University Hospital of Wales, Cardiff, UK.

Sandra Youngman PhD
Research Officer in Molecular Genetics,
Institute of Medical Genetics,
University of Wales College of Medicine and University Hospital of Wales, Cardiff, UK.

Presently Post-doctoral Research Fellow, Imperial Cancer Research Foundation, London, UK.

Series Editor's Foreword

This welcome addition to our series of volumes *Major Problems in Neurology* is to be greatly welcomed and represents in my view an outstanding achievement by Professor Peter Harper and his colleagues. Ever since Huntington's disease was so elegantly and concisely described by George Huntington, a young Long Island general practitioner, in 1872, this tragic condition has continued to exercise the interest of generations of neurologists worldwide. Tragic it certainly is, since neither the distressing and disabling involuntary movements that it produces nor the progressive intellectual decline that occurs eventually in most cases can be significantly influenced by any form of medical or surgical treatment, except temporarily, and the progressive nature of the affliction imposes a heavy and increasing burden upon other family members and upon society. The tragic nature of the condition is compounded by the fact that, except in the very rare juvenile and rigid form, the condition rarely develops until middle life and, being due to an autosomal dominant gene (so that each offspring of an affected individual has a 50:50 chance of being affected), most patients have passed through their reproductive life before being aware as to whether or not they are going to develop the disease.

However, as this book clearly demonstrates, advances in molecular biology have brought us much nearer to identification of the causal gene, and closely linked markers are already available which help in predicting whether or not an individual is likely to be carrying the gene, at least in many families.

I say that this book represents a remarkable achievement because of its sheer comprehensiveness and because of the outstanding clarity and sensitivity of the writing. As one would expect, the historical introduction by Peter Harper and Michael Morris is scholarly and complete, and the chapters on the clinical neurology, natural history, epidemiology, genetics and genetic counselling, in which Peter Harper has been either the sole author or a co-author, demonstrate to the full his encyclopaedic knowledge and experience of this disease gleaned over many years of careful study in South Wales. But in addition his colleagues, Oliver Quarrell, Michael Morris, Audrey Tyler, Duncan Shaw and Sandra Youngman, have contributed other chapters in which the psychiatric aspects of the condition, its neurobiology, its social aspects and its management are thoroughly and sensitively considered. Similar careful and comprehensive attention is paid to linkage, molecular genetics and predictive testing, with all of its psychological and social implications.

There have been other publications on Huntington's disease in the past, but having read this monograph, I believe it to be both unique and exceptional, not least in the range of topics relevant to the disease which have been given such careful and detailed consideration, but because of the fine sense of the art of medicine which shines through its pages. It is in my view the definitive work on Huntington's disease against which subsequent publications will inevitably be judged. Professor Harper and his colleagues deserve our congratulations and our gratitude.

Lord Walton of Detchant

Preface

This book had its origins almost 20 years ago, when one of us (PSH) returned from America to develop medical genetics in Cardiff, Wales, after 2 years working at Johns Hopkins Hospital, Baltimore, where colleagues were involved in a project attempting to map the gene for Huntington's disease (HD). Considerable numbers of referrals of HD families for genetic counselling in Cardiff soon made it clear that the disease was frequent in South Wales, an impression reinforced by a series of family records kept by a neurologist colleague in Cardiff, Dr Charles Wells, who suggested that there might be as many as 20 families in the area and encouraged a systematic survey.

Twenty years later we know of over 200 HD families in Wales and the number is probably still incomplete. If we had known the scale of the project we were undertaking when we were starting, we might well not have had the courage to begin! However, begin we did, and over the years a group of workers has evolved, with a combination of skills that has been able to make a significant contribution to our understanding of HD, as well as providing a service to our families locally.

The decision to write this book came from the realization that our group had a particularly wide range of expertise, including psychiatric and genetic counselling experience, as well as molecular genetic skills and knowledge of the social effects of HD. Four of us had written theses on the disorder, with a considerable amount of information gained and collected that we felt could be valuable to others. Urgency was given to the task by the realization that some of this expertise would soon disperse, with two authors moving to senior posts elsewhere; if a book was to be written, the opportunity had to be seized.

Although we have placed a particular emphasis on the genetic aspects of HD in view of our own work and interests, we have tried to make this a balanced book. The contents list of chapters indicates which authors have been principally responsible for writing the different sections but we should emphasize that they have all been written in close collaboration with the group as a whole, with criticism, additions and amendments often being made by others. This close consultation, only possible within a single working group, will we hope give a unity of style and content that is often lacking in a multi-author book.

We had nursed a hope that the HD gene might itself have been isolated in time for inclusion in the book, but while this appears very close at the time of writing (March, 1991), it has not yet happened and its description must await a future edition. In retrospect this is perhaps fortunate, as isolation of the gene will lead to a complete reassessment of much of the older work and will radically change both experimental strategies and the practical approach to genetic counselling and prediction. We hope that the present account will give readers an up-to-date assessment of our current knowledge and that it will prove of practical use to clinicians involved with HD families, whether neurologists, psychiatrists, or from other specialities, to geneticists and others in relation to genetic counselling and predictive testing, and to the many scientists in HD research who wish to know about the broader aspects of the disease they are studying.

Many thanks are owed for a book such as this, which encompasses work done over such a prolonged period. Here we can only mention a few of the most conspicuous.

First, to our clinical colleagues in Cardiff, especially those in the Departments of Neurology and Psychiatry, we owe a considerable debt for their close and continuing partnership over a long period. Within the Institute and Department of Medical Genetics there have been numerous major contributors over the years, notably Mrs Pat Jones, Nursing Sister, who has been the mainstay of our service to the HD families in South Wales, along with Mrs Morag Nordin and other fieldworkers. Dr David Walker undertook the original clinical study, while Dr Robert Newcombe was responsible for much of the statistical analysis of various studies up to the present, including the life-table data that have been re-analysed for this book. The HD register has seen the involvement of a number of colleagues, including Valerie McBroom, Mansoor Sarfarazi, Jeff Wolak, Iain Fenton and Lodewijk Sandkuijl. In the laboratory, Russell Snell, Shelley Rundle, Tracey Ford and Nicole Datson have all been closely involved with our molecular genetics research, while Laz Lazarou and Linda Meredith have been responsible for the diagnostic analyses in predictive testing.

The advice of Dr Benno Müller-Hill has been particularly valuable in handling the difficult and sensitive area of HD in Nazi Germany, while the help of Drs Manuela Koch, Kathy Davies and Nicole Datson in obtaining and translating material from the German literature is much appreciated.

A number of colleagues have been particularly helpful in reading and criticizing parts of the manuscript, including Drs Niall Quinn, David Turner, David Ball and Raymund Roos, Professors Mark Wiles, Peter McGuffin, Michael Conneally and Alan Richens, and Mrs Shirley Dalby. There is no doubt that their comments have improved the end result considerably, though it goes without saying that any remaining defects are our own responsibility.

A number of illustrations have been provided by colleagues where our own material seemed inadequate; they are mentioned individually in the text but we have greatly appreciated their generosity, as we do the permission of those who have allowed us to quote unpublished data. Dr Raymund Roos of Leiden and Dr J. Neal of Cardiff kindly provided and organized the neuropathological material on HD. Our own department of medical illustration under Professor Ralph Marshall, has been unfailingly helpful in turning our own efforts into high quality drawings and prints.

We thank our hard-pressed secretarial staff, especially Michelle Thomas and Karen Evans, for their help in producing the manuscript and in the correspondence involved; as well as the staff of Baillière Tindall and Saunders for their efficiency and their tolerance in allowing late alterations.

Our work has received financial support from many sources over the years, but the Huntington's Disease Association, Mental Health Foundation, Medical Research Council, Wellcome Trust, Department of Health and Welsh Office, and the Hereditary Disease Foundation deserve special mention.

Finally, this book could not have been written were it not for the HD patients and families in Britain, and especially in Wales, on whom our own studies have been based. Working with them over the years has been a privilege for us all, and it is appropriate that it should be to them that this book is dedicated.

1

Introduction: a Historical Background

INTRODUCTION

Huntington's disease (HD) is a disorder of the central nervous system and is thus rightly classified as a neurological disorder; yet it is in many ways a condition whose effects extend across many fields and which is encountered by clinicians in widely differing specialities. Research scientists in increasing numbers are also involved with HD as biochemists, neuropharmacologists and molecular biologists. In both clinical and basic science aspects this disorder serves as a model from which we can learn much about other progressive genetic disorders of the nervous system.

This book attempts to look at HD from a number of different angles and to be of use to both clinicians and scientists who are involved in work on this or allied conditions. Before introducing the disorder further, it is thus worth outlining how the rest of the book is organized.

Chapter 2 deals with the clinical aspects of HD as they may present to the practising clinical neurologist, describing the main patterns of neurological involvement, their investigation and their differential diagnosis. Most HD patients will first be seen by physicians in primary care or general internal medicine, so that in writing this chapter we have not assumed that all readers will have expertise in neurology. The following chapter is in many ways complementary, written from the viewpoint of the psychiatrist, but hopefully useful also to other clinicians who will need to know the range of potential psychiatric problems of patients under their care. While only a minority of HD patients have psychiatric symptoms of a degree that necessitates specialist care by a psychiatrist, there is no doubt that the involvement of a psychiatrist or psychologist with special interest and experience in the disorder can be of the greatest help to both patient and family. Early recognition of the psychiatric aspects of the disorder may also avoid or postpone the need for acute or long-term hospital care.

Chapter 4 describes the natural history of HD, particularly in relation to its range of age at onset and at death. We hope that future workers will no longer have to observe its unaltered course and that the extensive studies that have been done on the natural history of HD will be able to act as a comparison with the treated state, not, as at present, the unmodified course that we still see. In Chapter 5 we try to place the clinical observations of HD into the framework of what is known about normal and disordered structure and function of the brain, in particular, those areas of the brain most involved. It will be quite clear from our descriptions of the neuropathology, and from the extensive neurochemical and pharmacological research, that we have a long way to go before we can fully link the clinical features with their molecular and cellular basis, yet this is one of the areas where possibilities of progress seem greatest.

Two chapters (6 and 7) on the social aspects of HD and on its management deal with these areas in considerable detail. Some readers may feel that we have overemphasized these at the expense of the more medical problems; we would argue strongly that in terms both of what we can offer as help for HD patients and families, and the problems and unmet needs that they perceive, the social aspects are at present more important than the strictly medical ones. We would venture to suggest that neurologists have (with honourable exceptions) underestimated and at times, even ignored the social aspects of the disorders with which they are involved for far too long. Perhaps the social bias in this book may help to change the image and practice of the speciality in this regard!

Chapter 8 discusses the epidemiology of HD, a fascinating topic which in many ways mirrors the major migrations of people from European countries across the world, but which also points out that HD has a multifocal origin in the major populations studied. We expect soon to be able to revise this account in terms of the origin and spread of individual identifiable mutations in the HD gene.

The final four chapters deal with various aspects of the genetics of HD – here the authors can without hesitation be rightly accused of having a personal bias! However, by our close involvement in much of the work described, we hope that non-geneticists will share in our fascination with this topic. HD is still providing surprises in its inheritance pattern; the paternal transmission of juvenile cases promises to provide a model for non-Mendelian factors acting in single gene disorders, while the molecular genetics research towards isolating the gene provides a particular combination of excitement and frustration, though the latter will be dispelled once the goal has been reached.

Our own experience in providing a genetics service for HD families over many years has encouraged us to provide a full account of the issues involved in genetic counselling and in predictive testing. We have tried here to point out the pitfalls and problems involved, as well as indicating our concern for future (and past) abuse in this area. Once again HD can be regarded as providing an important model for many other progressive genetic disorders where the new techniques of molecular genetics will provide powerful applications.

Having outlined the topics that are to come in later chapters, we must now return to the beginning, more than a century ago, and place HD in its historical context. There can be no better way of doing this than by outlining the original description that gave the disorder its name, and which has served as the foundation for all subsequent studies.

GEORGE HUNTINGTON AND HEREDITARY CHOREA

The description by George Huntington in 1872 of the disease that has subsequently borne his name is one of the most remarkable in the history of medicine. Subsequent workers, from Osler through to the present, have remarked on its clarity, brevity and comprehensiveness. It was not the first description of the disorder, as will be seen, but it stands out as the first full delineation of the condition as a specific disease entity, quite separate from other forms of chorea.

Huntington's paper was given before the Meigs and Mason Academy of Medicine at Middleport, Ohio, on 15 February 1872 and published only 2 months later in the Philadelphia journal, *The Medical and Surgical Reporter* (Fig. 1.1); the first part dealt with chorea in general, and does not contain any particularly original information. The final part, occupying only a single page of printed text, is strikingly different in character; its vividness and authenticity of clinical detail come across as strongly today, more than a century later, as they did to contemporary readers such as Osler. One can do no better than quote it in full here.

THE

MEDICAL AND SURGICAL REPORTER.

No. 789.]　　　　PHILADELPHIA, APRIL 13, 1872.　　　　[Vol. XXVI.—No. 15.

ORIGINAL DEPARTMENT.

Communications.

ON CHOREA.

By GEORGE HUNTINGTON, M. D.,
Of Pomeroy, Ohio.

Essay read before the Meigs and Mason Academy of Medicine at Middleport, Ohio, February 15, 1872

Chorea is essentially a disease of the nervous system. The name "chorea" is given to the disease on account of the *dancing* propensities of those who are affected by it, and it is a very appropriate designation. The disease, as it is commonly seen, is by no means a dangerous or serious affection, however distressing it may be to the one suffering from it, or to his friends. Its most marked and char-

The upper extremities may be the first affected, or both simultaneously. All the voluntary muscles are liable to be affected, those of the face rarely being exempted.

If the patient attempt to protrude the tongue it is accomplished with a great deal of difficulty and uncertainty. The hands are kept rolling—first the palms upward, and then the backs. The shoulders are shrugged, and the feet and legs kept in perpetual motion; the toes are turned in, and then everted; one foot is thrown across the other, and then suddenly withdrawn, and, in short, every conceivable attitude and expression is assumed, and so varied and irregular are the motions gone through with, that a complete description of

Figure 1.1 The title page of George Huntington's 1872 paper in the *Medical and Surgical Reporter*.

And now I wish to draw your attention more particularly to a form of the disease which exists, so far as I know, almost exclusively on the east end of Long Island. It is peculiar in itself and seems to obey certain fixed laws. In the first place, let me remark that chorea, as it is commonly known to the profession, and a description of which I have already given, is of exceedingly rare occurrence there. I do not remember a single instance occurring in my father's practice, and I have often heard him say that it was a rare disease and seldom met with by him.

The *hereditary* chorea, as I shall call it, is confined to certain and fortunately a *few* families, and has been transmitted to them, an heirloom from generations away back in the dim past. It is spoken of by those in whose veins the seeds of the disease are known to exist, with a kind of horror, and not at all alluded to except through dire necessity, when it is mentioned as *'that disorder.'* It is attended generally by all the symptoms of common chorea, only in an aggravated degree hardly ever manifesting itself until *adult* or *middle* life, and then coming on gradually but surely, increasing by degrees, and often occupying years in its development, until the hapless sufferer is but a quivering wreck of his former self.

It is as common and is indeed, I believe, *more* common among *men* than women, while I am not aware that season or complexion has any influence in the matter. There are three marked peculiarities in this disease: 1. Its hereditary nature. 2. A tendency to insanity and suicide. 3. Its manifesting itself as a grave disease only in adult life.

1. Of its hereditary nature. When either or both the parents have shown manifestations of the disease, and more especially when these manifestations have been of a *serious* nature, one or more of the offspring almost invariably suffer from the disease, if they live to adult age. But if by any chance these children go through life *without* it, the thread is broken and the grandchildren and great-grandchildren of the original shakers may rest assured that they are free from the disease. This you will perceive differs from the general laws of so-called hereditary diseases, as for instance in phthisis, or syphilis, when *one* generation may enjoy entire immunity from their dread ravages, and yet in another you find them cropping out in all their hideousness. Unstable and whimsical as the disease may be in *other* respects, in *this* it is firm, it never skips a generation to again manifest itself in another; once having yielded its claims, it never regains them. In all the families, or nearly all in which the choreic taint exists, the nervous temperament greatly preponderates, and in my grandfather's and father's experience, which conjointly cover a period of 78 years, nervous excitement in a marked degree almost invariably attends upon every disease these people may suffer from, although they may not when in *health* be over nervous.

2. The tendency to insanity, and sometimes that form of insanity which leads to suicide, is marked. I know of several instances of suicide of people suffering from this form of chorea, or who belonged to families in which the disease existed. As the disease progresses the mind becomes more or less impaired, in many amounting to insanity, while in others mind and body both gradually fail until death relieves them of their sufferings. At present I know of two married men, whose wives are living, and who are constantly making love to some young lady, not seeming to be aware that there is any impropriety in it. They are suffering from chorea to such an extent that they can hardly walk, and would be thought, by a stranger, to be intoxicated. They are men of about 50 years of age, but never let an opportunity to flirt with a girl go past unimproved. The effect is ridiculous in the extreme.

3. Its third peculiarity is its coming on, at least as a grave disease, only in adult life. I do not know of a single case that has shown any marked signs of chorea before the age of thirty or forty years, while those who pass the fortieth year *without* symptoms of the disease, are seldom attacked. It begins as an ordinary chorea might begin, by the irregular and spasmodic action of certain muscles, as of the face, arms, etc. These movements gradually increase, when muscles hitherto unaffected take on the spasmodic action, until every muscle in the body becomes affected (excepting the

involuntary ones), and the poor patient presents a spectacle which is anything but pleasing to witness. I have never known a recovery or even an amelioration of symptoms in this form of chorea; when once it begins it clings to the bitter end. No treatment seems to be of any avail, and indeed nowadays its end is so well-known to the sufferer and his friends, that medical advice is seldom sought. It seems at least to be one of the incurables.

Dr. Wood, in his work on the practice of medicine, mentions the case of a man, in the Pennsylvania Hospital, suffering from aggravated chorea, which resisted *all* treatment. He finally left the hospital uncured. I strongly suspect that this man belonged to one of the families in which hereditary chorea existed. I know nothing of its pathology. I have drawn your attention to this form of chorea gentlemen, not that I considered it of any great practical importance to you, but merely as a medical curiosity, and as such it may have some interest.

It can be seen that all the cardinal features of HD are recognized in this description: the adult onset, progressive course and eventually fatal outcome; the choreic movements combined with mental impairment, and risk of suicide; even the pattern of inheritance, with 'the thread broken' once a person had gone through life without developing it. A description of this nature could only have been written by one whose observations were based on direct and continued contact with affected patients; George Huntington's role as a family doctor, following his father and grandfather to give a 78-year total period of observation, gave him this perspective to a unique degree.

The active involvement of the two older generations of the family was acknowledged by George Huntington himself (1909) and is attested by the presence of pencil notes and corrections by his father on the original manuscript (Winfield, 1908; De Jong, 1937). The paper was prepared while George Huntington was working in his father's practice before leaving for Ohio, so that it can indeed by regarded as a distillation of the observations of three generations of family doctors.

Huntington was more fortunate than many authors who have given classical descriptions of a disease; his paper was widely appreciated from the outset and soon became internationally recognized. Browning (1908a) discussed the reasons why this was so.

> There were good reasons why his paper succeeded in drawing general attention to this disorder and in securing for it permanent recognition.
> Huntington was the first to give definitely the location of his cases and thus positively establish a verifiable record.
> The abstracting of his original article by Kussmaul and Nothnagel in Virchow-Hirsch's 'Jahrbuch' for 1872.
> Thanks to the work of Friedreich on hereditary ataxia, as well as the growing interest in heredity, the time was ripe for its appreciation. Only work of unusual, incisive and wide reaching interest could attract such a share of attention.

A further factor was the interest and appreciation of Sir William Osler, Professor of Medicine at Philadelphia and then Johns Hopkins Hospital, Baltimore. Osler remarked that 'there are few instances in the history of medicine in which a disease has been more accurately, more graphically or more briefly described' (Osler, 1894).

GEORGE HUNTINGTON – LIFE AND BACKGROUND

The intimate connection between George Huntington's description of HD and his background in family practice gives a particular interest to knowing more about his life. Fortunately this had already become a subject of interest during his lifetime; the 'Huntington's number' of *Neurographs*, edited by William Browning (1908a,b,c,d) provides much detail, with a valuable biographical sketch by Winfield (1908). Huntington's ancestors came, as did those of the families he studied, from the East Anglia region of England. Simon Huntington, of Norwich, is recorded as sailing to America in 1633 with his wife and children; he died on the voyage, but a son settled in Connecticut, from where Abel Huntington, grandfather of George moved in 1797 to practise medicine in East Hampton, Long Island. Dr Abel Huntington was a distinguished physician and was the first on Long Island to perform the operation of lithotomy. He was also interested in infectious diseases and personally prepared and preserved the variola virus. His interests were not confined to medicine. He held several important public offices: in 1820, he was Presidential Elector; in 1821, he was elected New York City Senator; he was elected congressman for two terms; in 1845, he was appointed Collector of Customs for Sag Harbour; and in 1846 he was a member of the committee to revise the constitution of the State of New York. Dr Abel Huntington's son, George Lee Huntington, succeeded him

Figure 1.2 George Huntington as a young man (left) and in later life (right). Reproduced from the 'Huntington number' of *Neurographs* (Browning, 1908).

in his practice, and George Huntington the younger was born at East Hampton in 1850.

George Huntington's upbringing would seem to have been a quiet and stable one, in a respectable small town in rural surroundings. He accompanied his father on his rounds and began his medical studies with him before graduating at Columbia University, New York in 1871. In present day medical education, where originality is often stifled by excessive facts, it is salutary to note that he qualified aged 21 and wrote his paper at the age of 22.

Although George Huntington initially returned to East Hampton to practice, he soon moved to Pomeroy, Ohio, where he married, but apparently found Pomeroy 'abundantly supplied with Physicians', causing him to move again, first to Dutchess County, New York, then to Asheville, North Carolina. He had serious health problems during this time, principally asthma, but by 1903 was able again to return to Dutchess County, where he remained in practice until 1915. In this year he retired to live with his son not far away, dying in 1916 at the age of 65 years.

George Huntington never published further papers after his 1872 description, either on chorea or on other topics. This would seem to have been due not just to his health, nor to his practice commitments, but to his removal from the environment that had resulted in his single, but remarkable contribution to medicine – the presence of the patients studied by three generations of his family, suffering from the disorder which has so appropriately preserved the name of Huntington.

Perhaps the clearest insight into how George Huntington's background affected his description of the disorder is seen in an address that he gave in 1909 to the New York Medical Society.

> Over fifty years ago, in riding with my father on his professional rounds, I saw my first cases of 'that disorder', which was the way in which the natives always referred to the dreaded disease. I recall it as vividly as though it had occurred but yesterday. It made a most enduring impression upon my boyish mind, an impression every detail of which I recall today, an impression which was the very first impulse to my choosing chorea as my virgin contribution to medical lore. Driving with my father through a wooded road leading from East Hampton to Amagansett, we suddenly came upon two women, mother and daughter, both tall, thin, almost cadaverous, both bowing, twisting, grimacing. I stared in wonderment, almost in fear. What could it mean? My father paused to speak with them and we passed on. Then my Gamaliel-like instruction began; my medical education had its inception. From this point on my interest in the disease has never wholly ceased.

EARLY CONCEPTS OF CHOREA

The work of George Huntington and other early writers on hereditary chorea was rooted in a large body of pre-existing knowledge and tradition concerning chorea as a whole. Those who wish to follow the development of our knowledge of chorea in detail will find the early monographs on the subject, such as those of Osler (1904) and Huet (1889) valuable. The possible links between HD and the dancing mania of the middle ages have received much attention in previous works on HD (Bruyn, 1968; Hayden, 1981). The following note does not explore these areas in detail, but aims simply to set the later work on hereditary chorea in its context.

Little is known about movement disorders before the advent of writing. Undoubtedly, they existed in prehistorical times, but they were probably thought to be the result of evil spirits or godly retribution.

The term chorea originates from the Greek χορος, meaning 'dance'. In time, the word passed into Latin usage (*choreus*), but it was still used in a secular rather than a medical sense. It was not until the Renaissance that Paracelsus (Theophrastus Bombastus von Hohenheim, 1493–1541) classified chorea and first used the term to indicate organic disease.

Paracelsus, who was Professor at Basel, rejected Galenic psychophysiology and the theory of humours. This was revolutionary for the time and indeed the classical humours hypothesis was only eventually abandoned 200 years later in the eighteenth century. His approach to existing medical knowledge was critical and his writings were the result of direct observation of the patient.

Paracelsus classified chorea into three types. *Chorea naturalis* referred to chorea of organic or physiological aetiology and patients 'only felt an involuntary impulse to allay the internal sense of disquietude'. Although *chorea naturalis* cannot be identified with a specific neurological disorder, Bell (1934) considered that this group might have included patients with HD. Treatment for *chorea naturalis* 'must come from nature' and Paracelsus recommended a complex mixture which included *hyosciamus*.

Paracelsus believed that somatic symptoms and disease may be caused by a disturbance in the 'spiritus vitae'. He identified two types of chorea of non-organic or 'spiritual' aetiology but he concentrated on *chorea lasciva*, the term he used for the Dancing Mania of the Middle Ages.

The dancing mania was first noted in western Europe during the eleventh century and was characterized by excitement and overactivity. The most famous episode of the Dancing Mania occured at Aix-la-Chapelle in 1374. At this time the Black Death had ravaged Europe and the people lived in fear. On 24 June, which was the Feast of St John the Baptist, the peasants joined hands, 'lost all control and danced continuously (until) they fell to the ground in a state of exhaustion. They then complained of extreme oppression and groaned as if in the agonies of death, until they were swathed in cloths bound tightly around their waists, upon which they again recovered and remained free from complaint until the next attack. While dancing, they neither saw nor heard, being insensible to external impressions' (Heckler, 1844). At Strasburg in 1418, the dancing phenomenon reappeared and it was then that the name of St Vitus, a fourth century Christian martyr, was invoked to protect sufferers of the Dancing Mania. The afflicted went to a church dedicated to the saint in Zabern, Alsace.

The critical faculties of Paracelsus are much in evidence when he commented on St Vitus' Dance:

We do not wish to admit in this chapter that the saints can cause a plague or the diseases which eventually are named after them. In our opinion, such diseases have nothing to do with the works of the saints. There are so many who connect great sufferings with saints and attribute their sufferings to God rather than nature. This is idle talk. We dislike talk behind which there is no proof but mere belief. As this disease is known under the name of a saint, we do not intend to change the name; it should however be called *chorea lasciva* . . . (It) is a mere opinion and idea assumed by the imagination (Temkin *et al.*, 1941).

The cause of the Dancing Mania of the Middle Ages has not been fully explained, but Paracelsus is probably right in suggesting a significant psychological component in the aetiology. Similar outbreaks have occurred in other places subsequently. Paracelsus' main contribution was to recognize the different aetiologies of chorea. It was Sydenham who provided the first important clinical descriptions of choreic movement.

Thomas Sydenham (1624–1689) and William Harvey were the two great English physicians of the seventeenth century. So great is Sydenham's reputation that he has been called the 'English Hippocrates', 'the father of British medicine' and 'the father of chorea'. He was born in Dorset and after a military career and imprisonment, he was 'created' a Bachelor of Medicine in 1648 at Oxford. He was a great clinical observer and in 1666 published his first book *Methodus Curandi Febres*. He is mainly remembered today because of his concise description of chorea in childhood in his *Schedula Monitoria de Novae Febris Ingressa* (1686): 'He that is affected with this disease can by no means keep in the same posture for one moment . . . if a cup of drink be put into his hand, he represents a thousand gestures like jugglers before he brings it to his mouth'.

Confusion has arisen because Sydenham called chorea in childhood St Vitus' dance. The Dancing Mania was also known as St Vitus' dance so that all syndromes with chorea became known by this saint's name. It soon became necessary to separate the choreic syndromes; chorea in childhood was known as chorea minor and the Dancing Mania was known as chorea major. Nowadays, chorea minor is called Sydenham's chorea and is recognized to be associated with juvenile rheumatic fever and carditis.

DESCRIPTIONS OF HD BEFORE 1872

There can be few diseases in medicine where the description that has given the author's name to the condition is truly the first. HD is no exception, and there is no doubt that others had recognized and to some extent described the condition before George Huntington published his 1872 paper. Not surprisingly there have been a number of claims of such descriptions (De Jong, 1937; Bruyn, 1968; Stevens, 1972) but few stand up to critical examination. Thus Elliotson's (1832) mention of chronic adult chorea, noting that 'I have often seen it hereditary' might or might not have referred to HD, while Husquinet's interesting finding (1975) that the case described in 1873 by Landouzy could be traced to Belgium and linked to records from the previous century can hardly be regarded as a prior description.

The first definite record of HD was in a letter by Charles Oscar Waters (1816–1892), in 1841 and published by Dunglison in the first edition of his *Practice of Medicine* in 1842. It is worth reproducing this in detail, as it gives such a clear picture of the clinical features and natural history.

> It consists essentially in a spasmodic action of all, or nearly all, the voluntary muscles of the system, of involuntary and more or less irregular motions of the extremities, face and trunk. In these involuntary movements the upper part of the air passages occasionally participate as is witnessed by the 'cluckling' sound in the glottis and in a manifest impediment to the powers of speech. The expression of the countenance and general appearance of the patients are very much such as are described as characteristic of chorea.
>
> The disease is markedly hereditary, and is most common among the lower classes, though cases of it are not unfrequent among those who by industry and temperance, have raised themselves to a respectable rank in society. These involuntary movements of the face, neck, extremities and body cease entirely during sleep.
>
> This singular disease rarely, very rarely indeed makes its appearance before adult life, and attacks after forty-five years of age are also very rare. When once it has appeared, however, it clings to its suffering victim with unrelenting tenacity until death comes to his relief. It very rarely or never ceases while life lasts.
>
> The first indications of its appearance are spasmodic twitchings of the extremities generally of the fingers which gradually extend and involve all the voluntary muscles. This derangement of muscular action is by no means uniform; in some it exists to a greater, in others to a less extent, but in all cases it gradually induces a state of more or less perfect dementia.
>
> When speaking of the manifestly hereditary nature of the disease, I should perhaps have remarked that I have never known a case of it to occur in a patient, one or both of whose ancestors were not, within the third generation at farthest, the subject of this distressing malady.

In 1846, a further description occurred in the form of a thesis submitted to Jefferson Medical College, Philadelphia, by Dr Charles Gorman (1817–1896) entitled 'On a form of chorea, vulgarly called magrums'. Although the thesis is lost, it is mentioned in the third edition (1848) of Dunglison's book: 'an inaugural dissertation, presented before the Faculty of Jefferson Medical College of Philadelphia by Charles R. Gorman of Luzerne County, Pa., the writer states that this affection prevails also in other portions of the country. According to him, it seems to be circumscribed by neighbourhood boundaries, and to be confined to sections of the country, the inhabitants of which are intimately connected in their Social or Business relations'.

A valuable account of these early American descriptions of hereditary chorea is given by Browning (1908a,b,c,d) in the 'Huntington number' of *Neurographs*, along with the location of the families and biographical details of Waters, Gorman and other early workers.

The other independent early description that undoubtedly represented HD is that of Johan Christian Lund, written in Norwegian and only generally appreciated since relevant parts were translated by Ørbeck (1959), though recognized in Scandinavia before that. Lund was public health physician in the region of Saetersdal, Norway and wrote in his medical report for 1860:

> As recorded in the previous medical report, chorea St. Vitus (which is Lund's term for St. Vitus's Dance) seems to recur as an hereditary disease in Saetersdal. It is commonly known as the 'twitches', occasionally as the 'inherited disease'. It usually occurs between the ages of 50 and 60, generally starting with less obvious symptoms, which at times only progress slowly, without becoming violent, so that the patient's normal activities are not particularly hindered: but more often after a few years they increase to a considerable degree, so that any form of work becomes impossible and even eating becomes difficult and circuitous. The entire body, though chiefly the head, arms, and trunk, is in constant jerking and flinging motion, except during sleep, when the patient is usually motionless. A couple of the severely affected patients have during the last days of their lives become *fatui* (i.e. demented). The disease occurs in two families which are registered below. Information is not as complete as could be desired though enough to start with, as long as doctors in Saetersdalen are mindful of the disease in future. (Lund, 1860; quoted by Ørbeck, 1959.)

Lund gave further details on family members in later reports and according to Ørbeck descendants with HD still exist today. It seems that Lund was doubly unlucky in not gaining recognition, since not only was the original source restricted in readership by being written in Norwegian, but subsequent commentators on it confused the disorder with Parkinson's disease.

One further description of HD before 1872 is widely quoted, namely that of Lyon (1863) who reported three families from New York State with hereditary chorea under the title 'Chronic hereditary chorea' in the American Medical Times. Lyon's families do not seem to have been connected with those of Waters, but were also called 'megrim families' locally. Browning (1908d) traced the localization of the families to Bedford, on the New York–Connecticut border, but no subsequent clinical evaluation of the patients or their descendents has been carried out. However, a critical look at the description of these cases suggests that Lyon's families 1 and 2 (family 3 was not described in detail) are far from typical for HD and much more suggestive of benign familial chorea (see Chapter 2). So that readers can decide for themselves, these cases are quoted here.

> CASE I. – Mr A., residing in the town of ———, county of ———, N.Y., has well marked chorea, which is quite general; so that he is constantly, when awake, making irregular movements with the upper and lower extremities, facial muscles, and more or less with those of the body. This condition has existed for many years, but seems not to interfere materially with his general health, the vegetative functions being well performed. Mr. A. has two brothers and three sisters; the two brothers have themselves never had any choreal symptoms, but one of them has two children in whom well defined chorea has existed for many years; of the three sisters, two have had chorea for the most of their lives, being now past the middle age.
>
> The progenitors of Mr. A., on the male side, were perfectly free from chorea, but not so on the maternal side; his mother had well developed choreal manifestations from early life, which continued till her decease; she had also a brother who died during adult life from the severity of the disease; but to go still further, both the grandfather and great-grandfather of Mr. A., on the maternal side, had the same disorder which

we find in their children: whether collateral instances of the affection occurred in the families we are not advised.

II. – Mrs. K., of the town of ———, Ct., and a descendant from a family which has long been known and designated as migrim, had chorea for the most of her life, being about seventy-five years old at the date of death. She had a family of two sons and three daughters: of these one son and two daughters had chorea, with which disease they attained an advanced age; no satisfactory information can be readily obtained in relation to the offspring of the son and one of these daughters so affected; but the other daughter married, and had a son, who is now forty years of age, in whom chorea has exhibited itself from puberty.

In particular, the fact that the proband of family 1 is stated to have had chorea for many years, but that the condition had not interfered with his general health, while his two affected sisters had 'had chorea for most of their lives', and their mother from early life, is most unlike HD. So is the fact that the grandson in family 2, then 40 years old, had chorea since puberty. There is no mention of progression or of general deterioration apart from the one individual of family 1, who is stated to have 'died during adult life from the severity of the disease'.

Thus it would seem that Lyon's families should not be accepted as HD, but are rather characteristic of benign familial chorea. Such a situation has an interesting precedent in the myotonic disorders, where the non-progressive and much rarer myotonia congenita was described by Thomsen in 1876 considerably before the recognition of the commoner myotonic dystrophy by Steinert and by Batten and Gibb in 1909.

WILLIAM OSLER AND HUNTINGTON'S DISEASE

The rapid spread and wide distribution of knowledge concerning HD was in no small measure due to the interest that Sir William Osler (1849–1919) took in it. Osler had a lifelong interest in chorea and his monograph *On Chorea and Choreiform Affections* (1904) contains a wealth of personally collected data on rheumatic (Sydenham's) chorea. However, he included a separate chapter on hereditary chorea in this, as well as a brief section in his *Principles of Medicine* (1892) and dealt specifically with Huntington's chorea in several case reports and papers (1890, 1893, 1894). (Surprisingly and most atypically he misspells the eponym as 'Huntingdon' in some of these!)

Osler made attempts to reassess Huntington's original family during the summer of 1887 but Huntington wrote back indicating that they would not welcome this, though Osler was sent notes on these patients the following year; he was able to gain personal experience from patients seen at Johns Hopkins Hospital of both English and German origin; his case reports are a model of clarity and detailed observation and ring as true today as a century ago.

> When sitting in a chair, at ease, the arms and hands are in more or less constant irregular motion. The fingers are extended and flexed alternately; sometimes only one, sometimes the entire set. At other times the whole hand will be lifted, or there are

constant movements of pronation and supination. For half a minute or so they may be perfectly motionless. The head and trunk present occasional slow movements; in the latter more of a swaying character. The legs jerk irregularly and the feet are flexed or extended; but the movements are not so frequent as in the arms. The face in repose is usually motionless, but the lips are occasionally brought together more tightly and the chin elevated or depressed. There is an occasional movement of the zygomatic and of the frontal muscles. He puts out the tongue, with tolerably active associated movements of the face, and it is usually quickly withdrawn or rolled from side to side. It is impossible for him to hold it out for any length of time. There are no irregular movements of the palate muscles.

He walks with a curious irregular gait, displaying distinct incoordination, swaying as he goes, hesitating a moment in a step, keeping the arms out from the body and in constant motion. The legs are spread wide apart; the steps are unequal in length and he seems rather to drag the feet. He stands well with the heels close together and the eyes shut. (From Osler, 1894.)

Osler's description shows how the clinical picture of HD and its distinction from other causes of chorea were already accurately defined by this period, with recognition of both the psychiatric and neurological aspects, the progressive nature and the adult onset. Osler also recorded the pathology of the brain, but could find no very specific changes and did not note particular abnormalities of the basal ganglia.

THE SPREAD OF KNOWLEDGE ON HD

The rapid spread of awareness of George Huntington's 1872 paper has already been mentioned, but from the work on hereditary chorea discussed so far it could be imagined that most of the interest and activity was in America, not in Europe. Such an impression would be entirely wrong, for not only was the topic of chorea in general one of great interest among European physicians and neurologists, but many of the early reports and studies of HD in the last quarter of the nineteenth century were European in origin. The monograph of Petit (1970) and the review of Bruyn (1968) are especially valuable in giving appropriate recognition to these early European workers.

Table 1.1, based on the bibliography of Bruyn *et al*. (1974) gives a picture of how widely diffused information on HD soon became. Some of these reports (e.g. that of Landouzy in 1873) were described before being aware of Huntington's description, but international communication seems to have been remarkably rapid at that time (perhaps because the number of investigators and volume of literature were limited). Medicine, neurology and pathology had not yet become fully demarcated specialties, and in reading the early descriptions one gains the impression of an actively communicating body of workers whose interests ranged over a wide variety of disorders. The names of some of these workers who published on HD, including Landouzy (1873), Bourneville (1874), Golgi (1874), Déjerine (1886) and Hoffmann (1888) are now better remembered in relation to other neurological disorders than in connection with HD.

Table 1.1 The spread of information on Huntington's disease. First reports from different countries, based on Bruyn et al., 1974

Country	Date of report*	Author*
United States	(1842), 1872	(Waters), Huntington
Norway	(1862)	(Lund)
France	1873	Landouzy
Italy	1874	Golgi
Germany	1877	Meynert
United Kingdom	1880	Harbinson
Russia	1889	Kornilowa
Cuba	1890	Arostegui
Netherlands	1890	Beukers
Poland	1890	Biernacki
Brazil	1891	Couto
Denmark	1892	Friis
Argentina	1894	Costa
Czechoslovakia	1895	Ganghofner
Yugoslavia	1900	Gutschy
Australia	1902	Hogg
Canada	1904	Mackay

*Entries in parenthesis indicate a description before George Huntington's 1872 report.

It is also relevant to note that HD was described early in a number of countries, such as Cuba and Brazil, in which the disorder is not today recognized as a frequent occurrence, reinforcing the impression, discussed further in Chapter 8, that HD has been a widespread condition for a considerable time.

HD IN NEW ENGLAND – HISTORY AND ORIGINS

While the recognition of HD was spreading rapidly throughout the world during the last decades of the nineteenth century, laying the foundations of our detailed knowledge of the natural history of the disease, there was considerable activity also in tracing and attempting to connect the various families in the New England region that had been responsible for most of the original descriptions. Jelliffe (1908) and Davenport and Muncey (1916) were able to compile extensive pedigrees and to link them into groups which could be traced back to possible founding members. Most of these appeared to originate in the early seventeenth century from the East Anglia area of England (from which George Huntington's ancestors had also come), and which had seen extensive migration to the United States.

Unfortunately, the attempts to identify specific individuals as the source of the gene for HD all seem to have stretched the evidence far beyond what actually exists. Vessie (1932) claimed to have traced one large HD grouping back over a 300-year period to three individuals from the village of Bures in

Suffolk. He gave these individuals pseudonyms but the actual names were used by Critchley (1934, 1964, 1973) who extended the tracing in East Anglia, as did van Zwanenberg (1974); Vessie (1932, 1939) and later Maltsberger (1961) developed a further aspect of these families during the early colonial period in New England – their possible involvement in the notorious witchcraft trials of that time. Their papers, especially that of Maltsberger, are more notable for their lurid and exaggerated descriptions than for their scientific approach and do not actually show any direct involvement of likely HD patients in these episodes.

Subsequent investigation has shown that the whole of this work is based on inadequate and probably erroneous foundations. Caro and Haines (1975) and Caro (1979) who are themselves based in East Anglia, traced the genealogical evidence in both England and America, and found that the principal person named by Critchley as most likely to have introduced the HD gene did not actually exist; additionally they noted numerous errors and discrepancies in the previous accounts. The previous conclusions associating carriers of the HD gene with witchcraft and criminality were also invalidated. Despite this, Critchley (1984) was still giving essentially the original account 10 years later. In view of the frequent tendency to stigmatization of HD patients and families that has existed from the beginning and still remains a problem, it seems especially important to correct this much cited chapter of history and to exhort workers on HD to be as cautious in their historical conclusions as in their scientific work. When accurate mutational analysis becomes possible in the near future, it will become possible to document the descent and relationships of many groups of HD pedigrees with considerable accuracy; it will be of particular importance that any genealogical studies accompanying such work are not flawed.

Regardless of the precise details, New England has clearly played a pivotal role in receiving HD genes from Europe and distributing them throughout the United States. It is possible also that it may have been responsible for a wider spread of the disorder; as described in Chapter 8. It seems likely that the HD gene in some Pacific Island communities may have been brought by visiting New England whaling ships in the early nineteenth century (Scrimgeour, 1983). If this can be confirmed (a possibility with future molecular analysis) it will provide a particularly interesting example of how a genetic disorder can not only contribute to a chapter of history but also leave long-term effects that persist after the original events of the Pacific whaling industry are all but forgotten, except as literature.

EVOLUTION OF THE CLINICAL PICTURE

Compared with most original descriptions of a disease, that given by George Huntington in 1872 was remarkably complete, despite its brevity. This was largely due to the all-round view that he was able to obtain as a general practitioner and the length of time over which his own family had observed the patients for whom they cared. Subsequent work over the next three decades thus served mainly to give more detail to the neurological and psychiatric symptoms, to extend awareness of the range and variation seen in the disorder, and to attempt to analyse the pathological and genetic aspects.

The numerous case and family reports following the original description have already been mentioned and, in the case of Osler, quoted. The existence of HD as a specific entity of chronic, progressive, hereditary chorea of adult onset soon became widely accepted; the detailed clinical picture that emerged will be described in Chapter 2, but it is worth here examining some of the more important aspects that had not been fully appreciated in the initial descriptions already given.

Psychiatric syndromes

Huntington in his original paper recognized that mental deterioration was one of the characteristic features of the disorder that was to bear his name; he noted 'the tendency to insanity, and sometimes that form of insanity that leads to suicide'. Most of the subsequent case reports, however, concentrated on the chorea rather than the psychiatric manifestations of HD.

One of the more systematic of the early studies on mental disorder in HD was conducted by Phelps (1892) at Rochester, Minnesota. He found five cases of HD, or one in 600 admissions, that had been admitted to the Second Minnesota Hospital for the Insane. In addition, he wrote to the superintendents of 50 psychiatric institutions all over the United States asking for details on HD patients. Thirteen cases were reported to him in replies from 24 hospitals, so that he had the largest series up to that time. He described many of the major forms of psychiatric disorder, including psychotic symptoms, 'melancholia', suicide, irritability and 'steady mental degeneration'.

In the early literature on HD, there are clear descriptions of psychosis. Delusions of grandeur figure prominently in a number of clinical descriptions. One inpatient said that 'God is my lawyer' (Phelps, 1892), while another called her mental hospital a 'castle' (Eager and Perdrau, 1910). Delusions of royalty in HD have also been reported (Phelps, 1892; Eager and Perdrau, 1910). An interesting American case of delusions of special ability was described in a man who said he was getting 'a million dollars for standing back and allowing Harrison to be President instead of himself' (Phelps, 1892). Some of these cases may have had concurrent general paresis, which is a well-known cause of grandiose delusions.

Cognitive decline was described in the early reports (e.g. Osler, 1893) and indeed in the opinion of Sinkler (quoted by Mitchell, 1895) 'nearly all cases of Huntington's chorea terminate in dementia'. Perhaps the most perceptive of the early accounts of impaired intellectual functioning was provided by Edward Mapother (1911) from Dublin:

> As the disease progresses the most marked noteworthy and constant feature comes to be a failure of the power of sustained attention. Often there is no marked deficit of memory. There is no disorientation either in regard to place or time. Comprehension of speech is good and the capacity for simple judgements and deduction is unimpaired. But if one gives the patient a somewhat complex order involving the performance of several successive actions or the observation of a series of phenomena, his capacity for sustained attention is immediately manifested.

Bower and Mills (1890) also commented on higher mental functions in HD, especially abnormalities of speech and handwriting. Although dementia was thus well recognized in HD, early authors also observed that there were exceptions and that some patients remained 'mentally clear' (Fisher, 1906).

Juvenile HD and the rigid form

The paper ofCed (1863) is commonly quoted as the first description of childhood HD but, as mentioned earlier, the lack of clear progression and the early onset in his families make it likely that he was dealing with a separate disorder such as benign familial chorea. A number of other early childhood cases are discussed by Bruyn (1968) in his review, but the most detailed of these is the report of Hoffmann (1888). In this three-generation family with HD there were two daughters who had onset at 4 and 10 years, showed rigidity, hypokinesia and seizures, as well as choreic movements, both showing prolonged survival. This report showed clearly that not only could childhood onset occur within a typical HD family, but that the clinical features might be strikingly different to those normally seen in the adult disorder.

The recognition of juvenile HD was closely linked with the realization that chorea was not the only motor disorder in HD, and that some patients showed a predominance of rigidity and hypokinesia. The term 'Westphal variant' has often been used for this clinical presentation, but in describing his 18-year-old patient with these features, Westphal (1883) attributed them to a separate cause rather than to HD. The family of Hoffmann (1888) already mentioned, together with those reported by Curschmann (1908), Freund (1911) and many others, reinforced the existence of the rigid form of HD as an important clinical picture, usually in young adults or children, but occurring in the same families as cases with more typical chorea as the presenting feature.

Thus by the end of the nineteenth century, as a result of a very large number of careful and detailed descriptions, the clinical picture of HD was, if not complete, at least well established in terms of its main neurological features, its psychiatric involvement and its natural history. It was already quite clear that it was a specific disorder with well defined, though variable clinical features; the universal adoption of the title 'Huntington's chorea' or 'Huntington's disease' reinforced the degree to which the disorder was accepted as an entity, as well as the contribution that George Huntington had made to its original description.

NEUROPATHOLOGY

The search for neuropathological abnormalities in HD began early. Lewis (1876) was forced to conclude that 'no definite portion of the cerebrospinal system can at present be chosen as the site of lesions peculiar to chorea'. However, Meynert (1877) was more successful in detecting the postmortem changes in HD. He proposed that chorea could be explained by lesions in the corpus striatum, but this remained contentious for many years. Osler (1893) found no specific changes, as already mentioned, and could give no clear explanation for the clinical features. Some workers thought that brain inflammation was a neuropathological characteristic of HD (Oppenheim, 1887; Phelps, 1892; Sinkler, 1892) whereas others believed that the disorder was caused by a congenital malformation of the motor cortex (Stier, 1902; Müller, 1903). These reports were disputed by other investigators who thought that vascular sclerosis and the overgrowth of neuroglia were the pathological changes in HD (see Mapother, 1911).

Perhaps the most influential of the early neuropathological studies was that conducted by Jelgersma (1908). He described generalized shrinkage of the HD brain and atrophy of the caudate nucleus (reduced to one-third of its original volume). These findings were confirmed by later workers (Alzheimer, 1911; Pfeiffer, 1913), but it was not until the 1920s that there was general agreement that the brain changes in HD were primarily degenerative and atrophic, and that the caudate nucleus was preferentially involved in the process.

EARLY FORMS OF TREATMENT

At the time of Huntington's report, there was little in the therapeutic armamentarium for most diseases. None the less, several authors tried drug treatment for HD and some animal trials were also conducted. In the 1890s, strychnia was injected into dogs until they were 'violently choreic' (Wood, 1893). Quinine was found to 'arrest these movements' and in a clinical trial on a patient with pronounced chorea, the drug was reported to be effective (Wood, 1893). After administration of bromide of potassium, it was noted that 'twitchings decreased remarkably' in another case report (McFaren, 1874). Other drugs were less successful, such as hyoscamine (Lewis, 1876) and arsenic (Eager and Perdrau, 1910). McFaren (1874) was probably the first investigator to recommend a nutritional diet for weight loss in HD. None the less, admission to an asylum was the only significant intervention that could be offered.

INHERITANCE

The later part of this book is largely devoted to a detailed description of the genetic basis of HD (Chapters 9 and 10) and the practical applications of this knowledge in genetic counselling and prediction (Chapters 11 and 12). The main features of the inheritance pattern were already well recognized by George Huntington in his original description.

> When either or both parents have shown manifestations of the disease, and more especially when these manifestations have been of a *serious* nature, one or more of the offspring almost invariably suffers from the disease, if they live to adult age. But if by any chance these children go through life *without* it, the thread is broken, and the grandchildren and greatgrandchildren of the original shakers may rest assured that they are free from the disease. Unstable and whimsical as the disease may be in *other* respects, in this it is firm; it never skips a generation to manifest itself in another; once having yielded its claims, it never regains them. (From Huntington, 1872.)

Reading this passage today, it is as clear a description of Mendelian dominant inheritance as one could wish. However, though Mendel had published his work in 1865, it was only rediscovered in 1900, so no theoretical basis could be provided for the pattern that had been observed until after this time. It did not take long, however, for workers looking for possible human examples of Mendelian inheritance to recognize that HD provided a likely instance. In February 1908, Punnett, a close colleague of Bateson whose work was primarily on poultry, cited HD as likely to follow dominant inheritance, while later in the same year Jelliffe (who had read Punnett's paper) also mentioned this. Neither was very definite, both preferring to cite brachydactyly as a more conclusive example.

By 1911, Davenport was able to be much more confident in listing HD as an autosomal dominant disorder, along with other conditions illustrating autosomal recessive and sex-linked inheritance. Interestingly, he also cites other forms of chorea as being dominantly inherited. By the time of his later collection of material specifically on HD (Davenport and Muncey, 1916), the mode of inheritance was beyond doubt, and detailed analysis could begin.

In any discussion of the early development of our concepts of the genetics of HD, the work of Charles Davenport (Davenport, 1911; Davenport and Muncey, 1916) requires special discussion. Davenport's prejudiced attitude to the disorder, his advocacy of radical eugenic views (Davenport, 1911) and later political association with the German race-hygienists (see Chapter 11) make it difficult to assess his work objectively, as does his tendency to overinterpret Mendelian principles, especially in relation to other disorders involving intelligence and mental characteristics. Despite this, though, and bearing in mind the early time at which he was working, he must be acknowledged as the person who not only made it clear beyond doubt that HD followed Mendelian dominant inheritance, but also for his documentation of such important aspects as age of onset, variation between families, possible anticipation and its biases, and the reasons for apparent skipped generations. Clinically, too, he recognized the variability in degree and type of mental involvement and the variety of motor disturbance that might occur. This compilation of data on almost 1000 affected individuals, mainly in the New England area and descended from a small number of original progenitors, provided the foundation

Table 1.2 Landmarks in the study of Huntington's disease

1841	First definite description of HD (Waters)
1872	George Huntington's definitive description
1888	Juvenile HD clearly described (Hoffmann)
1908	Mendelian dominant inheritance recognized
1934	Systematic study of inheritance (Bell)
1958–1959	First detailed genetic-epidemiological survey in specific region (Michigan)
1967	Committee to Combat Huntington's Disease formed
1967	World Federation of Neurology research group formed
1972	Centennial Symposium, Columbus, Ohio
1983	Localization of HD gene
1987	First applications of DNA markers in prediction
199?	Isolation of HD gene

for much of the future work that was to come and should be recognized for this, even though many of the conclusions drawn may have been flawed.

THE DEVELOPMENT OF HD RESEARCH

Until recently, the history of research on HD has been one of gradual progress, rather than of sudden leaps. The main discoveries in the different fields of work are recorded in the specific chapters and have been shared by many different disciplines in addition to the neurosciences, genetics having been a major contributor from the beginning to the present.

A broad picture of the activity relating to the study of HD can be obtained from the bibliography of Bruyn *et al.* (1974). By 1890 up to 20 papers per year were being published on HD, with a peak just before World War I only reached again in the mid 1930s, falling again during World War II, but rising sharply and continuously thereafter. Understandably, the main topics have changed over the years, with studies of neuropathology prominent in the earlier phases, while pharmacological and biochemical topics have been more abundant in recent years, clinical and genetic reports remaining relatively constant, at least until 1972.

Table 1.2 lists some of the landmarks in the development of our understanding of HD. The major landmark anticipated in the very near future is the identification of the HD gene and its function, steps that will arguably represent the most important advance in the history of the disease. Several other less spectacular, but significant steps that do not readily relate to the topics covered by subsequent chapters are described here.

ONE HUNDRED YEARS OF HD – THE CENTENNIAL SYMPOSIUM AND BIBLIOGRAPHY

The anniversary of George Huntington's 1872 paper was marked by two events which have proved less evanescent in their effects than most such celebrations.

The published results of a symposium held in Columbus, Ohio brought together a wealth of material and summarized the state of knowledge at that time (Barbeau et al., 1973). Some parts of the volume, notably the section on biochemistry, now appear dated and the book reflects our lack of progress in some aspects during the century, as well as the very real progress in other ways. Nevertheless, despite this, it provides a real landmark and can serve as a valuable starting point for anyone embarking on research on HD.

An even more valuable resource has been provided by the 'Centennial Bibliography' on HD (Bruyn et al., 1974), a major achievement which has collected all known publications on HD up to 1972 (since updated to 1978). The critical approach taken, the extensive cross indexing and informative comment accompanying the bibliography not only make this the definitive collection of material on HD, but give a clear picture of the pattern of research, its progress and evolution into different subjects, as well as the geographical distribution of work on the disorder. It has been extensively used during the writing of this book as a source of reference for older papers (particularly those written in languages other than English) which would otherwise have been difficult to obtain. Computer-based abstracting services now ease the task of keeping in touch with the ever-expanding literature, but the centennial bibliography will remain an indispensable resource for people wishing to inform themselves on the work of the first century since Huntington's description.

An important focus and forum for discussion of new work on HD over the past 20 years has been the World Federation of Neurology Research Group on Huntington's chorea. Founded by Dr André Barbeau, this group first met at Montreal in 1967 and has since held a succession of valuable workshops at about 2-yearly intervals. It has helped to create a community of research workers involved with HD who have developed close and continuing links and collaborations spanning a number of different disciplines.

THE COMMITTEE TO COMBAT HUNTINGTON'S DISEASE AND THE DEVELOPMENT OF LAY SOCIETIES

A striking feature of the medical scene during the past 30 years has been the development of self-help groups run by sufferers and their families. Fundraising for research, improvement of services and greater public awareness have all been major aims, as discussed more fully in Chapter 7. In the case of HD, progress would undoubtedly have been slower had such societies not been developed, and it is unlikely that the disorder would have the high public profile internationally that it does today.

The initial development in this area arose from the illness of Woody Guthrie, the American folk-singer (Figure 1.3), who developed HD symptoms around 1952 and died in 1967 at the age of 55. The remarkable biography by Klein (1981) gives considerable insight into the interplay between personality, creativity

Figure 1.3 Woody Guthrie, the American folk singer affected with HD, whose illness was the catalyst for the founding of the first lay society involved with HD, the Committee to Combat Huntington's Disease. Courtesy of Woody Guthrie publications.

and disease in Guthrie's life. His widow Marjorie devoted the later part of her life to promoting all aspects of HD, and in 1967 the Committee (later the Association) to Combat Huntington's Disease was formed with objectives to provide services for families and to promote education and research (Figure 1.4).

Marjorie Guthrie was also largely responsible for the initiation of comparable organizations in other countries, resulting in 1978 in the International Huntington's Association, a body that now contains 27 member countries. Close links with medical groups, such as the WFN Research Group have ensured that the lay societies have played a constructive role in developing policies in such critical areas as predictive testing.

A further effect of lay pressure was the formation in 1975 of a Congressional Commission on HD, which reported 2 years later (US DHEW, 1977). While

HD COMMITTEE TO COMBAT HUNTINGTON'S DISEASE, INC.
Suite 1304 · 200 West 57 Street · New York, N.Y. 10019

NEWSLETTER

NUMBER 1   © 1968, Committee to Combat Huntington's Disease, Inc.   SPRING 1968

Purposes of the Committee to Combat Huntington's Disease, Inc.

The Committee's Charter contains a statement of the purposes for which the Committee was organized. Briefly, these purposes are:

Educational. To collect information about all aspects of Huntington's disease and distribute it to interested individuals, "for the purpose of increasing public awareness;"

Assistance. "To assist those afflicted with Huntington's disease and their families in meeting the social, economic, and emotional problems resulting from such affliction;"

Financial. The Committee will raise funds to help support "the advancement

The day-to-day work of the Committee to Combat Huntington's Disease, Inc. is carried out by six operating committees working within the framework of the larger organization. Any member is welcome to join the operating committee whose work interests him most. These committees are:

Financial Committee. Responsible for fund-raising activities and for dispensing the funds that are raised.

Membership Committee. Handles all questions of membership; eventually, this committee will be instrumental in contacting other groups interested in Huntington's disease to form national or international organizations.

Administrative Committee. Works with the Executive Secretary to perform various daily activities of the Committee.

Family Counseling Committee. Helps to determine the needs of families involved in Huntington's disease, and plans action to aid these families and individuals.

Publications and Publicity Committee. Prepares the Newsletter and other publications; makes sure that the work of the Committee receives attention in newspapers, magazines, etc.

Program Committee. Plans general meetings in advance; makes arrangements such as getting speakers, etc.

Figure 1.4   The initial newsletter of the Committee to Combat Huntington's Disease.

much of its weighty publication makes turgid reading there are some lively and controversial discussions included and there is no doubt that it made legislators aware of the disorder in a way that would not have happened otherwise.

THE VENEZUELA PROJECT

The remarkable concentration of HD patients living in the Zulia region of Venezuela, by the shores of Lake Maracaibo (Figure 1.5), represents the largest cluster of cases derived from a single ancestor that has remained geographically localized. This pedigree contains about 7000 members with over 100 living affected subjects. Some of the epidemiological aspects of this concentration are described in Chapter 8, but the wider significance of HD in Venezuela for the development and understanding of the disease deserves a special mention here.

The high frequency of the disorder in some of the small and isolated lakeside communities was first documented by Negrette (1963), and was brought to general attention by his colleague Avila-Giron (1973) who presented details at the centennial symposium on HD. It was then recognized that this isolate could be of special significance, particularly in studying possible homozygotes since several families had both parents affected; visits by Dr André Barbeau and

Figure 1.5 HD in Venezuela. The concentration of the disorder in the villages around Lake Maracaibo and the occurrence there of probable homozygotes for the disorder has led to a major longitudinal study of the disease and to studies of the HD gene. (Courtesy of Dr Nancy Wexler.)

others confirmed this, but the key development was the decision of the Hereditary Disease Foundation to mount a systematic study of the population. After an initial visit in 1979, comprehensive studies were carried out by an annual visiting team beginning in 1981; detailed pedigrees were drawn up, allowing a full genealogy of the different branches to be pieced together, while accurate clinical assessment of both affected members and relatives at risk was carried out, concentrating on those families with two affected parents containing possible homozygotes. Blood samples were taken for DNA analysis and cell lines set up allowing their long-term study. At the same time the workers tried to provide as much practical help as possible to these poor and deprived communities. The background to the work is well described in a number of general readership articles (Drake, 1984; Kolata, 1984; Pines, 1984; Steinmann, 1987), but a full account of this remarkable project has yet to be written.

The most spectacular result of the Venezuela project has been its crucial role in locating the HD gene, outlined below and described further in Chapter 10. Almost as important, however, have been the clinical studies, which are providing a unique longitudinal documentation of HD in a population without access to medication for the disease, as well as clearly showing that the likely homozygotes for HD are no more severely affected than are those with a single copy of the gene. The Venezuela project is a striking example of the value of a long-term clinically-based project and there is no doubt that this value will increase further when isolation of the gene and understanding of its function lead to therapy that can significantly alter the natural history of the disorder.

HD RESEARCH IN WALES

This book has been written to give a broad perspective on the major problems associated with HD and on the research, past and current, which has thrown light on some of these problems. Hopefully it provides an objective account of the situation and we have tried to avoid over-emphasizing our own views, or our own contributions to advances that have been made. Nevertheless, the consistent involvement of a single unit in HD care and research over a period approaching 20 years is unusual and it is likely that this sustained activity over a prolonged period has itself made possible studies that could not have been done otherwise. In particular, the close interaction between clinical and laboratory research has been fruitful, as has the consistent policy of trying to provide a high standard of services for the population we have been studying.

In this section we summarize the work done in Cardiff on HD over the past two decades and bring together widely different studies that are scattered in the literature. Details of the individual studies are given in the various other parts of this book as appropriate. Some general reviews of our work have been written previously (Harper, 1986; Harper et al., 1988) but the collected references are given chronologically at the end of the book (Appendix 1). We hope that this account will encourage others by showing what a relatively small and modestly funded group can do by basing its research on the population which it is trying to serve. Wales (Figures 1.6, 1.8) is a small country approximately 200 miles long and 100 miles wide. It borders England to the west and retains a clear identity from its larger neighbour, together with a considerable number of cultural, linguistic and to some extent genetic differences (Harper and Sunderland, 1984). Its population of almost 3 million is mainly concentrated in the industrial south-east part of the country and the northern coastal strip, the rest being almost entirely rural, mountainous and sparsely populated. Cardiff, the administrative capital, situated in the south-east corner, contains the only medical school, and is also the base for the regional medical genetics service, which covers the entire country.

A limited study of HD families in South Wales was first undertaken many years ago (Spillane and Phillips, 1937). Twenty four cases from six families were described, mainly from south-west Wales, and the authors gave a general review of the clinical features of the disorder. Unfortunately, the original records of this study were not preserved, so that it proved impossible to link these families with those studied subsequently.

In late 1971 one of us (PSH) was appointed to develop the speciality of medical genetics in Wales and was immediately impressed by the number of HD families referred for genetic counselling and by the extent of some of them. One kindred in particular (Harper, 1976, 1986), descended from a single immigrant to Wales born in 1845, contained 47 affected individuals and had over 100 living people at high risk (Figure 1.7). It soon became clear from these referrals and by records of the local neurology department that HD represented a major problem in south-east Wales and that a major initiative was required if genetic counselling and other service needs were to be met and if systematic research was to be undertaken. It was not until late 1973 that funding was obtained for a clinical research fellow and a nurse fieldworker, which allowed a core team to be formed for the Cardiff HD project. A systematic survey was carried out between 1974 and 1978. This study was in some respects very different from most other studies of HD in a geographical region, and these differences were deliberate. First, the survey was planned

Figure 1.6 Map of Wales and other parts of the British Isles, showing the original study area for the HD total ascertainment project.

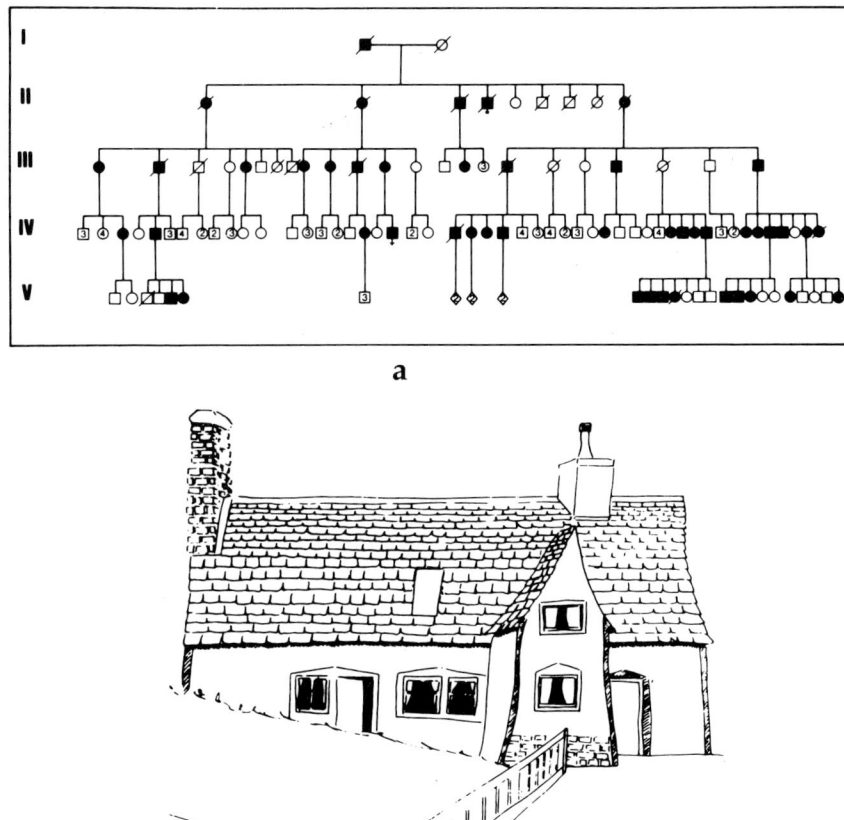

Figure 1.7 HD research in Wales. The development of a major long-term research programme relating to HD in Wales was based originally on the recognition of extensive kindreds, including the one shown here **(a)**. The ancestral affected member, born in south west England in 1845, migrated to the developing industrial area of south east Wales and lived in the isolated farmhouse shown above **(b)** (courtesy of Dr David Walker). Over 100 descendants at risk of HD remain in the immediate area today.

from the beginning as the initial phase of a long-term study which would subsequently maintain contact with families and would collect data prospectively. Second, the principal aim of the survey was to give an accurate basis for service provision, especially genetic counselling. It was felt that it was impossible to meet or predict service needs or to achieve any long-term preventive goals, without a full knowledge of the numbers of people affected by and at risk for HD, and that a detailed register was an essential part of this. Third, the study was restricted to a defined part of Wales, with intensive effort to obtain total ascertainment within this. The choice of the counties of Glamorgan and Gwent which essentially make up industrial South Wales, gave a compact area (see Figure 1.6) containing 1.7 million people dependent on a small number of hospitals and clinicians and easily accessible from Cardiff.

This initial survey (Harper *et al.*, 1979; Walker *et al.*, 1981) provided few real surprises but was the foundation for all subsequent work. HD proved to be much commoner than even the investigators suspected, with a prevalence of 7.61 per 100 000 (prevalence date 1971) and a heterozygote frequency of 1 in 5000. The prevalence was especially high in the county containing the two largest families, emphasizing the importance of using a large population base to avoid local bias produced by such families. Not a single new mutation was conclusively documented, though systematic genealogical searches to connect pedigrees were not feasible, owing to lack of old church records for most families and the relatively recent adoption of fixed surnames in Wales, so that a shared surname usually provided no evidence for kinship.

The study gave further evidence (Walker *et al.*, 1983) that there is an absolute as well as a relative increase in fertility of those with the HD gene (discussed further in Chapter 9) and it also confirmed the earlier onset with paternal transmission of the disease (Newcombe *et al.*, 1981). More practically it had produced a total of almost 1000 living relatives with an age-adjusted risk of greater than 10%, most of whom had previously had little information concerning HD, almost no formal genetic counselling, and very little practical support in relation either to the care of affected members or concerning their own fears and worries. The initial study took almost 5 years, and could at no stage be described as complete since information continued to arise on both existing and new families. The subsequent 5 years can best be described as a time of consolidation, with the work developing along three principal lines.

The first main task was to construct a systematic genetic register suitable for maintaining information in the long term. The initial study data had been collected with this in mind, but a major advance meanwhile had been the development of a life-table approach to risk estimation which had allowed a more accurate approach to age at onset analysis (Newcombe *et al.*, 1981) but was also relevant to genetic counselling. It was decided to build this into a computerized service-orientated HD register (Harper *et al.*, 1982) which could be used to monitor trends in families and also to ensure that by recording births into families, children growing up in HD families could be offered genetic counselling at an appropriate age. The evolution of this register has paralleled advances in computing. The original study was dependent on punched cards, the initial register on a main-frame computer, while its subsequent version is microcomputer based (Sarfarazi *et al.*, 1987) giving considerable advantages in ease of use and confidentiality (see Chapter 11). A further revision of the system is in progress.

The second main area of activity resulting from the survey was in genetic counselling and prevention. The study itself had shown a major unmet need in this area, but a general policy was only adopted after considerable thought and debate among colleagues, whose opinions varied considerably. Some were strongly directive in approach (Spillane and Phillips had earlier concluded that 'only legislative measures will succeed in eradicating the disease') while others expressed the view that anything one might do in informing families would be likely to cause more harm than good. The policy finally chosen was one of systematic but entirely non-directive counselling, based primarily on home visits. All adults at significant risk (over 10%) were contacted (after informing their family doctor where possible). A home visit with full explanation of the genetic risks and other aspects of HD was undertaken by an experienced genetics nurse–fieldworker and social worker, and a further appointment was offered for the genetic counselling clinic if requested. Importantly, a system of regular (usually annual or alternate year) contact was offered either as a visit or by telephone, and support was arranged either directly or through the appropriate agencies for such aspects as arranging invalidity benefit or other social welfare allowances, alterations in the home to help disabled sufferers, and temporary admissions to hospital for respite care.

The creation of a genetic register, with regular collection of data on new births, together with the systematic approach to genetic counselling in the population, made it possible to ask whether any trends in birth rate were occurring that might be associated with our genetic counselling programme. Close to total ascertainment of HD in the study population allowed a projection forwards to expected new cases based on the number of births at risk, and a comparison could be made with retrospective birth analyses. The initial study (Harper *et al.*, 1979) showed that there had been a relatively constant actual and predicted number of cases during the decades preceding our survey, but that subsequently a marked decline occurred (Harper *et al.*, 1981), though addition of further data showed this to be less steep than originally calculated (Quarrell *et al.*, 1987). The significance of these results in terms of prevention is discussed in Chapter 11, but this work shows the value of a completely ascertained population in conjunction with a prospectively maintained register, if changes are to be accurately monitored.

The emergence, as anticipated, of numerous social problems that had previously been neglected provided the focus for the third major area of work. This was a study of the social aspects of HD, in particular the incidence of family breakdown and also the economic burden of the disorder, based on an unselected series of 92 affected individuals (Tyler, 1982; Tyler *et al.*, 1982, 1983). This study largely provided the basis for the UK Office of Health Economics' report on HD (1980). At the same time this group of patients, their spouses and a random sample of 100 first degree relatives at risk were questioned regarding their knowledge of the inheritance of HD and their attitudes to childbearing, genetic counselling and future predictive tests (Tyler, 1982; Tyler and Harper, 1983). This study gave detailed and essential information that formed the basis not only of our genetic counselling service provision, but also for our subsequent evaluation of predictive testing.

The continued close monitoring of HD families, together with the extension of genetic counselling clinics to all districts of Wales, has allowed the prevalence estimates to be extended from the original study area to other parts of Wales. Quarrell *et al.* (1988) were able to show that the disorder was also common in north Wales, where no systematic study had previously been done, while a reassessment of the prevalence in South Wales after a 10-year interval showed that this had risen from 7.61 in 1971 to 8.85 per 100 000 in 1981, even though the original estimate had been thought to represent complete ascertainment. Figure 1.8 shows the detailed breakdown for the different parts of Wales.

As with any major and continued study, unplanned observations arose. Families with benign hereditary chorea were encountered and documented (Harper, 1978) and an assessment of the outcome of isolated cases coming to autopsy provided unexpected findings in terms of both positive and negative results (Quarrell *et al.*, 1986). However, the major development in HD research

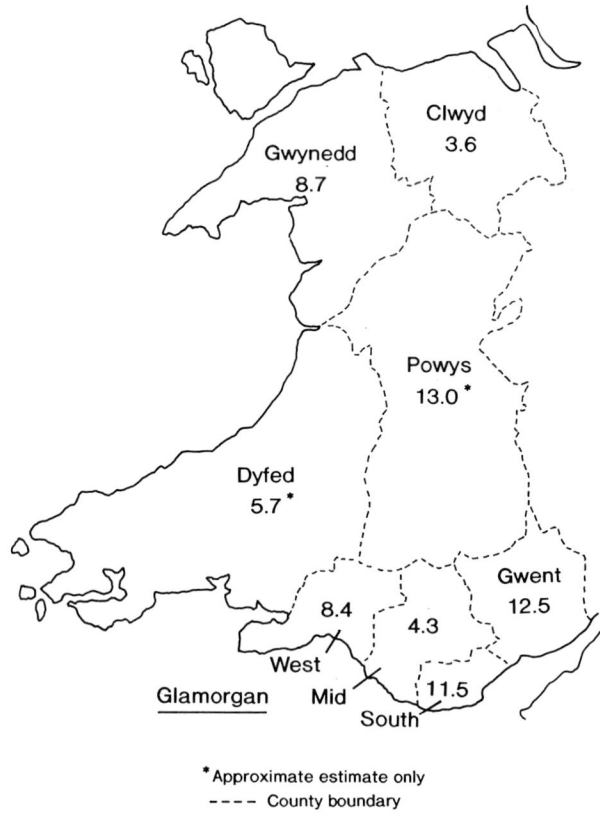

Figure 1.8 The prevalence of HD in different areas of Wales.

was foreshadowed when a genetic linkage study was carried out, based on the largest South Wales pedigrees, which clearly offered special opportunities for this.

The outcome of this collaborative study (Volkers et al., 1980) was negative; as were all the studies from this time based on classical genetic markers (see Chapter 10). However, in 1981 a visit by Dr David Housman alerted us to the study developing in Venezuela and the importance of collecting DNA from our own linkage families. During 1982–1983 this was progressively carried out, and cell lines were established in Dr James Gusella's laboratory in Boston, so that by the time of the unexpectedly rapid detection of linkage (Gusella et al., 1983), material was already available for detailed molecular analysis in both Boston and Cardiff.

The Cardiff research thus moved into its current molecular phase, in which our extensive clinical resources have also proved a particular strength. Confirmation of the genetic linkage on chromosome 4 was soon obtained on the south Wales families (Harper et al., 1985; Youngman et al., 1986), while benign hereditary chorea was shown to be non-allelic (Quarrell et al., 1988). A further step was the isolation of a probe 'D5' (D4S90) from a chromosome 4 DNA library which proved to be more telomeric than any other (Youngman et al., (1988) and which has proved particularly useful in the ordering of the HD gene in individual families and in the analysis of this chromosomal region in general (Youngman, 1989; Youngman et al., 1989). A separate approach to the localization of the gene has been that of linkage disequilibrium, detected by Snell et al. (1989) with markers more proximal than had previously been considered likely and discussed fully in Chapter 10.

Finally, and appropriately, the two main themes of HD research in Wales, clinical and molecular, have combined in their application to predictive testing. We were cautious in embarking on presymptomatic diagnosis, partly because the family structures were often unsuitable in our register population (Harper and Sarfarazi, 1985), but also because of the ethical problems surrounding this. We thus concentrated initially on 'exclusion testing' related to pregnancy, a form of testing more comparable to conventional genetic testing than is presymptomatic diagnosis of HD. As a result of this we have gained particular experience in this field as described in Chapter 12 (Quarrell et al., 1987b; Tyler et al., 1990). Our work on presymptomatic testing has largely involved families from outside Wales (Morris et al., 1988, 1989) and has particularly concentrated on documenting the practical and ethical problems involved, including the potential dangers of adoption and childhood testing (Morris et al., 1988; Harper and Clarke, 1990) and contributing to British and international guidelines on predictive testing drawn up by the World Federation of Neurology and International Huntington Association.

In all these studies a feature has been our close collaboration with other groups around the world involved with HD. The existence of these links has been particularly productive and has allowed us to utilize the resources of other units, as well as contributing to them ourselves. It is quite clear that such collaborations, informal as well as formal, will be essential for future advances, and to ensure that their results are used equitably to benefit HD families in all parts of the world.

REFERENCES

Alzheimer A (1911) Uber die anatomische Grundlage der Huntington'schen Chorea und der choreatischen Bewegungen uberhaupt. *Zeitschrift für die gesamte Neurologie und Psychiatrie*. **3**:566–567.
Avila-Giron R (1973) Medical and social aspects of Huntington's chorea in the State of Zulia, Venezuela. In A Barbeau, TN Chase and GW Paulson (eds). *Advances in Neurology*, pp. 261–266. New York: Raven Press.
Barbeau A, Chase TN and Paulson GW (1973) *Huntington's Chorea, 1872–1972*. New York, Raven Press.
Batten FE and Gibb HP (1909) Myotonia atrophica. *Brain* **32**:187–205.
Bell J (1934) Huntington's chorea. In: *Treasury of Human Inheritance* (ed RA Fisher). Cambridge: Cambridge University Press.
Bourneville DM (1874) De l'emploi thérapeutique du monobromure de camphre. *Progrès Médical* (Paris) **2**:456–459.
Bower JL and Mills CK (1890) Notes on some cases of chorea and tremor. *Journal of Nervous and Mental Disease* **15**:131–142.
Browning W (1908a) Huntington number. *Neurographs* **1**:1–164.
Browning W (1908b) Rev Charles Oscar Waters MD. I. Biographic sketch. II. Location of his cases. *Neurographs* **1**:137–144.
Browning W (1908c) Dr Charles Rollin Gorman. I. Personal sketch. II. His relation to the chorea question. *Neurographs* **1**:144–147.
Browning W (1908d) Irving Whitehall Lyon MD. I. Personal sketch. II. Location of his cases. *Neurographs* **1**:147–149.
Bruyn GW (1968) Huntington's chorea. Historical, clinical and laboratory synopsis. In: *Handbook of Neurology* (eds PJ Vinken and GW Bruyn). Amsterdam: Elsevier.
Bruyn GW, Baro F and Myrianthopoulos NC (1974) *A Centennial Bibliography of Huntington's Chorea 1872–1972*. The Hague: Martinus Nijhoff.
Caro A (1970) PhD Thesis. University of East Anglia.
Caro A and Haines S (1975) The history of Huntington's chorea. *Update* **11**:91–95.
Critchley M (1934) Huntington's chorea and East Anglia. *Journal of State Medicine* **42**:575–587.
Critchley M (1964) Huntington's Chorea: Historical and geographical considerations. In: *The Black Hole and other Essays*, pp. 210–219. London: Pitman Medical Publishing Company Ltd.
Critchley M (1973) Great Britain and the early history of Huntington's chorea. In: *Huntington's Chorea, 1872–1972* (eds TN Chase, GW Paulson, and A Barbeau). pp.115–122. New York: Raven Press.
Critchley M (1984) The history of Huntington's chorea. *Psychology and Medicine* **14**:725–727.
Curschmann H (1908) Eine neue Chorea-Huntingtonfamilie. *Deutsche Zeitschrift für Nervenheilkunde* **35**:293–305.
Davenport CB (1911) *Heredity in Relation to Eugenics*. New York: (re-issued by Arno Press, 1972).
Davenport CB and Muncey EB (1916) Huntington's chorea in relation to heredity and eugenics. *Eugenics Record Office Bulletin* **17**:195–222.
Déjerine J (1886) *L'hérédité dans les maladies du système nerveaux*. Paris: Thèse Aggreg.
De Jong RN (1937) George Huntington and his relationship to the earlier descriptions of chronic hereditary chorea. *Annals in Medical History* **9**:201–210.
Drake DC (1984) The curse of San Luis. *Philadelphia Inquirer* Aug 26.
Dunglison R (1842) *Practice of Medicine*, 1st edn. Philadelphia: Lee and Blanchard.
Dunglison R (1848) *Practice of Medicine*, 3rd edn. Philadelphia: Lee and Blanchard.
Eager R and Perdrau JR (1910) Notes on four cases of Huntington's chorea. *Journal of Mental Science* **56**:506–509.
Elliotson J (1832) St Vitus's dance. *Lancet* **1**:162–165.
Fisher ED (1906) A case of Huntington's chorea. *Journal of Nervous and Mental Disease* **33**:781.
Freund CS (1911) Zwei Brüder mit Huntingtonscher Chorea. *Berliner klinische Wochenschrift* **48**:735.

Golgi C (1874) Sulla alterazioni deglia organi centrali nervosi in uno caso di corea gesticulatoria assoziata ad alienazione mentale. *Rivista Clinicale Bologna* **4**:361.
Gorman CR (1848) In R Dunglison (ed) *Practice of Medicine*, 3rd edn, Vol 2, p. 218. Philadelphia: Lee and Blanchard.
Gusella JF, Wexler NS, Conneally PM *et al.* (1983) A polymorphic DNA marker genetically linked to Huntington's disease. *Nature* **306**:234–238.
Harper PS (1976) Genetic variation in Wales. *Journal of the Royal College of Physicians of London* **10**:321–332.
Harper PS (1978) Benign hereditary chorea, clinical and genetic aspects. *Clinical Genetics* **13**:85–95.
Harper PS (1986) The prevention of Huntington's chorea. The Milroy lecture 1985. *Journal of the Royal College of Physicians of London* **20**:7–14.
Harper PS and Sunderland E (1984) *Population and Genetic Studies in Wales*. Cardiff: University of Wales Press.
Harper PS and Sarfarazi M (1985) Genetic prediction and family structure in Huntington's chorea. *British Medical Journal* **290**:1929–1931.
Harper PS and Clarke A (1990) Should we test children for 'adult' genetic diseases? *Lancet* **335**:1205.
Harper PS, Walker DA, Tyler A, Newcombe RG and Davis K (1979) Huntington's chorea. The basis for long-term prevention. *Lancet* **ii**:411–413.
Harper PS, Tyler A, Smith S, Jones P, Newcombe R and McBroom V (1981) Decline in the predicted incidence of Huntington's chorea associated with systematic genetic counselling and family support. *Lancet* **ii**:411–413.
Harper PS, Tyler A, Smith S, Jones P, Newcombe RG and McBroom V (1982). A genetic register for Huntington's chorea in South Wales. *Journal of Medical Genetics* **19**:241–245.
Harper PS, Youngman S, Anderson MA *et al.* (1985) Genetic linkage between Huntington's disease and the DNA polymorphism G8 in South Wales families. *Journal of Medical Genetics* **22**:447–450.
Harper PS, Quarrell WJ and Youngman S (1988) Huntington's disease: Prediction and prevention. *Philosophical Transactions of the Royal Society, London* **319**:285–298.
Hayden MR (1981) *Huntington's Chorea*. Berlin: Springer Verlag.
Heckler JFC (1844) Epidemics of the Middle Ages. Translated in 1844 for the Sydenham Society from the German by BG Babington.
Hoffmann J (1888) Uber Chorea chronica progressiva (Huntingtonsche Chorea, Chorea hereditaria). *Virchows Archiv für pathologische Anatomie* **111**:513–548.
Huet E (1889) *De la Chorée chronique*. Paris.
Huntington G (1872) On Chorea. *Medical and Surgical Reporter* **26**:320–321.
Huntington G (1910) Recollections of Huntington's chorea as I saw it at East Hampton, Long Island, during my boyhood. *Journal of Nervous and Mental Disorders* **37**:255–257.
Husquinet H (1975) Premières descriptions de la chorée de Huntington en France et en Belgique. *Clio Medica* **10**:197–204.
Jelgersma G (1908) Die anatomische Veranderungen bei Paralysis agitans und chronischer Chorea. *Verhandlungen der Gesellschaft deutscher Naturforscher und Arzte* **2** (2):383–388.
Jelliffe SE (1908) A contribution to the history of Huntington's chorea: a preliminary report. *Neurographs* **I**:116–124.
Klein J (1981) *Woodie Guthrie. A Life*. London: Faber and Faber.
Kolata G (1984) Closing in on a killer gene. *Discover* March:83–87.
Landouzy LTJ (1873) Mouvements choréiques des membres inférieurs. *Gazette Médicale de Paris* **48**:329–330.
Lewis B (1876) A case of chorea associated with mania, terminating fatally by cerebellar apoplexy. *Medical Times and Gazette* **2**:280–282.
Lund JC (1860) Chorea St Vitus Dance in Saetersdalen. Report of Health and Medicine and Medical Conditions in Norway in 1860, p.137. (Quoted by Ørbeck, 1959.)
Lyon IW (1863) Chronic hereditary chorea. *American Medical Times* **7**:289–290.
MacFaren J (1874) A case of chorea. *Journal of Mental Science* **20**:97–99.
Maltsberger JT (1961) Even unto the twelfth generation – Huntington's chorea. *Journal of the History of Medicine and Allied Sciences* **16**:1–17.
Mapother E (1911) Mental symptoms in association with choreiform disorders. *Journal of Mental Sciences* **57**:646–661
Mendel G (1865) Versuche über Pflanzenhybriden. *Proceedings of the Natural History Society of Brunn* **4**:3–47. English translation reprinted 1965. Edinburgh: Oliver and Boyd.
Meynert T (1877) Discussion to Fritsch. *Psychiatry Clb* **4**:47.
Mitchell JK (1895) Huntington's chorea. *Journal of Nervous and Mental Disorders* **22**:395–397.

Morris M, Tyler A and Harper PS (1988) Adoption and genetic prediction for Huntington's disease. *Lancet* **ii**:1069–1070.

Morris M, Tyler A, Lazarou L, Meredith L and Harper PS (1989) Problems in genetic prediction for Huntington's disease. *Lancet* **ii**:601–603.

Müller LTR (1903) Uber drei Falle von Chorea chronica progressiva (Chorea hereditaria, Chorea Huntington). *Deutsche Zeitschrift für Nervenheilkunde* **23**:315–335.

Negrette A (1963) Corea de Huntington (Estudio de una sola familia investigade, través de varias generaciones. Talleros Graticos. University of Zulia, Maracaibo, Venezuela.

Newcombe RG, Walker DA and Harper PS (1981) Factors influencing age at onset and duration of survival in Huntington's chorea. *Annals of Human Genetics* **45**:387–396.

Oppenheim H (1887) Eine seltene Motilitatsneurose (chorea hereditaria?). *Berliner klinische Wochenschrift* **24**:309–310.

Ørbeck AL (1959) An early description of Huntington's chorea. *Medical History* **3**:165–168.

Osler W (1890) Hereditary chorea. *Johns Hopkins Hospital Bulletin* **I**:110.

Osler W (1892) *The Principles and Practice of Medicine*, pp. 944–945. Edinburgh and London: Young J. Pentland.

Osler W (1893) Remarks on the varieties of chronic chorea and a report upon two families of the hereditary form with one autopsy. *Journal of Nervous and Mental Disorders* **18**:97–111

Osler W (1894) Case of hereditary chorea. *Johns Hopkins Hospital Bulletin* **5**:119–129

Osler W (1904) On chorea and choreiform affections, pp. 96–112. Philadelphia: Blakiston and Son.

Osler W (1908) Historical note on hereditary chorea. *Neurographs* **I**:113–116.

Petit H (1970) La maladie de Huntington. In P Warot (ed) C.R.. 67e. *Congress de Psychiatre et Neurologie. Langue Franc.* 901–1058 (Masson, Paris).

Pfeiffer JAF (1913) A contribution to the pathology of chronic progressive chorea. *Brain* **35**:276–292.

Phelps RM (1892) A new consideration of hereditary chorea. *Journal of Nervous and Mental Disorders* **19**:765–776.

Pines M (1984) In the shadow of Huntington's disease. *Science* (May):32–39.

Punnett RC (1908) Mendelian inheritance in Man. *Proceedings of the Royal Society of Medicine* **1**:135–168.

Quarrell OWJ, Tyler A, Cole G and Harper PS (1986) The problem of isolated cases of Huntington's disease in South Wales. *Clinical Genetics* **30**:331–337.

Quarrell OWJ, Meredith AL, Tyler A *et al*. (1987) Exclusion testing for Huntington's disease in pregnancy with closely linked DNA markers. *Lancet* **i**:1281–1283.

Quarrell OWJ, Youngman S, Sarfarazi M and Harper PS (1988) Absence of close linkage between benign hereditary chorea and the locus D4S10 (probe G8). *Journal of Medical Genetics* **25**:191–194.

Quarrell OWJ, Tyler A, Jones MP, Nordin M and Harper PS (1988a) Population studies of Huntington's disease in Wales. *Clinical Genetics* **33**:189–195.

Sarfarazi M, Wolak G, Quarrell O and Harper PS (1987) An integrated microcomputer system to maintain a genetic register for Huntington's disease. *American Journal of Medical Genetics* **28**:999–1006.

Scrimgeour EM (1983) Possible introduction of Huntington's chorea into Pacific Islands by New England whalemen. *American Journal of Medical Genetics* **15**:607–613.

Sinkler W (1892) On hereditary chorea with a report of three additional cases and details of an autopsy. *Medical Record*. (NY) **41**:281–285.

Snell RG, Lazarou L, Youngman S. *et al*. (1989) Linkage disequilibrium in Huntington's disease: an improved localisation for the gene. *Journal of Medical Genetics* **26**:673–675.

Spillane J and Phillips R (1937) Huntington's chorea in south Wales. *Quarterly Journal of Medicine* **6**:403–425.

Steinert H (1909) Myopathologische Beitrage. 1. Uber das klinische und anatomische Bild des Muskelschwunds der Myotoniker. *Deutsche Zeitschrift für Nervenheilkunde* **37**:58–104.

Steinmann M (1987) In the shadow of Huntington's disease. *Columbia*:14–19.

Stevens DL (1972) The history of Huntington's chorea. *Journal Royal College Physicians* **6**:271–282.

Stier E (1902) Zur pathologischen Anatomie der Huntington'schen Chorea. *Münchener medizinische Wochenschrift* **49**:770.

Temkin CL, Rosen G, Zilboorg G and Higerist HE (1941). *Theophratus von Hohenheim called Paracelsus*. Baltimore: Johns Hopkins Press.

Thompson J (1876) Tonische kramote in Willkurlich beweglichen muskeln. *Archiv Psychiatrie Nervenheilkunde* **6**:702–718.

Tyler A (1982) The social, personal and economic burden of Huntington's chorea in South Wales. MSc Thesis. University of Wales.

Tyler A and Harper PS (1983) Attitudes of subjects at risk and their relatives towards genetic counselling in Huntington's chorea. *Journal of Medical Genetics* **20**:179–188.
Tyler A, Harper PS, Walker DA, Davis K and Newcombe RG (1982) The socioeconomic burden of Huntington's chorea. *Journal of Biosocial Science* **14**:379–389.
Tyler A, Harper PS, Davis K and Newcombe R (1983) Family breakdown and stress in Huntington's chorea. *Journal of Biosocial Sciences* **15**:127–138.
Tyler A, Quarrell OWJ, Lazarou, LP, Meredith AL and Harper PS (1990) Exclusion testing in pregnancy for Huntington's disease. *Journal of Medical Genetics* **27**:488–495.
United States Congress Commission (1977) Report: *Commission for the control of Huntington's disease and its consequences.* Washington: US Department of Health, Education and Welfare.
Van Zwanenberg F (1974) The geography of disease in East Anglia. *Journal Royal College of Physicians, London* **8**:145–153.
Vessie PR (1932) On the transmission of Huntington's chorea for 300 years. The Bures family group. *Journal of Nervous and Mental Disorders* **76**:553–573.
Vessie PR (1939) Hereditary chorea: St Anthony's dance and witchcraft in colonial Connecticut. *Journal of the Connecticut State Medical Society* **3**:596–600.
Volkers WV, Went LN, Vegter-Vlis M, Harper PS and Caro A (1980) Genetic linkage studies in Huntington's chorea. *Annals of Human Genetics* **44**:75–80.
Walker DA, Harper PS, Wells CGC, Tyler A, Davis K Newcombe RG (1981) Huntington's chorea in South Wales. A genetic and epidemiological study. *Clinical Genetics* **19**:213–221.
Walker DA, Harper PS, Newcombe RG and Davis K (1983) Huntington's chorea in South Wales. Mutation, fertility and genetic fitness. *Journal of Medical Genetics* **20**:12–17.
Waters CO (1842) In: *Practice of Medicine*, Vol. 2 (ed Dunglison R), p. 312. Philadelphia: Lee and Blanchard.
Westphal C (1883) Uber eine dem Bilde der cerebrospinalen grauen Degeneration-ähnliche Erkrankung des zentralen Nervensystems ohne anatomischen Befund, nebst einigen Bemerkungen über paradoxe Kontraktion. *Archiv für Psychiatrie und Nervenkrankheiten* **14**:87–96, 767–773.
Winfield JM (1908) A biographical sketch of George Huntington, M.D. *Neurographs* **1**:89–91.
Wood HC. The choreic movements (1893). *Journal of Nervous and Mental Disorders* **4**:241.
Youngman S (1989) Studies on Huntington's disease using recombinant DNA techniques. University of Wales PhD Thesis.
Youngman S, Sarfarazi M, Quarrell OWJ et al. (1986) Studies of a DNA marker (G8) genetically linked to Huntington's disease in British families. *Human Genetics* **73**:333–339.
Youngman S, Shaw DJ, Gusella JF et al. (1988) A DNA probe D5 (D4S90) mapping to human chromosome 4p16.3. *Nucleic Acids Research* **16**:1648.
Youngman S, Sarfarazi M, Bucan M et al. (1989) A new DNA marker (D4S90) is located terminally on the short arm of chromosome 4, close to the Huntington Disease gene. *Genomics* **5**:802–809.

2

The Clinical Neurology of Huntington's Disease

INTRODUCTION

Most clinical surveys of Huntington's disease have shown that between one half and three quarters of patients present initially with symptoms that are neurological in nature, principally those related to chorea. It is thus to the neurology clinic that most patients will be initially referred by their family doctor or other primary care physician. This chapter considers the range of neurological disturbance found in HD, the appropriate approaches to clinical assessment and investigation, and the differential diagnosis in a neurological context. Psychiatric symptoms, the other principal group of abnormalities, are discussed in the following chapter and it is clearly important to recognize that most patients will prove to have a mixed pattern of neurological and psychiatric disturbances, regardless of the speciality to which they present. The natural history of the disorder is covered in Chapter 4.

The clinical diagnosis of HD is dependent upon recognition of patterns of symptoms given in the history, the clinical signs present in the patient and on the family history. The various patterns which can be recognized were described in case reports during the late nineteenth century and the early part of this century and have been outlined in Chapter 1. Despite these variations, HD is still classically considered as a neurodegenerative disorder beginning in midlife, characterized by chorea and dementia, with death 15–20 years later. Patients who have features which depart from this description are often described as having 'atypical' HD; this may lead to diagnostic difficulty and error, particularly if it is not realized that chorea is not always a prominent sign and in some cases may be absent altogether. Furthermore, understanding of the clinical signs must take account of the fact that signs change during the course of the illness and that different patterns may be observed depending on the age of onset. Evidence that making the correct diagnosis is not always easy in HD is shown by the considerable number of misdiagnoses found in the systematic surveys of outcome, discussed later. Many neurologists' experience of HD comes from diagnosing early cases, after which there may be no subsequent follow-up. Consequently, they may be less familiar with the clinical picture of some patients with advanced disease.

CHOREA

Chorea is a characteristic feature of HD and until recently, the disorder was still commonly called Huntington's chorea, though the more general term, Huntington's disease, is more appropriate in view of the widespread nature of the neurological disturbance. Chorea has been defined by the World Federation of Neurology as 'A state of excessive, spontaneous movements, irregularly timed, randomly distributed and abrupt. Severity may vary from restlessness with mild, intermittent exaggeration of gesture and expression, fidgeting movements of the hands, unstable, dance-like gait to a continuous flow of disabling, violent movements'. (Lakke, 1981.)

Distinction between chorea and the various other types of abnormal movement is not always easy, but some of the main characteristics are listed in Table 2.1. Patients are often described as having involuntary movements, but as Bruyn and Went (1986) pointed out, many normal movements are involuntary (such as blinking, swallowing and arm swinging whilst walking); HD patients have additional unwanted and unnecessary movements. In a Croonian lecture, to the London College of Physicians, Wilson (1925) noted that chorea could be described as 'objectively purposeful but subjectively purposeless'. In the same lecture, Wilson described the disorganized muscle physiology, there being simultaneous contraction of agonist and antagonist muscles; this finding has been confirmed by subsequent EMG studies (Thompson et al., 1988). Surface EMG recordings of chorea show both brief, solitary contractions and also longer co-contractions. All chorea therefore includes some element of dystonia.

Table 2.1 Chorea and allied movement disorders

Disorder	Character of movement
Chorea	Irregular, generalized, flitting and flowing, often pseudo-purposive (see definitions in text)
Dystonia	Sustained muscle contractions, frequently causing twisting and repetitive movements, or abnormal postures
Athetosis	Peripheral movements of dystonic nature
Ballismus	Gross unilateral choreic movements; usually confined to limb(s)
Tics	Repetitive, stereotyped jerks of limited distribution (especially involving head, neck and shoulders) that can usually be markedly suppressed voluntarily, but at the expense of rising inner tension
Myoclonus	Brief, shock-like muscle jerks which originate in the central nervous system and often occur repetitively in the same muscles

Chorea in HD can vary from barely perceptible to extemely severe. In the early stages it may appear to be no more than an exaggeration of normal restlessness and may only be seen when the person is stressed emotionally or by being given a complex task. At this stage choreic movements can usually be suppressed by voluntary action, something that frequently happens during a formal neurological examination. Later the choreic movements become more obvious, can no longer be suppressed and are of greater amplitude.

All parts of the body are usually involved in choreic movements in HD, an important distinguishing feature from such disorders as tardive dyskinesias, where the movements have a predilection for face and jaw, and from hemiballismus, where a single limb or side is affected.

In the early stages of the disease the chorea of HD is indistinguishable from Sydenham's chorea, but the latter is self-limiting, and when well established the movements of HD involve more limb segments, as well as having repetitive and stereotypic elements. A pseudo-purposive appearance may be present, possibly due to the patient converting an initial involuntary movement into a voluntary one.

The visible effects of chorea will depend greatly on the part of the body affected (Figure 2.1). In the arms and hands the main symptomatic effect will be clumsiness in occupational or domestic duties, while in the lower limbs disturbance of gait will result; the patient may unjustly be accused of being drunk. Involvement of face, jaw and pharyngeal muscles will affect both speech and swallowing, as discussed below, while grunting and clicking sounds can occur, some of which may reflect respiratory movements against a partially closed glottis.

The severity of chorea tends to increase progressively during the initial years of the illness, later tending to plateau, in contrast to the continued progression of the other motor features of the disorder (Table 2.2). In the later stages chorea may be completely masked by rigidity and bradykinesia, while in the juvenile form and in adults presenting with rigidity, chorea may be minimal or even absent from the outset and even throughout their illness (Curran, 1930). The diagnostic value of chorea is somewhat lessened by the truly remarkable number of disorders in which it may be seen. The review of Padberg and Bruyn (1986) lists 143, most of which are exceptionally rare, posing no real diagnostic problems; some of the more important are discussed further in the section on differential diagnosis.

OTHER MOTOR ABNORMALITIES

While chorea is the most frequent and conspicuous feature in the majority of HD patients, especially early in the disorder, it is only part of a much more profound and generalized disturbance of motor function. Indeed it is these other aspects that are the most important in functional terms, as witnessed by

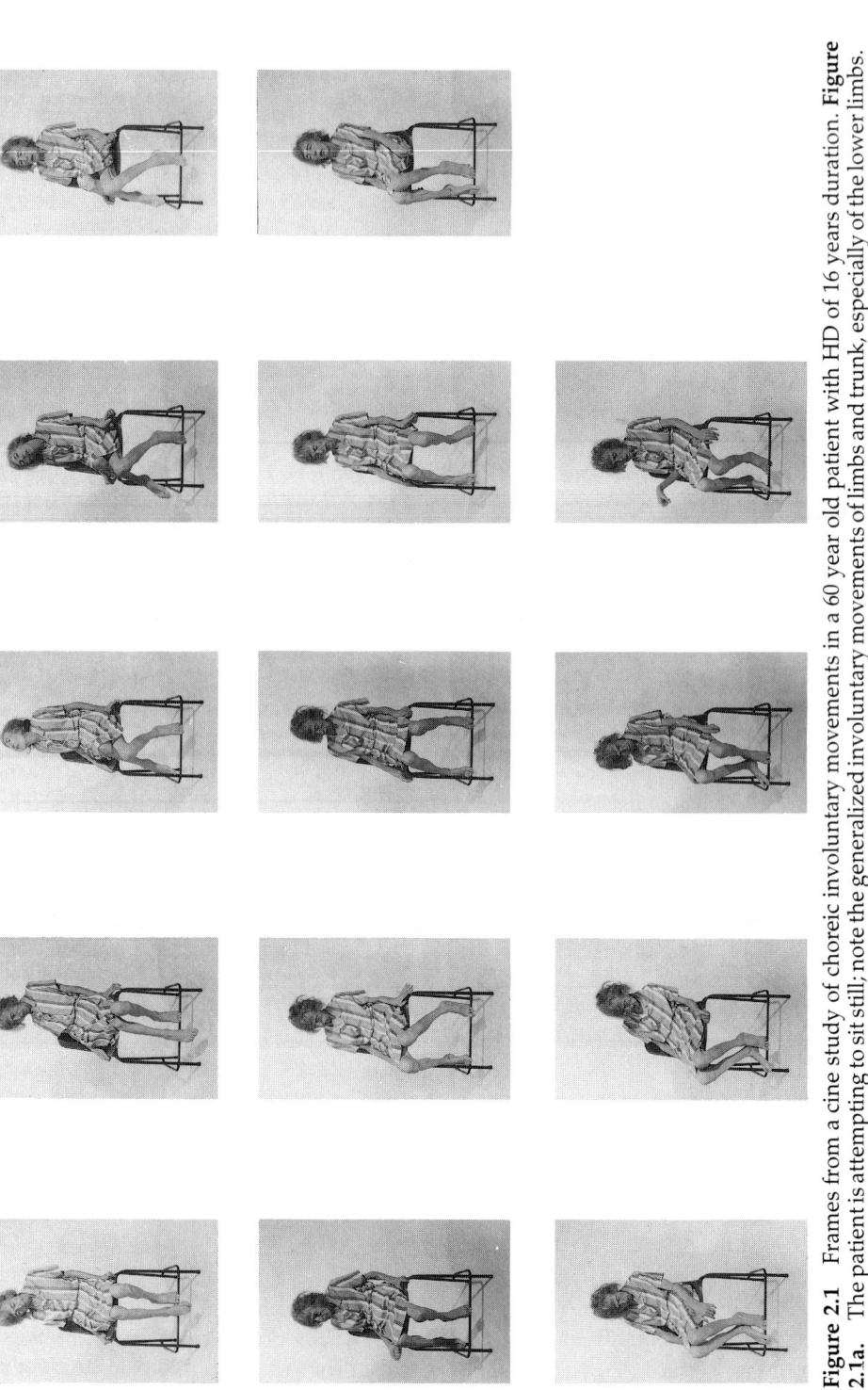

Figure 2.1 Frames from a cine study of choreic involuntary movements in a 60 year old patient with HD of 16 years duration. **Figure 2.1a.** The patient is attempting to sit still; note the generalized involuntary movements of limbs and trunk, especially of the lower limbs.

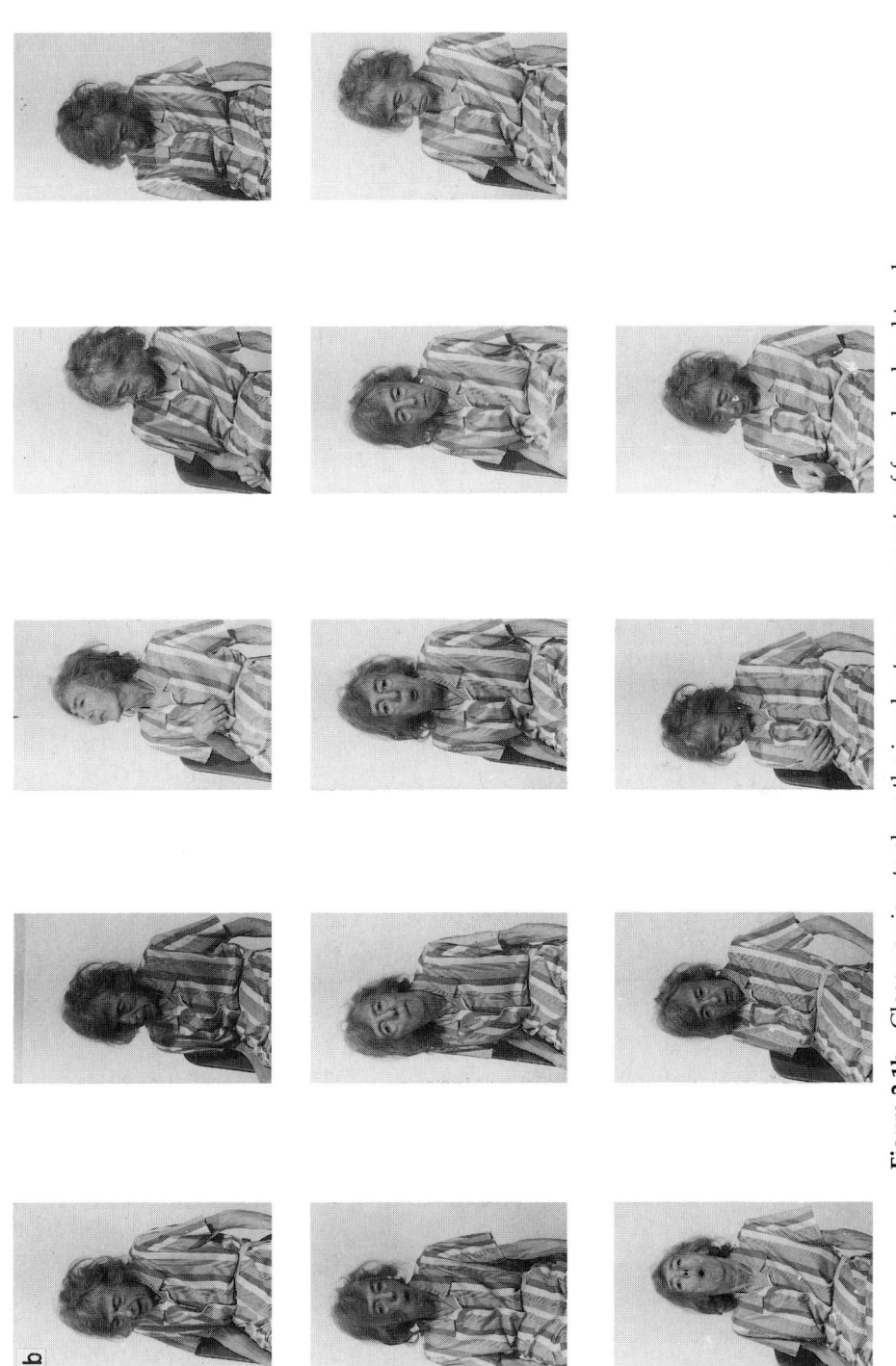

Figure 2.1b Close up view to show the involuntary movements of face, head and trunk.

Figure 2.1c The patient is attempting to hold the arms outstretched. Note the marked involuntary movements of the upper limbs.

the relative lack of serious disability produced by 'pure chorea' in such syndromes as benign familial chorea. Table 2.2 summarizes some of these aspects.

Dystonic movements are frequently seen in advanced HD, usually appearing well into the course of the disease and increasing in severity. They can be defined as 'slow, sustained muscle contractions which distort limbs, trunk, neck, face or mouth into characteristic postures' (Lakke, 1981) or 'a syndrome of sustained muscle contractions, frequently causing twisting and repetitive movements, or abnormal postures' (Fahn, 1988). In contrast to idiopathic dystonia, dystonic movements in HD are not seen in the absence of the other features of the disorder, nor do they show any particular spatial distribution.

Rigidity has already been mentioned as a major feature of most advanced cases of HD; it is a prominent feature of juvenile HD (see below) and of a minority of adult onset patients who may present with this as the major feature, the so-called 'Westphal variant'. Westphal (1883) described an 18-year-old girl with this type of movement disorder: she came from an HD family but Westphal himself did not recognize that her clinical features were part of the HD phenotype. Hamilton (1908) noted that two of the 27 patients he described progressed from choreic twitching to muscular rigidity. Bell (1934), in her review of 151 published pedigrees, noted that this feature had been documented in a further five families. Bittenbender and Quadfasel (1962) described eight patients with HD, in whom the predominant sign was rigidity, and reviewed 62 similar cases which had been reported in the literature. Patients were grouped as: starting rigid; starting with mixed features; starting with chorea; and miscellaneous onset. The authors noted that many of the patients starting with rigidity had an earlier age of onset; the mean was 22.2 years and 27 of the 70 patients were juvenile (onset before 20 years). In addition, they felt that their study supported the earlier suggestion by Denny-Brown (1960) that HD evolves towards an akinetic rigid state. Despite this wealth of literature Bittenbender and Quadfasel commented that 'among English-speaking neurologists, the limited classical concept of Huntington's chorea as a symptom triad of dementia, chorea and hypotonia still to some extent persists'.

Table 2.2 The range of motor abnormalities in HD

Frequent (at some point in disease course)
  Chorea
  Dystonic movements
  Eye movement abnormalities
  Rigidity
  Bradykinesia

Less common (except in juvenile HD)
  Cerebellar dysfunction
  Upper motorneurone abnormalities
  Epilepsy
  Myoclonus

Much of our information on the rigid form of HD comes from the early German literature (e.g. Vogt and Vogt, 1920; Entres, 1921a,b, 1940). The review of Bruyn (1968) provides a detailed discussion of these cases and of the relationship between the different types, but more recently the concept of separate juvenile, rigid and classical forms has progressively broken down in favour of a more general relationship between age at onset and pattern of clinical features.

Bradykinesia and akinesia are important and frequent features of HD. Not only may they be a prominent feature later in the course of the illness, but they may also be observed early in the course of the disease even when the predominant neurological sign is chorea. This has been demonstrated electrophysiologically by Hefter *et al.* (1987) and Thompson *et al.* (1988) with EMG recordings obtained during simple voluntary, rapid alternating and complex sequential movements. The results were independent of whether or not the patients were taking neuroleptic medication. Cognitive impairment was assessed in the study of Hefter *et al.* (1987); apart from a prolongation of reaction times, there was no significant correlation between bradykinesia and tests of cognitive function. These observations may in part explain some of the abnormalities of voluntary movement which may be observed clinically.

A quantitative clinical assessment of voluntary movement impairment in HD has been provided by Folstein *et al.* (1986) who reported the neurological abnormalities of 149 patients; all had some abnormality of voluntary movement. Among a subset of 34 patients with a duration of illness between 0 and 3 years, approximately 80% had abnormalities of rhythm and speed which included at least one of the following: alternating rapid tongue movements, finger thumb tapping and dysdiadochokinesis. These abnormalities were seen in 90–100% of patients with a longer duration of illness. Deficits in voluntary motor control worsen as the disease progresses (Penney *et al.*, 1990) in contrast, chorea worsens in the early course of the illness and then plateaus (Folstein *et al.*, 1983). In the final stages of the illness, patients become wheelchair bound, bed-ridden and eventually immobile.

The pattern of disease progression for many patients with onset in midlife is summarized as in Table 2.3. As with all generalizations there are exceptions; it has already been emphasized that individual patients may present predominantly with hypertonia and bradykinesia, in which case, chorea may be absent or only a minor component of the disorder. Progression of the disease may be

Table 2.3 Pattern of progression of HD (mid-life onset)

Stage of HD	Early	Mid	Late
Chorea	+++	++	++
Dystonia	+	++	+++
Bradykinesia/akinesia	+	++	+++
Decreased voluntary movement	+	++	+++

GAIT

Abnormalities of gait are obvious among those patients who have had signs and symptoms of HD for some years (Figure 2.1). HD patients in whom chorea is a predominant sign, walk with a wide based staggering gait. HD patients do fall but it is surprising that mobility is maintained despite seemingly most precarious arrangements of the limbs and trunk. Those patients with significant bradykinesia and hypertonicity may walk with a slow, stiff, unsteady gait; 'freezing', especially in confined spaces, may occur, and may precipitate falls. In a formal study of 13 male patients, in which stride abnormalities were recorded using ultrasound transducers strapped to the ankes, 12 showed significantly decreased and variable walking speed and stride length (Koller and Trimble, 1985). The same study also documented gait abnormalities in four out of five patients with symptoms of less than 5 years' duration. Increasing doses of haloperidol decreased choreic movements but did not improve gait or mobility, which can often worsen as a result of neuroleptic treatment.

In the early stages of the disease, abnormalities of gait can be provoked by asking the patient to walk away from the examiner and suddenly stop and turn through 180° on command. Similarly, abnormalities of balance and posture can be noted by asking the patient to stand with heels together for 15–20 s or to walk across the room in heel to toe fashion. Abnormalities of these manoeuvres are not specific to HD but are useful additional signs in those suspected of having developed the disorder.

EYE MOVEMENTS

Abnormalities of eye movement are seen in a number of disorders affecting the vestibular apparatus, cerebellum, pons, superior colliculus, basal ganglia and cortex. A detailed description of the normal control of eye movement has been given by Leigh and Zee (1983). The most significant abnormality in HD concerns the fast accurate saccadic movements which allow the foveae to move the visual focus rapidly to different objects.

Detailed measurements have been made of small groups of HD patients using a number of neurophysiological techniques: DC electro-oculography (Starr, 1967; Avanzini et al., 1979; Oepen et al., 1981; Beenen et al., 1986; Lasker et al., 1988); the scleral search coil technique (Leigh et al., 1983; Kirkham and Guitton, 1984; Collewijn et al., 1988); contact lenses with luminescent spots (Petit and Milbled, 1973) and the infra red reflectance method (Zangemeister and Mueller-Jensen, 1985; Bollen et al., 1986). Abnormalities have been noted irrespective of whether or not patients were taking neuroleptic medication.

Abnormalities in ocular movement occur early in the disease in a majority of patients and gradually worsen (Leigh et al., 1983; Folstein et al., 1986). Some authors (Oepen et al., 1981; Zangemeister and Mueller-Jensen, 1985; Beenen et al., 1986) have reported that a few patients show no ocular abnormalities but percentages need to be interpreted with caution since patients in these studies were selected on the basis of their ability to cooperate with the test.

HD patients have difficulty in initiating saccades (whether voluntary, in response to verbal commands, or reflex to a visual stimulus within the field of vision) and a decrease in saccadic velocity. Patients are often unable to suppress head movements or blinks during saccadic movements, (Starr, 1967; Leigh et al., 1983; Folstein et al., 1986). The accuracy of saccadic movements is often impaired with undershoot of the target (Leigh et al., 1983; Bollen et al., 1986; Lasker et al., 1988). Some authors have suggested that vertical saccades may be more severely affected than horizontal saccades (Petit and Milbled, 1973; Avanzini et al., 1979; Oepen et al., 1981; Leigh et al., 1983) but this finding was not confirmed by Collewijn et al. (1988), who suggested that their patients were less severely affected and that normal individuals have more difficulty with vertical saccades.

Gaze fixation abnormalities have been noted, the patient being unable to fix on a light source without the intrusion of small saccadic movements; similarly, smooth pursuit is interrupted by small jerky saccadic movements. HD patients cannot suppress reflex saccades to a visual stimulus in the field of vision despite having understood the instruction; this saccadic distractibility was a prominent finding in the study by Lasker et al. (1987).

Lasker et al. (1988) studied 10 patients with onset under 30 years and 10 with onset over 30 years. They suggested that a predominant early sign in those with a young age of onset was increased latency whilst those with older age of onset had decreased velocity as a predominant early sign. This observation was not confirmed by Collewijn et al. (1988).

Bedside examination of eye movements is possible. The patient is asked to look at the examiner's nose and then quickly flick his or her eyes towards objects held by the examiner in the periphery of the visual fields. In this way an impression may be gained of the patient's ability to move the eyes rapidly without head turns, blinks, undue delay or decreased velocity. Similarly, bedside impression of smooth pursuit and gaze fixation may be assessed. Although not pathognomic, abnormal eye movements reflect the underlying pathology in the basal ganglia and, provided other features are present, may be considered as evidence to support a diagnosis of HD.

DYSARTHRIA

Dysarthria is a common sign among HD patients. The rate and rhythm of speech may be disturbed early in the course of the disease (Folstein et al., 1983, 1986). As the disease progresses further the dysarthria worsens and speech may become unintelligible. In one study, eight patients were unable to sustain three vowel sounds without errors, which included reduced duration, low frequency elements and vocal arrests, suggesting involvement of the laryngeal and respiratory muscles in the movement disorder (Ramig, 1986). In addition to the mechanical problems, neuronal loss in the basal ganglia and cerebral cortex disrupts linguistic abilities resulting in reduced vocabulary and syntactic errors (Gordon and Illes, 1987; Troster et al., 1989). Some patients may become totally mute at a stage before motor disability is extremely severe. Speech disturbance represents an important aspect of management in advanced HD and is discussed further in Chapter 7.

DYSPHAGIA

Progression of HD is associated with dysphagia which, in some, may result in choking and asphyxia (Edmonds, 1966; Lanska et al., 1988). Dysphagia may involve multiple abnormalities of ingestion including inappropriate food selection and rate of eating (attributed to the dementing process), bite size and bolus formation, retention of food in the mouth after swallowing, aerophagia and regurgitation, and poor respiratory control (Leopold and Kagel, 1985). The implications for management are discussed in Chapter 7.

CACHEXIA

Cachexia is a frequent, but not invariable, observation in the later stages of HD. Oepen (1963) documented this in a large survey of 217 post mortem cases. Bruyn (1968) summarizes the situation in a characteristically succinct fashion. 'I have personally seen well over a hundred cases but never a fat one'. The generalized lack of muscle bulk is often obvious; a useful clinical sign is guttering of the dorsal aspects of the hands. Sanberg et al. (1981) demonstrated that weight loss occurred in HD patients despite an adequate diet and feeding. Hyperkinesia was not thought to be responsible for the observation since high calorie diets were provided to half the patients; in any case, chorea is less marked in the end stage of the disease and rigidity more common. Similar results were obtained in an anthropometric analysis of HD patients in the early- to mid-stages of the disease (Farrer and Yu, 1985). Highly significant differences were found between HD patients and normal controls for weight; body mass index; triceps and subscapular skinfold thickness; arm, chest and abdominal circumferences, and bizygomatic and bigonial breadths. The only dietary difference was increased calorie intake among the HD patients. In addition, the HD patients had a lower index of strenuous activity than the controls.

In conclusion, weight loss is an integral manifestation of the disease that is not fully understood; it is independent of the hyperkinesia and does not always respond to treatment with high calorie diets in the later stages of the disease, though they may be helpful in the early stages in preventing or even reversing weight loss.

SLEEP

A striking clinical observation, even in patients with severe or longstanding chorea is that choreic movements largely cease during sleep. In the mid- to late-stages of the disease it is not uncommon for the spouse or carer to complain that the patient was restless in bed, but even in such advanced cases hospital observation suggests the choreic movements are absent when the patient is deeply asleep. A recent quantitative study (Fish et al., 1990) of HD and other movement disorders has confirmed that movements are greatly reduced in the deeper sleep stages in all disorders studied, with a comparable reduction of semi-purposive movements in both normal and affected subjects. A further problem, probably related to the dementing process, is sleep reversal; the patient being somnolent during the day and restless at night which may be disruptive for family life.

Hansotia et al. (1985) studied sleep function in seven HD patients; two with mild and five with moderately severe chorea. There was no clinical abnormality in the mildly affected patients but those with moderately severe chorea had an increased latent period before the onset of sleep and increased interspersed wakefulness. More time was spent in Stage I and II sleep and correspondingly less time in Stage III, IV and REM sleep. Emser et al. (1988) studied 10 HD patients and found their sleep patterns were essentially normal; however, in contrast to Parkinson's disease patients, there was an increase in sleep spindle density during Stage II sleep. The authors attributed this to differences in dopamine levels within the basal ganglia.

INCONTINENCE

Urinary incontinence is not an uncommon problem for HD patients and despite the difficulties caused for their families it has received little attention. Hayden (1981) stated that approximately 20% of patients become incontinent in the terminal stages of the illness; however, the basis for this estimate was not stated. Although not an unexpected finding, it would be wise to exclude recurrent urinary infections if patients experience incontinence.

Wheeler et al. (1985) found that the causes of incontinence could be related to one or more of the following: dementia, depression, decreased mobility or detrusor hyperreflexia. In this study five of the six patients were female; their symptoms included frequency, urgency, nocturia and incontinence. Two patients had normal perineal muscle EMG and cystometry but four had decreased voluntary control of the perineal musculature and detrusor hyperreflexia, the latter being defined as the inability to suppress contraction greater than 15 cm water pressure when requested. The guarding reflex (the increase in striated sphincter activity during vesical filling), was normal. This pattern of results was interpreted to indicate a suprapontine lesion and was similar to that seen in other disorders such as Parkinson's disease and cerebrovascular disease (Siroky and Krane, 1982). The EMG studies also demonstrated involuntary movements of the perineal musculature which, in contrast to other muscles, ceased during micturition.

EPILEPSY

Epilepsy may occur in HD. It is a common feature of juvenile HD (see below). As with many of the clinical features, descriptions of epilepsy in HD date back to the late nineteenth century and early twentieth century. Notkin (1931) summarized 21 patients with HD and epilepsy from the literature; at least seven had juvenile HD. Several series have considered this feature and have shown a frequency of 1–3% in adults with HD (Oepen, 1963; Markham and Knox, 1965; Hayden, 1979). The prevalence of epilepsy among the general adult population is approximately 0.35–0.62 per 100 population (Hauser and Kurland, 1975), therefore, epilepsy occurs slightly more frequently in adult patients with HD. It is important to consider anoxia due to choking as a preventable cause of seizures in advanced HD.

Myoclonus is an occasional feature of HD. It is most often associated with juvenile cases (see below) but may also occur in young adults (Kereshi *et al.*, 1980). The authors have seen a similar case aged 26 years and have had correspondence concerning a 41-year-old woman, said to have had onset of symptoms aged 30 years. In both cases family history was undisputed and in the former there were neuropathological changes which supported the clinical diagnosis.

CEREBELLAR ABNORMALITIES

In general, macroscopic cerebellar atrophy is not a usual feature of HD (see Chapter 5). In describing the general diagnostic features of HD, Paulson (1979) considered abnormalities of finger-nose testing to result from sudden shifts in tone which could be misinterpreted as indicative of a cerebellar disburtance. However, post-mortem studies of some juvenile cases have shown abnormalities of the cerebellum; case 7 of Markham and Knox (1965) had loss of Purkinje cells and gliosis of the molecular layer. Case 4 of Jervis (1963) had considerable dysmetric and coarse tremors; pathologically there was cerebellar atrophy, absent Purkinje cells, and atrophy of the dentate nucleus; his brother, case 3, had less marked cerebellar atrophy. Byers *et al.* (1973) reported neuropathological findings in the cerebellum of four children but commented: 'the severe atrophy of the cerebellum stands in striking contrast to the absence of cerebellar symptoms and signs'.

The cerebellar abnormalities found in some of the larger systematic neuropathological studies, such as those of Rodda (1981) are discussed in Chapter 5, as are the results of quantitative studies (Jeste *et al.* 1984).

TENDON REFLEXES AND PYRAMIDAL TRACT ABNORMALITIES

Tendon reflexes are variable in HD; they will be normal in many cases, reduced in some and pathologically brisk in others. Areflexia, however, is not a feature of HD, and should raise the possibility of neuroacanthocytosis (p. 73), a condition whose clinical picture can often be otherwise indistinguishable from HD. Likewise, the plantar response is often flexor, but in some, particularly in the later stages when rigidity is a feature, it may well be extensor. Late in the course of the illness primitive reflexes (pout and grasp) may be observed; clearly these are non-specific but reflect the degree of neurodegeneration.

The percentage of patients with abnormal tendon reflexes varies with different studies: Heathfield (1967) reported brisk reflexes, clonus and extensor plantar responses in 22% of patients; Stevens (1976) reported 30% of patients to have marked and 18% to have mild pyramidal tract signs; Hayden (1979) reported that 26% had brisk reflexes, 12.3% had clonus and 7.6 had extensor plantar responses. In their study of rigid forms of HD, Bittenbender and Quadfasel (1962) found increased reflexes in 67% of their cases and clonus in 22%. These figures are interesting in themselves but have to be interpreted in the light of changes in phenotype in relation to duration of disease and age of onset.

HD IN THE ELDERLY

The remarkable range of age at both onset and death seen in HD may cause it to be unsuspected in the elderly in the same way as often occurs in childhood. Onset over the age of 70 years is exceptional; it is more common to find an elderly person who at the time of presentation is relatively little disabled but who in retrospect has a history of choreic movements present for many years, even decades. Survival into the 80s is not uncommon and patients living to over 90 years have been recorded. Age at onset and death frequently show familial correlation, as well as sex-related effects of parental transmission, as discussed in Chapters 4 and 9.

Myers et al. (1985) studied the phenotypic features of 25 patients with onset of neurological disorder after 50 years. They concluded that both chorea and cognitive impairment were slowly progressive but that symptoms could plateau for several years. Table 2.4 summarizes some of the more important clinical aspects that need to be considered in diagnosis and management.

Late onset cases commonly present with chorea and few other features of HD. Normal intellect, at least initially, may deter clinicians from making the diagnosis, while minor degrees of deterioration may pass unnoticed by the family who have attributed them to the effects of age. If the patient's heterozygous parent also had been destined to develop late onset HD, a positive family history may be lacking because their death preceded the onset of clinical disease. A member of an HD family who presents in old age with dementia may well prove to have another cause for it, such as Alzheimer's disease, and neuropathology will be of particular importance in resolving the diagnosis, though the brain of late onset cases of HD may have minimal or even no pathological abnormalities at autopsy, as discussed in Chapter 5.

In the elderly patient with suspected HD, tardive dyskinesia related to drug therapy for other disorders requires special consideration, but HD does not show the particular buccal and lingual distribution characteristic of this. Although the course of HD in the elderly is often relatively benign the mean survival is not increased in this group, largely because of the greater mortality from all causes. Death is commonly from vascular disease, or other unrelated disorders. Sometimes late-onset cases are only diagnosed when they are examined after one or more of their offspring has presented with chorea.

Relatives of an elderly HD patient whose course has been benign may reasonably wish to know whether this implies a correspondingly benign course for themselves, should they become affected. Attitudes to genetic counselling and reproduction are considerably influenced by the perceptions in the family of the disorder's severity. Unfortunately, despite the familial correlation for age at onset and death, variation within a kindred is still too great to give any adequate reassurance (see Chapter 9).

A final question to be raised is whether there really is a separate condition of 'senile chorea' (Alcock, 1936; Friedman and Ambler, 1990), either as a non-genetic disorder or as a separate inherited condition to HD. Most workers

Table 2.4 Huntington's disease in the elderly

Chorea	Prominent
Rigidity	Uncommon
Other motor abnormalities	Usually minor or absent
Mental changes	Often slight, may be absent
Course	Often prolonged; death from other causes frequent
Family history	Correlation for late onset, but still very variable
Autopsy changes	May be minimal

would now agree that senile chorea does not constitute a single syndrome entity. Most cases of chorea in the elderly without a clear cause are likely to be due to HD, though a minority are due to cerebrovascular disease.

JUVENILE AND CHILDHOOD HUNTINGTON'S DISEASE

The term 'juvenile Huntington's disease' is generally applied to those cases of HD with onset before 20 years of age, though this is an arbitrary division. This group has been estimated to make up between 3% and 10% of all HD in different surveys, as shown in Chapter 4. Particular interest has focused on juvenile HD on account of its unusual genetic transmission, with a preponderance of affected fathers, and also because the clinical features are often far from typical, giving rise to diagnostic difficulties.

Many clinicians are still unaware that HD may occur in young children, not just in adolescence; the group of HD patients with onset under 10 years is rare, probably no more than 0.5% of all cases, and the differences, both genetic and clinical, that are seen in the juvenile category overall, are still more striking when this restricted group of early childhood cases is considered. However, the clinical features, while applying particularly to children, are not exclusive to them and can also be seen in adolescents and in some young adults. They are described here in the context of childhood HD since these children will usually present not to neurologists but to general paediatricians, child psychiatrists or educational psychologists and other clinicians who will often not usually be considering HD in their range of expected diagnoses.

The history of the recognition of HD in childhood is closely bound up with that of the rigid form or 'Westphal' variant, since so many childhood cases fall into this clinical grouping. The family reported by Hoffmann (1888) contained a girl whose first symptom was seizures aged 4 years, followed by chorea and rigidity, along with other relatives showing early onset and features of the rigid form. Some of the other early reports of childhood HD (notably that of Lyon (1863), discussed in Chapter 1), may well have represented other disorders such as benign familial chorea. Over 30 cases with juvenile onset had been reported by the end of the century, so that by the time of the systematic studies of Entres (1921a,b) and Bell (1934), it was a well-recognized entity.

The studies of Bittenbender and Quadfasel (1962), Campbell *et al.* (1961), Jervis (1963) and Osborne *et al.* (1982) provide clear pictures of the clinical features while the descriptions of Byers and Dodge (1967) and Oliver and Dewhurst (1969) give details of young children with HD. The best summary of the early literature is that of Bruyn (1968), though his views on how childhood HD should be classified and its relation to the rigid form as a whole have changed subsequently (Bruyn and Went, 1986). Figure 2.2 illustrates the case history of a patient studied personally.

The most important clinical point to appreciate about juvenile HD is that most cases present with a clinical picture that differs markedly from that shown

Table 2.5 Juvenile HD – distinguishing features

	Juvenile	Adult
Mental deterioration	Severe	Variable
Chorea	Often not prominent even absent	Usually prominent
Rigidity	Marked	Usually in later stages
General motor abnormalities	Prominent	Usually in later stages
Upper motorneurone and cerebellar signs	Common	Uncommon or late
Epilepsy	Common	Uncommon
Time course	More rapid decline, especially mental; but survival can be prolonged	Variable; survival not age dependent
Inheritance	Father usually affected	Either parent affected

by most adult-onset cases of HD. Table 2.5 shows the most frequent presenting and early features; it is obvious that these are not only strikingly different from most cases of adult HD, but also that they could be caused by a wide variety of generalized neurodegenerative disorders.

Presentation of juvenile HD is commonly with symptoms of mental disturbance. Behavioural disorder in a previously normal child may cause school problems and lead to a psychological referral; it may not be recognized at this stage that an organic disorder is underlying these symptoms. Once a brain disorder is suspected, either because of the presence of neurological abnormalities, or because of the results of specific psychometric testing, it may still not be appreciated that a progressive disorder is present. The abnormalities may be regarded as mental handicap due to a perinatal event or other static factor, rather than as a progressive mental deterioration.

Even when the clinical picture is correctly recognized as a progressive neurodegenerative disorder, the associated features are often not sufficiently specific to give a clear indication of the cause. Extensive metabolic investigations may be undertaken searching for a suspected lysosomal storage disease or other metabolic disorder.

Seizures of a variety of types are common in juvenile HD (Brackenridge, 1980), occurring in around 30% of cases, in contrast to the low frequency of around 2% in adults. They may be *grand mal*, akinetic or myoclonic in type. The EEG frequently shows epileptiform features, again in contrast to adult HD, but it is not specific for juvenile HD.

Rigidity is a prominent feature of most cases with juvenile onset, at least when neurological abnormalities become established. The parkinsonian features may suggest a diagnosis of Wilson's disease of Hallevorden–Spatz disease. The rigidity may also be aggravated by spasticity, with a general

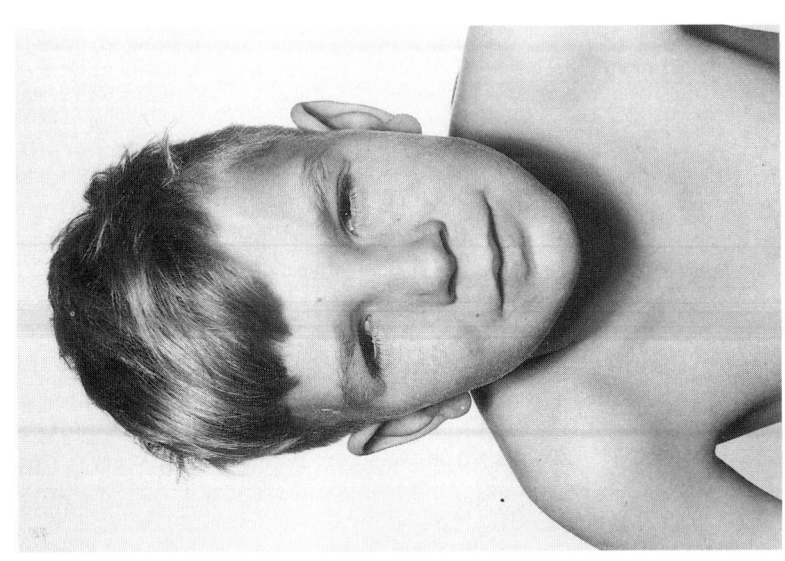

Figure 2.2

increase in upper motor neurone abnormalities, changes rarely seen in adult HD except at advanced stages.

Bradykinesia is an important clinical feature accompanying the rigidity, affecting general motor functions and especially gait, abnormality of which may be the presenting feature.

Chorea has not so far been mentioned in the clinical picture of juvenile HD, but it is frequently present. However, it is not usually a presenting or predominant feature as it is in adult HD, being dominated by other motor or mental features already described. The child who presents with chorea as the only or predominant feature is most likely to have a different cause, such as benign familial chorea. Nor is chorea the only movement disorder to be seen in juvenile HD; tremor may occur, again accentuating the parkinsonian picture when taken together with rigidity, but is usually of the action rather than resting variety; various dystonic movements and myoclonic jerks may be seen. Levodopa or dopamine agonist therapy in akinetic-rigid cases of HD usually have no effect in such patients.

Other points to be mentioned are the involvement of speech, due to any combination of extrapyramidal or pyramidal dysarthria, and mental deterioration; ataxia, which may have a cerebellar component; and disturbance of eye

---

Figure 2.2 Juvenile Huntington's Disease. This boy, from a large South Wales HD kindred, was initially seen in 1964 aged 12 years with a one year history of slowness of gait and slurred speech, together with progressive mental deterioration. The principal findings were an expressionless face, general slowness of all movements and increased tone in the limbs, particularly the arms. There was dysarthria and mild ataxia of the lower limbs. Reflexes were normal but the left plantar response was extensor. School and psychological assessment confirmed extremely poor school progress and a withdrawn and easily frightened attitude; formal assessment showed a reading and arithmetic age of around 7 years.

Full investigation failed to show any metabolic or other cause, including ophthalmological assessment for Wilson's disease. Air encephalogram was normal. During the subsequent 13 years he slowly deteriorated, becoming totally dependent and bedridden. At the age of 22 he started to have recurrent generalized epileptic seizures, following one of which he developed bronchopneumonia and died.

Autopsy confirmed the clinical diagnosis of HD.

The patient's sister developed intellectual and motor deterioration aged 19 and was diagnosed as having HD aged 22. She showed marked rigidity with a coarse tremor which was not significantly affected by a wide variety of drug treatment. Eye movements were severely restricted. She deteriorated progressively and died aged 31 years.

Autopsy showed generalized cerebral atrophy, especially marked in the caudate nucleus. Microscopy showed gliosis and almost total absence of neurones in the caudate nucleus.

These two patients were members of a sibship of four. One brother committed suicide aged 25; there was no clinical features of HD at the time, nor did autopsy show features of the disorder. Another brother remains well aged 45. The father died aged 40 after a clinical diagnosis of multiple sclerosis; study of the brain at autopsy showed typical changes of HD in the caudate nucleus. Numerous paternal relatives are also known to be affected with HD.

movements, a prominent feature in juvenile and rigid cases of HD in general. Vision itself is not affected, a useful distinguishing feature from some other neurodegenerative diseases such as Batten's disease.

The course of childhood HD is usually a slow but relentlessly downhill one. The rate of decline has been shown to be somewhat more rapid than in adult HD, but survival may be prolonged and there is some disagreement as to whether overall survival is shorter than in the adult form (see Chapter 4). Van Dijk *et al.* (1986) analysing 195 cases of juvenile HD in the literature, found some shortening even when correcting their data for cases still alive, while Farrer and Conneally (1985) found that only those cases with onset under 14 years showed reduced survival, there being no general relationship between survival and age at onset. Van Dijk *et al.* found two other relevant points in their study; survival of those juvenile cases showing the rigid form was shorter than those showing juvenile onset with chorea, while the minority of juvenile cases that were maternally transmitted generally did not show features of the rigid form.

Laboratory investigations of childhood HD, often undertaken in great detail to exclude metabolic disorders, are uniformly negative, while radiology confirms the widespread and severe involvement of the brain, atrophy being seen widely in the basal ganglia, cerebral cortex and sometimes cerebellum. The neuropathological changes, described in Chapter 5, are again more severe and generalized than in most adult cases, but do not differ in any specific manner; the increased frequency and degree of cerebellar involvement has been noted earlier.

The unusual genetic aspects of juvenile HD have been the subject of numerous studies over the past 20 years since the initial observations of Bruyn (1968) and Merrit *et al.* (1969) that most cases were transmitted by an affected father. For those cases with onset under 10 years almost all cases are paternally transmitted, while for juvenile cases overall the proportion is 80–90%. The genetic aspects are discussed more fully in Chapter 9, but it should be noted here that the evidence is strongly against juvenile HD representing a separate disorder since it occurs in families where the majority of affected members show the typical adult onset disease.

The paternal transmission of childhood HD can cause practical problems in diagnosis. The father may be mildly affected, may deny symptoms, refuse examination and in some cases actively conceal the family history of HD. This, together with the generalized and initially non-specific nature of the clinical features in an affected child, may result in the diagnosis of HD not being made, most commonly because it has never been considered as a possibility. One of the values of a systematic genetic register for HD is that the diagnosis is more likely to be considered when a known genetic risk is present.

Should molecular genetic studies be undertaken as a diagnostic test for HD in childhood? Hammer *et al.* (1987) have reported such a case where DNA analysis was done to confirm a suspected clinical diagnosis and showed a high risk of the HD gene being present. The question of childhood testing for *prediction* of HD and other late-onset disorders is a controversial one (Harper and Clarke, 1990) and is discussed in Chapter 12, but while it might seem that the diagnostic use of this approach is free from these problems, this is not entirely so (Craufurd *et al*., 1990). In particular it must be recognized that the fact that the HD gene may have been inherited by the child does not necessarily mean that current symptoms are the result of HD. It would certainly not seem justified to embark on such testing when the clinical features are vague (e.g. non-specific behavioural problems in the absence of neurological abnormality) or when the expected risk is low (e.g. when the mother is the affected parent).

CLINICAL ASSESSMENT OF THE SUSPECTED HD PATIENT

The neurological abnormalities characteristic of HD, as well as those that are found less frequently, will have already given the clinician a clear indication of what features should be particularly sought in the neurological history and examination. Table 2.6 summarizes these, but it is worth emphasizing some of the points relevant to the examination.

Many patients with suspected early HD will show almost no abnormality on routine examination apart from minor choreic movements. Even these, as noted earlier, may be suppressed during the examination itself and only present when the patient is stressed, when entering the room or rising from a chair, or during taking of the history. Thus the examination of the patient must be considered as encompassing the entire consultation, not just the formal examination. Detailed neurological examination should focus on those aspects likely to be abnormal early in the course of the disease. Problems with voluntary movement may be detected by asking the patient to make rapid tongue movements, finger thumb tapping or testing for dysdiadochokinesia. Similarly, clinical assessment of eye movements may reveal abnormalities of saccadic movements. Dysarthria may have been noted during history taking but can be more formally assessed by asking the patient to repeat simple words or syllables. In most cases a confident clinical diagnosis may be made; however, there will be an occasional individual with some abnormal features for whom repeated examinations at annual intervals should be considered before the diagnosis is established.

Assessment of higher mental functions will frequently be abnormal, as discussed in the next chapter; however, in older patients, this may be normal, even on detailed testing. Again the observer may be alerted to problems by evidence of disturbed mood or affect during the general interview.

On examining the limbs, chorea is most likely to be the predominant abnormality, with bradykinesia and dystonia in the later stages; these may be seen in younger patients, especially children, at the time of presentation. Tremor and myoclonus may also be observed in this latter group. In assessing chorea in the limbs it is helpful to observe the fingers when the arms are outstretched. Chorea may also be detected by the 'milkmaid sign'; the patient is asked to hold the examiner's index and middle fingers, when variable grip can

Table 2.6 The neurological examination in HD; points of particular diagnostic relevance

General interview	Be alert for minor involuntary movements, evidence of altered mental state, abnormal gait or posture
General examination	No systemic abnormality expected
Mental state	See Chapter 3 for details. May be entirely normal
Cranial nerves	
Fundi	Abnormality suggests different or additional diagnosis
Eye movements	Sensitive early abnormality, especially saccades
Hearing	Normal
Facial muscles	Often give early evidence of choreic movements
Tongue	Inability to stay protruded; chorea and dyspraxia
Speech	Rarely markedly altered early in disease, but severely affected in later stages
Limbs	
General observation	Choreic movements, early in hands and feet. No early wasting
Power	Normal, but may seem to fluctuate because of superadded chorea
Tone	Increased early in rigid cases; otherwise normal initially
Reflexes	Usually normal early except for juvenile form; may be affected by choreic movements ('hung-up' jerks) often increased later with extensor plantars in some cases. Impaired/absent in neuro-acanthocytosis
Coordination	Specific cerebellar signs rare
Sensation	Normal
Gait	Usually normal early; later affected by chorea and dystonia; may be very bizarre

be felt easily if movements are present. Inability to maintain the tongue protruded may reflect both chorea and dyspraxia; the patient should be asked to protrude the tongue for 20 s, care being taken that it is not gripped by the patient's teeth.

A helpful measure, which will also accentuate any choreiform movements, is to ask the individual to close their eyes. Similarly, a trained professional observer also present in the room but who is not partly occupied by the interactions of the clinical interview, may be better placed to detect mild chorea.

Assessment of muscle strength will usually be normal in early cases but may be affected by any significant bradykinesia, dyspraxia or general motor disturbance. Tone will likewise usually be normal initially; hypotonia is not so characteristic of HD cases with marked chorea as it is in Sydenham's chorea, although it may seem variable because of superimposed chorea. Rigidity is rarely as marked at early stages as in Parkinson's disease. The tendon reflexes will be affected by choreic movements, but will generally be normal in early disease. The 'hung-up' knee jerk is produced by the interposition of a choreiform contraction of knee extensors upon the normal course of the reflex.

Pyramidal signs may develop later, while again they may be present early in juvenile HD.

Incoordination of the limbs is generally a reflection of the movement disorder; intention tremor and signs specifically suggestive of cerebellar involvement are rare.

On bedside examination external ocular movements are clinically normal in schizophrenia and tardive dyskinesia. Thus a significant disturbance of eye movements is a valuable clinical sign in neuroleptic-treated subjects where there is a suspicion of HD.

It can be seen from this description that while the neurological features characteristic of HD are relatively few, a large number of abnormalities can occur. It is thus worth asking whether there are findings which, if present, would strongly contraindicate HD as a diagnosis. There are in fact very few such findings, but among them are optic atrophy (or other abnormalities of the fundus), sensory changes of any kind, a strongly focal or otherwise restricted distribution of the abnormalities, or involvement of organs outside the nervous system. Such findings should strongly suggest an alternative diagnosis or, if present in a patient with undoubted HD, the presence of additional separate pathology.

QUANTITATIVE ASSESSMENT SCALES

As mentioned earlier these are widely used by clinical research workers, but they have a valuable place in the serial assessment of HD patients, especially in judging the response to therapeutic agents. They can also be valuable in assessing relatives at risk, particularly if a suspicion of early HD has been raised by the patient or by another observer. While at first sight they may seem complex and time consuming, this is far from being the case, especially if used regularly, and they provide much more valuable documentation than can usually be obtained from a regular neurological record.

The quantitative neurological examination devised by Folstein *et al.* (1983) and used by them in their extensive study of HD patients in Maryland (Folstein, 1989), is particularly valuable in allowing detection and assessment of motor abnormalities other than chorea. It has proved especially useful in the assessment of applicants prior to predictive testing, in order to exclude early clinical features of HD in asymptomatic individuals. Three useful subsets of information can be obtained from it: the *motor impairment scale*, the *chorea scale* and the *eye movement scale*. A separate scale for assessment of chorea has been described by Marsden and Quinn (1981).

The *functional assessment scale* of Shoulson and Fahn (1979) (Appendix 2) provides a score of handicap as relevant to the patient. By concentrating on what the patient can or cannot do, as opposed to the severity of neurological impairment perceived by the observer, this type of scale is particularly suited to trials of therapy, where the improvement in a particular aspect of the condition claimed by medical staff may not be reflected in overall reduction in handicap. The scale is also valuable in large scale comparisons of natural history, as discussed in Chapter 4.

Comparable scales are useful in assessment of the mental state in HD; these are discussed in the next chapter.

The isolated case

The patient showing clinical features suggestive of HD but with no family history of the disorder is quite commonly seen. Thirty-two such patients were encountered among a total of 192 cases of HD over a 10 year period in the study of Quarrell *et al.* (1986). The serious dilemma that this poses for genetic counselling is discussed in Chapter 11, but the clinical aspects deserve consideration here.

To start with, the most important point is to maintain a strong attitude of scepticism about the isolated nature of the case. If this is done and efforts to trace details are exhaustive, it is surprising how often a clear history of HD in a previous generation or collaterals emerges. As mentioned in Chapter 11 this requires a combination of a particularly indefatigable temperament and the skills of a detective as much as those of a clinician.

While such a search is in progress, it should go without saying that the clinical assessment and investigation of the patient should be especially thorough, so that there can be no doubt that the disorders considered later under differential diagnosis have been excluded. When this has been done it should be clear whether the case is typical of HD apart from the absent family history, or whether it is a puzzling neurological disorder, not fitting any precise diagnostic category, and where HD is simply one of a number of possibilities. In both situations, the authors' policy has been to use diagnostic labels that clearly indicate the uncertainty and need for further confirmation. The apparently typical case is classed as 'probable Huntington's disease, not proven', whereas that where HD is less likely would be classed differently, e.g. 'movement disorder; possible Huntington's disease'. The precise way in which this is done is of less importance than that it should clearly indicate that a firm diagnosis has not been reached and that further assessment, particularly detailed neuropathological study in the event of death, is essential. DNA should be isolated from a blood sample and banked for future study; brain or other tissue can be used if blood has not been taken prior to death. A videotape recording of the movement disorder is also valuable.

Do truly non-genetic cases of clinically typical HD really exist? The rarity and difficulty in proving a case to represent a new mutation are discussed in Chapter 9; the same problems are present in proving that a case is non-genetic, but there is no convincing evidence that an isolated patient with typical HD, confirmed neuropathologically, can result from a non-genetic cause. Mutational analysis of the HD gene will hopefully resolve this problem in the not too distant future.

THE CLINICAL AND LABORATORY INVESTIGATION OF SUSPECTED HD

If the diagnosis of HD is not definite after a full history and neurological examination have been undertaken, along with an assessment of the patient's mental state and a carefully documented pedigree, then it is unlikely that the studies listed in Table 2.7, or any others, short of study of the brain at autopsy, will make the diagnosis more certain. This makes it all the more important that the clinical assessment is thorough and especially that no stone is left unturned in obtaining details on relatives that might provide definite information on HD in a family member.

Nevertheless this should be no excuse for not undertaking full investigations as one would for any serious neurological problem. The gravity of the diagnosis is such that few patients or their relatives will accept it unless a clinical assessment has been backed up by other studies. Even when the patient knows the risk and likely diagnosis, coming from a large and well-established family, a

Table 2.7 Principal investigations in the diagnosis of Huntington's disease (selection determined by clinical features and family history)

Investigation	Changes in HD	Value
Blood film (fresh thick wet film)	Normal	Excludes neuro-acanthocytosis; may suggest alcoholic encephalopathy
CPK	Normal	Raised in neuro-acanthocytosis
Nerve conduction studies	Normal	Usually abnormal in neuro-acanthocytosis
Red-cell indices	Normal	Exclude polycythaemia
Copper studies	Normal	Exclude neurological Wilson's disease
Slit-lamp examination		
Thyroid function tests	Normal	Exclude thyrotoxicosis
Anti-nuclear factor	Normal	Exclude systemic lupus erythematosus
Serological tests for syphilis	Normal	Exclude syphilis
EEG	Normal or non-specific	Doubtful value
CT scan	Loss of caudate; general cortical atrophy	Distinguishes other characteristic patterns and focal disease
CSF analysis	Normal	Doubtful value
Neuropathology (in fatal cases only)	Usually characteristic, but occasionally appears normal in early stages	Distinguishes other neurodegenerative disorders (see text)
DNA analysis and storage	Detection of specific gene defect likely in future	May exclude other specific disorders (e.g. prion disease)

brief period of investigation will provide an opportunity for a step-by-step coming to terms with the diagnosis, and will make it easier for those involved to disclose it in the least devastating manner possible. Most of the value of the tests listed lies in their exclusion of other neurological disorders, a few of which (e.g. Wilson's disease) may be amenable to treatment, while others have either a different prognosis or different (even absent) genetic basis.

Fortunately the investigations of most value are now non-invasive ones. Pneumoencephalography has been replaced by CT and MRI scanning techniques. Brain biopsy of the cortex (Tellez-Nagel *et al.*, 1973; Drymiotis *et al.*, 1973) is of no value in establishing a positive diagnosis of HD, or excluding it. Even lumbar puncture is now an investigation of marginal significance unless there is positive blood serology for syphilis, or there are atypical features to suggest multiple sclerosis.

Electroencephalography (EEG)
The routine use of EEG is of doubtful value. Scott *et al.* (1972) reviewed the EEG findings of 95 patients; 31 had definite but non-specific abnormalities, low voltage and lack of alpha rhythm. These findings occurred more often in institutionalized patients with cortical atrophy. With regard to early or pre-symptomatic diagnosis, Bruyn and Went (1986) considered that 'accumulated experience showed it (EEG) to have no diagnostic contributory value'. Other electrophysiological techniques are more appropriately considered as research than clinical investigations, and are considered as such in Chapter 5.

DNA analysis
In the very near future DNA analysis in HD is likely to be of major clinical significance in diagnosis as well as in prediction. Currently its diagnostic role is limited, though its use should be considered in the patient with an apparently distinct neurological disorder who is also at risk for HD. Meanwhile all suspected HD patients should have a blood sample taken for DNA isolation and storage, pending isolation of the HD gene and the possibility of identifying specific mutations. Most regional genetics centres in North America and Britain now maintain DNA banks set up specifically for the storage of samples from patients with genetic disorders where future developments are likely to result in new diagnostic possibilities, with correspondingly important implications for their relatives.

Structural imaging – CT and MRI
The pneumoencephalogram was one of the earliest imaging techniques to become available. It was possible to demonstrate loss of the convex bulge of the caudate nucleus into the anterior horn of the lateral ventricle and broadening of the sulci over the frontal and parietal lobes (Blinderman *et al.*, 1964; Gath and Vinje, 1968). This technique did not have wide application and has been superseded by computerized tomography (CT) of the brain. An example of the CT scan from a patient with HD is shown in Figure 2.3.

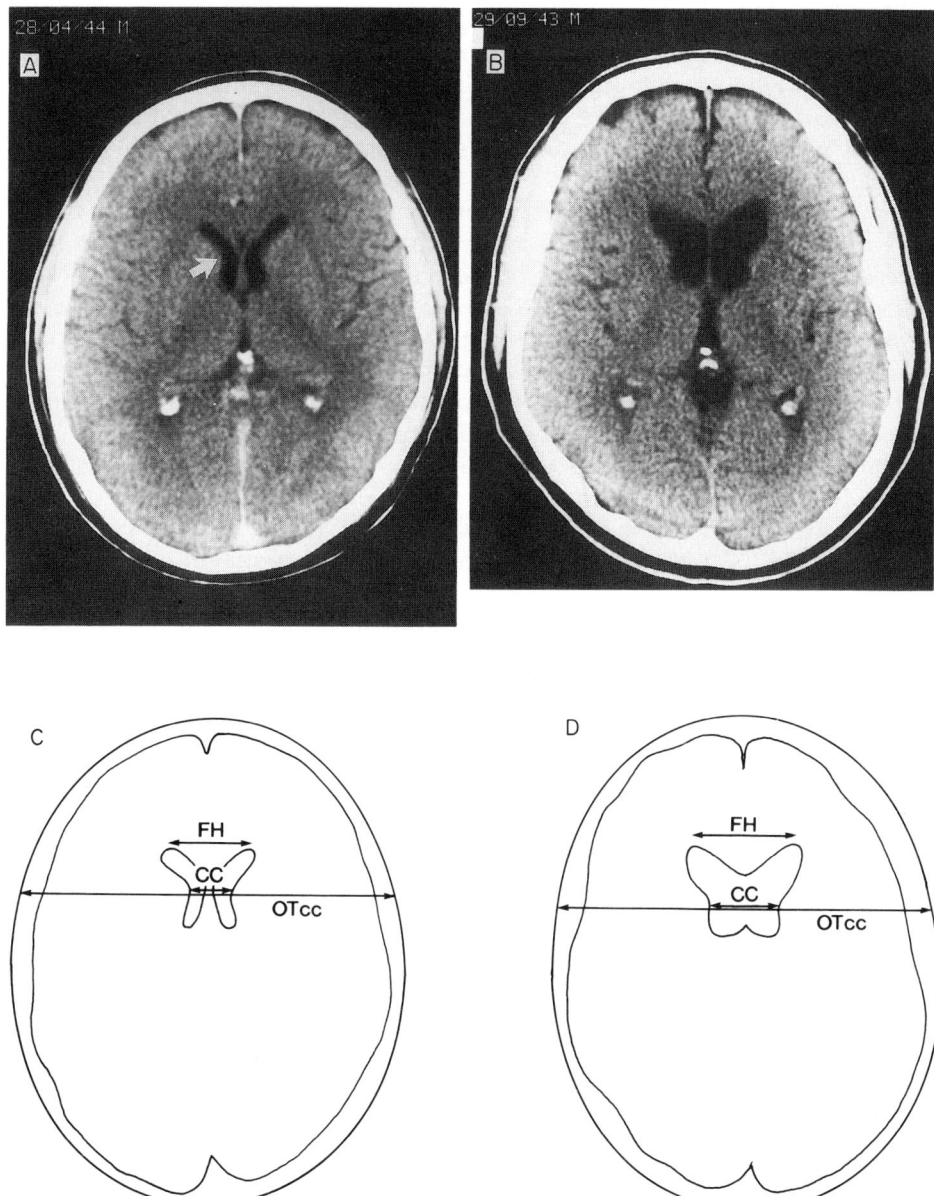

Figure 2.3 The CT scan in HD. **(A)** CT scan of a 45 year old male showing the normal concave contour of the lateral ventricle which is adjacent to the caudate nucleus (arrowed). **(B)** CT scan of a male of similar age affected by HD. Note the enlargement of the lateral ventricle due to atrophy of the caudate nucleus. **(C)** Diagrammatic representation of scan A. **(D)** Diagrammatic representation of scan B. Figures C and D illustrate some of the measurements which may be made to quantify the degree of caudate nucleus atrophy. See text for detailed explanation.

In the early stages of the disease the CT scan may be normal. Hayden et al. (1986) described 10 patients with symptoms of less than 5 years' duration who had FH/CC ratios (see Figure 2.3) within 2 standard deviations of the normal range. Starkstein et al. (1989) suggested that measurements of subcortical atrophy correlated with duration of symptoms; whereas cortical atrophy correlated with the patient's age rather than assessment of dementia.

A number of studies have been undertaken to define CT parameters which discriminate between HD patients and normal controls. Measurements have varied between studies, making direct comparisons difficult. Commonly, subcortical atrophy has been assessed by measuring the shortest distances between the indentations of the caudate nuclei (the bicaudate diameter, CC) combined with another measure such as the widest distance between the frontal horns in the same plane (FH), or the distance between the outer tables of the skull along the same line as the bicaudate measurement (OTcc) (Barr et al., 1978). The result is often expressed as a ratio: FH/CC or CC/OTcc. A number of reports have demonstrated that HD is associated with observable atrophy of the caudate nuclei as demonstrated by a significant rise in the CC/OTcc ratio or fall in the FH/CC ratio. Care is required in the interpretation of individual results since the bicaudate diameter and frontal horn measurements increase with age but at different rates; consequently the normal FH/CC ratio decreases with age.

The CT scan should be considered as a useful confirmatory aid allowing the exclusion of other causes of chorea such as tumours or infarcts. In practice its use is limited by the fact that a normal scan in the early stages of the disease does not preclude the diagnosis of HD and patients presenting late in the course of the disease may be unable to cooperate with the investigation. Moreover, an abnormal scan is not diagnostic of HD. However, in a suspected patient from a known HD family, selective caudate atrophy on CT could be highly suggestive of the diagnosis. The magnetic resonance image (MRI) scan provides similar information but there is improved definition of the caudate and putamen nuclei (Simmons et al., 1986).

Functional imaging

Tanahashi et al. (1985) noted that decreased cortical blood flow correlated with tests of cognitive function and occurred prior to cortical atrophy as judged by CT scans.

More recently research interest has focused on metabolic assessments of brain function using positron emission tomography (PET). Local cerebral glucose metabolic rates may be measured using the tracer ^{18}F-2-fluorodeoxyglucose (^{18}FDFG). Following intravenous injection, this substance is converted by brain hexokinase to ^{18}FDG-6-phosphate which remains fixed with little further metabolism; glucose utilization may then be assessed from a combination of the PET scan and arterial glucose and ^{18}FDG levels.

Kuhl et al. (1982) were the first to demonstrate that HD patients have reduced glucose metabolism in the caudate and putamen nuclei. This observation was confirmed even in the early stages of the disease when CT measurements were normal (Garnett *et al.*, 1984; Hayden *et al.*, 1986; Young *et al.*, 1986). Reports of decreased glucose metabolism in the caudate nuclei of asymptomatic individuals at risk for HD have been inconsistent and conflicting (Kuhl *et al.*, 1982; Hayden *et al.*, 1987; Mazziotta *et al.*, 1987; Young *et al.*, 1987); the possible applications in prediction are considered in Chapter 12.

Caudate nucleus hypometabolism is not pathognomic for HD. Similar observations have been noted in benign familial chorea, neuro-acanthocytosis, and Lesch Nyhan syndrome (Palella *et al.*, 1985; Suchowersky *et al.*, 1986; Phillips *et al.*, 1987). However caudate glucose hypometabolism is not a necessary correlate of chorea, since it was absent in four patients with systemic lupus chorea in studies by Guttman *et al.* (1987). These conditions are considered later in this chapter.

Reid *et al.* (1988) have been able to demonstrate reduced blood flow in the area of the caudate nucleus in six HD patients. Imaging was undertaken using the single photon emission tomography technique (SPECT) and the tracer hexamethylpropylenamineoxime complexed with ^{99}technetium. The amount of tracer taken up by cerebral tissue is related to blood flow. The authors suggested that reduced blood flow in the caudate nucleus may be an early feature as changes were noted in a seventh patient who had psychiatric features suggestive of HD but who had not developed neurological features which would be diagnostic. More recently a patient with definite features of HD had normal uptake of tracer in the region of the basal ganglia and a second patient had reduced uptake unilaterally (Waters and Chen, 1990). Therefore, the value of this imaging technique requires further assessment.

Neuropathology
The detailed structural changes in the brain seen in patients with HD, both macroscopic and histological, are discussed in Chapter 5, but the clinical aspects of this subject need considering at this point. Even though brain biopsy is no longer performed in HD and full study of the brain is only possible after death, neuropathology is one of the most important investigations to be considered in the suspected patient with HD. It must be remembered that making a diagnosis of HD will have immediate and long-term implications for the entire family, and that obtaining or failing to obtain neuropathology may be the determining factor that supports or makes impossible a definite diagnosis or risk assessment for them. In the same way it may be possible to use previous clinical assessments and neuropathological reports on other family members in the investigation of the patient under one's care.

Table 2.8 summarizes some of the main situations in which the availability of neuropathological evidence is of particular importance. It cannot be too strongly emphasized that detailed neuropathological study of at least one affected member of a kindred is important even in an apparently typical family with HD, in view of the other brain disorders that may mimic it. Neuropathology has become increasingly essential now that relatives in the younger

Table 2.8 Neuropathology in Huntington's disease – indications and importance

Indication	Information provided
Clinically definite HD with positive family history but no previous autopsy	Provides certainty for relatives in genetic counselling and predictive tests. May provide DNA for analysis
Clinically definite HD, but isolated case or uncertain family history	As above
Atypical or unlikely HD in a known HD family	Will resolve question of whether diagnosis is HD or other disorder
Suspected HD in early stage, not confirmed (e.g. suicide) in a known family	May give definite evidence as to whether HD present or not but it should be remembered that clinical signs can precede pathological changes
Elderly patient with established family history who may have Alzheimer's disease	May provide evidence that HD and Alzheimer's disease coexist

generations may request presymptomatic testing. While most centres would not rule out doing such tests in its absence, clear neuropathological support of the clinical diagnosis of HD is of the greatest value in allowing testing to be carried out on firm diagnostic foundations, without the worry that the diagnosis may be queried at a later point. Stored brain material may also be the only available source of DNA in an older deceased patient (Upadhyaya et al., 1985), though DNA previously isolated from blood or other tissue and stored as DNA is more reliable in this respect.

The follow-up study of Quarrell et al. (1986), already mentioned in the context of the isolated case, shows not only the value of detailed neuropathology, but sadly how often clinicians fail to obtain it. In part this problem arises because death most often occurs long after hospital investigations have ceased and in a situation where carrying out neuropathological studies may be difficult.

The variation in the clinical features of HD may result in the situation where a member of a known HD family develops a neurological disorder that is thought to be something distinct, but yet could be an atypical presentation of HD. Again detailed neuropathology at autopsy may be the only way to resolve the dilemma.

Perhaps the group of patients where the opportunity for neuropathology is most likely to be missed is those where death is not directly due to HD. Patients in an HD family who commit suicide or die accidentally may be suspected as showing early symptoms; unless the forensic pathologist or whoever is responsible for the acute autopsy is aware of the possibility of HD and the need for the brain to be fully studied, the opportunity may be lost. Carrasco and Mukherji (1986) have recorded an accidental death in an apparently normal 48-year-old man at risk of HD where marked caudate atrophy was present at autopsy. However, caution should be exercised in the interpretation of a normal result, since it is well established that neuropathological features may only occur after the onset of signs and symptoms (Vonsattel et al., 1985). Neuropathological investigation should also be sought when an elderly person, at risk for HD, develops clinical features which could be consistent with a diagnosis of Alzheimer's disease. Again, interpretation of the result needs to take account of all available clinical and pathological evidence. Moreover, it should be remembered that features of HD and Alzheimer's disease or Parkinson's disease may coexist in the same patient.

In conclusion, the number of clinical diagnoses supported by neuropathological examination needs to be increased. As with all investigations the neuropathological examination should not be considered in isolation; there is no single pathognomic feature of HD, the pathologist relying upon recognition of a pattern of cell loss in the context of information from the history and clinical examination. Molecular analysis will in future be an integral part of the assessment and blood or tissue should always be retained for this.

DIFFERENTIAL DIAGNOSIS

The differential diagnosis of HD falls into four main areas:

(1) The various causes of chorea.
(2) The distinction between other neurological disorders affecting the basal ganglia which may overlap with HD in various ways.
(3) Distinction from the various causes of dementia and from the major psychoses.
(4) Huntington's disease in childhood.

The last of these categories has already been considered in the discussion of juvenile HD, while the psychiatric differential diagnosis is dealt with in the next chapter. Categories (1) and (2), however, contain a number of important disorders that must be considered and carefully excluded if errors are not to be made. That the correct diagnosis of HD is often far from easy is evidenced by surveys showing a high frequency of misdiagnosis. In the Maryland study (Folstein, 1989) 24 of 217 patients (11%) proving to have HD had been notified to the study with a different diagnosis, including alcoholism (seven cases), a variety of psychiatric diagnoses, 'senile chorea' and Parkinsonism. Conversely 31 out of 212 patients notified as HD (15%) actually proved to have a different diagnosis, tardive dyskinesias (eight cases) being the most frequent, followed by Alzheimer's disease (four cases). Neuropathological studies have shown a comparable misdiagnosis rate; thus in the series of Bird et al. (1977) 7% of patients with a clinical diagnosis of HD proved to have a different neuropathological disorder, while Quarrell et al. (1986) found a number of other disorders in their survey of the outcome in isolated cases, 15 out of 37 proving to have disorders other than HD. Tardive dyskinesia was again the most common (six cases).

There are specific situations where the diagnosis of HD can be particularly difficult. Childhood HD has already been mentioned, as has the problem of the isolated case. Some patients may only be referred for investigation late on in the course of the disease, with inadequate documentation of the early features, while the diagnosis in early cases can often hinge on the existence of other affected family members who may be at a remote distance or unwilling to undergo tests, or whose illness may appear to be different from that shown by the patient under investigation.

The causes of chorea are numerous and some of these are summarized in Table 2.9. More extensive lists are to be found in the reviews of Padberg and Bruyn (1986) and Shoulson (1986); the former lists over 140 possible causes but in most of these, chorea is either an exceptionally rare occurrence, or else an incidental one in a disorder where additional features can readily be distinguished from HD, such as systemic lupus or polycythaemia vera. Among the various exogenous causes, drug-induced tardive dyskinesia is the most important group. The various other hereditary disorders characterized by chorea are discussed here individually more because clinicians are unaware of their existence than because of their similarity to HD. There is still a widespread assumption that any hereditary chorea must be Huntington's chorea.

Sydenham's chorea

Until a generation ago this was by far the commonest cause of chorea in children and young adults; indeed to the writers of the last century the term 'chorea' was synonymous with Sydenham's chorea unless stated to the contrary. George Huntington's classical description of HD was essentially an appendage to a more general discussion of Sydenham's chorea, while Osler, who took a keen interest in all forms of chorea, devotes only one page of his

Table 2.9 Some causes of chorea relevant to the diagnosis of HD

Mendelian	Inheritance
Huntington's disease	(AD)
Benign familial chorea	(AD)
Neuro-acanthocytosis	(AR)
Paroxysmal choreoathetosis	(AD)
Wilson's disease	(AR)
Lesch Nyhan disease	(XR)
Neuronal lipofuscinosis	(AD or AR)
Familial basal ganglion infarcts	(AD when familial)
DRPLA	(AD)
Non-Mendelian	
Sydenham's chorea	
Pregnancy/oral contraceptives	
Systemic lupus erythematosus	
Thyrotoxicosis	
Polycythaemia	
Drug induced (various – especially tardive dyskinesia)	

Principles and Practice of Medicine (1892) to Huntington's chorea, as compared to eight pages for Sydenham's chorea. Likewise his monograph on chorea (1904) is mainly concerned with Sydenham's chorea, though hereditary chorea received a chapter.

The current situation is now totally different, with Sydenham's chorea having declined dramatically in all developed countries over the past 30 years, along with other manifestations of streptococcal infection such as acute rheumatic fever. Indeed it has come so close to complete disappearance that childhood chorea is now more likely to have a different underlying cause, usually metabolic or genetic. The closely allied adult cases of chorea related to pregnancy and oral contraceptives may also have declined comparably; in view of the universal nature of both pregnancy and the use of oral contraceptives, it is most unwise to attribute chorea to either unless very careful exclusion of other causes, including HD, has been made.

Clinical confusion of HD with Sydenham's chorea is likely only in the early stage of HD, where the movements are slight and other features of HD absent. Even here the finer and more rapid nature of the movements in most (though not all) cases of Sydenham's chorea should allow distinction, while the disorder is usually self-limiting and never relentlessly progressive. In HD families it is not uncommon to find a family history of childhood 'St Vitus Dance' in individuals later developing HD or in healthy relatives, reflecting the commonness of Sydenham's chorea in previous generations. In the rare cases where there remains any doubt, observation over time will clarify the distinction between the two disorders.

Tardive dyskinesia

The well-recognized association of involuntary movements with prolonged administration of chlorpromazine and other neuroleptic drugs (Tanner, 1986) can create a difficult diagnostic problem in relation to HD, and has been a frequent cause of misdiagnosis in the studies already mentioned; the use of dopa and dopa agonists are additional causes to be considered. A variety of choreic and dystonic movements may occur; they may be generalized, but initially predominantly involve face and mouth. A particularly difficult situation is posed by the person at risk for HD who has received phenothiazines for psychiatric symptoms; if movements later develop, are they the result of the therapy, or of HD? Apart from the distribution of the movements, other helpful distinguishing features are remission or lack of progression after stopping the drug (but some with tardive dyskinesia often worsen initially, or even continue to progress), lack of the organic and progressive mental impairment seen in HD, normal eye movements and absence of caudate atrophy or other evidence of progressive cerebral degeneration. Many patients are elderly, with movements first seen at an age where onset of HD is uncommon. However, none of these findings provide absolute distinction, and all the facts, including a possible family history of HD and the strength of evidence for any drug-related causation, will need to be carefully considered.

Benign familial chorea

This disorder (also termed benign hereditary chorea, familial essential chorea) is a well-recognized condition, whose distinction from HD is of considerable importance (Table 2.10). First fully described in 1967 by Haerer *et al.* in a large Black American kindred and independently by Pincus and Chutorian (1967), there have been numerous subsequent reports which have amplified the clinical features and inheritance pattern (Chun *et al.*, 1973; Sadjadpour and Amato, 1973; Burns *et al.*, 1976; Foerster and Foerster, 1978; Fisher *et al.*, 1979). Reviews have been provided by Bird *et al.* (1976), Harper (1978) and Sleigh and Lindenbaum (1981) with a notably complete account of the disorder being given by Bruyn and Myrianthopoulos (1986). The probability that the patients described by Lyon (1863) had benign familial chorea, not HD, has been discussed in Chapter 1.

The choreic movements may be indistinguishable from those of HD (Figure 2.4) though varying considerably in severity and often much milder. The main distinction lies in the history. In most families onset is clearly recognized from infancy or early childhood and is essentially non-progressive throughout life. Patients survive with their chorea little changed into old age and show none of the other progressive motor disturbances of HD. Mentality is likewise usually normal, though considerable psychological upset may occur, particularly in relation to teasing at school or from prolonged investigations and inappropriate approaches to management (Sleigh and Lindenbaum, 1981; Loosmore and Wood, 1988). Leli *et al.* (1984) found some evidence of mental impairment in one kindred but this seems to be the exception. The term 'benign' is not always appropriate since some patients have had considerable motor delay in childhood and continuing disability from their movement disorder affecting their later ability to work.

Inheritance is clearly autosomal dominant, but with considerable variation in severity within a family and lack of penetrance in some gene carriers, especially

Table 2.10 Benign familial chorea. Distinguishing features from Huntington's disease

	Benign familial chorea	Huntington's disease
Onset	Infancy or childhood	Usually adult
Course	Static through adult life	Progressive, fatal
Generalized motor disorder	Absent, except elements of dystonia	Frequent, progressive
Mental deterioration	Absent; mild retardation recorded in few cases	Frequent, progressive
Inheritance	Autosomal dominant; not at same locus as HD	Autosomal dominant

females. Harper (1978) estimated that only 75% of females carrying the gene were clinically affected. This probably explains early reports suggesting recessive inheritance on the grounds of neither parent being affected. The locus has been shown by Quarrell et al. (1988) to be distinct from that for HD. Harper (1978) estimated a frequency of 1 in 500 000 but it could well be more frequent than this.

It is obviously essential that families with benign familial chorea are not erroneously labelled as having HD if major errors in prognosis and implications for genetic counselling are to be avoided. In early childhood most confusion has been with 'cerebral palsy' and various epileptic disorders (Sleigh and Lindenbaum, 1981). Many of the patients reported appear to have been subject to repeated or prolonged unnecessary investigations, as well as fruitless orthopaedic procedures. Awareness of the disorder should at least spare the patients this, even though no active measures seem to be helpful. Most approaches to drug treatment have shown no change, though a patient reported by Robinson and Thornett (1985) showed improvement of chorea on several occasions when his asthma was treated by steroids.

Paroxysmal choreoathetosis
The distinctive disorder, well reviewed by Buruma and Roos (1986), probably contains a number of separate conditions, all extremely rare. Mount and Reback (1940) described the first family, in which episodes were triggered by coffee and other beverages. Stress and sudden movement have subsequently been found to be more frequent precipitants (Williams and Stevens 1963; Jung et al., 1973). The attacks consist of sudden episodes of generalized abnormal movements, which may be choreic, dystonic, athetoid or mixed. Episodes generally last a few minutes only, and in between them the individual is entirely healthy. Lance et al. (1977) distinguished kinesigenic and dystonic forms according to the nature of the attacks. The disorder is essentially non-progressive with onset in infancy or childhood and persistence through life. Improvement with such drugs as phenytoin or carbamazepine supports the concept that the condition may represent a form of subcortical epilepsy. Clear autosomal dominant inheritance is present in most families. From the viewpoint of distinction from HD it can be seen from this brief description that there should be no danger of confusion provided the existence of other dominantly inherited basal ganglion disorders characterized by chorea is recognized.

Chorea-acanthocytosis (neuro-acanthocytosis)
Unlike the above two disorders, chorea-acanthocytosis is progressive, so that confusion with HD is a greater possibility. First described independently in American and British families (Critchley et al., 1967; Cederbaum et al., 1971; Aminoff, 1972), the condition has been most extensively documented in Japan, where it may well be more common than HD (Shimizu et al., 1974, 1980; Shibasaki et al., 1982). Bruyn (1986) has provided a valuable review. The clinical features are rather characteristic, though showing some variation between

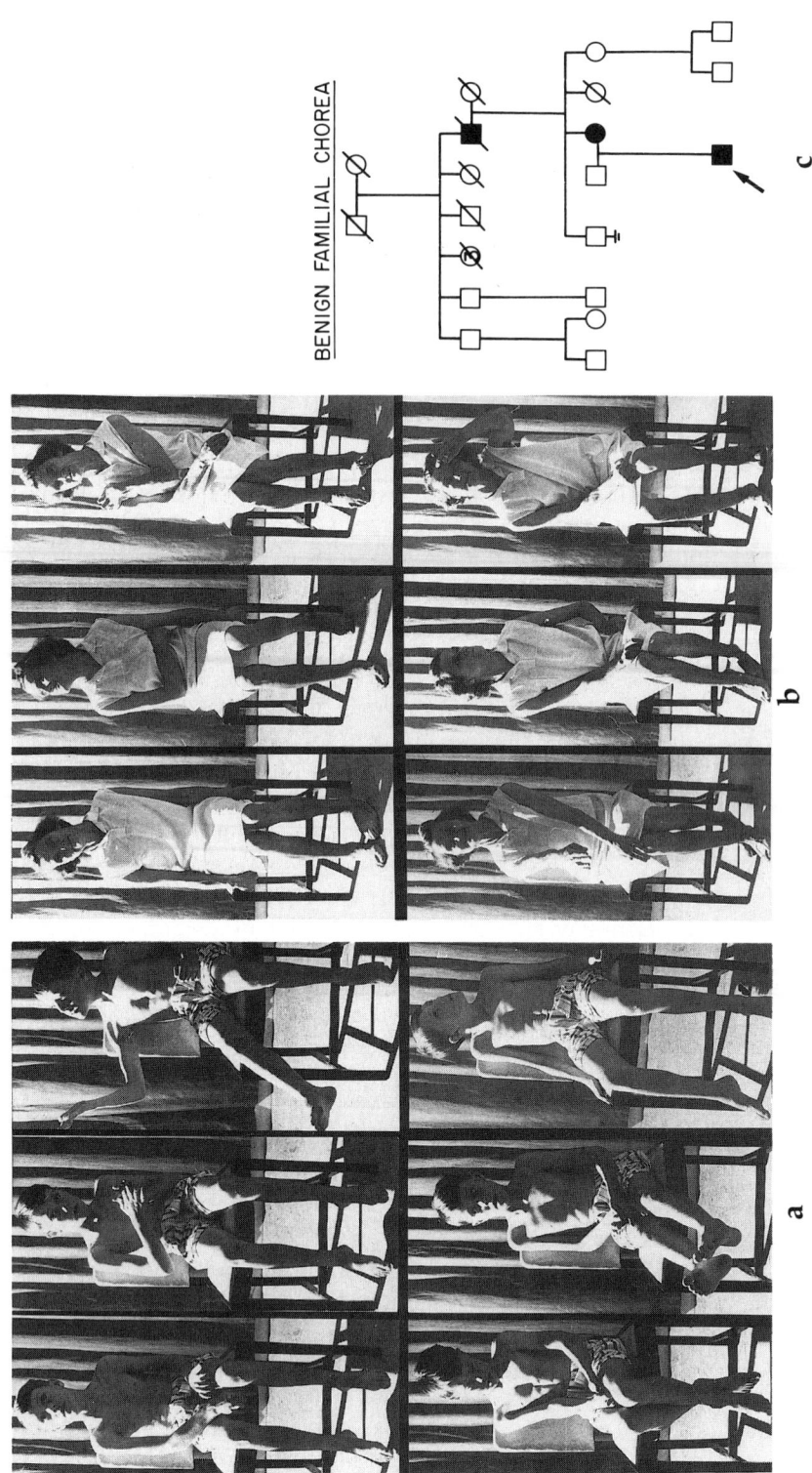

Figure 2.4

families. Onset is in early adult life with abnormal movements involving jaw, tongue and face, which may cause self mutilation. Generalized chorea and dystonic movements subsequently develop, but are accompanied by clinical and electrophysiological signs of a motor neuropathy, with muscle wasting and loss of reflexes. Epileptic attacks and mild intellectual decline are frequent, but prominent dementia is uncommon. All these features should allow a distinction from HD, but the specific diagnostic test is the finding of acanthocytes on a thick wet blood film, the metabolic significance of which has not yet been elucidated. The serum creatine kinase is also elevated. Neuropathology, illustrated in Chapter 5, is often characteristic (Sato *et al.*, 1984) with changes confined to the basal ganglia and complete sparing of the cortex, in contrast to HD. In most families, notably in Japan, inheritance has been autosomal recessive, with a high rate of consanguinity and multiple affected offspring born to healthy parents.

Figure 2.4 Benign familial chorea. (Family originally reported by Harper, 1978.) **(a)** Propositus, aged 9 years. During the first days of life the mother noticed unusual writhing movements of arms and legs and suspected he was affected in view of the similar disorder in herself and her father. Motor landmarks were delayed, and he did not walk until 2.5 years old. Mental development was normal, with first words around 1 year. The involuntary movements remained unchanged throughout infancy and early childhood, and caused little disability apart from general clumsiness and slight unsteadiness of gait. He attended normal school from the age of 5 years; progress was initially normal, and he was able to play games. Assessment of intelligence at 5 years showed IQ was 115, and at 9 years IQ was 108. Behavioural disturbances developed at 9 years of age, and were attributed to awareness of his handicap and to teasing at school. The movements have remained unchanged into adult life. He was initially assessed medically when aged 5 years, and familial cerebellar ataxia with athetosis was diagnosed at this time. The diagnosis was revised to familial choreoathetosis following further investigation when aged 9 years. Examination at this time showed generalized choreiform movements of moderate amplitude, affecting face, trunk and limbs, with unsteadiness of gait and irregular speech rhythm. Neurological examination was normal apart from the movements; in particular, no rigidity or cerebellar incoordination could be demonstrated, while cranial nerves were normal. General examination was entirely normal. Movements were accentuated by worry, but were diminished by performing specific actions. **(b)** Mother of propositus. This patient was stated to have had involuntary movements since shortly after birth, and not to have walked unaided until she was 11 years old. She had at no time received medical treatment for the movements, which had persisted unchanged into adult life, being unaffected by her only pregnancy. General health had been good; she had attended a normal school. When examined at the age of 44 years, she showed marked choreic movements of limbs and trunk, with facial grimacing and disturbance of speech; gait was moderately affected by the movements, but neurological examination was otherwise normal. Her mental state was, in particular, entirely normal. The pedigree of the family is shown in **(c)**. The maternal grandfather of the propositus died from tuberculosis aged 53 years. He was stated by his daughter to have had lifelong involuntary movements similar to those of other affected family members, but he had never received medical attention for these.

A systematic and repeated search for acanthocytes is essential in the isolated case of HD, or where recessive inheritance appears to be operating, even if the clinical features are typical.

Dentato-rubro-pallido-luysian atrophy (DRPLA)
Despite its rarity this disorder probably offers more likelihood of confusion with HD than the conditions discussed so far. Indeed a 'pseudo-Huntington' type has been described in Japan, where the condition seems to be commoner than elsewhere. (Sakamoto *et al.*, 1971; Kobayashi *et al.*, 1975; see Iizuka *et al.* (1984) for review.) Most patients have cerebellar-type ataxia as the initial feature, with later development of chorea and particular involvement of eye movements. In the Japanese families the chorea soon becomes more prominent than the ataxia and is accompanied by progressive mental deterioration, clearly indicating the possibility of diagnostic confusion with HD. Inheritance is also autosomal dominant.

Neuropathology provides the only definite distinguishing feature, with especial involvement of the red and dentate nuclei, cerebellum and posterior columns – a very different pattern to that of HD. Despite the rarity of DRPLA in most countries, its phenotype illustrates the importance of full neuropathological studies being done on at least one affected individual in any HD kindred. DRPLA has recently been confirmed to be determined by a different genetic locus to HD (Kondo *et al.*, 1990).

Other hereditary neurological disorders
Numerous degenerative disorders of the nervous system exist, often still poorly understood and classified, which may affect the basal ganglia and which may show some clinical features that overlap with those of HD. Disorders so far confined to one or a few families and whose status remains incompletely defined include familial basal ganglion infarcts and familial cystic degeneration of the basal ganglia. When the history, genetic aspects and detailed neurological assessment are all taken into account, there should rarely be serious doubt; when doubt remains, then it is preferable, as stated earlier, to leave the diagnosis open rather than to attempt a definite and possibly erroneous conclusion which the facts do not justify.

Parkinson's disease should not be a serious possibility for confusion, except possibly in juvenile HD and cases presenting with rigidity. Despite this it is surprising how often a family history in a previous generation of an HD patient is erroneously given by relatives as Parkinson's disease. It is possible that this is because the latter is not just a more familiar term to many families, but also more socially acceptable. Choreic movements may result from excessive l. dopa therapy in this disorder.

Wilson's disease is a more serious possibility for misdiagnosis, again particularly in juvenile and rigid cases, but copper studies and search for the Kayser–Fleischer ring should resolve the situation; the possibility of therapy makes this particularly important.

Hereditary ataxias of various types generally show a predominant cerebellar and posterior column involvement in the entire kindred, though some of the olivo-ponto-cerebellar group (e.g. Machado Joseph disease) may also show pronounced movement disorder and some mental deterioration. This extensive and heterogeneous group of disorders is beginning to be resolved more clearly by gene-mapping studies. A recent study has shown that movement disorders, including chorea, may also occur in mitochondrial cytopathies (Truong et al., 1990).

In torsion dystonia, again a heterogeneous group, the different character of the movement disorder and absence of mental deterioration should avoid confusion. As stated earlier, dystonic movements do not occur alone in HD and are rare in the early stages. However, many patients with dystonia may have additional jerky movements (myoclonic dystonia), and some families show autosomal dominant inheritance.

In conclusion, despite the very large number of causes of chorea and the numerous other disorders which can at times resemble HD, it should be possible to make a firm diagnosis in the great majority of cases. One can only reinforce the advice, given at the beginning of this section, to document a full and accurate history and clinical assessment (including the mental state), to expend the greatest possible effort in establishing a family history of HD, and to obtain full neuropathology where the patient or an affected relative dies. If, despite these efforts, the diagnosis remains uncertain, it is better to admit this and to make continued efforts to confirm it. Most current diagnostic uncertainty results from lack of detailed documentation in the past; a conscientious and thorough approach will help to ensure that such problems are not handed on to our successors.

REFERENCES

Alcock NS (1936) A note on the pathology of chronic progressive chorea. *Brain* **59**:376–387.
Aminoff MJ (1972) Acanthocytosis and neurological disease. *Brain* **95**:749–760.
Avanzini G, Girotti F, Caraceni T and Spreafico R (1979) Oculomotor disorders in Huntington's chorea. *Journal of Neurology, Neurosurgery and Psychiatry* **42**:581–589.
Barr AN, Heinze WJ, Dobben GD, Valvassori GE and Sugar O (1978) Bicaudate index in computerised tomography of Huntington disease and cerebral atrophy. *Neurology* **28**:1196–1200.
Beenen N, Bttner U and Lange HW (1986) The diagnostic value of eye movement recordings in patients with Huntington's disease and their offspring. *Electroencepth. Clinical Neurophysiology* **63**:119–127.
Bell J (1934) Huntington's chorea. In: *The Treasury of Human Inheritance, Vol. IV*, Part I (ed RA Fisher). Cambridge: Cambridge University Press.
Bird TD, Carlson CB and Hall JG (1976) Familial essential ('benign') chorea. *Journal of Medical Genetics* **13**:357–362.
Bittenbender JB and Quadfasel FA (1962) Rigid and akinetic forms of Huntington's chorea. *Archives of Neurology* **7**:275–288.
Blackburn R (1988) On moral judgements and personality disorders: The myth of Psychopathic personality revisited. *British Journal of Psychiatry* **153**:505–512.
Blinderman EE, Weidner W and Markham CH (1964) The pneumoencephalogram in Huntington's chorea. *Neurology* **14**:601–607.

Bollen E, Reulen JPH, Den Heyer JC, Van der Kamp W, Roos RAC and Buruma OJS (1986) Horizontal and vertical saccadic eye movement abnormalities in Huntington's chorea. *Journal of Neurological Sciences* **74**:11–22.

Brackenridge CJ (1980) Factors influencing dementia and epilepsy in Huntington's disease of early onset. *Acta Neurologica Scandinaviea* **62**:305–311.

Bruyn GW (1968) Huntington's chorea. Historical, clinical and laboratory synopsis. In: *Handbook of Clinical Neurology*, Vol. 16 (eds PJ Vinken and GW Bruyn). Amsterdam: North Holland.

Bruyn GW (1986) Chorea – acanthocytosis. In: *Handbook of Clinical Neurology*, Vol. 49 (eds PR Vinken, GW Bruyn and HL Klawans), pp.327–334. Amsterdam: Elsevier.

Bruyn GW and Myrianthopoulos NC (1986) Chronic juvenile hereditary chorea (benign hereditary chorea of early onset) In: *Handbook of Clinical Neurology*, Vol. 49 (eds PR Vinken, GW Bruyn and HL Klawans). Amsterdam: Elsevier.

Bruyn GW and Went LN (1986) Huntington's chorea. In: *Handbook of Clinical Neurology*, Vol. 49 (eds PR Vinken, GW Bruyn and HL Klawans). Amsterdam: Elsevier.

Burns J, Neuhauser G and Thomas IL (1976) Benign hereditary chorea. Non-progressive chorea of early onset. Clinical genetics of the syndrome and report of a new family. *Neuropädiatrie* **7**:431–438.

Buruma OJS and Roos RAC (1986) Paroxysmal choreoathetosis. In: *Handbook of Clinical Neurology*, Vol. 49 (eds PR Vinken, GW Bruyn and HL Klawans). Amsterdam: Elsevier.

Byers RK and Dodge JA (1976) Huntington's chorea in children. *Neurology* **17**:587–596.

Byers RK, Gilles FH and Fung C (1973) Huntington's disease in children. Neuropathologic study of four cases. *Neurology* **23**:561–569.

Campbell AMG, Corner B, Norman RM and Urich H (1961) The rigid form of Huntington's disease. *Journal of Neurology, Neurosurgery and Psychiatry* **24**:71–77.

Carrasco LH and Mukherji CS (1986) Atrophy of corpus striatum in normal male at risk of Huntington's chorea. *Lancet* **i**:1388–1389.

Cederbaum S, Heywood D, Aigner R and Motulsky A (1971) Progressive chorea, dementia and acanthocytosis: a genocopy of Huntington's chorea. *Clinical Research* **19**:177.

Chun RWM, Daley RF, Manshein BJ and Wolcott GJ (1973) Benign familial chorea with onset in childhood. *Journal of the American Medical Association* **225**:1603–1607.

Collewijn H, Went LN, Tamminga EP and Vegter-Van der Vlis M (1988) Oculomotor defects in patients with Huntington's disease and their offspring. *Journal of Neurology Science* **86**:307–320.

Craufurd D, Donnai D, Kerzin-Storrar L *et al.* (1989) Testing of children for 'adult' genetic disease (letter). *Lancet* **i** 1406.

Critchley EMR, Clarke DB and Wikler A (1967) An adult form of acanthocytosis. *Transactions of the American Neurology Association* **92**:132–137.

Curran D (1930) Huntington's chorea without choreiform movements. *Journal of Neurology and Psychopathology* **10**:305–310.

Denny-Brown D (1960) Diseases of the basal ganglia and their relation to movement. *Lancet* **ii**:1099–1105 and 1155–1162.

Drymiotis A, Whittier JR and Korenyi C (1973) Brain biopsy and Huntington's chorea: a critical review of the literature. In: *Huntington's Chorea, 1872–1972* (eds A Barbeau, TN Chase and GW Paulson), pp. 439–452. New York: Raven Press.

Edmonds C (1966) Huntington's chorea, dysphagia and death. *Medical Journal of Australia* **53**:273–274.

Emser W, Brenner M and Stober Schimrig K (1988) Changes in nocturnal sleep in Huntington's and Parkinson's disease. *Journal of Neurology* **235**:177–179.

Entres JL (1921a) Uber Huntington'sche Chorea. *Zeitschrift für die gesamte Neurologie und Psychiatrie* **73**:541–551.

Entres JL (1921b) Zur klinik und vererbung der Huntington'schen chorea. *Monographien des Gesamtgebietes der Neurologie und Psychiatrie* **27**. Berlin: Springer.

Entres JL (1940) Der erbvitststanz Huntingtonsche Chorea: erbbiologischer Teil. In *Gutt A Handbuch der Erbkrankheiten, Vol III*. Leipzig: Thieme.

Fahn S (1988) Concept and classification of dystonia. *Advances in Neurology* **50**:1.

Farrer LA and Conneally PW (1985) A genetic model for age at onset in Huntington's disease. *American Journal of Human Genetics* **37**:350–357.

Farrer LA and Yu P-L (1985) Anthropometric discrimination among affected, at-risk, and not-at-risk individuals in families with Huntington disease. *American Journal of Medical Genetics* **21**:307–316.

Fish DR, Sawyers D, Allen P *et al.* (1990) The effect of sleep on the dyskinetic movements of Parkinson's disease, Gilles de la Tourette's syndrome, Huntington's disease and torsion dystonia. *Archives of Neurology* (in press).

Fisher M, Sargent J and Drachman D (1979) Familial inverted Choreoathetosis. *Neurology* **29**:1627–1631.
Foerster K and Foerster G (1978) Benigne hereditare, nicht progressive Chorea. *Nervenarzt* **49**:724–725.
Folstein SE (1989) *Huntington's Disease: a Disorder of Families*. Baltimore: The Johns Hopkins University Press.
Folstein SE, Jensen B, Leigh JR and Folstein MF (1983) The measurement of abnormal movement: methods developed for Huntington's disease. *Neurobehaviour Toxicology and Teratology* **5**:605–609.
Folstein SE, Leigh JR, Parhad IM and Folstein MF (1986) The diagnosis of Huntington's disease. *Neurology* **36**:1279–1283.
Friedman JH and Ambler M (1990) A case of Senile chorea. *Movement disorders* **5**:251–253.
Garnett ES, Firnau G, Nahmias C, Carbotte R and Bartolucci G (1984) Reduced striatal glucose consumption and prolonged reaction time are early correlates in Huntington's disease. *Journal of Neurological Sciences* **65**:231–237.
Gath I and Vinje B (1968) Pneumoencephalographic findings in Huntington's chorea. *Neurology* **18**:991–996.
Gordon WP and Illes J (1987) Neurolinguistic characteristics of language production in Huntington's disease: a preliminary report. *Brain and Language* **31**:1–10.
Guttman M, Lang AE, Garnett ES et al. (1987) Regional cerebral glucose metabolism in SLE chorea: Further evidence that striatal hypometabolism is not a correlate of chorea. *Movement Disorders* **2**:201–210.
Haerer AF, Currier RD and Jackson JF (1967) Hereditary nonprogressive chorea of early onset. *New England Journal of Medicine* **276**:1220–1224.
Hamilton AS (1908) A report of twenty-seven cases of chronic progressive chorea. *American Journal of Insanity* **64**:403–475.
Hammer J, Machler M, Schmid W and Schomig-Spingler M (1987) Linked DNA markers in clinical diagnosis of juvenile Huntington's disease. *Lancet,* **7 Nov**: 1088.
Hansotia P, Cleeland CS and Chun RWM (1968) Juvenile Huntington's chorea. *Neurology* **18**:217–224.
Hansotia P, Wall R and Berendes J (1985) Sleep disturbances and severity of Huntington's disease. *Neurology* **35**:1672–1674.
Harper PS (1978) Benign hereditary chorea, clinical and genetic aspects. *Clinical Genetics* **13**:85–95.
Harper PS and Clarke A (1990) Should we test children for 'adult' genetic diseases? *Lancet* **335**:1205–1208.
Hauser WA and Kurland LT (1975) The epidemiology of epilepsy in Rochester, Minnesota, 1935 through 1967. *Epilepsia* **16**:1–66.
Hayden MR (1979) Huntington's chorea in South Africa. PhD Thesis, University of Cape Town.
Hayden MR (1981) *Huntington's Chorea*. Berlin: Springer-Verlag.
Hayden MR, Martin WRW, Stossel AJ et al. (1986) Positron emission tomography in the early diagnosis of Huntington's disease. *Neurology* **36**:888–894.
Hayden MR, Hewitt J, Martin WRW, Clarke C and Amman W (1987) Studies in persons at risk for Huntington's disease. *New England Journal of Medicine* **317**:382–383.
Heathfield KWG (1967) Huntington's chorea. Investigation into the prevalence of this disease in the area covered by the North East Metropolitan Regional Hospital Board. *Brain* **90**:203–232.
Hefter H, Homberg V, Lange HW and Freund H-J (1987) Impairment of rapid movement in Huntington's disease. *Brain* **110**:585–612.
Hoffmann J (1888) Uber Chorea chronica progressiva (Huntingtonsche Chorea hereditaria). *Virchows Archiv. A. Pathological Anatomy and Histopathology* **111**:513–548.
Iizuka R, Hirayama K and Maehara K (1984) Dentato-rubro-pallido-luysian atrophy: a clinicopathological study. *Journal of Neurology, Neurosurgery and Psychiatry* **47**:1288–1298.
Jervis GA (1963) Huntington's chorea in childhood. *Archives of Neurology* **9**:244–257.
Jeste DV, Barban L and Parisi J (1984) Reduced purkinje cell density in Huntington's disease. *Experimental Neurology* **85**:78–86.
Jung S-S, Chen K-M and Brody JA (1973) Paroxysmal choreoathetosis. Report of a Chinese case. *Neurology* **23**:749–755.
Kereshi S, Schlagenhauff RE and Richardson KS (1980) Myoclonic and major seizures in Huntington's chorea: case report and electro-clinical findings. *Clinical Electroencephalography* **11**:44–47.
Kirkham TH and Guitton D (1984) A quantitative study of abnormal eye movements in Huntington's chorea using the scleral search coil technique. *Neuro-ophthalmology* **4**:27–38.

Kobayashi H, Kosaka K, Hoshino C and Shibayama B (1975) An autopsy case of the characteristic degeneration of the dentate nucleus with choreic movement and psychic symptoms. *Clinical Neurology (Tokyo)* **15**:724–730.

Koller WC and Trimble J (1985) The gait abnormality of Huntington's disease. *Neurology* **35**:1450–1454.

Kondo I, Ohta H, Yazaki M, Ikeda J-E, Gusella JF and Kanazawa I (1990) Exclusion mapping of the hereditary dentatorubropallidoluysian atrophy gene from the Huntington's disease locus. *Journal of Medical Genetics* **27**:105–108.

Kuhl DE, Phelps ME, Markham CH, Metter J, Reige WH and Winter J (1982) Cerebral metabolism and atrophy in Huntington's disease determined by ^{18}FDG and computed tomographic scan. *Annals of Neurology* **12**:425–434.

Lakke PWF (1981) Classification of extrapyramidal disorders. Proposal for an international classification and glossary of terms. *Journal of Neurological Sciences* **51**:313–327.

Lance JW (1977) Familial paroxysmal dystonic choreothetosis of Mount and Reback and its differentiation from related syndromes. *Transactions of American Neurological Association* **102**:46–48.

Lanska DJ, Lanska MJ, Lavine L and Schoenberg BS (1988) Conditions associated with Huntington's disease at death: a case control study. *Archives of Neurology* **45**:878–880.

Lasker AG, Zee DS, Hain TC, Folstein SE and Singer HS (1987) Saccades in Huntington's disease: initiation defects and distractibility. *Neurology* **37**:364–370.

Lasker AG, Zee DS, Hain TC, Folstein SE and Singer HS (1988) Saccades in Huntington's disease: slowing and dysmetria. *Neurology* **38**:427–431.

Leigh JR and Zee DS (1983) *The Neurology of Eye Movement*. Philadelphia: FA Davis.

Leigh RJ, Newman SA, Folstein SE, Lasker AG and Jensen BA (1983) Abnormal ocular motor control in Huntington's disease. *Neurology* **33**:1268–1275.

Leli DA, Furlow TW and Falgout JC (1984) Benign familial chorea: an association with intellectual impairment. *Journal of Neurology and Neurosurgery and Psychiatry* **47**:471–474.

Leopold NA and Kagel M (1985) Dysphagia in Huntington's disease. *Archives of Neurology* **42**:57–60.

Loosmore SJ and Wood K (1988) Benign hereditary chorea: a case report. *British Journal of Psychiatry* **152**:131–134.

Lyon IW (1863) Chronic hereditary chorea. *American Medical Times* **7**:289–290.

Markham CH and Knox JW (1965) Observations of Huntington's chorea in childhood. *Journal of Paediatrics* **67**:46–57.

Marsden CD and Quinn NP (1981) Appendix 6. In: *Methods in Clinical Pharmacology – Central Nervous System* (eds MH Lader and A Richens). London: Macmillan.

Mazziotta JC, Phelps ME, Pahl JJ et al., (1987) Reduced glucose metabolism in asymptomatic subjects at risk for Huntington's disease. *New England Journal of Medicine* **316**:357–362.

Merrit AD, Conneally PM, Rahman NF and Drew AL (1969) Juvenile Huntington's chorea. In: *Progress in Neurogenetics 1* (eds A Barbeau and JR Brunette). Amsterdam: Excerpta Medica Found.

Mount LA and Reback S (1940) Familial paroxysmal choreoathetosis: preliminary report on a hitherto undescribed clinical syndrome. *Archives of Neurology* **44**:841–846.

Myers RH, Sax DS, Schoenfeld M et al. (1985) Late onset of Huntington's disease. *Journal of Neurology, Neurosurgery and Psychiatry* **48**:530–534.

Notkin J (1931) Convulsive manifestations in Huntington's chorea. *Journal of Nervous Mental Disorders* **74**:149–160.

Oepen G, Clarenbach P and Thoden U (1981) Disturbance of eye movements in Huntington's chorea. *Archives Psychiatrie Nervenkrankheiten* **229**:205–213.

Oepen H (1963) Uber 217 Korpersektionsbefunde bei der Huntington'schen chorea. *Beiträge zur pathologischen Anatomie und zur Allgemeinen Pathologie* **128**:12–24.

Oliver J and Dewhurst K (1969) Childhood and adolescent form of Huntington's disease. *Journal of Neurology, Neurosurgery and Psychiatry* **32**:455–459.

Osborne JP, Munson P and Burman D (1982) Huntington's chorea. Report of 3 cases and review of the literature. *Archives of Disease in Childhood* **57**:99–103.

Osler W (1892) *The Principles and Practice of Medicine*, pp. 944–945. Edinburgh and London: Young J. Pentland.

Osler W (1904) *On Chorea and Choreiform Affections*. Philadelphia: Blakiston and Son.

Padberg G and Bruyn GW (1986) Chorea: differential diagnosis. In: *Handbook of Clinical Neurology, Vol. 49* (eds PR Vinken, GW Bruyn and HL Klawans). Amsterdam: Elsevier.

Palella TI, Hichwa RD, Ehrenkaufer RL et al. (1985) 18-F Fluorodeoxyglucose PET scanning in HPRT deficiency. *American Journal of Human Genetics* **37**:A70.

Paulson GW (1979) Diagnosis of Huntington's disease. *Advances in Neurology* **23**:177–184.
Penney JB, Young AB, Shoulson I, Starosta-Rubenstein S, Snodgrass SR and Sanchez-Ramos J (1990) Huntington's disease in Venezuela: 7 years of follow-up on symptomatic and asymptomatic individuals. *Movement Disorders* **5**:93–99.
Petit H and Milbled G (1973) Anomalies of conjugate ocular movements in Huntington's chorea. *Advances in Neurology* **1**:287–294.
Pincus JH and Chutorian A (1967) Familial benign chorea with intention tremor: a clinical entity. *Journal of Paediatrics* **70**:724–729.
Quarrell OWJ, Tyler A, Cole G and Harper PS (1986) The problem of isolated cases of Huntington's disease in South Wales. *Clinical Genetics* **30**:331–337.
Quarrell OWJ, Youngman S, Sarfarazi M and Harper PS (1988) Absence of close linkage between benign hereditary chorea and the locus D4S10 (probe G8). *Journal of Medical Genetics* **25**:191–194.
Ramig LA (1986) Acoustic analyses of phonation in patients with Huntington's disease. *Annals of Otology, Rhinology and Laryngology* **95**:288–293.
Reid IC, Besson JAO, Best PV, Sharp PF, Gemmell HG and Smith FW (1988) Imaging of cerebral blood flow markers in Huntington's disease using single photon emission computed tomography. *Journal of Neurology, Neurosurgery and Psychiatry* **51**:1264–1268.
Robinson RO and Thornett CEE (1985) Benign hereditary chorea – response to steroids. *Developmental Medicine and Child Neurology* **27**:814–816.
Rodda RA (1981) Cerebellar atrophy in Huntington's disease. *Journal of the Neurological Sciences* **50**:147–157.
Sadjadpour K and Amato K (1973) Hereditary non-progressive chorea of early onset – a new entity? *Advances in Neurology* **1**:79–91.
Sakamoto H, Matsushita M and Nabano Y (1971) An autopsy case diagnosed as Huntington's chorea with severe degenerative changes in the dentate nucleus. *Advances in Neurological Science (Japan)* **15**:794–795.
Sanberg PR, Fibiger HC and Mark RF (1981) Body weight and dietary factors in Huntington's disease patients compared with matched controls. *Medical Journal of Australia* **1**:407–409.
Sato Y, Ohnishi A, Tateishi J et al. (1984) An autopsy case of chorea-acanthocytosis. *No To Shinkei* **36**:105–111.
Scott DF, Heathfield KWG, Toone B and Margerison JH (1972) The EEG in Huntington's chorea: a clinical and neuropathological study. *Journal of Neurology, Neurosurgery and Psychiatry* **35**:97–102.
Shibaski H, Sakai T, Nishimura H, Sato Y, Goto I and Kuroiwa Y (1982) Involuntary movements in choreo-acanthocytosis: a comparison with Huntington's chorea. *Annals of Neurology* **12**:311–314.
Shimizu T and Kamakura K (1980) Choreo-acanthocytosis. *Journal of Clinical Neurology (Tokyo)* **20**:1056–1058.
Shimizu T, Inoue K and Sugita H (1974) Self-mutilation, choreo-acanthocytosis, muscular hypotonia, absence of deep tendon reflexes and normouricemia. A report of an adult case. *Neurological Medicine of Tokyo* **1**:135–136.
Shoulson I (1986) In: *Diseases of the Nervous System* (eds AK Asbury, CM McKhann and WI McDonald), pp. 1258–1267. Philadelphia: WB Saunders.
Shoulson I and Fahn S (1979) Huntington disease: clinical care and evaluation. *Neurology* **29**:1–3.
Simmons JT, Pastakia B, Chase TN and Shults CW (1986) Magnetic resonance imaging in Huntington disease. *American Journal of Neuroradiology* **7**:25–28.
Siroky MB and Krane RJ (1982) Neurologic aspects of detrusor sphincter dyssynergia, with reference to the guarding reflex. *Journal of Urology* **127**:953–957.
Sleigh G and Lindenbaum RH (1981) Benign (non-paroxysmal) familial chorea. Paediatric perspectives. *Archives of Disease in Childhood* **56**:616–621.
Starkstein SE, Folstein SE, Brandt J, Pearlson GD, McDonnell A and Folstein M (1989) Brain atrophy in Huntington's disease. *Neuroradiology* **31**:156–159.
Starr A (1967) A disorder of rapid eye movements in Huntington's chorea. *Brain* **90**:545–564.
Stevens DL (1976) Huntington's chorea: a demographic, genetic and clinical study. MD Thesis, University of London.
Suchowersky O, Hayden MR, Martin WRW, Stossel AJ, Hildebrand AM and Pate BD (1986) *Movement Disorders* **1**:33–44.
Tanahashi N, Myers JS, Ishikawa Y et al. (1985) Cerebral blood flow and cognitive testing correlate in Huntington's disease. *Archives of Neurology* **42**:1169–1175.
Tanner CM (1986) Drug-induced movement disorders (tardive dyskinesia and dopa-induced dyskinesia. In: *Handbook of Clinical Neurology, Vol. 49* (eds PJ Vinken, GW Bruyn and HL Klawans). Amsterdam: Elsevier.

Telez-Nagel I, Johnson AB and Terry RD (1973) Ultrastructural and histochemical study of cerebral biopsies in Huntington's chorea. In: *Huntington's Chorea, 1872–1972* (eds A Barbeau, TN Chase and GW Paulson), pp. 387–398. New York: Raven Press.

Thompson PD, Berardelli A, Rothwell JC *et al.* (1988) The coexistence of bradykinesia and chorea in Huntington's disease and its implications for theories of basal ganglia control of movement. *Brain* **111**:223–244.

Troster AI, Salmon DP, McCullough D and Butters N (1989) A comparison of the category fluency deficits associated with Alzheimer's and Huntington's disease. *Brain and Language* **37**:550–513.

Truong DD, Harding AE, Scaravilli *et al.* (1990) Movement disorders in mitochondrial myopathies. *Movement Disorders* **5**: 109–117.

Upadhyaya M, Reynolds GP and Harper PS (1985) Recombinant DNA studies on stored necropsy brain samples from patients with Huntington's chorea. *Journal of Clinical Pathology* **38**:1093–1095.

Van Dijk JG, Van der Velde EA, Roos RAC and Bruyn GW (1986) Juvenile Huntington disease. *Human Genetics* **73**:235–239.

Vogt C and Vogt O (1920) Zur Lehr der Erkrankungen des striaten Systems. *Journal für Psychologie und Neurologie* **25**:627–846.

Vonsattel J-P, Myers RH, Stevens TJ *et al.* (1985) Neuropathological classification of Huntington's disease. *Journal of Neuropathology and Experimental Neurology* **44**:559–577.

Waters C and Chen D (1990) SPECT-IMP in Huntington's disease. *Movement Disorders* **5**(supp.1):38.

Westphal CFO (1883) Uber eine dem Bilde der cerebrospinalen grauen Degeneration ähnliche Erkrankung des centralen Nervensystems ohne anatomischen Benfund, nebst einigen Bemerkungen über paradoxe Kontraction. *Archive Psychiatrie Nervenkrankheiten* **14**:87–95 and 767–773.

Wheeler JS, Sax DS, Krane RJ and Siroky MB (1985) Vesico-urethral function in Huntington's chorea. *British Journal of Urology* **57**:63–66.

Williams J and Stevens H (1963) Familial paroxysmal choreo-athetosis. *Pediatrics* **31**:656–661.

Wilson SAK (1925) The Croonian lectures on some disorders of motility and of muscle tone, with special reference to the corpus striatum. *Lancet* **ii**:169–179.

Young AB, Penney JB and Starosta-Rubenstein S (1986) PET scan investigations of Huntington's disease: cerebral metabolic correlates of neurological features and functional decline. *Annals of Neurology* **20**:296–303.

Young AB, Penney JB, Starosta-Rubenstein S *et al.* (1987) Normal caudate glucose metabolism in persons at risk for Huntington's disease. *Archives of Neurology* **44**:254–267.

Zangemeister WH and Mueller-Jensen A (1985) The co-ordination of gaze movements in Huntington's disease. *Neuro-ophthalmology* **3**:193–206.

3

Psychiatric Aspects of Huntington's Disease

INTRODUCTION

Psychiatric manifestations, which were recognized in the earliest descriptions of Huntington's disease (HD) are wide-ranging but non-specific. Schizophrenia, paranoia, depression, suicide, and personality disorder have been documented in sufferers of HD but a large number of these early descriptions were case reports (e.g. MacFaren, 1874; Lewis, 1876; Lind, 1927) rather than systematic studies. One of the problems of reviewing the literature on the psychiatric aspects of HD is the great variation in the meaning and definition of 'mental' symptoms. At the time of Huntington's original report (1872), depression was not generally distinguished from cognitive decline and so for several decades most psychiatric symptoms were grouped together rather than categorized. There were some exceptions such as the early description of the cognitive impairment (Hallock, 1898) and the attempted identification of 'biotypes' (Davenport and Muncey, 1916).

Psychiatric features of HD will be considered first in relation to assessment, initial symptoms and the problem of misdiagnosis. Each of the major psychiatric syndromes in HD will be considered, in the context of recent international concepts of classification. This is followed by a discussion of the cognitive aspects of HD and finally its differential diagnosis in relation to other psychiatric disorders.

ASSESSMENT

The overall clinical assessment and investigation of the HD patient have been discussed in Chapter 2, but a note on psychiatric assessment is added here.

Many HD patients, especially those in the early stages, are able to give a good account of themselves. They can describe the onset and course of their symptoms and how they are affected in functional terms. Before diagnosis, they may have been 'symptom-searching' and can often recognize the onset or worsening of psychological symptoms such as irritability. However, there are difficulties with taking a history from HD patients. Firstly, as the disease progresses, dysarthria may become more pronounced, thus impairing communication. Secondly, memory problems can lead to distortion of events. Thirdly, HD patients are not infrequently uncooperative and may resent being interviewed. They often fail to attend hospital appointments but usually agree to being interviewed in their homes. Time spent establishing rapport is crucial to the success of the interview.

Mental status examination may support details obtained from the history (see *Handbook on Mental State Examination* produced by the Institute of Psychiatry, 1986). Chorea and weight loss may be evident. Other features such as the level of self-care should also be noted and whether this is the result of the patient's or the carer's efforts. Some HD patients have diminished awareness of their environment, as may be evident from wandering, but misidentification of key relatives is rare. Several types of mood disturbance may occur in HD, including depression, irritability and, in later stages, emotional blunting. Among abnormalities of speech are slurring and sudden and unexpected pauses. In HD, delusions (especially paranoid delusions) are more common than abnormal perceptions (e.g. hallucinations). Cognitive functioning, in particular orientation and memory, should be assessed in every HD patient. Contrary to what is often believed, not all HD patients suffer from intellectual decline.

History from other informants is very important, especially in the case of patients with advanced HD or those with denial. Relatives can usually date the onset of HD or its psychiatric manifestations more accurately than the affected patient and can often provide useful information on how HD affects social functioning and family life. The comments of relatives are often coloured by their emotional attitude to the affected patient and it is wise for the clinician not to take sides in any family disputes. Another useful informant is the general practitioner who should be contacted for first-hand knowledge of the family.

There is no consensus on the role of psychometric assessments in HD. Some of the problems are that they offer a limited cross-sectional view of a longitudinal disorder and that factors such as poor cooperation will affect test results. Chorea itself will adversely affect performance tests (e.g. assembling objects from their parts and reproducing designs with blocks). Many cognitive tests were standardized on the general population and so are unsuitable in the assessment of HD patients, for whom tests to identify specific cognitive impairment should be used. Cognitive tests may have to be conducted over several sittings owing to the HD patient's reduced attention span. The main usefulness of cognitive assessment in HD is to indicate areas for rehabilitation and to provide a baseline for later reference.

INITIAL MANIFESTATIONS OF HD

A substantial body of research has shown differing estimates of the initial symptoms of HD. Some authors (e.g. Brothers, 1964) reported that the first

Table 3.1 Initial symptom of affected individuals

Author	Year	Neurological (%)	Psychiatric (%)	Both
Minski and Guttmann	1938	21	79	—
Oepen	1963	34	24	42
Brothers	1964	59	27	14
Heathfield	1967	54	46	—
Bolt	1970	54	27	19
Mattsson	1974	22	48	30
Stevens	1976	65	35	
Hayden	1979	39	51	10
Walker et al.	1981	66	34	—

features of HD are mainly neurological whereas others (e.g. Minski and Guttmann, 1938) presented evidence to the contrary (Table 3.1). Mattsson (1974) stated that personality change (including 'alcoholism and criminal behaviour') is the most common psychiatric presentation of HD and that disorders of consciousness are rarely seen as presenting symptoms.

Certain factors may explain these different rates. Were only living patients taken into account? Were all patients personally examined? Was another informant interviewed? The training of the investigator would also appear to be relevant. Neurologists would be likely to identify higher rates of neurological abnormalities than psychiatrists and vice versa.

In a survey in South Wales (Walker, 1979; Walker et al., 1981) where home visits were made to collect full clinical information, about two-thirds of living HD patients initially presented with neurological manifestations and one-third with psychiatric changes (Table 3.2). Information was obviously less accurate for deceased patients and those with advanced HD, in whom roughly equal numbers had presented with neurological and psychiatric symptoms.

Table 3.2 Initial manifestations of HD in a survey in South Wales (From Walker et al., 1981)

First symptom	Alive	Dead	Not known	Total
Chorea	80	104	1	185
Mental	41	96	1	138
Rigidity	—	1	—	1
Other	1	—	—	1
Not known	6	85	2	93
Total	128	286	4	418

Table 3.3 Initial diagnoses in studies of HD

Diagnosis	Dewhurst (1970) No. = 102 (%)	Bolt (1970) No. = 334 (%)	Mattsson (1974) No. (n = 162) (%)
HD	38	61	33
Schizophrenia/psychoses	7	8	17
Mood disorder	9	5	15
Personality disorder	10	4.5*	24
Neurosis	11	—	—

*Bolt categorized personality disorder and neurosis together.

MISDIAGNOSIS

Several epidemiological studies have found significant rates of initial misdiagnosis of HD, especially in the early stages of the disorder (Table 3.3). Bolt (1970) found that more than two-thirds of the misdiagnoses were ascribed to primary psychiatric illnesses, of which schizophrenia and paranoid psychoses were the commonest. Shokeir (1975), in a prevalence study of HD in the Canadian prairies also found significant rates of misdiagnosis. Many of his HD patients were initially diagnosed as having Parkinson's disease, premature senile dementia, Pick's disease, dystonia, cerebral atherosclerosis and psychosis. He added that 'in some instances these diagnostic labels were retained . . . until another member of the family was diagnosed as having Huntington's disease'. In a large epidemiological study of 801 HD patients in Michigan (Chandler et al., 1960) a similarly wide range of disorders mistaken for HD was reported. Twenty-four (11%) out of 217 patients with HD were misdiagnosed as having a variety of neurological and psychiatric disorders (e.g. parkinsonism, movement disorder due to trauma, alcoholism) in a careful study in Maryland (Folstein, 1989). This study also found that 31 patients (15%) thought to have HD were found on investigation to suffer from another disorder (Folstein, 1989). Diagnostic problems in a HD survey in South Wales have been reviewed by Quarrell et al. (1986) (see Chapter 2).

PREVALENCE OF PSYCHIATRIC DISORDER IN HD

Reported prevalence rates of psychiatric symptoms in HD vary widely (see Table 3.4) but many of these studies provide only limited descriptions of the nature of observed psychiatric phenomena. Some studies have used standardized questionnaires but only rarely have standardized diagnostic criteria been used. Depression has been the most studied psychiatric state in HD and other aspects of psychiatric status such as anxiety have been relatively neglected. Over the past decade research workers, particularly the group at Johns

Table 3.4 Prevalence of psychiatric disorder in HD (not including dementia)

Authors	Year	Number of HD patients	Number with psychiatric disorder	%
Bickford and Ellison	1953	21	14	(66%)
Heathfield	1967	80	40	(50%)
Oliver	1970	100	94*	—
Saugstad and Odegard	1986	199	71	(35%)
Folstein et al.	1987	186	130	(73%)

*Some patients were counted more than once.

Hopkins University (Folstein et al., 1979; Folstein, 1989), have been using both questionnaires and standardized semi-structured interviews, which offer greater flexibility when assessing psychiatric symptoms and have thus permitted a more detailed description of the psychiatric profile of sufferers of HD.

GENERAL CLASSIFICATION OF PSYCHIATRIC ILLNESS

Considerable controversy surrounds the concept of psychiatric illness, not least because so little is known about aetiology (Wootton, 1959; Clare, 1976). This is particularly relevant when classification is attempted. Some writers, such as Menninger et al. (1963) have argued against classification, on the grounds that since every patient is unique so must be their illness. Yet it is generally accepted that psychiatric classification, with all its shortcomings, is necessary for useful communication. There are currently two main systems of classification in psychiatric practice: the *International Classification of Diseases* (World Health Organization, 1978, 1987) and the *Diagnostic and Statistical Manual* which is produced by the American Psychiatric Association (1980, 1987).

The first time mental disorders (Section V) were incorporated into the *International Classification of Diseases* in its sixth revision (ICD-6) published in 1948. This system was only officially recognized by five countries and so not widely used but subsequent revisions have been formally adopted by almost every country. The ninth and current revision (ICD-9), which has been criticized for being a compromise, has nonetheless been valuable in encouraging greater uniformity of classification around the world (Saugstad and Odegard, 1985). The World Health Organization has prepared the tenth revision of the *ICD* which will come into international use in 1991–1992. Psychiatric symptoms will be classified under Chapter V (F) which will be entitled 'Mental, Behavioural and Developmental Disorders' (World Health Organization, 1987; Cooper, 1989; Table 3.5).

In the United States, alternative classification was introduced in an attempt to overcome the perceived weaknesses of ICD-6. The first edition of the

Table 3.5 Classification of mental diseases (ICD-10)

F0	Organic, including symptomatic mental disorders
F1	Mental and behavioural disorders due to psychoactive substance use
F2	Schizophrenia, schizotypal and delusional disorders
F3	Mood (affective) disorders
F4	Neurotic, stress-related, and somatoform disorders
F5	Behavioural syndromes and mental disorders associated with physiological dysfunction and hormonal disturbances
F6	Disorders of adult personality and behaviour
F7	Mental retardation
F8	Developmental disorders
F9	Behavioural and emotional disorders with onset usually occurring in childhood and adolescence

Diagnostic and Statistical Manual (DSM-I) appeared in 1952 and was revised in 1965. The third edition (DSM-III) (American Psychiatric Association, 1980), in contrast to its predecessors, had five axes (psychiatric syndromes, personality disorders, physical disorders, severity of psychosocial stressors and highest level of adaptive functioning). It contained explicit operational criteria for making diagnoses, thus improving its reliability. DSM-III and its most recent revision (DSM-III-R), have been widely used not only in the United States but also in many other countries.

The following sections describe the major psychiatric syndromes seen in HD. The numbers in brackets refer to their ICD-10 classification.

SCHIZOPHRENIA, SCHIZOTYPAL AND DELUSIONAL DISORDERS (F2)

Schizophrenia (F20)

The nature of schizophrenia is probably best understood by taking a historical perspective. Until about a century ago, many influential psychiatrists believed that all mental disorders were expressions of a single entity which was thought to begin with depression, become a psychosis and end as a dementia (Griesinger, 1845). Emil Kraepelin (1856–1926), a German psychiatrist, disagreed and in 1898 proposed his concept of dementia praecox, thus laying the groundwork for contemporary views of schizophrenia (Berrios and Hauser, 1988). His dementia praecox included three subgroups, namely hebephrenia, catatonia and paranoia. In 1903, he added the group of dementia simplex. He clearly separated dementia praecox from manic depressive illness. Eugen Bleuler (1857–1939), a Swiss psychiatrist, widened the diagnostic boundaries of what he felt was a group of disorders and introduced the term schizophrenia (Bleuler, 1911). It thus became necessary for psychiatrists to agree what must be present to make a diagnosis of schizophrenia and in 1959, Kurt Schneider (1887–1967) proposed his well-known first-rank symptoms (FRS) which would differentiate schizophrenia from other conditions. This was a pragmatic approach and was not evolved from theoretical concepts. It was widely adopted in Britain and has been the basis of numerous research studies on schizophrenia.

First rank symptoms of schizophrenia include passivity experiences in which thoughts, emotions, impulses or actions are believed by the individual to be under external, alien control. Auditory hallucinations in the third person, in which the patient hears a voice or several voices discussing him in the third person, are also a diagnostic symptom of schizophrenia in the absence of physical disease. These symptoms are used in the DSM-III and DSM-III-R classifications of schizophrenia. However, in contrast to cross-sectional definitions, these criteria require 'continuous signs of disturbance for at least 6 months'. This longitudinal requirement has been criticized because there are patients who fulfil DSM-III criteria for days or weeks but not for as long as 6 months (Birley, 1990).

The relationship between schizophrenia and HD will be examined under four headings: HD misdiagnosed as schizophrenia, the prevalence of schizophrenic symptoms in established HD, whether there is any difference in the symptomatology of a schizophrenia-like disorder when it occurs in HD and whether a schizophrenia-like disorder can precede chorea.

Misdiagnosis
Panse (1942) reported that the most common misdiagnosis of HD was as schizophrenia. Twenty-three (5%) of his series of 461 HD patients had a diagnosis of schizophrenia and the true diagnosis of HD was only uncovered by investigating the family history.

Dewhurst *et al.* (1970) in their study of HD in Oxfordshire found that seven out of 102 patients were initially diagnosed as having schizophrenia. Similarly, Bolt (1970) found that about 8% of 334 HD patients in Scotland whose case notes were available were initially misdiagnosed as having schizophrenia or paranoid psychosis. Unfortunately, few other details about these patients (e.g. nature of symptoms) are given. In the Maryland survey, Folstein (1989) reported that schizophrenia was diagnosed incorrectly in two patients (out of 217 HD patients) who were later found to have HD.

Prevalence of schizophrenic symptoms in HD (Table 3.6)
It is difficult to be certain about the rate of schizophrenic symptoms in HD because of differences in study design, whether patients were personally interviewed and whether biased HD populations were studied (e.g. HD patients in a mental hospital). Oliver (1970) in his series of 100 HD patients did not see any cases of typical schizophrenia. However, he described three patients with 'schizophreniform symptoms' of 'queerness, irrational behaviour, writing odd letters, etc.', symptoms that must be regarded as non-specific. Mattsson (1974) reported that about 12% of a large series of HD patients had a history of schizophrenia but the main weakness of this research is that it was a

Table 3.6 Frequency of schizophrenia in HD

Authors	Year	No. with schizophrenia	Total HD cases	Frequency %	Standardized diagnostic criteria
Minski and Guttmann	1938	3	50	6	No
Brothers	1964	16	312	5	No
Heathfield	1967	9	96	10	No
Streletzki	1961	67	1200	5.7	No
Oltman and Friedman	1961	4	57	7	No
Bolt	1970	58	334	8	No
Dewhurst	1970	7	102	7	No
Oliver*	1970	12	100	12	No
Mattsson	1974	—	—	12	No
Stevens	1976	12	106	11	No
Folstein	1989	3	88	3.4	Yes

*'Schizophreniform symptoms'.

case record study. Data were collected by inviting all mental hospitals and departments of psychiatry and neurology to report cases of HD and whether they had any psychiatric illness. The results provide a useful guide to the frequency of schizophrenia in HD but, under ordinary clinical conditions, there are considerable diagnostic discrepancies among psychiatrists (Ward et al., 1962; Cooper, 1967). In a survey of the west of Scotland, Bolt (1970) found that eight HD patients out of 124 personally examined had features of schizophrenia but she did not state which diagnostic criteria were used. Folstein (1989), using a standardized diagnostic instrument, found three HD patients out of 88 had a history of schizophrenia in a survey in Maryland. Despite the various methodological problems, it appears that the association of schizophrenia and HD exceeds chance expectation.

The classical subtypes of schizophrenia have all been described in HD, especially paranoid schizophrenia (Naef, 1917; Panse, 1942; Heathfield, 1967) but also hebephrenic and catatonic subtypes (Panse, 1942; Brothers, 1964).

The course of schizophrenia in HD depends on many factors, including antipsychotic treatment. Streletzki (1961) noted that schizophrenic symptoms could remit after 1–3 years when cognitive decline supervened. Others have reported that a 'schizophrenic' personality change supervenes (Minski and Guttmann, 1938; Panse, 1942).

Nature of the association between HD and schizophrenia: the 'organic schizophrenias'
Davison and Bagley (1969) have pointed out that various forms of 'organic' or 'symptomatic' schizophrenia have the same range of symptoms and fulfil the same operational diagnostic criteria as spontaneously occurring schizophrenia (Davison and Bagley, 1969; Davison, 1983). Thus there are no convincing distinguishing features between 'organic' and 'true' schizophrenia. Streletzki (1961) thought that schizophrenic symptoms in HD are indistinguishable from

'endogenous schizophrenia'. Cummings (1985) suggested that lesions in the subcortical structures are the commonest sites for the genesis of organic schizophrenia.

There are conflicting opinions about whether organic schizophrenia tends to develop *de novo* or occurs more frequently in those with a family history of schizophrenia (Nicotra, 1938; Tyskiewicz, 1960). Panse (1942) reported three sisters with a psychotic presentation of HD while Heathfield (1967) found 'identical' schizophrenic symptoms in two affected siblings and the same psychiatric disorder in a third who was unaffected with HD. In other HD families, psychotic presentation has been reported in three generations (Oppler, 1933; Meierhofer, 1938). Despite these familial aggregations, Davison (1983) believes that there is a lack of genetic predisposition in organic schizophrenia.

It can be concluded that the occurrence of schizophrenic symptoms in patients with organic disease, including HD, could lead to greater understanding of 'idiopathic' schizophrenia and is evidence of a syndromal rather than single disease entity concept of schizophrenia.

Schizophrenia as a precursor of HD
Several authors have suggested that schizophrenia-like disorder may be a 'forme fruste' of HD. One of the earliest to suggest this was Evrard (1936) who described schizophrenia as a 'phenotypic polymorphism' of HD. Other authors were Tusques and Feuillet (1937), Laane (1951) and Panse (1942) who found that the psychosis could precede the onset of chorea by 1–35 years. More recently, Brothers (1964) in his Australian series noted that 16 of his series were diagnosed as having schizophrenia several years before the onset of chorea. This finding was supported by Heathfield (1967) and Bolt (1970). Garron (1973) urged caution in interpreting these results and recommended systematic longitudinal studies of those at risk.

Persistent delusional disorders (F22)
This group of disorders includes a variety of conditions in which longstanding delusions constitute the only, or the most conspicuous, clinical characteristic, and which cannot be classified as organic, schizophrenic or affective.

Paranoia
The clinical concept of 'paranoia', which is derived from the Greek word meaning madness, has been subject to controversy for over a century and remains unresolved (Lewis, 1970; Hart, 1990). Kraepelin (1912) stated that paranoia was a condition which developed insidiously and was initiated by a single unshakeable delusion not necessarily of persecutory content; secondary delusions usually arise later culminating in a complex delusional system. The other mental functions, intellectual and emotional, remained unimpaired. According to Post (1966) paranoid is now used almost exclusively to describe 'any delusional form of self-reference concerned with persecution, litigation,

love, envy, hate, honour or the supernatural'. Examples of paranoid psychosis in HD encountered by us are described below.

A 53-year-old man studied by the authors had an 8-year history of HD and was admitted to a pyschiatric hospital under the Mental Health Act 1983 because of paranoid symptoms. He believed that the British Broadcasting Corporation was secretly recording his movements. He acted on his delusional belief by digging up his garden looking for wires and 'bugging devices'. He believed his neighbours were involved in a conspiracy against him and smashed their windows. This patient had clear paranoid symptoms which took the form of delusions of persecution. He believed that individuals and an organization were trying to harm him and damage his reputation. HD was the main aetiological factor although he had a premorbid suspicious personality.

Another patient, who was aged 45 and had a 10-year history of HD, was admitted to hospital as an emergency. She refused to eat because she believed that her daughter was poisoning her food and had already poisoned the family dog. Her explanation was that her daughter wanted to inherit the house prematurely. She was unable to explain how she knew her home food was poisoned but was prepared to accept hospital meals. She was treated with antipsychotic medication and her symptoms improved.

Paranoid psychoses may occur in organic mental states arising from any cause such as primary degeneration, trauma, and infection. Rosenbaum (1941), in a retrospective study of the case-notes of 46 HD patients admitted to St Laurence State Hospital, New York, found that 32 patients (70%) had paranoid delusions which were in general directed against relatives and friends. Bolt (1970) found that one-third of HD patients ascertained in the west of Scotland had 'often poorly systematized' paranoid ideas. Considering that the term paranoid is so variably used, it is a pity that these patients were not described more fully.

Othello syndrome
In the Othello syndrome, or the syndrome of pathological jealousy (*Eifersuchtswahn*), one partner (usually the husband) becomes convinced that the spouse is unfaithful (Shepherd, 1961; Tarrier *et al.*, 1990). The delusion of infidelity is the predominant symptom and it is often preceded by increasing suspiciousness over several months. Not infrequently, the husband acts on his delusion by accusing, cross-examining and following his wife. Incidental occurrences are misinterpreted as providing definite proof of extramarital liaisons. The jealous husband spends hours trying to extract a 'confession' and may murder or seriously assault his wife. The onset is typically in the fourth decade.

There are no reliable studies of the frequency of delusional jealousy in HD. It is occasionally mentioned in epidemiological studies such as in the early South Wales survey by Spillane and Phillips (1937). Geraud *et al.* (1970) reported in some detail the Othello syndrome in a 49-year-old man with HD. Bigelow *et al.* (1959) described six female affected patients 'freely expressing delusions of infidelity'. We have observed the following case.

A 52-year-old man at risk for HD was referred as an emergency because of his disturbed behaviour. While holidaying in Spain he began to suspect his wife's fidelity and accused her of having sexual relations with numerous partners. On the first week of their holiday, he was convinced she had intercourse with 30 men nightly and in the second week, it was 40 men nightly. He accumulated trivial incidents to support his case. Whenever his wife left the hotel-room, he believed she went to the nearest lift and approached the first man (especially if he was alone) for sex. He further believed that each 'propositioned' man immediately accepted the offer of her favours. When asked to explain this improbable behaviour of his wife, he returned that she was 'good at persuading' and was dressed in her night-clothes to save time. His wife's denial of these events led to persistent quarrelling so that since their return they have been sleeping apart. He had a history of alcohol misuse and during his holiday he took his first alcoholic drink for 6 years. On examination he had choreiform movements and the diagnosis of HD was confirmed on inpatient investigation. This patient had the pure form of the Othello syndrome. Alcohol was an important factor in the aetiology of this man's symptoms together with early signs of his organic neurological disease.

Ekbom's syndrome
The characteristic symptom of Ekbom's syndrome (to be distinguished from the neurological disorder of 'restless legs' also named after Ekbom) is the delusional conviction of being infested with small organisms, such as fleas or mites (Ekbom, 1938). The description of delusional infestation has been subject to imprecise terminology. 'Acarophobia', 'parasitophobia' and 'entomophobia' are some of the misleading terms that have been applied to the condition. The majority of patients are not strictly suffering from a phobia, or morbid fear; rather they present with an unshakeable false belief that they are infested. Ekbom (1938) rightly described the phenomenon as a delusion ('dermatozoenwahn' or delusion of animal life in the skin).

Ekbom's syndrome has a wide variety of underlying aetiologies, such as affective psychosis, schizophrenia, and pellagra. It is recognized in several neurological disorders, e.g. brain tumours, cerebral infarction and Alzheimer's disease. Three cases of Ekbom's syndrome have been ascertained from our HD register of 800 affected patients.

A 66-year-old married woman had a 3-year history of delusions of infestation by fleas, snails and mice. The onset was 5 years after formal diagnosis of HD. She constantly scratched herself, especially in the pubic region, in a vain attempt to relieve the symptoms of the alleged infestation. She ascribed swellings on her legs to wasp bites. She often refused to eat because she believed her food was poisoned by 'mouse droppings'. Her husband was particularly exasperated when she woke him during the night about these abnormal beliefs. She insisted that he bought insecticidal agents and washed the bed-clothes daily. Her family, general practitioner and inspectors from the Environmental Health Department failed to confirm the presence of insects or rodents. She eventually accepted treatment with an antipsychotic and her delusions remitted.

Induced delusional disorder (F24)

An induced psychosis is a delusional system which appears to have developed in a person as a result of a close relationship with another person who already has an established and similar delusional system (Enoch and Trethowan, 1979). Lasegue and Falret (1877) in their seminal paper described seven cases of *folie à deux* which means insanity or psychosis of two. It is a very rare phenomenon even when two individuals are involved but rarer still when three (*folie à trois*) or four (*folie à quatre*) people are involved. It may also occur in an entire family (*folie à famille*). Induced delusional disorder has rarely been reported in HD families.

Sims *et al*. (1977) described *folie à quatre* which affected the proband, his wife, sister and brother. The proband believed that an industrial organization had put 'bugging' devices in the wall of his brother's house. The proband's wife, who had an a priori risk of 50% of developing HD, corrobated his account by claiming that the bugging devices had been placed to find out what disablement category her brother-in-law was in for compensation purposes and alleged that the post office had confirmed her findings. In this family, the proband had the primary psychosis and his wife may be described as having a *folie imposée*. Unfortunately, she has not been followed up long enough to find out whether she has developed HD.

AFFECTIVE (MOOD) DISORDERS (F3)

The affective disorders are so called because an abnormality of mood – either depression or elation – is a prominent feature. Depression is a word which is employed in many different contexts and senses. It is used to describe the normal feeling of unhappiness in times of adversity. It is also used to describe a symptom which is a component of many psychiatric and physical disorders. Depression can be used to describe a syndrome, in which depressed mood is the central symptom, but other features of the syndrome may include negative thinking, lack of enjoyment and anergia.

There has been little consensus about how depression should be classified (Kendell, 1976). Since the 1920s there has been much controversy about the merits of classifying depression into 'reactive' and 'endogenous'. The reactive depression was thought to be relatively mild with less disturbance of vegetative functions such as appetite, sleep and libido, while it was also considered understandable in terms of adverse circumstances. Guilt, loss of weight and appetite, constipation, reduced libido, early morning waking and high risk of suicide were said to characterize endogenous depression. The terms 'reactive' and 'endogenous' suggest aetiology but it is likely that life events and constitutional factors contribute to both depressions. There are also doubts about the validity of the two syndromes and it is now generally considered that the symptomatology of depression forms a continuum.

ICD-9 has two main groups for depression – manic depressive psychosis and neurotic depression. Neurotic depression includes mild depressive disorders in which disproportionate depression has usually followed a distressing experience. Over the past decade, operational criteria for depression have become widely used, although the problem of validity has not been surmounted (Farmer and McGuffin, 1989). In ICD-10, depression will be classified as single depressive episode (F31), depression occurring as a current illness within the context of bipolar disorder (F32) and recurrent depression (F33). The classification of affective disorders in DSM-III and DSM-III-R uses different terminology and does not employ familiar terms such as manic depressive, neurotic, endogenous and reactive. Instead the main categories are 'major depressive episode' and 'organic affective syndrome' which are operationally defined. The organic affective syndrome is characterized by a predominantly dysphoric mood over at least a 2-week period accompanied by at least two of the following: appetite or weight change; insomnia or hypersomnia; psychomotor agitation or retardation; loss of interest in daily activities; fatigue; feelings of worthlessness; diminished concentration, and recurrent thoughts of death or suicide. No evidence of significant dementia or alteration in consciousness should be present and the mood disturbance must be judged to be related to an underlying disorder. The major depressive episode is distinguished by the absence of a causative or related 'organic' disorder and includes four of the symptoms described above. Both forms of depression may occur in HD but these criteria have not been used in most studies of depression in HD.

Mild depressions are classified under 'dysthymic disorder' (Burton, 1990), but Murphy (1991) has criticized this term as ill-defined and non-specific.

Depression in HD

Affective symptoms were recognized by Huntington (1872) himself when he said that 'In all the families or nearly all in which the choreic taint exists, the nervous temperament greatly preponderates'. In the opinion of Davenport and Muncey (1916), manic-depressive illness 'is far and away the commonest type of insanity that is associated with chorea'. Descriptions of affective symptoms in HD are also found in the German literature. Symptoms of depression, including suicidal ideation, and anxiety in HD patients were reported by Entres (1921) and Meggendorfer (1923). Panse (1942) in a major study (see also Chapter 11) considered the relationship between depression and HD. He concluded that depressive episodes in HD were short-lived and better described as 'moroseness' or 'acute dysphoric attacks' rather than as 'endogenous' depression. Koehler and Sass (1984) argue convincingly that many of Panse's cases had 'true' manic-depressive illness but that he tried to 'explain away' and reduce the importance of these symptoms which were then interpreted in terms of personality deviation.

Table 3.7 Prevalence of depression in HD

Author(s)	Year	No. depressed	(%)	Controls
Rosenbaum	1941	13/46	28	No
Heathfield	1967	10/80	12.5	No
Dewhurst et al.	1970	15/102	14.7*	No
Bolt	1970	84/334	25	No
Oliver	1970	9/100	9†	No
Mattsson	1974		16	No
Folstein et al.	1983	36/88	41	No
Mindham et al.	1985	12/27	44	Yes

*Also includes 'suicide attempts and self-starvation'.
†Another 24% had 'depressive symptoms' including apathy, tiredness, inability to cope and insomnia.

Depression is said to be frequent in Huntington's disease and in various studies, prevalence has varied from 9 to 44% (Table 3.7). Most of these studies have methodological shortcomings. Sample sizes were often small, many sufferers were inpatients in psychiatric hospitals (Saugstad and Odegard, 1986) and not all cases were personally interviewed (Mattsson, 1974). Other limitations were the failure to state that definition of depression was used and lack of a control group.

Folstein, in one of the major studies of depression in HD (Folstein et al., 1983a,b; Folstein, 1989) found that 41% of a consecutive series of 88 patients had a history of major depressive episodes. These patients had an established family history of HD and each patient was personally interviewed using a structured interview (Diagnostic Interview Schedule). A spouse was used as an additional informant and there was a review of all obtainable case-notes.

In a controlled study, Mindham et al. (1985) compared the frequency of depression in HD and Alzheimer's disease. Twenty-seven patients from each of the two diagnostic groups were interviewed and a history of depression was ascertained used the Diagnostic Interview Schedule. It was found that HD patients showed twice the prevalence of major affective disorder. The limitations of the study, as the authors point out, were that the patients were not accurately matched for age and the numbers were relatively small.

Familial association of affective disorder

Familial association of affective disorder was studied in 10 established HD families (Folstein et al., 1983a, b). In five families, the proband had HD but not affective disorder and in the other five families, the proband had both HD and affective disorder. In the first five families (i.e. proband without depression), one out of 25 at risk individuals had at least one episode of depression while the other five families, five out of 20 at risk individuals had such a depressive episode. The authors suggested a number of interpretations for this significant finding, which needs to be replicated; they concluded in favour of genetic heterogeneity in HD being responsible for the familial correlation.

This study also provided evidence that affective disorder may precede chorea. It was found that affective symptoms preceded the neurological and dementing features by an average of 5 years in 66% of individuals who had both HD and affective disorder. Those retrospective data on early manifestation of affective symptoms have been supported by long-term prospective follow-up of at-risk individuals. In addition, psychiatric investigation of those who have had high-risk presymptomatic tests has shown that some of them have developed affective disorder before the onset of neurological symptoms (see Chapter 12). However, these intriguing results which suggest that depressive symptoms could be predictive for subsequent general symptoms of HD have yet to be confirmed in an independent series using similarly clear diagnostic criteria for affective disorder.

Racial variation in the prevalence of depression
Folstein *et al.* (1987) in a systematic survey of HD in Maryland reported the unexpected finding of a relatively large number of black patients (50 cases out of 186). They also found significantly lower rates of affective disorder in the black HD population compared with the white. Affective disorder was found in 10% of the blacks and in 41% of the white HD cases. Other clinical differences among the blacks were the lower average age of onset and more pronounced bradykinesia. The authors suggest a number of hypotheses to explain these findings, including the paternal transmission effect, the possibility of two alleles at the HD locus or the possibility of an unlinked modifier of a single HD allele that is more common in the black population than the white. The true explanation is currently unknown (see Chapter 9).

Nature of the association between depression and HD
Coincidental hypothesis
According to this hypothesis, there is no difference between the rates of depression in HD and non-HD patients, a view that does not agree with most data.

Psychological (reactivity) hypothesis
This theory regards depression as a psychological response to the development of HD. However, it is often unclear what is meant by 'psychological' or whether the individual is reacting to the actual loss of health or to the threat of further loss of health. Reactivity is often viewed in terms of linear relationship between aspects of the disease (e.g. severity, disability) and depression scores; the finding of a positive correlation is taken as evidence of the psychological hypothesis. Although the reactivity hypothesis may have face validity, there is very little empirical evidence to support it. Folstein (1989) listed the main reasons against: the depressive symptoms precede the onset of neurological symptoms; about 10% of the Baltimore series had hypomania which would not

be expected; the familial association of depression is higher in certain families; and the depression responds to antidepressant medication.

Organic hypothesis
According to this view, depression results from neuropathological changes that are the same as those causing the neurological symptoms. This model can explain the appearance of depression both during the pre-neurological and the neurological stages of HD. However, it is uncertain if depression and the neurological symptoms of HD share the same organic substratum. Folstein and Folstein (1987) hypothesized that the neuronal loss in the medial anterior caudate is responsible for both the neurological and emotional symptoms. Even if this were so, the expression of these symptoms is likely to be controlled by many intervening variables and no one-to-one correspondence between lesions and neurological deficits can be assumed.

SUICIDE

In his original paper, Huntington (1872) drew attention to 'that form of insanity which leads to suicide'. Several other authors have supported his statement. However, there are different concepts of suicide and the legal one is particularly important when considering suicide statistics.

Table 3.8 Suicide rates of patients with HD

Country	Author	Year of publication	No. of suicides and total deaths	Percentage of all deaths
Norway	Ørbeck and Quelprud	1954	1/117	0.85
United States	Rosenbaum	1941	2/46	4.3
Australia	Parker	1958	4/65	6.4
United States	Reed et al.	1958	—	7.8
	Chandler et al.	1960		(males)
				6.4
				(females)
Tasmania	Brothers	1964	3/312	0.96
Holland	Bruyn	1968	—	7.0
South Africa	Hayden	1979	—	3.35
Australia	Chiu and Alexander	1982	3/182	1.6
United States	Schoenfeld	1984	—	12.7
United States	Farrer	1986	25/440	5.7
United States	Haines	1986	5/252	2.0
Norway	Saugstad and Odegard	1986	1/199	0.5
South Wales	Unpublished data	1974–1990	4/201	2.0

Suicide is one of four legal categories of the mode of death; the others are homicide, accident and natural causes. In law, to commit suicide means to kill oneself intentionally, when one is of sane mind and has reached years of discretion. Suicide can refer to the act and the victim. The term means not only 'the act of taking one's own life' but also 'one who dies from his own hand'. Beck et al. (1974) have usefully defined suicide as 'a wilful self-inflicted life-threatening act which has resulted in death'.

Suicides have been registered in Europe and North American since the early nineteenth century. In the UK, about 4400 suicides are recorded annually by the Registrar-General. These statistics are based on verdicts reached by coroners in England and Wales and the Crown Office in Scotland. Coroners' verdicts are also the source of suicide figures in the US. These officials observe the legal definition of suicide by seeking evidence of self-inflicted death.

Many psychiatrists argue that this definition is too restrictive and that therefore, official suicide rates tend to understate the truth. A well-known study of suicide in Dublin (McCarthy and Walsh, 1966) has shown a marked disparity between the numbers included by 'legal' and 'psychiatric' definitions. A Scottish study has revealed that the legalistic approach of the Crown Counsel has led to an under-reporting of suicide by 40% when compared with a 'psychiatric' approach. In the US, Dublin (1963), a leading authority on suicide, estimates that the number of suicides is probably higher by one fourth to one third than recorded.

Is suicide among HD patients commoner than that of the general population? The answer is probably yes, with several published studies showing high rates of suicide in HD. In the United States, the National Huntington Disease Roster was used to study causes of death in a large series of HD patients (Farrer, 1986). Out of 452 deceased patients, information on cause of death was available on 440 patients. It was found that 25 patients (5.7%) had committed suicide. Age at onset was not significantly lower among individuals who committed suicide (36.7 years) compared with non-suicidal deaths (36.9 years). Those who committed suicide died 7 years earlier and lived with their illness 5.8 years less than those who died of natural causes. It is a pity that the mode of suicide (self-poisoning, drowning, hanging) was not presented. A rate of 5.7 suicides per 100 deaths of HD sufferers was almost four times higher than the reported rate of 1.5 for the US Caucasian population in 1979. Suicide was found to be the third or fourth cause of death among those with HD while nationally, suicide is the tenth leading cause of mortality. A similar study in New England revealed a very high rate of suicide (12.7%) but one of the methodological faults was that causes of death were ascertained for only 157 out of 506 patients.

It is difficult to make valid international comparisons in suicide rate for HD as there is a great variety of methods and criteria for certification of suicide. For valid international comparisons in suicide rate to be made, uniform certification procedures and operational criteria for case-finding are necessary. Nevertheless, there appear to be lower rates of suicide in HD in the European studies.

Saugstad and Odegard (1986), using data from the Norwegian National Case Register of Serious Mental Disorder, have recently described a series of 199 HD patients who were admitted to mental hospitals from 1916 to 1975. Of 92 HD patients who died in a psychiatric hospital, only one committed suicide. In a total of 117 deaths, Ørbeck and Quelprud (1954) found one suicide and two probable ones. In Britain, Dewhurst et al. (1970) followed up 102 HD patients for 16 years and found that one patient during this time committed suicide. It was added that 'in some pedigrees there was a high incidence of suicide in other family members' but precise figures were not presented. Official underestimation of suicide may be explained by coroners who record 'death by misadventure' when they are unable to decide about 'intent'. Suicide may also be concealed by relatives and family doctors because the verdict of suicide is considered a disgrace.

In the South Wales series, there have been 4 suicides (3 males, 1 female) as the cause of death of 8201 HD patients over the past 16 years. Two, who were first cousins, jumped in front of a train. They were both depressed; one killed himself the weekend before a Monday psychiatric outpatient appointment and the other committed suicide while on weekend leave from a psychiatric hospital. The third suicide cut her wrists in the bath during an admission to a psychiatric ward for treatment of depression. The fourth was taunted by his friends for his incoordination and 'going the way of his brother'; after taking an overdose of anticonvulsants, he walked into the countryside and the following morning he was found to have died of exposure.

A lower rate of suicide than in the US has also been reported for HD patients in Australia and Tasmania. In his study of 312 HD patients (half of whom had been admitted to psychiatric hospitals), Brothers (1964) identified only three established suicides.

Several factors such as presence of depression, stage and severity of HD, and level of social supports will influence the suicide rate in HD. The timing of suicide in relation to stage of illness attracted the attention of the early authors who believed it could occur in the early stages of HD when chorea is not disabling and dementia is not pronounced (Ladame, 1911; Oltman and Friedman, 1961). This was supported by empirical work by Schoenfeld et al. (1984) who showed that more than half of the suicides occurred in the early stages of the disease and some of them had not been formally diagnosed. However Brothers (1964) found that two of the three suicides (all of whom were depressed) in his series occurred in the late stages of HD.

None of these studies has attempted to examine what proportion of suicides had been mentally ill, what social factors were related to the act and what stressful events preceded the act. Barraclough et al. (1974) in a general study of 100 consecutive suicides found that a high proportion of suicides had been mentally ill (93%). Eighty per cent were seeing a doctor and 80% were prescribed psychotropic drugs. Similar research among HD patients who commit suicide would be valuable and might lead to some of these suicides being prevented.

DELIBERATE SELF HARM

Problems of definition beset discussion of 'attempted suicide'. Until the 1950s, it was believed that those who committed suicide and those who survived it had an essential degree of suicidal intent. Those who survived were considered to be failed suicides.

These ideas were challenged in an important study by Kessel (1965) in Edinburgh. He argued that the term 'suicide attempt' was misleading; instead he suggested the term 'deliberate self-injury and self-poisoning'. Four-fifths of his patients 'performed their acts in the belief that they were comparatively safe; aware even in the heat of the moment that they would survive their overdosage and be able to disclose what they had done in good time to ensure rescue. What they were attempting to do was not suicide'. Kessel was therefore describing a pattern of behaviour and properly did not try to make a judgement on motivation. Other authors have since suggested similar behavioural terms to describe what was previously known as 'a suicide attempt'. Kreitman (1967) proposed the term 'parasuicide' to describe 'any act deliberately undertaken by a patient which mimics the act of suicide but which does not result in a fatal outcome'. Another term, 'deliberate self-harm' has been suggested by Morgan (1979).

There is a dearth of studies concerning self injury in sufferers of HD. Dewhurst *et al.* (1970) found that out of 102 HD patients followed up for 16 years, 'there were 10 cases of attempted suicide' but more precise details are not given. Similarly, in South Wales, there were 10 cases of self-harm in a series of 92 HD patients who were personally interviewed (Tyler, 1982). These tended to occur in the early stages of HD. The precipitating circumstances were following the birth of a baby (three), following the death of spouse (three), after being made redundant (two), after being pressed to attend hospital (one) and 'because people were staring at me' (one). In a much larger study of 831 patients from the United States Huntington's Disease Roster (Farrer, 1986), it was found that 27.6% 'attempted suicide at least once'. This is a very high rate. The information was obtained by questionnaire and the true rate might have been even higher if the patients had been interviewed personally because some of them may have concealed this part of their history from a questionnaire.

Areas for future research in the study of self-harm among HD sufferers would be social class, amount of planning, precipitating factors and concurrent use of alcohol. Many motivations for self-harm may coexist such as the distressing nature of HD, but also a blind reaction to escape an unendurable state of tension (Shneidman, 1964) and an attempt to influence family members after the failure of conventional means of doing so.

MANIA

Kraepelin (1921) brought mania and depression together into the single entity of manic-depressive psychosis. He described four forms of mania: hypomania (the mildest form with predominant euphoria, overactivity, disinhibition); acute mania (more severe, with transient grandiose delusions, a more labile mood, and at times incoherent talk); delusional mania (more persistent grandiose delusions and occasional hallucinations), and delirious mania (disorientation, vivid visual hallucinations; very labile mood). More recently, the concept of manic-depressive psychosis has been questioned by Leonhard (1962) and independently by Angst (1966) in Switzerland and Perris (1966) in Sweden. These European authors have suggested an altered classification dividing patients into those who have 'unipolar depression' (i.e. depressive disorder only), 'unipolar mania' (i.e. mania only) and 'bipolar' illness when the patient has had both depressive disorder and mania. Currently, the term unipolar mania is not used and the term bipolar has been extended to include patients who have had an episode of mania on the grounds that virtually all patients who have a manic illness will at some stage have a depressive disorder.

The literature on mania in HD is beset by inadequately defined terms. It can be said that severe mania occurs rarely in HD (Table 3.9). Bolt (1970) found that three out of 88 HD patients had a history of elevated mood while 11 had grandiose ideation. Four cases of 'hypomania with delusions of grandeur' were reported by Heathfield (1967) in his series of 65 patients. Tamir et al. (1969) noted that in a series of 13 male and 19 female patients, none of the males developed 'euphoria' while four females were 'euphoric'. No other details such as age of onset of elevated mood, or concurrent symptoms were given in this series. Cummings (1985) in a case report described a 59-year-old man with a 9-year history of HD who 'developed grandiose delusions, believed that he had special powers, could control the Federal Bureau of Investigation and several armies'.

Dewhurst et al. (1970) did not use the terms mania or hypomania in their psychiatric evaluation of 102 patients. There were no clearly defined cases of mania in a psychiatric description of 100 cases of HD in Northamptonshire although 'excitement' was found in two patients (Oliver, 1970). Similarly,

Table 3.9 Mania/hypomania in HD

Study		Total patients studied	Number with mania/hypomania	Percentage
Heathfield	1967	65	4	6
Tamir	1969	32	4	12
Bolt	1970	88	3	3
Dewhurst	1970	102	None	—
Oliver	1970	100	2*	2
Saugstad and Odegard	1986	71	None	—
Folstein	1989	186	18	10

*'Excitement'.

Saugstad and Odegard (1986) did not report any defined cases of mania/ hypomania in 71 HD patients admitted to a psychiatric hospital.

These low or absent findings of hypomania in HD are not supported in a Baltimore series (Folstein, 1989). Hypomania, described as increased activity level (frantic house cleaning, pacing about) uncharacteristic general cheerfulness and pressured speech was found in about 10% of HD patients by interviewing them and relatives in detail. These spells were difficult to observe directly because most of the episodes were short-lived. The following case history from our South Wales series provides an example.

A 39-year-old newsagent, who had a 2-year history of HD, was admitted to a psychiatric hospital as an emergency because of unproductive overactivity, irritability and almost incessant speech. He stayed up at night, assaulted his wife and was abusive to neighbours. His lifestyle was extravagant and because of overspending his 'flourishing' business went into liquidation. His lack of insight was again exemplified when he said he was '100% fit' and attempted to take part in marathons and rugby matches despite arthritis and early HD. He was overfamiliar with his medical and nursing attendants by backslapping and addressing female members of staff as 'baby'. He had pressure of speech and spoke unnecessarily loudly. His content of thought was expansive when he considered himself 'businessman of the year'.

IRRITABILITY

Irritability is a common abnormality of mood but has received much less attention from nosologists and researchers than depression or hypomania. Snaith (1985) defines it as 'a state of poor control over aggressive impulses directed towards others and manifested verbally as shouting or snapping and physically by such acts as throwing objects, slamming doors, or directly assaulting others. It is recognized that irritability may be directed more towards oneself than to others'.

Irritability and aggression have been noted in many epidemiological studies of HD. Heathfield (1967) found that 15 of his 80 patients had 'aggression, irritability, including attacking people and throwing things' and another two were 'abusive and swearing'. Bolt (1970) reported a higher proportion with these symptoms. She found that about 50% (53% of males and 48% of females) of her series of 334 cases had 'some degree of ill-humour' including 'irritability, aggressiveness and rage'. Goodman et al. (1966) in their detailed psychological evaluation of three affected patients reported 'chronic irritability, bitterness and cynicism'.

It is unusual for HD patients themselves to complain of irritability and it is another informant who draws this symptom to professional attention. This should be expected because as Weissman and Paykel (1974) point out, irritability is more likely to be directed towards individuals who are in a close social relationship with the patient. A 'continuum of intimacy' has been proposed in which the expression of irritability is most pronounced towards first-degree relatives, then members of the extended family and is least pronounced towards non-relatives and people in authority. Irritability will rarely be manifest in the clinic.

Irritability may be precipitated by drug therapy (it is a recognized side-effect of benzodiazepines) and alcohol consumption. It may also be a symptom of depressive illness. In women, it is likely to be cyclical as a partial manifestation of the pre-menstrual syndrome. Irritability may also be a longstanding personality trait.

DISORDERS OF ADULT PERSONALITY

An individual's personality refers to long-standing ways of behaving in a wide variety of situations. The characteristic ways of behaving are called traits and particular groups of traits may be called personality types. Personality disorder refers to characteristics of individuals that cause them to suffer or which cause others in society to suffer. According to the International Classification of Disease (World Health Organization, 1978), personality disorders are 'deeply ingrained maladaptive patterns of behaviour generally recognizable at the time of adolescence or earlier and continuing throughout most of adult life although often becoming less obvious in middle or old age'. Many psychiatrists regard the diagnosis of personality disorder as highly unsatisfactory in a nosological sense because of its relatively poor reliability and value-laden judgements. In the opinion of Blackburn, psychopathic personality is 'little more than a moral judgement masquerading as a clinical diagnosis'.

Despite the problems of defining both normal and abnormal personality, the concept remains useful, particularly if personality is considered as a separate axis of classification (Rutter, 1987). There has been considerable effort expended recently in an attempt to improve the assessment of abnormal personality so that acceptable levels of reliability can be achieved (Shepherd and Sartorius, 1974; Siever and Klar, 1986).

PERSONALITY IN HD

Personality in HD will be considered in a historical context. In many studies it is not clear whether personality change is being discussed or a possible association between HD and premorbid personality.

Early studies

Personality disturbance in patients with HD has been noted since the early literature. Davenport and Muncey (1916) in a study of 962 choreics in New England identified 'biotypes' (see Chapter 1) in certain families 'which show a specific complex of symptoms'. Personality featured in several of the biotypes but since the descriptions such as 'eccentricity', and 'religious mania' were not further defined, it is understandable that subsequent work did not confirm these findings.

'Choreopathy'
In Weimar and Nazi Germany, the bias in diagnosing personality disturbance instead of depression has already been mentioned (page 93). This bias sprang from Kehrer who proposed the term 'choreopathy' (Kehrer, 1928). He stated that choreopathy covered two aspects of personality: first, any early personality changes occurring in affected individuals and second, premorbid lifelong personality abnormalities supposed to occur in HD. The concept of choreopathy was accepted by many German psychiatrists, such as Bertha and Kolmer (1940) and Hochheimer (1936) who described the impaired 'Steuerung der Denkablaufe' (guidance of thought processes) in the thinking of choreics. However, it is Panse (1942) who not only is most associated with the term but also altered its meaning (Koehler and Sass, 1984). He concentrated on the first of Kehrer's meanings and excluded from his definition longstanding abnormal personality characteristics and such characteristics occurring in other family members of HD patients. He found that 20% of his 700 HD patients presented with personality changes before the onset of neurological symptoms. These patients were subdivided into two subgroups: the 'callous–instinctual–irritable' subgroup (which was easily the larger – 95% – and included cases of alcohol misuse and those convicted of criminal behaviour) and a miscellaneous subgroup with a wide variety of mental symptoms (depression, anxiety, asthenia). Panse's work must be viewed in the context of his close involvement with the Nazi political philosophy (see Chapter 11).

Although the term choreopathy is rarely used now, it is interesting to note that as recently as 1981, Huber expanded the term and described choreopathy in 81% of his HD cases. His biggest subgrouping was 'neurasthenic choreopathy' (31% of all HD patients) and included symptoms such as irritability, anergia and impaired concentration. In Panse's series, irritability would have included his 'callous' subgrouping and neurasthenia was only a part of his small miscellaneous subgroup.

Recent studies
A small number of studies have applied standardized tests to sufferers of HD. Boll *et al.* (1974) for example applied the Minnesota Multiphasic Personality Inventory to nine patients with HD and a control group of nine patients with mixed brain damage. They found that three clinical scales (hypochondriasis, depression and schizophrenia) were elevated in both groups and the authors concluded that the personality characteristics in HD are little different from those seen in patients with nonspecific forms of brain damage.

Saugstad and Odegard (1986) found that 40 out of 71 HD patients had 'personality disorder' but they did not define the term in an operational way. Dewhurst *et al.* (1970) reported a smaller number of HD patients (10%) with personality disorder (10 out of 100 HD patients) but again, the term was not defined or subclassified. Bolt (1970) reported that a similar proportion (10.2%) of male HD patients (15 out of 148) had an altered personality.

In chronic organic brain disorders such as HD, it is preferable to refer to personality change rather than personality disorder. Change of personality may be an early sign of HD. Patients may have diminished awareness of the needs of others, or may act out of character, for example by stealing or being involved in sexual indiscretions. McHugh and Folstein (1975) noted apathetic change in personality resulting in diminished interest in work and personal appearance and later in withdrawal, marked self-neglect and mutism. In Oliver's opinion (1970), personality changes in 20% of HD patients were 'emotional instability, expressed as "moodiness"'.

Antisocial (psychopathic) personality

Perhaps of all the personality disorders, the definition of the antisocial (or psychopathic) is the most controversial. The concept of psychopathic behaviour began with Pinel (1806) who described *manie sans délire* in violent patients who were not deluded. Many attempts to define and classify psychopathy have failed (Lewis, 1974). He writes elsewhere that 'it is misconceived to equate ill health with social deviation or maladjustment' (Lewis, 1953).

Several studies have noted antisocial personality characteristics in HD. Oliver (1970) subdivided antisocial behaviour among his 100 patients into five types; violence (five cases), lying, thieving and acting as a 'con-man' (four cases), cruelty to young children (three cases) and two groups of sexual disorders. Tamir *et al.* (1969) studied the case records of 32 HD patients at the Creedmoor Institute for Psychobiologic Studies, New York and found sex differences in the presentation of HD. Males were more likely to show 'aggressive and homicidal activity' whereas females were more likely to show depression. However, it is not stated what the authors included as aggression (e.g. both verbal and physical aggression) and whether the five patients showing 'homicidal activity' actually killed someone.

Many HD patients said to have psychopathic personalities and irritability were reported to have committed criminal offences. Criminal behaviour will be considered in Chapter 6.

At-risk individuals

Much research has been devoted to the assessment of abnormal personality in those at risk as a means of identifying those who would later develop HD. Minski and Guttmann (1938), suggested that traits such as irritability, quick-temperedness, and stubbornness may predict potential HD sufferers. Personality tests as a possible means of presymptomatic testing are discussed in Chapter 12 but it can be stated here that no clear correlation has been confirmed between pre-existing personality characteristics and later development of the disorder.

ALCOHOL

The term alcoholism has so many different meanings that it is probably best avoided. According to Kellner (1962), 'the definition of alcoholism has long been marked by uncertainty, conflict and ambiguity'. The World Health Organization (1952) defines alcoholics as 'those excessive drinkers whose dependence upon alcohol has attained such a degree that it shows a noticeable mental disturbance or an interference with their bodily and mental health, their interpersonal relations and their smooth social functioning; or those who show prodromal signs of such developments'.

Conflicting opinions exist on alcohol consumption in HD. A Swedish study of psychiatric symptoms in HD did not mention alcohol misuse (Mattsson, 1974). In Michigan, only one family out of 32 was reported to have excessive alcohol consumption (Hughes, 1925). In their South Wales survey, Spillane and Phillips (1937) thought that this problem was uncommon. On the other hand, some workers have thought that alcohol misuse is common in HD (Hattie, 1909; Chandler et al., 1960; Hans and Gilmore, 1968). However, these studies did not use explicit criteria for 'alcoholism'. A more systematic study was conducted by King (1985) who interviewed 42 HD patients (25 male; 17 female) in Maryland using questions derived from the Diagnostic Interview Schedule (Robins et al., 1981). Information was also obtained from other informants. Seven individuals (six male; one female) (16%) fulfilled DSM-III criteria for alcohol abuse and five of the seven were classified as alcohol dependent. These rates corresponded to the rates of alcohol misuse (13.7% lifetime alcohol abuse) in a concurrent community survey in east Baltimore (Robins et al., 1984). In HD, alcohol misuse tended to be associated with depressive symptoms. The complex relationship between mood disorder and alcohol has been reviewed elsewhere.

BULIMIA

Various reports have drawn attention to the 'ravenous appetite' in HD (Forrest, 1957; Podolsky and Leopold, 1974). Whittier (1976) considered that bulimia is a feature of HD and that this will aggravate feeding problems. Appetite and weight loss in HD have been discussed in Chapter 2.

SEXUAL DISORDERS

The sexual disorders may be divided into two groups, the dysfunctions and the deviations. Sexual dysfunction may be defined as an absolute or relative deficit in sexual performance and therefore includes conditions such as erectile dysfunction in the male, vaginismus in the female and low sex drive in both sexes. The sexual deviations are a group of sexual behaviours that are unacceptable to a given culture and generally include disorders of sexual orientation such as paedophilia and exhibitionism and gender role abnormalities such as transvestitism and transexualism.

There are no well-conducted studies of sexual disorders in HD. There are methodological problems in such studies because of the wide range of normal sexual functioning. Changing cultural and social attitudes towards sexuality render the comparison of studies over time more difficult. Classification has also changed; for example, some studies (Dewhurst et al., 1970) have included pathological jealousy among disorders of sexuality.

In a series of 334 living and dead HD patients, Bolt (1970) found that 20 HD patients (13 men and seven women) had sexual disorders. The sexual dysfunction took the form of increased libido and the deviations took the form of exhibitionism and 'an abnormal sexual interest in children'. Dewhurst et al. (1970) found that 30% of HD patients had sexual disorders. Twelve per cent of males had hypersexuality, meaning 'demanding an inordinate amount of sexual fulfilment at odd times or in inappropriate places and whenever these desires were rebuffed they became vindictive, abusive and frequently violent'. A lower rate of hypersexuality was found among affected females (7%). Hyposexuality was also found, affecting 7% of male and 4% of female patients. As men are responsible for most sexual initiatives in Western cultures, hyposexuality in an affected female patient may remain unrevealed. Dewhurst et al. described a wide range of sexual deviations in HD including exhibitionism, homosexual assault, and voyeurism but presumably these conditions only occurred in single or small numbers of patients.

In a Northamptonshire study, Oliver (1970) found that six out of 100 patients had the following deviations: 'uncontrolled sexual advances or displays', sodomy, and masturbating in front of own children. Heathfield (1967) only reported two cases of sexual disorder in his series of 80 patients: one had 'promiscuity' and another had exhibitionism. In a social study (which included interviews with spouses) of 92 HD patients in South Wales, 23% of males and 9% of females had sexual disorders (Tyler, 1982). Hyposexuality was reported in seven cases and hypersexuality in six. Diminished sexual desire was often welcomed by the wife but several men suffered severely when the conjugal relationship ceased and found it difficult to control their aggression. Regarding deviations, three men were reported to have made sexual advances to their daughters, one man to his son, while another male was suspected of sodomy. No convictions for sexual offences have occurred among this group. Fertility in HD is discussed in Chapter 9.

PSYCHODYNAMICS

There has been a dearth of studies on psychodynamics in HD. Martindale (1987) discussed a wide range of pathological defence mechanisms in HD families. Among the more important of these are denial, reaction formation and projection. An example of rationalization is given by Whittier *et al.* (1972) when a parent puts forward reasons like 'she's too young' and 'I'm frightened her engagement will be broken off' for not informing a daughter that she is at 50% risk.

Werner and Folk (1968) in a detailed case report described premorbid inability 'to express dependency needs', 'sublimated aggression' and 'poorly formed sexual identity' in a 43-year-old patient with a 9-year history of HD. After the onset of HD, dependency needs became more manifest and activated neurotic conflicts. The authors suggested various ways of dealing with altered dependency needs in organic disease.

Pathological defence mechanisms may also be adopted by health care professionals who are involved with HD families. These mechanisms, including avoidance and displacement of responsibility, are well described by Martindale (1987).

DEMENTIA

As dementia has been regarded as one of the three cardinal features of HD (the other two being abnormal movements and autosomal dominant inheritance) it is necessary to consider definitions and diagnostic criteria for a fuller understanding of the literature. Roth (1980) offered one of the most concise definitions of dementia: 'the global deterioration of the individual's intellectual, emotional and conative faculties in a state of unimpaired consciousness'. The Royal College of Physicians (1981) defines it as 'the global impairment of higher cortical functions including memory, the capacity to solve the problems of day-to-day living, the performance of learned perceptuo-motor skills, the correct use of social skills and control of emotional reactions, in the absence of gross clouding of consciousness. The condition is often irreversible and progressive' (Report of the Royal College of Physicians by the College Committee on Geriatrics, 1981). More recently in the United States, the Department of Health and Human Services (McKhann *et al.*, 1984) have proposed the following criteria for dementia:

> 'Dementia is the decline of memory and other cognitive functions in comparison with the patient's previous level of function as determined by a history of decline in performance and by abnormalities noted from clinical examination and neuropsychological tests. A diagnosis of dementia cannot be made when consciousness is impaired by delirium, drowsiness, stupor or coma or when other clinical abnormalities prevent adequate evaluation of mental status. Dementia is a diagnosis based on behaviour and cannot be determined by computerised tomography, electroencephalography or other laboratory instruments although specific causes of dementia may be identified by these means'.

The American Psychiatric Association (1980, 1987) has produced diagnostic criteria for dementia but these have been criticized for being too broad and so leading to unreliability of diagnosis (Jorm and Henderson, 1985). Another criticism is that dementia is treated as a categorical rather than dimensional disorder (Jorm and Henderson, 1985).

From the above definitions, it is apparent that certain elements contribute to the modern concept of dementia (Mahendra, 1984). Firstly, dementia is an acquired condition. Cognitive and behavioural changes present from childhood are usually included with mental handicap. Secondly, dementia is global (as emphasized in Roth's definition) and not merely a collection of focal deficits. Korsakoff's syndrome, which is characterized by a severe memory impairment is therefore not a dementia. Dementia involves impairment of memory, intellectual functions and personality. Thirdly, dementia is a condition which occurs in the absence of impairment of consciousness.

Prevalence of dementia in HD (Table 3.10)
Comparing studies concerned with the prevalence of dementia in HD is difficult because study designs, diagnostic criteria for dementia and methods of assessment vary widely. Other confounding factors are stage of disease, the coexistence of other psychiatric disorder or symptoms and background intellectual level of subjects tested.

The early clinical and epidemiological surveys in HD did not use psychometric tests when evaluating dementia and the above concepts were not fully considered. Several reported symptoms such as poor concentration allow considerable latitude of interpretation. Bolt (1970) found that 70% of 334 HD patients (living and dead) had evidence of organic brain disease, evidence of which was 'impairment of memory, judgement, or intellect, emotional lability or euphoria, a tendency to wander, perseveration and disorientation'. She reported that impairment of memory was the commonest of these signs, occurring in 28% of patients. In the 124 patients personally examined, she found that memory impairment was apparent in 50% of patients. However, the term 'memory impairment' is somewhat vague and it is well known that patients with depression are likely to complain of memory impairment.

Oliver (1970) in a sereis of 100 HD patients in Northamptonshire found that 15 cases had 'symptoms characteristic of dementia', such as impairment of memory and concentration and slowness in thinking. Heathfield (1967) found a higher rate of dementia in HD patients. All but four of the 82 HD patients in his series were reported to have had dementia but he appears to have used disturbance of behaviour ('aggression, irritability, including attacking people and throwing things, being abusive and swearing, selfish and demanding, volatile moods') as evidence of dementia.

Table 3.10 Prevalence of dementia in HD

Author	Year	No. of patients	No. with dementia
Heathfield	1967	82	78
Bolt	1970	334	234
Oliver	1970	100	15

Table 3.11 Wechsler Adult Intelligence Scale (WAIS) scores in HD patients

Authors		Verbal	Performance	Full scale
Boll et al.	1974	95.54	92.90	92.81
Aminoff et al.	1975			74.4
McHugh and Folstein	1975	99.7	77.8	89.4
Lyle and Gottesman	1977			
Early HD		95.7	94.4	95.1
Late HD		93.1	87.8	89.7
Josiassen et al.	1983	101.2	91.8	97.0
Sax et al.	1983	96	87	91
Brandt et al.	1984			
Early HD		98.3	92.8	
Late HD		88.3	78.9	

In the last 20 years, it has become recognized that the investigation of dementia in HD has been hampered by a lack of standardized techniques for its assessment. One of the most widely used of the short assessments is the Mini-Mental State Examination (Folstein et al., 1975). It is an 11-item cognitive screening test with scores ranging from 0 to 30 and is a useful aid in rapid assessment. However, one of the problems of the MMSE is that it is not a diagnostic aid. The results of such tests may be affected by the presence of depression and by acute confusional states. It is particularly useful in assessing change over time. In one of the few longitudinal studies of dementia in Huntington's disease, Folstein (1989) has shown that scores on the MMSE show a gradual decline with time. The Clifton Assessment Procedures for the Elderly (CAPE) (Pattie and Gilleard, 1975, 1979) is also frequently used for the assessment of dementia but again it is not a diagnostic instrument.

The Wechsler Adult Intelligence Scale (WAIS) (Wechsler, 1955) a standardized instrument for the measurement of intelligence, can identify organic brain damage. Brain damage is reflected in discrepancies between the verbal and performance IQs. Several studies of the WAIS in Huntington's disease have shown a decrement in both verbal and performance subtests and that the performance subtests are more affected than the verbal (see Table 3.11). This is likely to reflect motor impairment. Similarly, in Parkinson's disease, a discrepancy has been shown between verbal (mean 115) and performance (mean 95) scores (Loranger et al., 1972). It should be remembered that all the performance subtests involve time limits whereas only one of the verbal subtests is timed. The application of the WAIS as a diagnostic instrument has been criticized. A number of factors can explain variation, including educational and cultural background as well as disease states. The WAIS is too nonspecific for the study of the many different causes of organic disorder. Its main value is that low scores on subtests can be pointers for verification by more specialized testing.

The Halstead–Reitan Battery (Halstead, 1947; Reitan, 1966) is another comprehensive instrument used for the detection of brain damage. Using this instrument in 11 HD patients, Boll *et al.* (1974) found they showed greater impairment, especially in motor skills, problem-solving, memory and concentration, than the nine controls with cerebrovascular disease (six) and traumatic head injury (three). The choice of controls may be questioned and the results would have been more useful if they had all had the same disorder.

A shortcoming of early studies of intelligence in HD is that patients in all stages of the disease were grouped together with little attempt to distinguish early and late neuropsychological profiles. A number of studies have now addressed this aspect and stratified patients according to length of illness. Patients with advanced HD had significantly lower scores on the Digit Symbol and Object Assembly subtests than HD patients diagnosed within the previous year (Butters *et al.*, 1978). Another study (Moses *et al.*, 1981) examined three groups of HD patients (less than 1 year, 3–5 years and over 6 years since diagnosis) and showed that generalized cognitive decline occurred early in HD and that memory and visuospatial deficits were the most prominent. Brandt *et al.* (1984) also studied duration of symptoms but found that motor impairment at the time of testing was a better correlate of cognitive decline. Further research, especially longitudinal studies on the same patients, is necessary before the natural history of cognitive impairment in HD is fully understood.

Subcortical dementia
For nearly 20 years the controversial concept of subcortical dementia has been used to describe a clinical syndrome of acquired intellectual impairment. Albert *et al.* (1974) in a seminal paper proposed the term subcortical dementia to describe the mental defects in progressive supranuclear palsy (Steele–Richardson syndrome). They reviewed the 42 cases reported in the literature and added detailed neuropsychological evaluations of five cases of their own. They identified four characteristic features of the syndrome: forgetfulness, slowness of thought processes, altered personality with apathy or depression and impaired ability to manipulate acquired knowledge. It was suggested that higher cortical functioning was intact and that subcortical dementia was primarily a defect in timing and activation of mental processes. Albert (1978) subsequently proposed the identification of subtypes of subcortical dementia and that these could be related to impairment of specific neurochemical pathways. Jackson *et al.* (1983) described their experience of 16 patients, the largest reported series of progressive supranuclear palsy, and agreed with Albert *et al.*'s (1974) findings that in progressive supranuclear palsy, dementia is a consistent finding and there is a subcortical degenerative process involved.

In subcortical dementia, the neuropathology involves the deep grey-matter structures including the thalamus, basal ganglia and related brain stem nuclei. In progressive supranuclear palsy, for example, the pathological changes are found in the globus pallidus, red nucleus, and substantia nigra. McHugh and Folstein at a meeting of the American Academy of Neurology in 1973 were the first to suggest that the cognitive changes in HD reflected subcortical dysfunction.

Table 3.12 Features that distinguish cortical and subcortical dementias (modified from Cummings and Benson, 1984)

Characteristic	Subcortical	Cortical
Mental status		
Language	No aphasia	Aphasia
Memory	Forgetful (difficulty retrieving learned material)	Amnesia (difficulty learning new material)
Mood	Affective disorder common	Normal
Motor system		
Speech	Dysarthric	Normal
Motor speed	Slow	Normal
Gait	Abnormal	Normal
Movement disorder	Common	Absent
Anatomy		
Cortex	Largely spared	Involved
Basal ganglia, thalamus	Involved	Largely spared
Neurotransmitters		
	HD: GABA Parkinson's disease: dopamine	Alzheimer's disease: acetylcholine

The concept of subcortical dementia has come under critical scrutiny. Recent studies of progressive supranuclear palsy do not consider subcortical dementia as an important feature of the condition (Jackson et al., 1983; Kristensen, 1985). In the opinion of Mayeux et al. (1983), the concept of subcortical dementia lacks clinical validation and has 'only a questionable pathological basis'. Their study included 123 consecutive patients of whom 57 had Parkinson's disease, 46 Alzheimer's disease and 20 HD. They were subdivided into three functional disability stages according to their degree of impairment in activities of daily living. There were no significant differences in intellectual impairment, assessed using neuropsychological testing, among the three groups. The methodological problems of this study, especially small sample size, have been pointed out by Brandt et al. (1988) who attempted to determine whether large cohorts of patients with HD and Alzheimer's disease differ in their cognitive features on psychological testing. In an analysis of 145 patients with Alzheimer's disease and 84 with HD, they found that there were qualitative differences in cognitive impairment and suggested that these results supported the differentiation of cortical and subcortical dementias (Table 3.12).

Memory
Experimental psychologists distinguish between a primary (or short-term) memory of very limited capacity which holds information for a few seconds only and a secondary (or long-term) memory of much greater capacity and

durability. Secondary memory therefore encompasses all material recalled beyond a period of a few seconds and items within it have occurred within the remote or recent past. There have been various attempts to investigate the subcomponents of primary and secondary memory. Secondary memory for example has been considered in terms of episodic (personal) memory and semantic (conceptual) memory.

For clinical purposes memory may be classified into 'immediate', 'recent' and 'remote' memory. The immediate memory span, which represents the functioning short-term storage mechanisms, is tested by asking the subject to reproduce a brief digit sequence. Intact registration is shown by accurate recall of a digit sequence. Recent memory is reflected in ability to acquire and retain new knowledge and requires a process of consolidation as well as registration. It is tested by asking the subject to repeat simple information after a period of several minutes. Remote memory is tested by ability to recall personal or public events which have occurred many years previously. It therefore reflects a process of retrieval of material which has been held in long-term storage.

Aminoff et al. (1975) reported one of the earliest systematic studies of memory function in HD. Eleven HD patients completed intelligence tests and nine of them completed supplementary memory tests, including a digit span test, an immediate recall of 14 objects displayed visually for 14 s and immediate recall of 12 common objects shown as drawings. There was no significant impairment of immediate memory compared with normal controls of advanced age. The finding of normal immediate memory was supported by Caine et al. (1977) who found that initial registration and immediate recall of a digit span was not significantly impaired.

Recent memory has been shown to be impaired in several reports. Compared with normal controls HD patients had impairment of recall of word lists after a 2 min delay (Caine et al., 1977; Moss et al., 1986). Evidence is accumulating that the recent memory deficit in HD is due to failure of retrieval rather than of encoding (Weingartner et al., 1979) or storing (Butters et al., 1978) as previously thought.

A small number of studies have addressed remote memory in HD. Caine et al. (1978) in their series of 18 HD patients reported that remote memory was impaired. Patients were unable to recall information 'on command'. For simple factual questions, they found that the presentation of multiple-choice answers markedly improved recall. They found that although patients could remember important personal facts, there was loss of finely detailed memories. A controlled and more systematic study of long-term factual memory was conducted by Brandt (1985) who assessed general knowledge. By demonstrating a high correlation between the feeling of knowing an answer and its recognition on a multiple-choice test, he was able to provide evidence that remote memories were still present but could not be retrieved upon demand.

Language

Language is a communication skill involving both the ability to express oneself and to comprehend others in speech as well as in writing. Advanced dementia is usually associated with a severe impairment of all aspects of language functioning. Although language dysfunction has not been used as a central element in the various definitions of dementia, more recently it has been stated that some form of language impairment is invariably present in the dementing process (Bayles, 1982).

Regarding motor aspects of speech (see Chapter 2), dysarthria, which occurs even in the early stages of HD, leads to loss of verbal fluency and impairs communication (Albert, 1978; Butters et al., 1978). In more severe dysarthria, speech will be disjointed or even unintelligible. A number of studies have encountered perseveration in HD (Kleist, 1922; Panse, 1942; Muller, 1982).

In a comprehensive study of language in 12 HD patients and 24 at-risk controls using the Boston Diagnostic Aphasia Examination (Goodglass and Kaplan, 1983), Gordon and Illes (1987) showed major abnormalities among the affected group (Table 3.13). Regarding lexical production, they found that the number of words produced during 4-min speech samples were significantly reduced in HD patients (120.5 words/min for the at risk group and 89.1 words/min for the affected group). The HD group produced a significantly higher rate of mumbled words than the normal control group. A substantially higher number of unfilled pauses longer than 5 s were found in the speech of affected individuals. HD patients also had a diminished level of syntactical complexity. Cortical speech abnormalities including paraphrasic errors and word-finding difficulties were reported. The authors hypothesized that subcortical degeneration is sufficient to produce these linguistic defects and that there is a 'disconnection of cortical–neostriatal pathways that are ordinarily used in language'. A German study (Wallesch and Fehrenbach, 1988) compared the spontaneous speech of 18 HD patients with 15 controls who were suffering from Friedrich's ataxia. It was found that HD patients showed reduction in syntactical complexity but using the Aachen Aphasia Test (Huber et al., 1983, 1984), which is specifically designed for the investigation of aphasia, the study did not produce evidence that language abnormalities in HD were aphasic in nature.

Table 3.13 Speech profile ratings in affected and at-risk (AR) individuals (using the Boston Diagnostic Aphasia Examination). From Gordon and Illes (1987)

Scale type	Mean of HD	Mean of AR	F Ratio	p
Phrase length	3.82	6.45	15.80	0.0003
Articulatory agility	2.93	6.17	7.85	0.008
Grammatical form	4.50	6.66	13.57	0.0007
Paraphasias	5.28	6.79	6.34	0.0161
Word finding	2.71	3.66	7.23	0.0106

Although production of speech is impaired in HD, comprehension is remarkably well preserved.

Prosody
Speedie et al. (1990) reported an interesting study of prosody (the variations in stress and pitch that convey shades of meaning separate from the actual words said) in a controlled study of HD patients, unilateral stroke patients and controls. Two types of prosody were assessed, namely affective prosody when the patient was asked to indicate whether the tone of a recorded voice was happy, sad or angry, and propositional prosody when the subject was asked whether the tone was a question, command or statement. It was found that HD patients were impaired in comprehension of both affective and propositional prosody and that this failure to appreciate rhythm and changes of tone in normal speech may contribute to the social impairments of HD patients.

Writing
Writing is clearly affected in HD, and both motor disturbance and impaired cognition can be responsible for abnormalities. Patients write slowly and with great effort. Hochheimer (1936) described 'choreatic macrographia' in HD, the size of patient's writing being enlarged. Other disturbances are sudden stops during the act of writing, strokes across the paper due to chorea, while in some cases, chorea is so pronounced that patients are unable to write (Podoll et al., 1988). In late stages of HD, dysgraphic errors, such as omission of letters or entire words, may occur (Podoll et al., 1988). Perseveration of letters (Podoll et al., 1988) and words have been reported (Hochheimer, 1936).

Visuospatial praxis
Spatial disorientation in patients with Huntington's disease has not received adequate quantitative study. Tests of visuospatial praxis include the Bender–Gestalt test (Bender, 1938, 1946), the Benton Visual Retention Test (Benton, 1955, 1963), the Minnesota Percepto-Diagnostic Test (Fuller, 1967; Crookes and Coleman, 1973) and the Rey–Osterrieth Test (Rey, 1941; Osterrieth, 1944). Josiassen et al. (1983) who administered the Benton Visual Retention Test (a test which involves the reproduction of geometrical designs from memory) to a series of 16 HD patients found visuospatial deficits in patients in later stages of the disease. Potegal (1971) conducted a controlled study of spatial localization among seven patients suffering from HD and seven suffering from Parkinson's disease. They were first asked to remember the position of a single black dot on a sheet of paper and then after being blindfolded, they were asked to mark the point where the dot had been. In a variation of this test, they were then asked to take a single step sideways so the the subject would then be marking the sheet from a different position. It was found that there was no impairment of localization of the target dot for both groups if the patient did not move. However, for the HD group, there was significant impairment of target

localization after sideways movement. Potegal hypothesized that compensation of self-produced movement is a function of the basal ganglia and that the deficit observed in HD is explained by caudate damage.

Impairments in visuospatial ability may have prognostic significance. In a group of institutionalized women suffering from dementia, it was found that visuospatial dysfunction was associated with a high mortality (McDonald, 1969). Of those with a high error rate, 26% were dead within the following 6 months compared with 4% in the group with a low error rate. No such studies have been performed in HD and there is also a need for prospective research on the progression of visuospatial impairments in HD.

Concept formation

Concept formation and abstract problem-solving in HD have received insufficient attention. Bettner *et al.* (1971) found that a test for abstraction (Stroop Colour Word Test) yielded performance decrements in elderly cognitively impaired subjects (not due to HD). Impaired abstraction identified by the Stroop Test has also been found in HD (Fisher *et al.*, 1983).

Mental set

A mental set is defined as 'a state of brain activity which predisposes a subject to respond in one way when several alternatives are possible' (Flowers and Robertson, 1985). Cools *et al.* (1984) provide another definition: 'the ability to reorganise behaviour according to the requirements of a task'. This phenomenon may be detected by the Wisconsin Card Sorting Test which is found to be particularly useful in studies of patients with frontal lobe disease (Milner, 1963; Robinson *et al.*, 1980). In this test, the subject must discover by which criterion (e.g. colour, number) the examiner wants him to sort 64 cards. The examiner indicates whether a response is correct or incorrect and normal people soon discover which criterion for sorting the examiner is using. The examiner can shift the criterion for sorting without warning, thus requiring the patient to identify the new criterion and hence shift his frame of reference. Josiassen *et al.* (1983) found that patients with HD easily identified the first sorting criterion but when the sorting strategy was changed, they had difficulty in changing conceptual sets and made perseverative errors.

The Trail Making Test (Reitan, 1958) is another useful instrument to assess mental set and also spatial analysis. The test consists of two parts. In the first, the subject is asked to draw a line between 25 numbered circles as quickly as possible. In part B, the subject is required to alternate between numbers and letters when joining the circles. Starkstein *et al.* (1988) found that mental processing time was prolonged in 16 patients with HD.

Insight

Insight into cognitive disabilities is preserved in HD (Caine *et al.*, 1978). Sixteen of their 18 patients complained of their inability to perform mental functions at the level they had attained previously. At a more subjective level, the degree of insight shown by many HD patients into their illness and prognosis, even at a relatively advanced stage, stands in striking contrast to the early loss of insight in other dementing illnesses such as Alzheimer's disease.

HD and the differential diagnosis of dementia (Tables 3.14, 3.15)
The problems of making an accurate diagnosis of the underlying causes of dementia are well known. In a series of 106 patients who were given a diagnosis of presenile dementia, 16 were found not to be demented on follow-up. Similarly Nott and Fleminger (1975) found that half of a series of 50 patients diagnosed as having presenile dementia were found to have recovered on subsequent examination.

There are many dementing conditions which may be confused with HD but only a small number of these can be described here. Several of these diseases have also been described as coexisting with HD. The more general differential diagnosis of HD has been discussed in Chapter 2.

Alzheimer's disease
Alzheimer's disease is the most common type of dementia and it especially affects the parietal and temporal lobes. There is generalized involvement of cortical functions with aphasia, apraxia and agnosia. The aetiology is unknown but it is generally believed to be a form of accelerated aging. The familial transmission of Alzheimer's disease was first shown nearly 40 years ago (Sjögren *et al.*, 1953) and these pioneering findings have since been widely replicated. Many of these studies (Heston *et al.*, 1981; Breitner and Folstein, 1984) suggest autosomal dominant inheritance in a proportion of kindreds especially those with early onset. More recently, molecular studies have shown linkage between familial Alzheimer's disease and DNA markers from chromosome 21 in four large pedigrees (St George-Hyslop *et al.*, 1987). Later studies have confirmed the finding of highly significant linkage in early onset families (Goate *et al.*, 1989), while very recently a specific mutational defect has been

Table 3.14 Classification of dementia

1. Degenerative senile dementia, Alzheimer's, Pick's, Creutzfeldt–Jacob, normal pressure hydrocephalus, multiple sclerosis, Parkinson's, Schilder's, Wilson's progressive supranuclear palsy, progressive myoclonic epilepsy
2. Space-occupying lesions, cerebral tumour, subdural haematoma
3. Trauma, post-traumatic dementia
4. Infection, general paresis, subacute and chronic encephalitis
5. Vascular
6. Metabolic uraemia, liver failure
7. Endocrine myxoedema, Addison's disease
8. Toxic 'alcoholic dementia', Korsakoff's psychosis
9. Anoxia, anaemia, congestive cardiac failure, chronic pulmonary disease
10. Vitamin lack, lack of thiamine, nicotinic acid, B12, folic acid

Table 3.15 Main features of the presenile dementias

Characteristic	HD	Alzheimer	Pick
Decade(s) of onset	40–50	40–60	50–60
Sex	Equal	Female 2 : 1	Female 2 : 1
Inheritance	Dominant	Dominant (some families)	? Dominant (some families)
Course	Progressive	Progressive	Progressive
Cognition: dysphasia	Late	Early	Late
Apraxia/agnosia	Late	Early	Late
Memory impairment	Later	Early	Relatively late
Cerebral atrophy	Generalized	Generalized	Generalized
Specific site	Caudate	Frontal	Frontal
Microscopic	Neuronal loss/ gliosis	Senile plaques neurofibrillary tangles	'Balloon' cells

found in the beta amyloid locus on chromosome 21 in one kindred (Goate *et al.*, 1991).

There have been several reports of the coexistence of Alzheimer's disease and HD (McIntosh *et al.*, 1978; Reyes and Gibbons, 1985; Vonsattel *et al.*, 1985). This may be coincidence, since Alzheimer's disease is so common, but it illustrates the importance of not assuming that dementia in a person at risk for HD is always the result of HD itself.

Several authors have attempted to distinguish the cognitive features of HD and Alzheimer's disease. Brandt *et al.* (1988) in a study of 145 patients with Alzheimer's disease and 84 with HD reported qualitative differences in cognitive impairments using Mini-Mental State profiles. In particular they found that memory and the naming of objects were more significantly impaired in Alzheimer's disease. These findings were supported in a subsequent study by Salmon *et al.* (1989) who used the Dementia Rating Scale, another brief mental status examination.

Pick's disease

This disease, which was first described by Arnold Pick in 1892 as a form of presenile dementia, appears to be transmitted as an autosomal dominant trait, at least in some families. The disease affects the frontal and temporal lobes, thus accounting for the clinical features of a frontal lobe syndrome. The early manifestations are changes of personality and behaviour rather than cognitive decline. In particular there is disinhibition, tactlessness and an indifference to others. In later stages of the disease, memory and intellectual impairment become evident and eventually, the global cognitive deterioration becomes indistinguishable from other advanced dementias. The characteristic pathology is the balloon-like Pick cell with argyrophilic inclusions.

General paresis
General paresis was formerly one of the commonest of the organic psychoses of mid-life but is now one of the rarest. It has been occasionally described in association with HD and it is hardly surprising that most of the reports are from the early part of this century (Lowry and Smith, 1918; Pagliano and Avierinos, 1922; Urechia and Rusdea, 1922). Mersky (1958) described a case that was missed because the patient had obvious HD and concluded that treatable physical illnesses should be sought in all cases of HD.

Atypical motor neurone disease and dementia
It has been increasingly recognized that there may be an association between motor neurone disease and dementia (Hudson, 1981). In a recent investigation, Neary et al. (1990) describe a profound and rapidly progressive dementia occurring in patients with clinical features of motor neurone disease. On neurological examination these patients had muscle wasting and cognitive examination showed impaired frontal lobe function. There was a family history of dementia.

Other
Farmer et al. (1989) reported a newly defined hereditary disease which affected five generations of a North Carolina family. The clinical features included choreiform movements, ataxia, seizures, progressive dementia with onset usually between 15–30 years. Death ensued on average 15–25 years after the onset of illness. The condition was inherited in an autosomal dominant fashion. Neuropathology showed neuronal loss of the dentate nucleus, microcalcification of the globus pallidus, neuroaxonal dystrophy of the nucleus gracilis and demyelination of the centrum semiovale. The authors distinguish their syndrome from HD on account of the radiological investigations which did not show pathological findings of caudate atrophy. Other neurodegenerative disorders that may be confused with HD are discussed in Chapter 2.

CONCLUSION

The findings described in this chapter show that an exceptionally wide range of disorders of mental function may occur in Huntington's disease. None of these are in themselves diagnostic or specific, but the pattern is often characteristic, giving strong support to the diagnosis of HD in an individual already known to be at risk. Mental disturbance may occur at any time during the course of the disease; awareness of the possibility of affective disorder or, less commonly schizophreniform psychosis, occurring prior to the existence of any neurological abnormality or evidence of dementia, is of particular importance, even if a definite diagnosis of HD cannot be made at this stage.

An accurate assessment of the specific psychiatric disturbances that may be present in an individual HD patient is essential if therapy and management are to be satisfactory. While HD itself is at present incurable, some of its psychiatric manifestations may respond to appropriate treatment, especially those occurring early in the course of the disease, before severe organic neurologic deterioration has occurred. Even when specific therapy is of no help, the burden on the patient and family can be lightened if the nature of the particular psychiatric symptom can be recognised, explained and adapted to as far as is possible, rather than dismissed as part of the inevitable deterioration seen in the disease.

In the future we may be able to explain the different forms of mental disturbance in HD in terms of specific neurochemical functions of different areas of the brain. As will be seen, in Chapter 5, our current understanding of the neurobiological basis of the neurological features of HD is still extremely limited, and these limitations apply even more to the abnormalities of mental function. It is likely that our understanding will be greatly increased when the HD gene is isolated and we understand its primary function. When this goal is reached we should be in a position not only to interpret satisfactorily many of the unresolved psychiatric aspects of the disorder, but also to use these abnormalities to understand better the functioning of the normal brain in relation to mental processes.

REFERENCES

Albert ML (1978) Subcortical dementia. In: *Alzheimer's Disease: Senile Dementia and Related Disorders* (eds Katzman R, Terry RD and Bick KL), pp. 173–179. New York: Raven Press.

Albert ML, Feldman RG and Willis AL (1974) The subcortical dementia of progressive supranuclear palsy. *Journal of Neurology, Neurosurgery and Psychiatry* **37**:121–130.

American Psychiatric Association (1980) *Diagnostic and Statistical Manual of Mental Disorders* (3rd edn) (DSM-III). Washington, DC: American Psychiatric Association.

American Psychiatric Association (1987) *Diagnostic and Statistical Manual of Mental Disorders* (3rd edn) (DSM-III-R). Washington DC: American Psychiatric Association.

Aminoff MJ, Marshall J, Smith EM and Wyke MA (1975) Pattern of intellectual impairment in Huntington's chorea. *Psychological Medicine* **5**:169–172.

Angst J. (1966) *Zur Atiologie und Nosologie endogener depressiver Psychosen. Monographien aus dem Gesamtgebiete der Neurologie und Psychiatrie.* Berlin: Springer.

Barraclough B, Bunch J, Nelson B and Sainsbury P (1974) A hundred cases of suicide: clinical aspects. *British Journal of Psychiatry* **125**:355–373.

Bayles KA and Boone DR (1982) The potential of language tests for identifying senile dementia. *Journal of Speech and Hearing Discord* **47C22**:210–217.

Beck AT, Resaik HLP and Lettie DJ (eds) (1974) *The Prediction of Suicide.* Maryland: Charles Press.

Bell TJ, Heaton R and Rectan RM (1974) Neuropsychological and emotional correlates of Huntington's chorea. *Journal of Nervous and Mental Disease* **158**:61–69.

Bender L (1938) *A Visual Motor Gestalt Test and its Clinical Uses.* New York: American Orthopsychiatric Association.

Bender L (1946) *Instructions for the Use of Visual Motor Gestalt Test.* New York: American Orthopsychiatric Association.

Benson DF, Cummings JL and Tsai SY (1982) Angular gyrus syndome simulating Alzheimer's disease. *Archives of Neurology* **39**:616–620.

Benton AL (1955) *The Revised Visual Retention Test: Clinical and Experimental Applications.* New York: The Psychological Corporation.

Benton AL (1963) *The Revised Visual Retention Test: Clinical and Experimental Applications*. New York: The Psychological Corporation.
Berrios GE and Hauser R (1988) The early development of Kraepelin's ideas on classification: a conceptual history. *Psychological Medicine* **18**:813–821.
Bertha H and Kolmer H (1940) Uber psychopathologische Erscheinungen bei der Chorea Huntington (Choreophrenie). *Dtsch Z. Nervenheilkd* **151**:26–46.
Bettner LG, Jarvik LF and Blum JE (1971) Stroop color-word test, non-psychotic organic brain syndrome and chromosome loss in aged twins. *Journal of Gerontology* **26**:458–469.
Bickford JAR and Ellison RM (1953) The high incidence of Huntington's chorea in the Duchy of Cornwall. *Journal of Mental Science* **99**:291.
Bigelow N, Roizin L and Kaufman MA (1959) Psychoses with Huntington's chorea. In: *American Handbook of Psychiatry* (ed S. Arietti), pp. 1248–1259. New York: Basic Books.
Birley JLT (1990) DSM-III: from left to right or from right to left? *British Journal of Psychiatry* **157**:116–118.
Bleuler E (1911) *Dementia Praecox or the group of Schizophrenias*. English translation. New York: International Universities Press.
Boll TJ, Heaton R and Rectan RM (1974) Neuropsychological and emotional correlates of Huntington's chorea. *Journal of Nervous and Mental Disease* **158**:61–69.
Bolt JMW (1970) Huntington's chorea in the west of Scotland. *British Journal of Psychiatry* **116**:259–270.
Brandt J (1985) Access to knowledge in the dementia of Huntington's disease. *Developmental Neuropsychology* **1**:335–348.
Brandt J, Strauss ME, Larus J, Jensen B, Folstein SE and Folstein MF (1984) Clinical correlates of dementia and disability in Huntington's disease. *Journal of Clinical Neuropsychology* **6**:401–412.
Brandt J, Folstein SE and Folstein MF (1988) Differential cognitive impairment in Alzheimer's disease and Huntington's disease. *Annals of Neurology* **23**:555–561.
Breitner JCS and Folstein MS (1984) Familial Alzheimer dementia: a prevalent disorder with specific clinical features. *Psychological Medicine* **14**:63–80.
Brothers, CRD (1964) Huntington's chorea in Victoria and Tasmania. *Journal of the Neurological Sciences* **1**:405–420.
Bruyn GW (1968) Huntington's chorea. Historical, clinical and laboratory synopsis. In : *Handbook of Neurology* (eds PJ Vinken and GW Bruyn), **16**:298–378. Amsterdam: North Holland Publishing.
Burton, A (1990) *Dysthymic Disorder*. London: Gaskell.
Butters N, Tarlow S, Cermak LS and Sachs D (1976) Comparison of the information processing deficits of patients with Huntington's Chorea and Korsakoff's Syndrome. *Cortex* **12**: 134–144.
Butters N, Sax D, Montgomery K and Tarlow S (1978) Comparison of the neuropsychological deficits associated with early and advanced Huntington's disease. *Archives of Neurology* **35**:585–589.
Caine ED, Ebert MH and Weingartner H (1977) An outline for the analysis of dementia: the memory disorder of Huntington's disease. *Neurology* **27**:1087–1092.
Caine ED, Hunt RD, Weingartner H and Ebert MH (1978) Huntington's dementia. Clinical and neuropsychological features. *Archives of General Psychiatry* **35**:377–384.
Chandler JH, Reed TE and DeJong RM (1960) Huntington's chorea in Michigan. III. Clinical observations. *Neurology* **10**:148–153.
Chiu E and Alexander L (1982) Causes of death in Huntington's disease. *Medical Journal of Australia* **1**:153.
Clare AW (1976) *Psychiatry in Dissent: Controversial Issues in Thought and Practice*. London: Tavistock.
Cools AR, van den Bercken JHL et al. (1984) Cognitive and motor shifting aptitude disorder in Parkinson's disease. *Journal of Neurology, Neurosurgery and Psychiatry* **47**:443–453.
Cooper JE (1967) Diagnostic change in longitudinal study of psychiatric patients. *British Journal of Psychiatry* **113**:129–142.
Cooper JE (1989) An overview of the prospective ICD-10 classification of mental disorders. *British Journal of Psychiatry* **154**(**suppl.4**):21–23.
Crookes TG and Coleman JA (1973) The Minnesota percepto-diagnostic test (MPD) in adult psychiatric practice. *Journal of Clinical Psychology* **29**:204–206.
Cummings JL (1985) Organic delusions: phenomenology, anatomical correlations and review. *British Journal of Psychiatry* **146**:184–197.
Cummings JL and Benson DF (1984) Subcortical dementia: Review of an emerging concept. *Archives of Neurology* **41**:874–879.
Cummings JL, Gosenfeld LF, Houlihan JP et al. (1983) Neuropsychiatric disturbances associated with idiopathic calcification of the basal ganglia. *Biological Psychiatry* **18**:591–601.

Davenport CB and Muncey EB (1916) Huntington's Chorea in relation to heredity and eugenics. *Bulletin of the Eugenics Records Office*, Number 17. Washington: Carnegie Institute.
Davison K. (1983) Schizophrenia-like psychoses associated with organic cerebral disorders: a review. *Psychiatric Developments* **1**:1–34.
Davison K (1989) Organic cerebral concomitants of schizophrenia: association greater than chance. In: *Contemporary Themes in Psychiatry* (eds K Davison and A Kerr), pp. 395–409. London: Gaskell.
Davison K, and Bagley CR (1969) Schizophrenia-like psychoses associated with organic disorders of the CNS. In: *Current Problems in Neuropsychiatry, British Journal of Psychiatry Special Publication No. 4* (ed RN Herrington), pp. 113–184. Ashford: Headley Bros.
Dewhurst K. (1970) Personality disorder in Huntington's disease. *Psychiatria Clinica* **3**:221–229.
Dewhurst K, Oliver JE and McKnight AL (1970) Socio-psychiatric consequences of Huntington's disease. *British Journal of Psychiatry* **116**:255–258.
Dublin L (1963) *Suicide: a Sociological and Statistical Study*. New York: Raven Press.
Dynan NJ (1914) The physical and mental states in chronic chorea; summary of nineteen cases of chronic progressive chorea with the postmortem findings in eight cases. *American Journal of Insanity* **70**:589–636.
Ekbom KA (1938) Der prasenile Dermatozoenwahn. *Acta Psychiatrica et Neurologica.* **3**:227–59.
Enoch MD and Trethowan WH (1979) *Uncommon Psychiatric Syndromes* (2nd ed). Bristol: Wright.
Entres JL (1921) *Zur Klinik und Vererbung der Huntingtonschen Chorea. Monographien aus dem Gesamtgebiete der Neurologie und Psychiatrie*, Vol 27. Berlin: Springer.
Evvard EM (1936) Maladie de Huntington et schizophrenie. *Rev Neurol* **66**:421–426.
Farmer A and McGuffin P (1989) The classification of the depressions: contemporary confusion revisited. *British Journal of Psychiatry* **155**:437–443.
Farrer LA (1986) Suicide and attempted suicide in Huntington's disease: implications for preclinical testing of persons at risk. *American Journal of Medical Genetics* **24**:305–311
Fisher JM, Kennedy JL, Caine ED and Shoulson I (1983) Dementia in Huntington's disease: a cross-sectional analysis of intellectual decline. *Advances in Neurology* **38**:229–238.
Flowers K and Robertson C. (1985) The effect of Parkinson's disease and the ability to maintain a mental set. *Journal of Neurology, Neurosurgery and Psychiatry* **48**: 517–529.
Folstein S (1989) *Huntington's Disease: A Disorder of Families*. Baltimore: Johns Hopkins Press.
Folstein SE and Folstein MF (1987) Diseases of the Caudate as a model for a manic depressive disorder. *International Conference on the Basal Ganglia*. Abstract. University of Leeds.
Folstein MF, Folstein SE and McHugh PR (1975) Mini-mental state. *Journal of Psychiatric Research* **12**:189–198.
Folstein SE, Folstein MF and McHugh PR (1979) Psychiatric syndromes in Huntington's disease. *Advances in Neurology* **23**:281–290.
Folstein SE, Franz ML, Jensen BA, Chase GA and Folstein MF (1983) Conduct disorder and affective disorder among offspring of patients with Huntington's disease. *Psychological Medicine* **13**:45–52.
Folstein SE, Abbott MH, Chase GA, Jensen BA and Folstein MF (1983) The association of affective disorder with Huntington's disease in a case series and in families. *Psychological Medicine* **13**:537–542.
Folstein SE, Chase GA, Wahl WE, McDonnell AM and Folstein MF (1987) Huntington's disease in Maryland: clinical aspects of racial variation. *American Journal of Human Genetics* **41**:168–171.
Forrest AD (1957) Some observations on Huntington's chorea. *Journal of Mental Science* **103**:507–513.
Fuller GB (1967) *Revised Minnesota Percepto-Diagnostic Test*. New York: The Psychological Corporation.
Garron DC (1973) Huntington's chorea and schizophrenia. *Advances in Neurology* **1**:729–734.
Geraud J, Moron P, Bertrand J-C, Escande M and Conte F (1970) Délire de jalousie et maladie de Huntington. In *CR 67e Congr Psychiat Neurol Langue Franc* (ed P Warot), pp. 1132–1134. Paris: Masson.
Goate AM, James LA, Owen MJ et al. (1989) Predisposing locus for Alzheimer's disease on chromosome 21. *Lancet* **i**:352–355.
Goate A, Chartier-Harlin M-C, Mullan M et al. (1991) Segregation of a missense mutation in the amyloid precursor protein gene with familial Alzheimer's disease. *Nature* **349**:704–706.
Goldstein NP, Ewart JC, Randall RV et al. (1968) Psychiatric aspects of Wilson's disease (hepatolenticular degeneration): results of psychometric tests during long-term therapy. *American Journal of Psychiatry* **124**:1555–1561.
Goodglass H and Kaplan E (1983) *The Assessment of Aphasia and Related Disorders* 2nd Edition. Philadelphia: Lea and Febiger.

Goodman RM, Holl CL, Terango L, Perinne GA and Roberts PL (1966) Huntington's chorea: a multidisciplinary study of affected parents and first generation offspring. *Archives of Neurology* **15**:345–355.

Gordon WP and Illes J (1987) Neurolinguistic characteristics of language production in Huntington's disease: a preliminary report. *Brain and Language* **31**:1–10.

Griesinger W. (1845) *Die Pathologie und der Therapie der Psychosen Krankheiten*. Berlin.

Guberman A and Stuss D (1983) The syndrome of bilateral paramedian thalamic infarction. *Neurology* **33**:540–546.

Haines JL and Conneally PM (1986) Causes of death in Huntington's disease as reported on death certificates. *Genetic Epidemiology* **3(6)**:417–423.

Hallock FK (1898) A case of Huntington's chorea with remarks upon the propriety of naming the disorder 'dementia choreica'. *Journal of Nervous and Mental Disease* **25**:851–864.

Halstead H (1947) *Brain and Intelligence. A Quantitative Study of the Frontal Lobes*. Chicago: University of Chicago Press.

Hans MB and Gilmore TH (1968) Social aspects of Huntington's chorea. *British Journal of Psychiatry* **114**:93–98.

Hart JJ (1990) Paranoid states: classification and management. *British Journal of Hospital Medicine* **44**:34–37.

Hattie WH (1909) Huntington's chorea. *American Journal of Insanity* **66**:123–127.

Hayden MR (1979) Huntington's chorea in South Africa. University of Cape Town: PhD thesis.

Hayden MR, MacGregor JM and Brighton PH (1980) The prevalence of Huntington's chorea in South Africa. *South African Medical Journal* **58**: 193–196.

Heathfield KWG (1967) Huntington's chorea. Investigation into the prevalence of this disease in the area covered by the North East Metropolitan Regional Hospital Board. *Brain* **90**:203–232.

Heston LL, Mastri AR, Anderson VE and White J (1981) Dementia of the Alzheimer type: clinical genetics, natural history and associated conditions. *Archives of General Psychiatry* **38**:1085–1090.

Hochheimer W (1936) Kritisches zur medizinischen Psychologie dargestellt an Chorea-literatur. *Fortschritte der Neurologie, Psychiatrie* **8**:455–470.

Hochheimer W (1936) Zur Psychologie des Choreatikers. *Journal für Psychologie und Neurologie (Leipzig)* **47**:49–115.

Huber W, Poeck K, Weniger D and Willmes K (1983) *Der Aachener aphasie Test*. Göttingen: Hogrefe.

Huber W, Poeck K and Willmes K (1984) The Aachen Aphasia Test. *Advances in Neurology* **42**:291–304.

Hudson AJ (1981) Amyotrophic lateral sclerosis and its association with dementia, Parkinsonism and other neurological disorders: a review. *Brain* **104**:217–247.

Hughes EM (1925) Social significance of Huntington's chorea. *American Journal of Psychiatry* **4**:537–574.

Huntington GW (1872) On Chorea. *Medical and Surgical Reporter* **26**:317–321.

Jackson JA, Jankovic J and Ford J (1983) Progressive supranuclear palsy: clinical features and response to treatment in 16 patients. *Annals of Neurology* **131**:273–278.

Jorm AF and Henderson AS (1985) Possible improvements to the diagnostic criteria for dementia in DSM-III. *British Journal of Psychiatry* **147**:394–399.

Josiassen RC, Curry LM and Mancall EL (1983) Development of neuropyschological deficits in Huntington's disease. *Archives of Neurology* **40**:791–796.

Kehrer F (1928) Erblichkeit und Nervenleiden. I. Ursachen und Erblichkeitskreis von Chorea, Mykoklonie und Athetose. *Monographien aus dem Gesamtgebiete der Neurologie und Psychiatrie*, Vol. 50. Berlin: Springer.

Kendell RE (1976) The classification of depressions: a review of contemporary confusion. *British Journal of Psychiatry* **129**:15–28.

Kellner M (1962) The definition of alcoholism and its prevalence. In *Society, Culture and Drinking Patterns* (eds DJ Pittman and CR Snyder). New York: Wiley.

Kessel N (1965) Self-poisoning. *British Medical Journal* **ii**:1265–70 and 1671–1672.

King M (1985) Alcohol abuse and Huntington's disease. *Psychological Medicine* **15**:815–819.

Kleist K (1922) Die psychomotorischen Störungen und ihr Verhältnis zu den Motilitätsstörungen bei Erkrankungen der Stammganglien. *Monatsschrift für Psychiatrie und Neurologie* **52**:253–302.

Koehler K and Sass H (1984) Affective psychopathology in Huntington's disease: the Johns Hopkins hypothesis and German psychiatry. *Psychological Medicine* **14**:733–737.

Kraepelin E (1912) Uber paranoide Erkrankungen. *Zentralbl Gesamte Neurologie Psychiatrie* **11** 617–638.

Kraepelin E (1921) Manic depressive insanity and paranoia. In *Psychiatrie, ein Lehrbuch für Studierend und Artze*. 8th edition. Translated by RM Barclay. Edinburgh: Livingstone.

Kreitman N (1967) *Parasuicide*. London: Wiley.
Kristensen MO (1985) Progressive supranuclear palsy. *Acta Neurologica Scandinavica* **71**:177–189.
Laane CL (1951) Den tidlige diagnose av chorea Huntington. *Nordisk Medicin* **45**:835.
Ladame P (1911) Suicide et chorée de Huntington. *Encéphale* **6**:422–429.
Lasegue C and Falret J (1877) *Ann Med Psychol (Paris)* **18**:321.
Leonhard K, Korff I and Schulz H (1962) Die Temperamente und den familien der monopolaren and bipolaren Phäsischen. *Psychosen. Psychiatrie und Neurologie* **143**:416–434.
Lewis A (1953) Health as a social concept. *British Journal of Sociology* **4**:109–124.
Lewis A (1970) Paranoia and paranoid: a historical perspective. *Psychological Medicine* **1**:2–12.
Lewis A (1974) Psychopathic personality: a most elusive category. *Psychological Medicine* **4**:133–140.
Lewis B (1876) A case of chorea associated with mania terminating fatally by cerebellar apoplexy. *Medical Times and Gazette* **2**:280–282.
Lind WAT (1927) Mental symptoms and post mortem appearances in Huntington's chorea. *Medical Journal of Australia* **2**:53–56.
Loranger AW, Goodell H, McDowell FH *et al.* (1972) Intellectual impairment in Parkinson's syndrome. *Brain* **95**:405–412.
Lowrey LG and Smith CE (1918) Degenerative chorea (Huntington's type) with the serology of general paresis. Report of two cases, one with autopsy. *American Journal of Syphilogy* **2**:453–461.
Lyle OE and Gottesman II (1977) Premorbid psychometric indicators of the gene for Huntington's disease. *Journal of Consulting and Clinical Psychology* **45**:1011–1022.
MacFaren J (1874) A case of chorea. *Journal of Mental Science* **20**:97–99.
Mahendra B (1984) *Dementia*, pp. 1–18. Lancaster: MTP Press.
Martindale B (1987) Huntington's chorea: some psychodynamics seen in those at risk and in the response of the helping professions. *British Journal of Psychiatry* **150**:319–323.
Martone M, Butters M, Payne JT, Becker JT and Sax DS (1984) Dissociations between skill learning and verbal recognition in amnesia and dementia. *Archives of Neurology* **41**:965–970.
Mattsson B (1974) Huntington's chorea in Sweden. II. Social and clinical data. *Acta Psychiatrica Scandinavica* **Suppl 255**:221–235.
Mayeux R, Stern Y, Rosen J and Benson DF (1983) Is 'subcortical dementia' a recognizable clinical entity? *Annals of Neurology* **14**:278–283.
McCarthy PD and Walsh D (1976) Suicide in Dublin. I. The under-reporting of suicide and the consequences for national statistics. *British Journal of Psychiatry* **126**:301–308.
McDonald C (1969) Clinical heterogeneity in senile dementia. *British Journal of Psychiatry* **115**:267–271.
McHugh PR and Folstein MF (1973) Subcortical dementia. Address to the American Academy of Neurology, Boston, MA, April 1973.
McHugh PR and Folstein MF (1975) Psychiatric syndromes of Huntington's chorea. In: *Psychiatric Aspects of Neurological Disease* (eds DF Benson and D Blumer), pp. 267–286. New York: Grune & Stratton.
McIntosh GC, Jameson HD and Markesbery WR (1978) Huntington disease associated with Alzheimer disease. *Annals of Neurology* **3**:545–548.
McKhann G, Drachman D, Folstein M, Katzman R, Price D and Stadlan EM (1984) Clinical diagnosis of Alzheimer's disease: report of the NINCDS–ADRDA Work Group under the auspices of the Department of Health and Human Services Task Force on Alzheimer's disease. *Neurology* **34**:939–944.
Meggendorfer F (1923) Die psychischen Störungen bei der Huntingtonschen Chorea. Klinische und genealogische Untersuchungen. *Zeitschrift für die gesamte Neurologie und Psychiatrie* **87**:1–49.
Meierhofer M (1938) Atypische Psychosen in einer Chorea–Huntington Familie. *Monatsschrift für Psychiatrie und Neurologie* **97**:13–60.
Merskey H (1958) A clinical and psychometric study of the effects of procaine amide in Huntington's chorea. *Journal of Mental Science* **104**:411–420.
Milner B (1963) Effects of different brain lesions on card sorting. *Archives of Neurology* **9**:90–100.
Mindham RHS, Steele C, Folstein MF and Lucas J (1985) A comparison of the frequency of major affective disorder in Huntington's disease and Alzheimer's disease. *Journal of Neurology, Neurosurgery and Psychiatry* **48**:1172–1174.
Minski L and Guttmann E (1938) Huntington's chorea: a study of 34 families. *Journal of Mental Science* **84**:21–96.
Mitchell A (1973) What's on your label? *Mind and Mental Health Magazine* 32–35.
Morgan HG (1979) *Death Wishes: the Understanding and Management of Deliberate Self-harm*. London: Wiley.

Moses JA, Golden CJ, Berger PA and Wisniewski AM (1981) Neuropsychological deficits in early, middle and late stage Huntington's disease as measured by the Luria–Nebraska Neuropsychological Battery. *International Journal of Neuroscience* **14**:95–100.

Moss MB, Albert MS, Butters N and Payne M (1986) Differential patterns of memory loss among patients with Alzheimer's disease, Huntington's disease and alcoholic Korsakoff's syndrome. *Archives of Neurology* **43**:239–246.

Muller WK (1982) Der fall bei eine kasuistik zur diagnostik, pathogenese und behandlung der chorea major. In: *Die Therapie Extrapyramidal-Motorischer Erkrankungen* (ed. JC Aschoff), pp. 41–58. Stuttgart: Schattauer.

Murphy DGM (1991) The classification and treatment of dysthymia. *British Journal of Psychiatry*. **158**:106–109.

Naef ME (1917) Uber Psychosen bei Chorea. *Monatsschrift für Psychiatrie und Neurologie* **41**:65–88.

Neary D, Snowden JS, Mann DMA, Northen B, Goulding PJ and MacDermott N (1990) Frontal lobe dementia and motor neurone disease. *Journal of Neurology, Neurosurgery and Psychiatry* **53**:23–32.

Nicotra A (1938) Demenza precoce e corea degenerativa di Huntington. *Cervello* **17**:219–230.

Nott PN and Fleminger JJ (1975) Presenile dementia: the difficulty of an early diagnosis. *Acta Psychiatrica Scandinavica* **51**:210–217.

Oepen H (1963) Paroxysmale Störungen bei der Huntingtonschen Chorea. *Archiv für Psychiatrie und Nerven Krankheiten* **204**:245–261.

Oepen G, Mohr U, Willmes K and Thoden U (1985) Huntington's disease: visumotor disturbance in patients and offspring. *Journal of Neurology, Neurosurgery and Psychiatry* **48**:426–433.

Oliver JE (1970) Huntington's chorea in Northamptonshire. *British Journal of Psychiatry* **116**:241–253.

Oltman JR and Friedman S (1961) Comments on Huntington's chorea. *Diseases of the Nervous System* **22**:313–319.

Oppler W (1933) Erbbiologische Nachkommenuntersuchungen bei einem Fall von Huntingtonscher Chorea mit schizophren-gefärbter Psychose. *Zeitschrift für Neurologie und Psychiatrie* **144**:769–783

Ørbeck AL and Quelprud T (1954) Setesdalsrykka (Chorea progressive hereditaria). *Norske videnskap Acad Schr* **1**:1–125.

Osterrieth P-A (1944) Le test de copie d'une figure complexe. *Archives de Psychologie* **30**:206–353.

Pagaliani V and Avierinos F (1922) Un cas de chorée familiale très Ameliorée par le traitement antisyphilitique. *Marseille-medicin* **59**:78–79.

Panse F (1942) *Die Erbchorea: eine klinisch-genetische Studie. Sammlung Psychiatrischer und Neurologischer Einzeldarstellungen 18.* Leipzig: Thieme.

Parker N (1958) Observations on Huntington's chorea based on a Queensland survey. *Medical Journal of Australia* **45**:351–359.

Pattie AH and Gilleard CJ (1975) A brief psychogeriatric assessment schedule – validation against psychiatric diagnosis and discharge from hospital. *British Journal of Psychiatry* **127**:489–493.

Pattie AH and Gilleard CJ (1979) *Manual for the Clifton Assessement Procedures for the Elderly.* Sevenoaks: Hodder and Stoughton.

Perris C (1966) A study of bipolar (manic-depressive) and unipolar depressive psychoses. *Acta Psychiatrica Scandinavica* **42(suppl. 194)**:1–199.

Pick A (1892) On the relation of senile atrophy and aphasia. *Prager Medizinisches Wochenschrift* **17**:165–167.

Pinel P (1806) *A treatise on insanity.* Sheffield: Cadell and Davies (translated by DD Davis).

Podoll K, Caspary P, Lange HW and Noth J (1988) Language functions in Huntington's disease. *Brain* **111**:1475–1503.

Podolsky S and Leopold NA (1974) Growth hormone abnormalities in Huntington's chorea: effect of L-DOPA administration. *Journal of Clinical Endocrinology and Metabolism* **39**:36–39.

Post F (1966) *Persistent Paranoid States in the Elderly.* New York: Pergamon.

Potegal M (1971) A note on spatial-motor deficits in patients with Huntington's disease; a test of hypothesis. *Neuropsychologia* **9**:233–235.

Quarrell OWJ, Tyler A, Cole G and Harper PS (1986) The problem of isolated cases of Huntington's disease in South Wales 1974–1984. *Clinical Genetics* **30**:433–439.

Reed TE (1958) Population genetics. *Lancet* **2**:519.

Reed TW, Chandler JH, Hughes EM and Davidson RT (1958) Huntington's chorea in Michigan. I. Demography and genetics. *American Journal of Human Genetics* **10**:207–225.

Reitan RM (1958) Validity of the trail making test as an indicator of organic brain damage. *Perceptual and Motor Skills* **8**:271–276.

Reitan RM (1966) A research program on the psychological effects of brain lesions in human beings. *International Review of Research in Mental Retardation* (ed. NR Ellis) New York: Academic Press.

Report of the Royal College of Physicians (1981) Organic mental impairment in the elderly. *Journal of the Royal College of Physicians (London)* **15**:141–167.

Rey A. (1941) L'examen psychologique dans les cas d'encéphalopathie traumatique. *Archives de Psychologie* **28**:286–340.

Reyes MG and Gibbons SG (1985) Dementia of the Alzheimer type and Huntington's disease. *Neurology* **35**:273–277.

Robins LN, Helzer JE, Croughan J and Ratcliff KS (1981) National Institute of Mental Health Diagnostic Interview Schedule. *Archives of General Psychiatry* **38**:381–389.

Robins LN, Helzer JE, Weissman MM et al. (1984) Lifetime prevalence of specific psychiatric disorders in three sites. *Archives of General Psychiatry* **41**:949–958.

Robinson AL, Heaton RK, Lehman RAW et al. (1980) The utility of the Wisconsin Card Sorting Test in detecting and localising front lobe lesions. *Journal of Consulting and Clinical Psychology* **48**:605–614.

Rosenbaum D (1941) Psychosis with Huntington's chorea. *Psychiatric Quarterly* **15**:93–96.

Roth M (1980) The diagnosis of dementia in late and middle life. In: *The Epidemiology of Dementia* (ed. JA Mortimer). Oxford: University Press.

Rutter M (1987) Temperament, personality and personality disorder. *British Journal of Psychiatry* **150**:443–458.

Salmon DP, Kwo-on-Yuen P. Heindel WC, Butters, N and Thal, LJ (1989) *Archives of Neurology* **46**:1204–1208.

Saugstad L and Odegard O (1985) In defence of international classification. *Psychological Medicine* **15**:1–2.

Saugstad L and Odegard O (1986) Huntington's chorea in Norway. *Psychological Medicine* **16**:39–48.

Sax DS, O'Donnell B, Butters N, Menzer L, Montgomery K, Kayne HL (1983) Computed tomographic, neurologic, and neuropsychological correlates of Huntington's disease. *International Journal of Neuroscience* **18**:21–36.

Schoenfeld M, Myers RM, Cupples A, Berkman B, Sax DS and Clark E (1984) Increased rate of suicide among patients with Huntington's disease. *Journal of Neurology, Neurosurgery and Psychiatry* **47**:1283–1287.

Shepherd M (1961) Morbid jealousy: some clinical and social aspects of a psychiatric syndrome. *Journal of Mental Science* **107**:687–704.

Shepherd M and Sartorius N (1974) Personality disorder and the international classification of diseases. *Psychological Medicine* **4**:141–146.

Shneidman ES (1964) Suicide, sleep and death. *Journal of Consulting Psychology* **28**:95–106.

Shokeir MH (1975) Investigations on Huntington's disease in the Canadian Prairies. *Clinical Genetics* **7**:345–348.

Siever LJ and Klar H (1986) A review of DSM-III criteria for the personality disorders. In: *American Psychiatric Association Annual Review*, vol 5 (eds AJ Frances and RE Hales). Washington DC: American Psychiatric Press.

Sims A, Salmons P and Humphreys P (1977) Folie à quatre. *British Journal of Psychiatry* **130**:134–138.

Sjögren T, Sjögren H and Lundgren AGH (1953) Morbus Alzheimer and morbus Pick: a genetic, clinical and pathoanatomical study. *Acta Psychiatrica Neurologica Scandinavica* **Suppl 92**.

Snaith, RP and Taylor PT (1985) Irritability: definition, assessment and associated factors. *British Journal of Psychiatry* **147**:127–136.

Speedie LJ, Brake N, Folstein SE, Bowers D and Heilman KM (1990) Comprehension of prosody in Huntington's disease. *Journal of Neurology, Neurosurgery and Psychiatry* **53**:607–610.

Spillane J and Phillips R (1937) Huntington's chorea in south Wales. *Quarterly Journal of Medicine* **6**:403–423.

St George-Hyslop PH, Tanzi RE, Polinsky RJ et al. (1987) The genetic defect causing familial Alzheimer's disease maps on chromosome 21. *Science* **235**:885–890.

Starkstein SE, Brandt J, Folstein SE et al. (1988) Neuropsychological and neuroradiological correlates in Huntington's disease. *Journal of Neurology, Neurosurgery and Psychiatry* **51**:1259–1263.

Stevens DL (1976) Huntington's chorea: a demographic, genetic and clinical study. MD thesis, University of London.

Streletzki F (1961) Psychosen im Verlauf der Huntingtonscher Chorea unter besonderer Berücksichtigung der Wahnbildungen. *Archiv für Psychiatrie und Nerven Krankheiten* **202**:202–214.

Tamir A, Whittier J, and Korenys C (1969) Huntington's chorea: a sex difference in psychopathological symptoms. *Diseases of the Nervous System* **30**:103.

Tarrier N, Beckett R, Harwood S and Bishay N (1990) Morbid jealousy: a review and cognitive-behavioural formulation. *British Journal of Psychiatry* **157**:319–326.

Tusques J and Feuillet C (1937) Troubles psychiques et chorée de Huntington. *Annales Médico-psychologiques* **95**:109–114.

Tyler A (1982) The social and economic burden of Huntington's disease. University of Wales: MSc Thesis.

Tyler A (1983) Sharing the diagnosis of Huntington's chorea. *Community Care*.

Tyskiewicz M (1960) Some cases of Huntington's disease, chronic schizophrenia and other psychiatric diseases in members of one family. *Neurol Neurochir Psychiat Pol* **10**:121–124.

Urechia C and Rusden N (1922) Chorée chronique syphilitique. *Review of Neurology*. **38**:513–517.

Vonsattel JP, Myers RH, Stevens TJ et al. (1985) Neuropathological classification of Huntington's disease. *Journal of Neuropathology and Experimental Neurology* **44**:559–577.

Wallesch C-W and Fehrenbach RA (1988) On the neurolinguistic nature of language abnormalities in Huntington's disease. *Journal of Neurology, Neurosurgery and Psychiatry* **51**:367–373.

Walker DA (1979) Huntington's chorea in South Wales. MD Thesis, University of Liverpool.

Walker DA, Harper PS, Wells CEC, Tyler A, Davies K and Newcombe RG (1981) Huntington's chorea in South Wales: a genetic and epidemiological study. *Clinical Genetics* **19**:213–221.

Ward CH, Beck AT, Mendelson M, Mock JE and Erburgh JK (1962) The Psychiatric Nomenclature: reasons for diagnostic disagreement. *Archives of General Psychiatry* **7**:198–205.

Wechsler D (1955) *Manual for the Wechsler Adult Intelligence Scale*. New York: Psychological Corporation.

Weingartner H, Caine ED and Ebert MH (1979) Imagery, encoding, and retrieval of information from memory: some specific encoding – retrieval changes in Huntington's disease. *Journal of Abnormal Psychology* **88**:52–58.

Weissman MM and Paykel ES (1974) *The Depressed Woman: a Study of Social Relationships*. Illinois: University of Chicago Press.

Werner A and Folk JJ (1968) Manifestations of neurotic conflict in Huntington's chorea. *Journal of Nervous and Mental Disease* **147**:141–147.

Whittier JR (1976) Asphyxiation, bulimia and insulin levels in Huntington disease (chorea). *Journal of the American Medical Association* **235**:1423–1424.

Whittier JR, Heimler A and Korenyi C (1972) The psychiatrist and Huntington's disease (chorea). *American Journal of Psychiatry* **128**:96–100.

Wilson RS and Garron DC (1979) Cognitive and affective aspects of Huntington's disease. In: *Huntington's Disease. Advances in Neurology* (eds TN Chase, NS Wexler and A Barbeau), Vol 23, pp. 193–202. New York: Raven Press.

Wootton B (1959) *Social Science and Social Pathology*, pp. 203–226. London: George Allen and Unwin.

World Health Organization (1978) *Mental Disorders: Glossary and Guide to their Classification in Accordance with the Ninth Revision of the International Classification of Diseases*. Geneva: World Health Organization.

World Health Organization (1952) Expert Committee on Mental Health, Alcoholism Subcommittee, Second Report. Technical Report Series No. 48. Geneva: World Health Organization.

World Health Organization (1987) *ICD-10. 1986 Draft of Chapter V. Mental, Behavioural and Developmental Disorders*. Geneva: World Health Organization.

4

The Natural History of Huntington's Disease

For many disorders advances in prevention and therapy have removed the possibility of observing their unaltered natural history, but for Huntington's disease (HD) the absence of any effective measures that alter its course means that the disease we see today is essentially unchanged in its onset, progress and termination to that originally observed by George Huntington over a century ago. This sad fact does at least give the opportunity of documenting the natural history of HD accurately, so that any future measures claimed to be effective can be tested against a known background. The information summarized here has come from studies in HD carried out in numerous countries over a considerable period of time; it is remarkable how closely most of these studies agree, most variation probably reflecting the special features of the investigation and investigator rather than indicating true differences relating to the disease. The genetic implications of age of onset and death are considered in Chapters 9 and 11.

AGE AT ONSET

Anyone working with HD patients is aware of the difficulty in pinpointing a precise onset of the disease. Gradual behavioural changes may only become recognized as abnormal over a period of years, while choreic movements may go unnoticed for prolonged periods. Should onset be defined as when the first possible alteration was noted, or when definite abnormality is finally documented? Should such documentation be medical or can the evidence of the patient or a relative be accepted? Most studies have taken as the point of onset the first definite abnormality, whatever its nature, recorded by a reliable witness (commonly a spouse).

It might be considered that the inherent variability in any age at onset estimate would make it an observation of little value, but this is far from being the case. Provided that the study is conducted by a single observer using well-defined criteria, the results for an HD population show a clear and consistent pattern, whatever the difficulty may be in defining onset in an individual

Table 4.1 Mean or median age at onset and death for HD in different studies*

Study	Country	Mean age at onset (years)	Mean age at death (years)
Bell (1934)	Collected literature	35.5 (460)	53.1 (349)
Panse (1942)	Germany	36.2 (446)	52.1 (473)
Reed et al. (1958)	Michigan, USA	35.3 (262)	53.7 (262)
Brothers (1964)	Victoria, Australia	37.2 (206)	51.4 (123)
Bolt (1970)	Scotland	42.5	56.7 (269)
Dewhurst et al. (1970)	Northampton, England	39.0 (102)	
Husquinet (1970)	Belgium		54.1 (231)
Venters (1971)	Scotland	38.8 (123)	56.3 (155)
Brackenridge (1971)	Collected literature	33.8 (344)	51.7 (403)
Wendt and Drohm (1972)	Germany (remote data)	44.0 (762)	56.5 (246)
Wallace (1972)	Queensland, Australia	39.5 (144)	57.2 (106)
Stevens (1976)	Yorkshire, England	43.9 (298)	56.9 (256)
Walker et al. (1981)	Wales	41.6 (333)	56.5 (284)
Farrer and Conneally (1985)	US Huntington roster	38.0 (569)	
Myers et al. (1985)	USA (Boston)	40.9 (243)	
Folstein et al. (1987)	USA (Maryland)	40.3	
Adams et al. (1988)	Georgia, USA	38.7 (611)	
Births before 1920		43.4 (311)	

*Figures in parenthesis show sample size.

patient. Table 4.1 summarizes the results of the main studies of age at onset, while Figure 4.1 shows the characteristic distribution in two of them. Bell (1934), in the earliest study of this type, noted the close fit of the age at onset to the normal distribution curve, while the data of Wendt (Wendt et al., 1959; Wendt and Drohm, 1972), probably the largest individual study ever performed, give a clear picture of the detailed distribution.

Most studies show a mean age at onset of between 35 and 44, but biases exist which make it likely that the higher estimates are closer to reality. Surveys based on inpatients and primary cases will tend to favour those with earlier onset, as will studies based on cases reported in the literature. Likewise the use of case material close in time to the study will omit those later onset gene carriers still to develop the disease.

There are two possible ways of overcoming this bias. For recent data one can use a life table analysis, as done originally by Newcombe (1981), which applies a correction for those cases not ascertained because they have not yet developed the disorder. When this is done it gives a corrected age at onset around 4–7 years later than the directly observed value. Thus in the South Wales study the observed median age at onset was 41.6 years (Walker, 1979) but the corrected value using alternative methods of analysis was between 46 and 48.9 years (Newcombe, 1981). A similar analysis on HD families in Georgia, USA (Adams et al., 1988) showed a corresponding 5-year difference, being 38.7 years

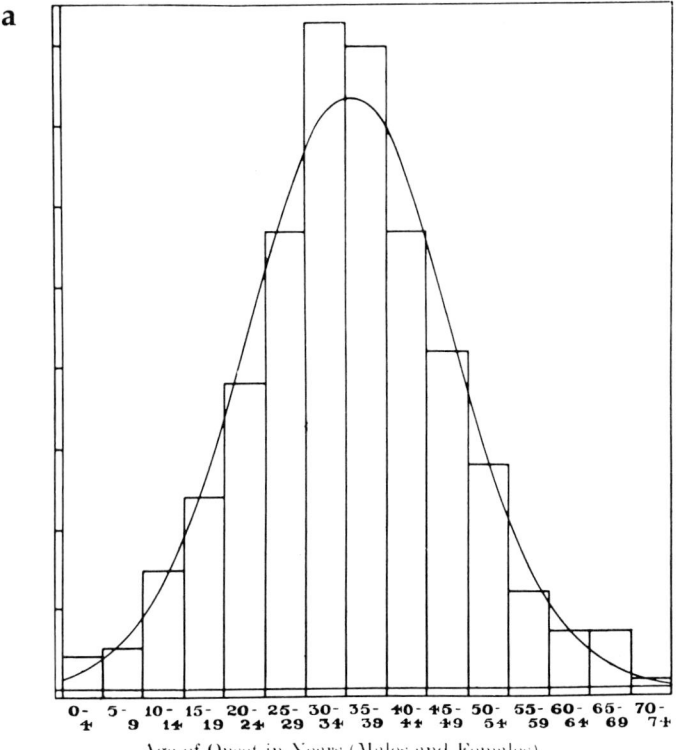
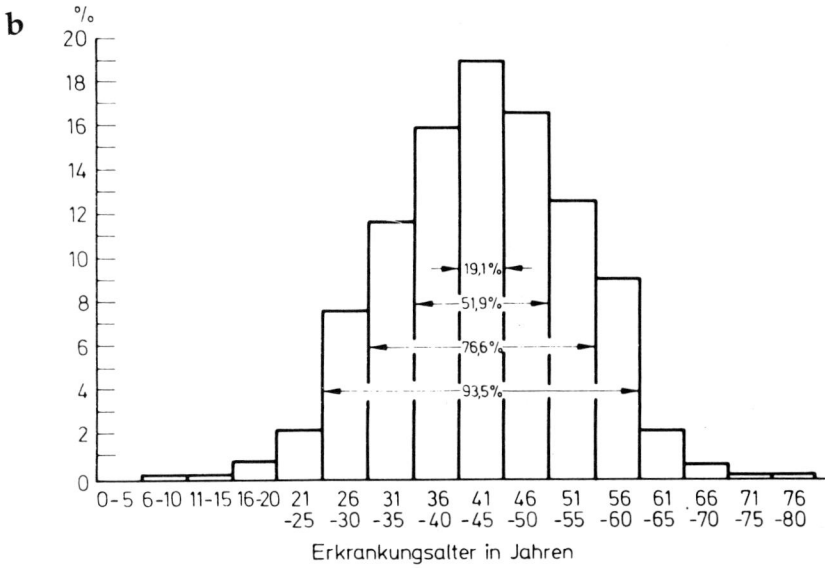

Figure 4.1 Early studies of the distribution of age of onset in HD. (a) From Bell (1934), the first systematic study of quantitative aspects of HD. (b) From Wendt and Drohm (1972), probably the largest individual study of genetic aspects of HD.

for the directly observed median age at onset and 44.0 years for the corrected estimate.

An alternative approach to removing bias is to use family data that are sufficiently remote from the time of study to be certain that all those who are going to develop HD will have done so. In this situation no correction should be necessary. This approach was taken by Wendt (Wendt et al., 1959; Wendt and Drohm, 1972), whose estimate of 43.4 years for age at onset in patients born between 1870 and 1900 was considerably later than previous estimates using more recent data. Unfortunately this approach is not free from problems either. Data from such early times may well be less accurate and complete, while early onset cases (occurring earlier in the period) may be preferentially lost. Stevens (1976) could find no difference in age at onset of his data when splitting them by decade of onset, but Adams et al. (1988) found a median onset of 43.7 years for those born before 1921, agreeing closely with the life table estimate. Similarly Pridmore (1990) in a reanalysis of the Tasmanian HD population, found a mean age at onset of 48.3 years for births before 1930, compared with 42.4 overall. This unusually late onset is based mainly on a single very large kindred.

Even more important perhaps than the mean age at onset are the extremes of the distribution curve. Those cases with onset under 20 years are usually defined as juvenile HD, while those with onset over 60 years can be considered as HD in the elderly. Both groups have a different and at times characteristic distribution of clinical features, discussed fully in Chapter 2, so that it is

Table 4.2 Early and late onset of HD in different studies*

Study	Number in total study	% With onset under 20 years (juvenile)	% With onset over 60 years (elderly)
Bell (1934)	460	10	3.2
Panse (1942)	456	7.2	3.4
Reed et al. (1958)	262	12	1
Brothers (1964)	206	7	3.9
Cameron and Venters (1967)	194	4.7	10.9
Oliver (1970)	115	9.5	3.5
Wendt and Drohm (1972)	762	2.2	4.9
Wallace (1972)	144	3.5	6.3
Stevens (1977)	298	1	1.9
Walker et al. (1981)			
Uncorrected	333	4	7.5
Life table analysis		3	23
Adams et al. (1988)			
Uncorrected	611	4.9	4.3
Life table analysis		3.2	16.6

*Studies prior to 1930 and of under 100 individuals omitted.

relevant to know how common such onsets are. Table 4.2 shows the data from the main surveys; there are probably greater biases here than in the estimates for mean age at onset, partly due to the small numbers but also due to the ascertainment used. The South Wales study (uncorrected data) showed 4.5% with onset under 20 and 8% over 60 years (Figure 4.2 and Table 4.3), figures agreeing closely with most other recent studies except for that of Stevens, who had few patients in either group.

The use of life table analysis has a particularly marked effect on the late onset end of the distribution curve, as shown in Figure 4.3. Thus in the Georgia data of Adams *et al.* (1988), only 4.3% had onset over 60 years using the total uncorrected data, whereas the figure was 16.6% when analysed by the life table method. For the South Wales data the value is changed from 8% to 23%. The importance of using appropriate age at onset curves for risk estimation in genetic counselling is discussed further in Chapter 11, but it must be strongly emphasized here that uncorrected age at onset curves are not suitable for this purpose.

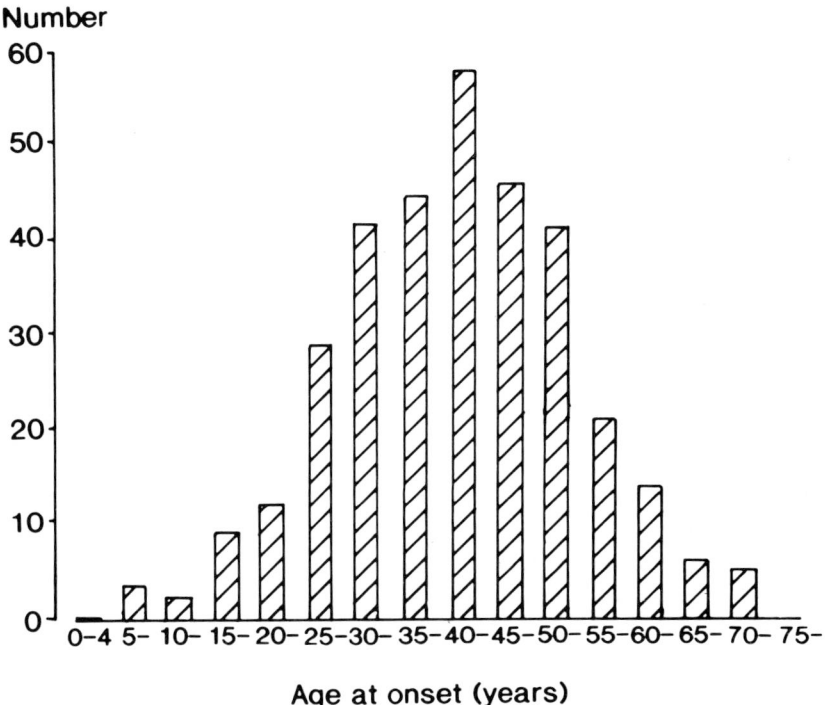

Figure 4.2 Distribution of age of onset and death from HD in the South Wales study (uncorrected by life table analysis). From Walker, 1979.

Table 4.3 Distribution of age at onset in HD. South Wales data, based on Walker (1979)*

Age group (years)	Number of onsets	%
0–4	0	0
5–9	3	0.9
10–14	2	0.6
15–19	9	2.7
20–24	12	3.6
25–29	29	8.7
30–34	42	12.6
35–39	45	13.5
40–44	58	17.4
45–49	46	13.8
50–54	21	12.3
55–59	21	6.3
60–64	14	4.2
65–69	6	1.8
70–74	5	1.5
75+	0	0

*Total 333.

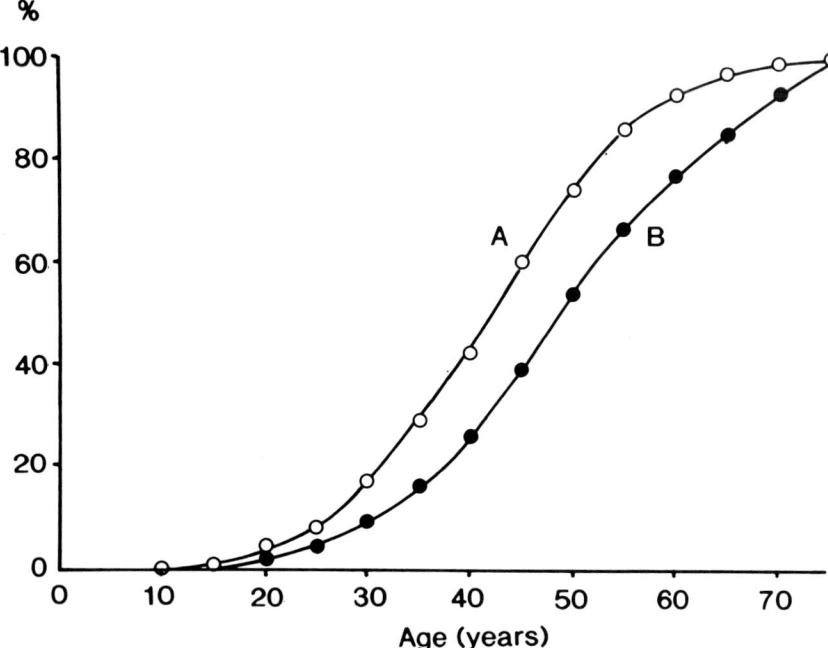

Figure 4.3 Age at onset in HD (cumulative frequency and curve). South Wales data, to show the difference between **A** uncorrected and **B** life table estimates. Courtesy of Dr Robert Newcombe.

AGE AT DIAGNOSIS

This lags considerably behind the age at onset, Brothers (1964) and Heathfield (1967) estimating a 10-year delay to be frequent. The South Wales study confirmed this, as can be seen from Figure 4.4, where the curve for age at diagnosis is shifted considerably to the right compared with the age at onset curve. Overall, this gives an 8-year interval between onset and diagnosis (Walker, 1979). The limitation of using age at diagnosis in charting the natural history of HD is further reflected in the finding that 20% of cases in this study were first diagnosed by the survey itself.

That the difficulty in defining a clear diagnosis of HD is a real one and not simply a question of oversight is confirmed by the recent study of Penney *et al.* (1990), based on an 8-year longitudinal study of the Venezuela isolate. They found that it was rare for patients to move from a clearly normal examination to one that allowed a definite diagnosis to be made. Rather there was an intervening period of several years in which suspicious minimal signs were present, which progressively evolved into definite HD. Their concept of a 'zone of onset' rather than a point is an important one and corresponds to what most workers have previously recognized intuitively but have not expressed quantitatively. It is of equal importance practically in allowing those with totally normal assessments a confident assurance that HD is unlikely to develop in the immediate future. Penney *et al.* (1990) estimated that such an individual would have a chance of only 3% of developing definite HD within the next 3 years.

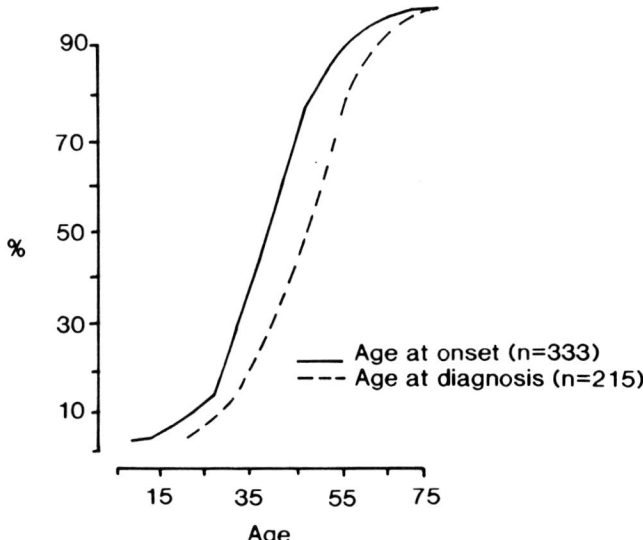

Figure 4.4 Distributions of age at onset and age at diagnosis in the south Wales study (based on Walker, 1979). There is an overall lag of eight years between onset and diagnosis.

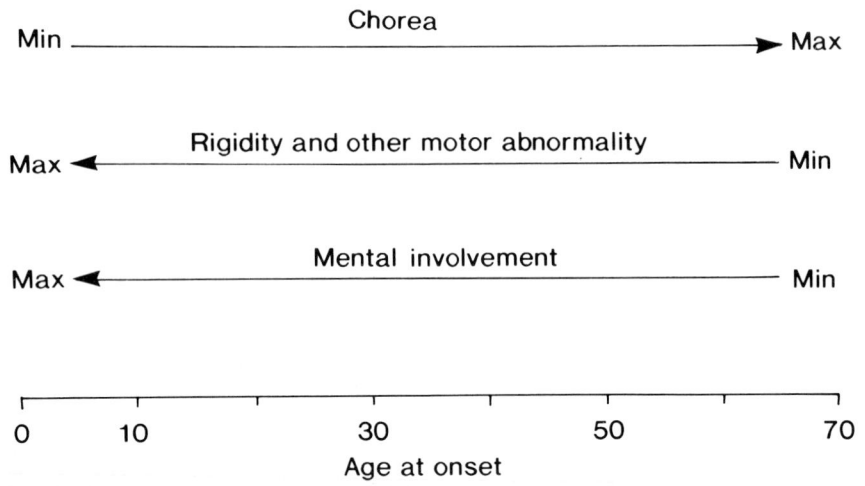

Figure 4.5 Presenting symptoms in HD in relation to age at onset.

CLINICAL FEATURES IN RELATION TO AGE AT ONSET

The tendency for juvenile HD to present with severe mental deterioration and rigidity, while HD in old age more commonly shows a relatively pure movement disorder, has already been mentioned in Chapter 2. It has led to the view that in classifying the various clinical types of HD the specific clinical features are largely secondary, reflecting patterns of brain involvement characteristic of the age at onset of the disease (Farrer and Conneally, 1987). Figure 4.5 shows this schematically, though in an over-simplified fashion; numerous individual exceptions to this pattern may occur. Nevertheless this scheme is of some diagnostic help – thus in the patient without a clear family history of HD presenting with chorea as the only significant feature, HD will be a likely diagnosis if the patient is aged 65, but unlikely if aged 15, even though both ages are towards the extremes for onset in HD.

AGE AT DEATH

This is the one variable in the natural history of HD that can be defined precisely, but in practice there are biases which affect it as much as the age at onset. Clearly the use of recent data will mean that the longer survivors in a cohort will still be alive and be excluded, thus giving an underestimate. The close correspondence of the major series, however, can be seen from Table 4.1, which ranges only between 51 and 57 years. Considering the span of half a century over which these studies have taken place it is remarkable how little relationship there is with time and with the country concerned. The recent study from Tasmania, restricted in numbers and based mainly on one large, late onset kindred, gives a mean age at death of 63 years (Pridmore 1990b). It is of interest that this HD population is derived from Somerset, England, where a similar late age of death and onset has been documented in two large families (Glendinning, 1975), suggesting that the properties of a particular kindred may have been retained in the Antipodes.

SURVIVAL

HD is a relentlessly progressive disorder, and the likely duration of the disease is one of the most important practical questions that patients, and even more their relatives, wish to know.

Survival has been estimated in most of the major studies of age at onset and death already quoted. Table 4.4 lists the results, which range between 10 and 17 years from the age at onset (not age at diagnosis). There is little difference between the sexes nor, importantly, is duration related clearly to age at onset; Conneally (1984), analysing 227 cases from the US HD roster, found a diminution in those with onset under 10 years or over 70 years, but relatively constant at all ages in between. The South African series of Hayden (1979, 1981) also found an apparent shortening of survival in juvenile-onset patients, but only included data on those patients who had died, inevitably biasing the data by exclusion of survivors. Van Dijk et al. (1986) found that juvenile cases in the literature did show a shorter survival even when a correction was made for this, but noted the very wide variation. The South Wales data (Newcombe et al., 1981) showed no difference in survival of juvenile cases, nor was there any relation to the sex of the affected parent; any tendency for a more benign course in those with late onset was likewise counteracted by the greater overall mortality in this age group. As with age at death, there does not appear to have

Table 4.4 Survival in HD (duration from onset to death)

Study	n	Duration (years)
Bell (1934)	204	13.7
Panse (1942)	446	13.4
Reed et al. (1958)	153	15.8
Brothers (1964)	97	12.2
Cameron and Venters (1967)	127	10.6
Petit (1969)	65	14.4
Bolt (1970)	176	14.6
Brackenridge (1971)	191	11.9
Stevens (1976)	180	13.6
Walker (1979)	204	15.3
Hayden (1979)	94	14.1
Conneally (1984)	227	17.1
Folstein (1989)	70	14.8

Table 4.5 Frequency distribution of duration of symptoms (onset to death) in 204 South Wales patients with Huntington's disease. From Walker, 1979.

Duration of symptoms (years)	n
0–4	8
5–9	34
10–14	57
15–19	61
20–24	22
25–29	10
30–34	7
35–39	4
40+	1

been any noticeable increase in survival in recent years, something that clearly indicates a lack of any effect of drug or other therapy on the outcome of the disorder.

The similar estimates for duration of the illness in different surveys conceal the marked variation seen between individuals; survival of up to 40 years was found in the South Wales study, as seen in Table 4.5.

CLINICAL STAGING AND RATE OF DECLINE

While onset and death provide the fixed points from which survival in HD can be measured, it is equally important to be able to assess the different stages in between and to assess quantitatively the rate of decline. In broad terms most workers have divided the time course of HD into three clinical phases, each corresponding approximately to a 5-year period of the overall 15-year course, though with much variation and overlap for individual patients.

As seen in Table 4.6, the HD patient in Stage 1 is able to lead a largely independent life. Continued work may be possible, depending on the responsibility and physical complexities involved; chorea is unlikely to cause serious disability, while rigidity and other motor problems are not yet prominent except in juvenile and some younger adult patients. However this relative independence may itself cause serious problems to family members, especially in those cases that present with serious psychiatric disturbances or are aggressive and impulsive.

In Stage 2 physical disability becomes severe, with chorea often marked and the generalized motor aspects becoming conspicuous. The patient is now increasingly dependent on others for help with mobility and overall care, as well as with taking on responsibility for general decision making and organization. This progressive disability may paradoxically make things easier in those

Table 4.6 The principal clinical stages of HD

Stage 1	Presentation with initial neurological or psychiatric symptoms Main features remain similar to those at presentation Chorea more prominent than other motor abnormalities Patient largely independent for most activities Burden on family mainly result of psychiatric problems Death rare except for suicide
Stage 2	Motor disorder more generalized Physical disability becomes major Patient dependent on others for many activities Burden on family both physical and psychological Death often from unrelated causes
Stage 3	Severe generalized motor disorder Physical disability severe to total Patient completely dependent for all aspects of care Burden on carers mainly physical Death frequent at any point

patients who previously have had mainly psychiatric problems from their disease, though these aspects may now arise for the first time in other patients whose initial phase was one largely of motor problems.

The third phase, Stage 3, is one of largely complete dependence; how it is managed and where will depend very much on family and social circumstances as discussed in Chapters 6 and 7. Death may occur at any time during this stage, commonly from the causes described below, though in older patients unrelated disorders are often the cause.

While the recognition of the differing neurological features characteristic of the main stages of HD is important for diagnostic purposes, a functional assessment is more relevant in both the assessment of therapy and the planning of services. The assessment scale of Shoulson and Fahn (1979) given in Appendix 3, has proved itself to be a valuable tool for this, while more detailed disability scores are useful in assessing these aspects (see Chapter 6). The Shoulson and Fahn scale is particularly useful in comparative studies. Thus a longitudinal assessment of the Venezuela HD population has shown a similar rate of decline to that seen in the US patients (Young et al., 1986; Penney et al., 1990). Functional decline of juvenile patients using this scale has been shown to be more rapid than that of adults.

CAUSES OF DEATH IN HD

From what do patients with HD actually die? This is a question often asked and not easy to answer, though it may appear unnecessary in the context of a severely disabled and debilitated person with a disorder such as HD. The use of death certificates to assess the frequency of the disease is described in Chapter

8, but they can also provide, though inadequately, information on the causes of death. Haines and Conneally (1986) have examined death certificates for 252 patients on the US National HD roster and showed, not surprisingly, that pneumonia and cardiovascular disease were the commonest causes, both as primary and contributory factors, accounting for 42% and 33% respectively, whereas no other cause (excluding HD itself) reached 10%.

While pneumonia is likely to have occurred in a severely terminally ill and immobile patient, cardiovascular disease could include individuals dying of unrelated causes, such as myocardial infarct, at any stage of the disease.

Significantly this study did not find suicide to be common (3%); the authors attribute this to reluctance to place this on the certificate, but this may not be entirely the case; suicide in HD has already been discussed in Chapter 3. Some deaths grouped as 'accidents' (2%) might belong in this group. Cancer (not sub-divided) was only given as a cause in 3%, a surprisingly low figure considering the number of elderly individuals and the high cigarette consumption of many HD patients. An Australian series (Chiu and Alexander, 1982) based on 182 deaths over a 27-year period, has given very similar results to the American study.

These data essentially confirm what those caring for HD patients have long recognized; there are no complications that are specific to HD, but the combination of immobility, weight loss, tendency to aspirate food and general debility leaves the patient vulnerable to any intercurrent disease. HD itself is the main cause of death, whatever may appear on the certificate.

REFERENCES

Adams P, Falek A and Arnold J (1988) Huntington's disease in Georgia: age at onset. *American Journal of Human Genetics* **43**:695–704.

Bell J (1934) Huntington's chorea. In: *Treasury of Human Inheritance* (ed. RA Fisher). Cambridge: University Press.

Bolt JMW (1970) Huntington's chorea in the West of Scotland. *British Journal of Psychiatry* **116**:259–270.

Brackenridge, CJ (1971) A genetic and statistical study of some sex-related factors in Huntington's disease. *Clinical Genetics* **2**:267–286.

Brothers CRD (1964) Huntington's chorea in Victoria and Tasmania. *Journal of the Neurological Sciences* **i**:405–420.

Cameron D and Venters GA (1967) Some problems in Huntington's chorea. *Scottish Medical Journal* **12**:152–156.

Chiu E and Alexander L (1982) Causes of death in Huntington's disease. *Medical Journal of Australia* **1**:153.

Conneally PM (1984) Huntington's disease: genetics and epidemiology. *American Journal of Human Genetics* **36**:506–526.

Dewhurst K, Oliver JE and McKnight AL (1970) Socio-psychiatric consequences of Huntington's disease. *British Journal of Psychiatry* **116**:255–258.

Farrer LA and Conneally PM (1985) A genetic model for age at onset in Huntington's disease. *American Journal of Human Genetics* **37**:350–357.

Farrer LA and Conneally PM (1987) Predictability of phenotype in Huntington's disease. *Archives of Neurology* **44**:109–113.

Folstein S (1989) *Huntington's Disease. A Disorder of Families*. Johns Hopkins University Press, Baltimore.

Folstein SE, Chase GA, Wahl WE, McDonnell, AM and Folstein MF (1987) Huntington disease in Maryland: clinical aspects of racial variation. *American Journal of Human Genetics* **41**:168–179.

Glendinning N (1975) A study in Huntington's chorea. University of London: MD thesis.

Haines JL and Conneally PM (1986) Causes of death in Huntington's disease as reported on death certificates. *Genetic Epidemiology* **3**:417–423.

Hayden MR (1979) Huntington's chorea in South Africa. University of Cape Town: PhD thesis.

Hayden MR (1981) *Huntington's Chorea*. Berlin: Springer Verlag.

Heathfield KWG (1967) Huntington's chorea. Investigation into the prevalence of this disease in the area covered by the North East Metropolitan Regional Hospital Board. *Brain* **90**:203–232.

Husquinet H (1970) La chorée de Huntington dans les 4 provinces belges. In: *CR 67e Congr Psychiat Neurol Langue franc* (ed. P Warot), pp. 1079–1118. Paris: Masson.

Myers RH, Sax DS and Schoenfeld M (1985) Late onset of Huntington's disease. *Journal of Neurology, Neurosurgery and Psychiatry* **48**:530–534.

Newcombe RG (1981) A life table for onset of Huntington's chorea. *Annals of Human Genetics* **45**:375–385.

Newcombe RG, Walker DA and Harper PS (1981) Factors influencing age at onset and duration of survival in Huntington's chorea. *Annals of Human Genetics* **45**:387–396.

Oliver JE (1970) Huntington's chorea in Northamptonshire. *British Journal of Psychiatry* **116**:241–253.

Panse F (1942) Die Erbchorea; eine klinisch-genetische Studie. *Samml Psychiat Neurol Einzeldarst 18*. Leipzig: Thieme.

Penney JB, Young AB and Shoulson I (1990) Huntington's disease in Venezuela: 7 years of follow-up on symptomatic and asymptomatic individuals. *Movement Disorders* **5**:93–99.

Petit H (1970) La maladie de Huntington. In: *CR 67e Congr. Psychiat. Neurol. Langue franc* (ed P Warot), pp. 901–1058. Paris: Masson.

Pridmore SA (1990a) Age of onset of Huntington's disease in Tasmania. *Medical Journal of Australia* **153**:135–137.

Pridmore SA (1990b) Age of death and duration in Huntington's disease in Tasmania. *Medical Journal of Australia* **153**:137–139.

Reed TE, Chandler JH, Hughes EM and Davidson RT (1958) Huntington's chorea in Michigan: 1 demography and genetics. *American Journal of Human Genetics* **10**: 201–225.

Shoulson I and Fahn S (1979) Huntington's disease: clinical care and evaluation. *Neurology* **29**:1–3.

Stevens DL (1976) Huntington's chorea: a demographic genetic and clinical study. University of London: MD thesis.

van Dijk, van der Velde EA, Roos RAC and Bruyn GW (1986) Juvenile Huntington's disease. *Human Genetics* **73**:235–239.

Venters G (1971) Epidemiology of Huntington's chorea. PhD Thesis, Edinburgh.

Walker DA (1979) Huntington's chorea in South Wales. University of Liverpool: MD thesis.

Walker DA, Harper PS, Wells CGC, Tyler A, Davies K and Newcombe RG (1981) Huntington's chorea in South Wales. A genetic and epidemiological study. *Clinical Genetics* **19**:213–221.

Wallace DC (1972) Huntington's chorea in Queensland. A not uncommon disease. *Medical Journal of Australia* **1**:299–307.

Wendt GG, Landzettel I and Unterreiner I (1959) Erkrankungsalter bei der Huntingtonschen Chorea. *Acta Genetica* **9 (Basel)**:18–32.

Wendt GG and Drohm D (1972) *Die Huntingtonsche Chorea. Eine Populations Genetische Studie*. Stuttgart: Thieme.

Young AB, Shoulson I and Penney JB (1986) Huntington's disease in Venezuela: neurologic features and functional decline. *Neurology* **36**:244–249.

5

The Neurobiology of Huntington's Disease

INTRODUCTION

The preceding chapters have attempted to describe and analyse the clinical features of Huntington's disease (HD) from the viewpoint of the neurologist and psychiatrist. The range of symptoms and variability in the natural history of the disease have been discussed, along with its relationship to other allied neurological and psychiatric disorders. In this chapter we examine what is known of the biological basis of these features, describe experimental studies that have increased neurological understanding of the disorder, and attempt to relate normal structure and function with the abnormalities seen in HD. As will be seen, this is no easy task and knowledge is still very far from complete. While it is known that the changes in HD result primarily from the mutation of a specific gene on chromosome 4, it is currently impossible to trace the chain of events that leads from this to disturbed neuronal structure and function and hence to the clinical features of HD.

The earliest experimental studies to be undertaken in HD were observations on the pathology of the diseased brain and comparisons with its normal counterpart. Initially, as discussed in Chapter 1, there was considerable confusion as to the areas of the brain involved and the types of change that occurred; however, the early part of the twentieth century saw a recognition of the particular involvement of the basal ganglia and the essentially atrophic rather than inflammatory nature of the changes. Quantitative and ultrastructural studies of the neuropathology, together with careful correlative studies between pathology and clinical features, have added important information.

More recently, the emphasis has shifted towards analysis of the physiology and pharmacology of the basal ganglia, with experimental animal studies providing the majority of the information. Our knowledge of the neurochemistry of the basal ganglia and of the specific neurotransmitters and receptors involved, has now reached the stage where suggestions can be proposed to explain some of the changes in HD; as yet we are unable either to confirm or refute these.

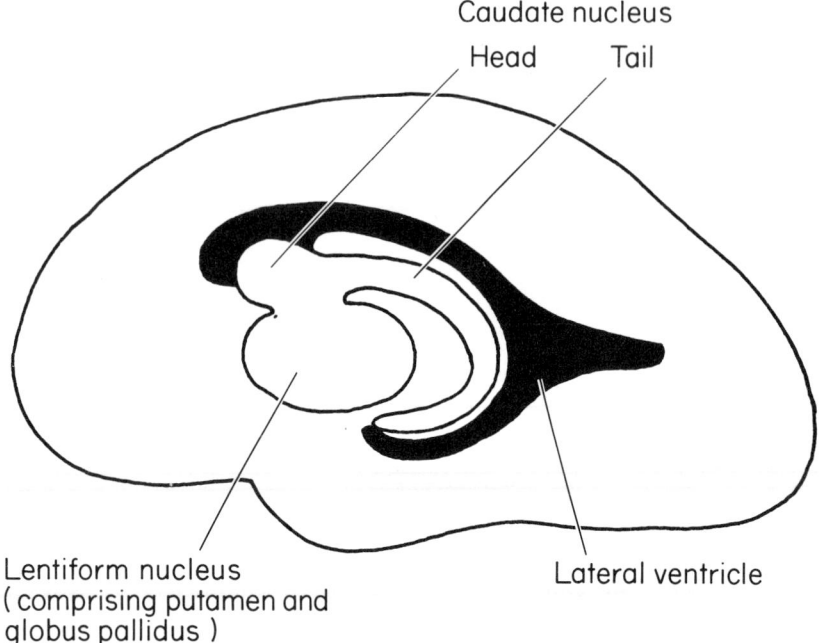

Figure 5.1 Longitudinal section of the developing human brain to show the 'C'-shaped development of the caudate nucleus and relative position of the lentiform nucleus (putamen and globus pallidus). Reproduced from Moore, *The Developing Human*, WB Saunders 1988, with permission.

Some of the experimental electrophysiological studies will be summarized; however, a final area of research that will not be discussed at any length is the study of peripheral tissues. Most of this work has proved either negative or contradictory and has reinforced the view that whilst the HD gene is present in all tissues its principal effect lies within the brain. Likewise it is to the disordered brain that one must look for an explanation of the abnormalities that are characteristic of HD.

NEUROANATOMY OF THE BASAL GANGLIA

The caudate nucleus, which arises from the floor of the expanding cerebral hemispheres, becomes 'C'-shaped during embryonic development and classically is divided into head and tail regions (Figure 5.1). The use of several terms describing the same structure may cause confusion; the corpus striatum includes the caudate nucleus, putamen and globus pallidus; the striatum or neo-striatum refers to the caudate nucleus and putamen. The ventral anterior part of the striatum, where the caudate nucleus and putamen fuse, is termed the nucleus accumbens. The term lentiform nucleus is sometimes used to describe the globus pallidus and putamen together.

The caudate nucleus and putamen may be considered as similar structures separated by white matter. The globus pallidus is often divided into medial and lateral (or internal and external) segments. The medial globus pallidus and substantia nigra pars reticulata contain a very similar distribution of neurones and may be considered as parts of the same structure.

PATHOLOGICAL CHANGES

Some of the historical aspects of the pathological changes described in HD have been discussed in Chapter 1. A good historical review was provided by Stone and Falstein (1939); in addition to their own six cases they were able to review the neuropathological findings of 159 published cases. Their last sentence remains a very appropriate summary: 'Huntington's chorea is an hereditary degenerative disease due to premature atrophy of the cerebral cortex and neo-striatum'.

A valuable summary of the pathological changes that occur in HD has been provided by Roos (1986) who commented that '... the diagnosis of Huntington's chorea can be posed with a fair degree of certainty on the basis of the morphological changes. Clinical and family history are indispensable to confirm the diagnosis of Huntington's chorea'.

Macroscopic inspection of the brain shows the leptomeninges to be thickened and slightly opaque. The brain of an advanced HD patient is smaller than that of age-matched controls (Figure 5.2), the weight being reduced by 200–300 g. Atrophy of the frontal lobes is often more apparent than in other areas. Coronal section through the anterior horns of the widened lateral ventricles shows the most striking macroscopic abnormalities with the caudate nucleus often being reduced to a rim of tissue; the putamen and globus pallidus are also atrophic. These changes are illustrated in Figures 5.3–5.5. Although cortical grey matter and subcortical white matter is lost, there is general shrinkage of brain tissue so cortical atrophy is less apparent macroscopically. Ventricular dilatation occurs to a greater extent than can be explained solely on the basis of striatal loss (de la Monte *et al.*, 1988); consequently, the macroscopic appearance of striatal loss is more striking.

The pattern of abnormalities seen in Figure 5.6 would, by itself, be suggestive of HD, but not diagnostic. Demonstration by light microscopy of neuronal loss in the striatum with gliosis together with an appropriate clinical history and absence of Pick bodies would be essential in order to confirm the diagnosis (Figures 5.7–5.9). Whilst atrophy of the caudate and putamen is the most characteristic feature (Figures 5.5 and 5.6), the pathological process is not restricted to the striatum, a point perceptively made by Bruyn, 1973: '... Huntington's chorea is the result of a pathological process involving not only the striatum, but to no less impressive degree, the cortex, the white matter, interhemispheric fibre systems, association fibre systems, and diencephalic nuclei'. The findings for various relevant areas of the brain are listed in Table

5.1, which is based largely on the text of Roos (1986). Some of the important quantitative studies are described in more detail below.

McCaughy (1961) commented that the number of striatal neuronal cells per high power field fell from the usual 30 to less than five and noted a greater loss of small cells, but more detailed quantitative analysis was not established until the 1970s. Two groups, Dom *et al.* (1973) and Lange *et al.* (1976) undertook detailed cell counts of various nuclei involved in the pathological process; this represented a landmark bridging qualitative and quantitative studies. The techniques are time consuming so it is not surprising that there were relatively

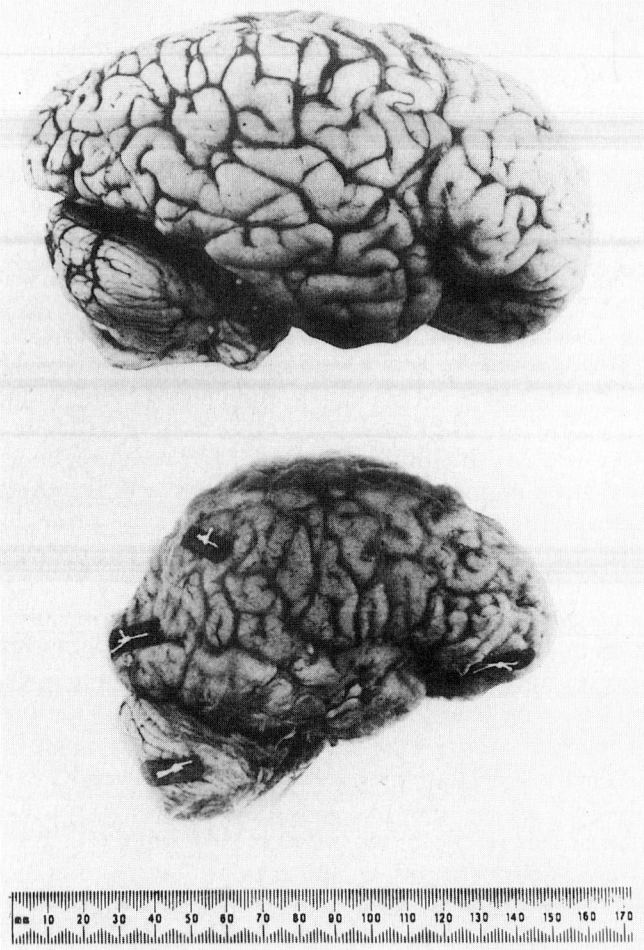

Figure 5.2 Lateral view of a 39-year-old female normal brain (above) compared with the same view of a brain from a female patient of the same age with advanced HD. Note the cortical atrophy in the frontal lobes in the HD brain (below). Courtesy of Dr Adrian Caro, Norfolk.

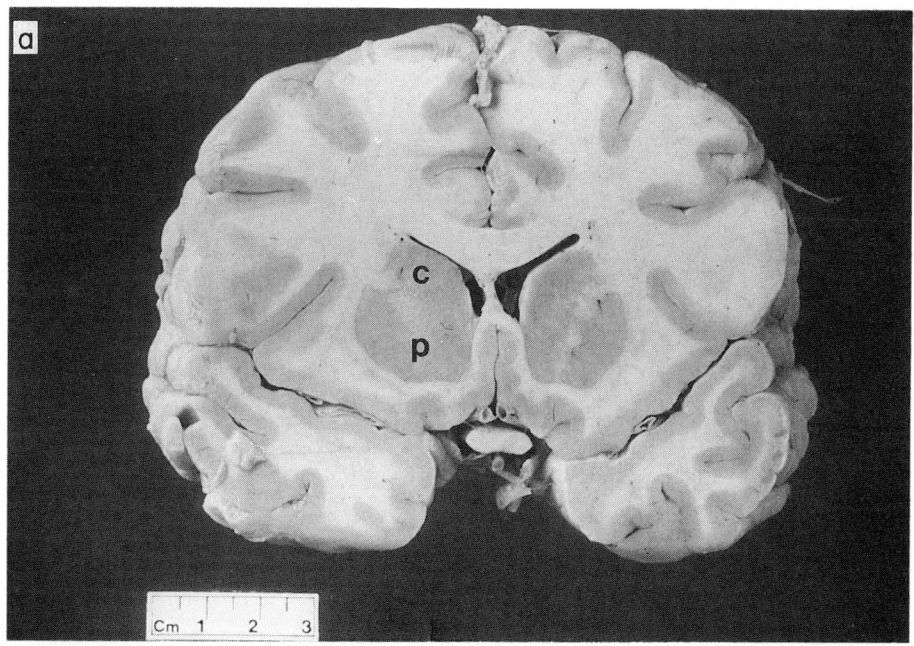
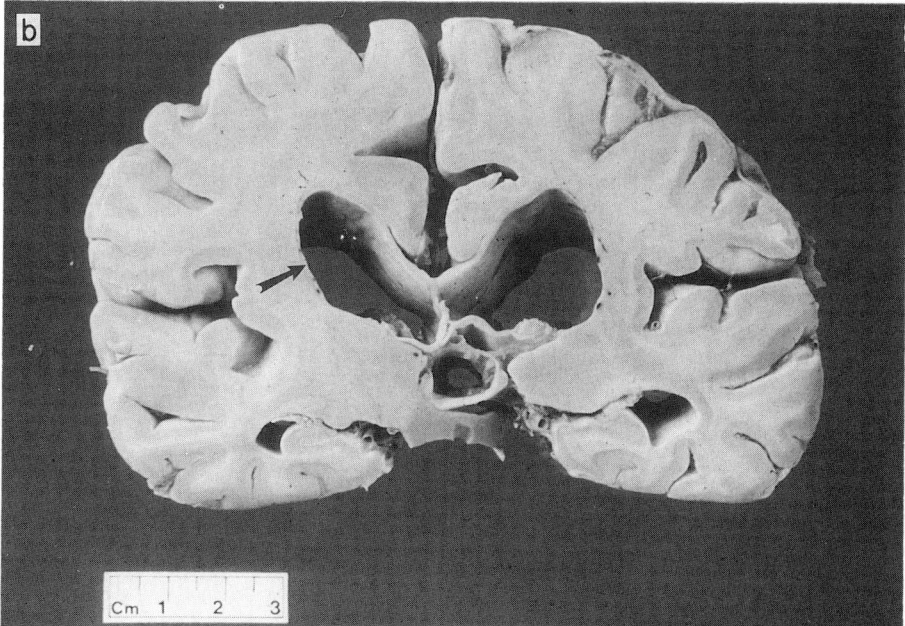

Figure 5.3 (a) Coronal section through the frontal lobe of a normal brain to show the caudate nucleus (C) and putamen (P). (b) Coronal section through HD brain, Note the degree of atrophy in both the caudate nuclei (arrow) and the cerebral cortex. Courtesy of Drs J Neal and G Cole, Cardiff.

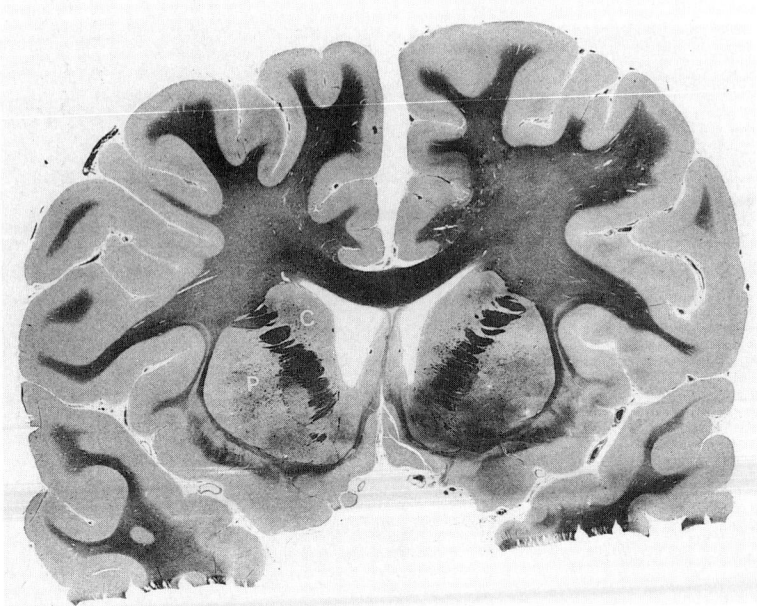

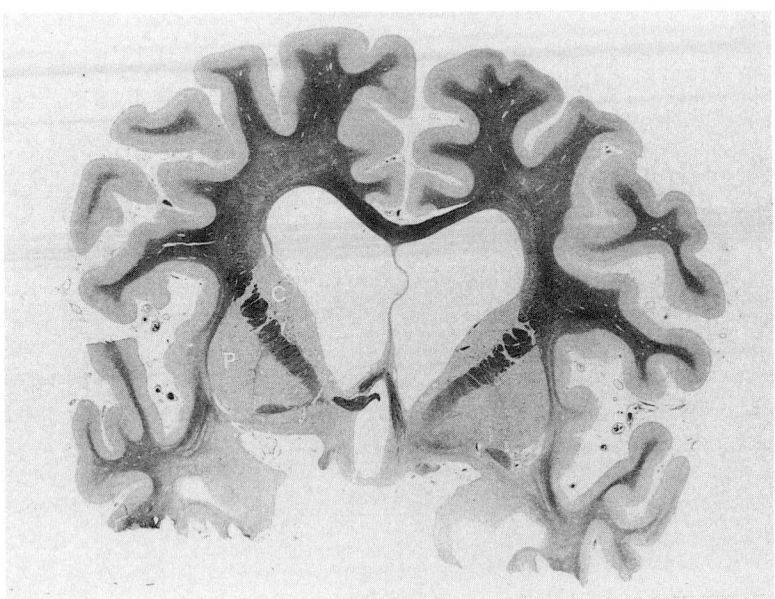

Figure 5.4 (a) and (b) Coronal sections through a normal brain. (a) is taken through the frontal lobe, while (b) is taken at a more posterior level. In both sections the caudate nuclei (C) and putamen (P) are shown. Phosphotungstic acid-haematoxylin (PTAH) stain. Courtesy of Drs J Neal and G Cole, Cardiff.

The Neurobiology of Huntington's Disease   147

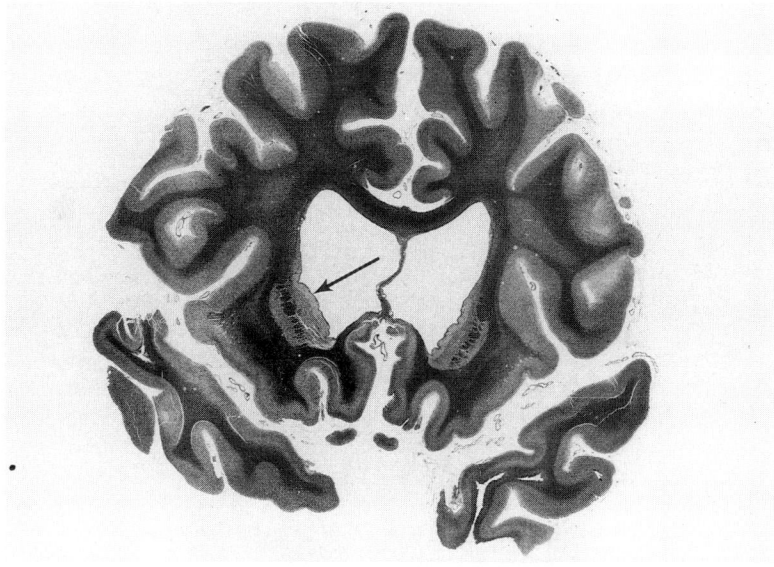

Figure 5.5 (a) and (b) These two photomicrographs show coronal sections from a brain with HD at different antero-posterior levels. (a) is a coronal section through the frontal lobe of the brain and shows atrophy of both caudate nuclei at this level. (b) is taken at a more posterior level and shows atrophy of the caudate nucleus (arrow). Compare these figures with the normal appearance as shown in Figures 5.4(a) and (b). Courtesy of Drs J Neal and G Cole, Cardiff.

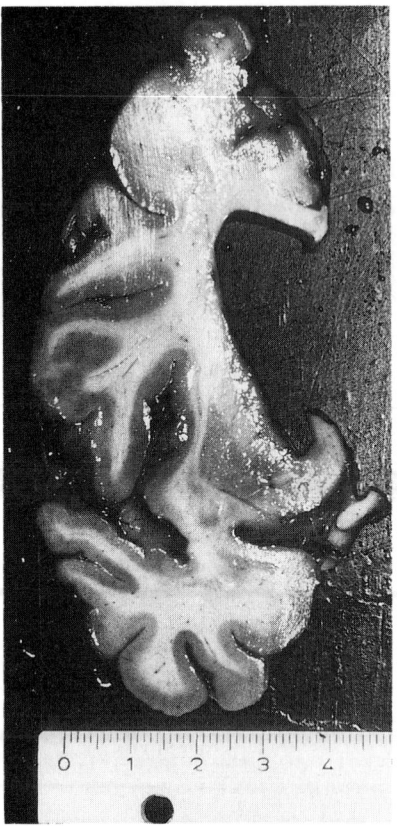

Figure 5.6 Coronal section through a brain from a patient with HD. Note the degree of atrophy in the caudate nucleus. (Courtesy of Dr Raymund Roos, Leiden.)

few samples in each study and that cell counts are not normally available as part of neuropathological service.

Dom *et al.* (1973) studied the putamen from four normal brains; the ratio of large to small neurones was 1:145. In the five cases of adult-onset HD there was a reduction in cell density for both large and small neurones, with much greater loss of the latter. Consequently the ratio of large to small neurones increased to 1:30 for those with bradykinesia as a prominent sign and 1:26 for those with chorea as the principal feature. Only one case of juvenile HD was studied; the loss of both large and small neurones was greater than that seen in the adult cases but the large to small neurone ratio was less dramatically altered, being 1:127.

Lange *et al.* (1976) undertook a more extensive study; they examined 13 normal brains and six cases with HD. They reported their results for the striatum as a whole and noted a 58% reduction in its volume in HD. The decrease in the absolute number of small neurones (70–80%) was more

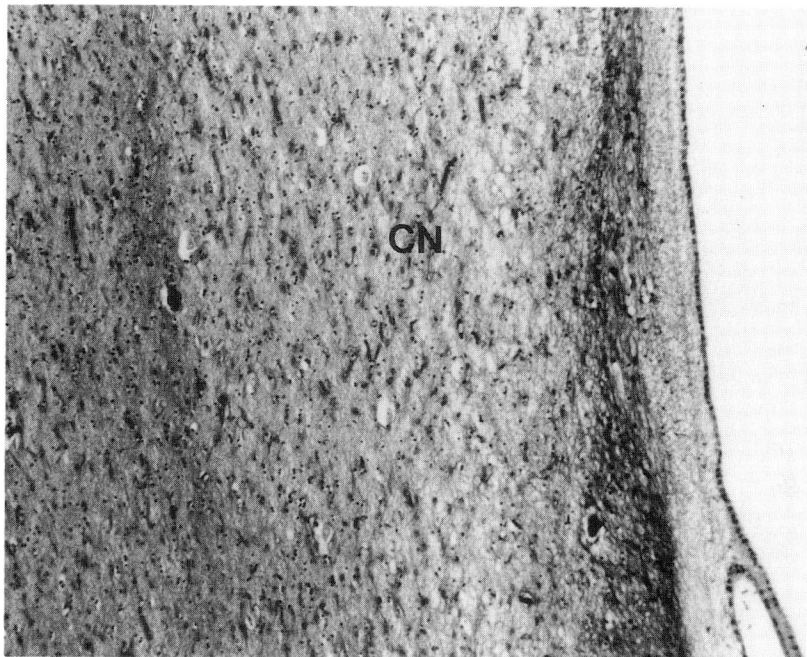

Figure 5.7 Photomicrograph to show the appearance of the caudate nucleus (CN) in HD; there is evidence of increased gliosis, together with some vacuolation of this nucleus. (PTAH stain × 250.) Courtesy of Drs J Neal and G Cole, Cardiff.

dramatic than the reduction of cell density (36–50%) because of shrinkage of the striatum. They also reported that the absolute number of large cells was not decreased, so the numerical cell density of this group increased; however, the large to small cell ratio increased from 1:145 in normal brains to 1:40 in those affected by HD, a figure in close agreement with the earlier study of Dom et al. (1973).

Lange et al. (1976) also reported a reduction in the absolute number of glial cells within the striatum but because of the greater reduction in neurones the glial index (number of glial cells per nerve cell) was increased to 169% of normal. This finding is significant in demonstrating that the usual finding of gliosis in the striatum of the HD brain is a relative rather than absolute phenomenon (Figures 5.7–5.10).

The effect of HD on the large cell population of the striatum has received further attention; Oyanagi and Ikuta (1987) noted that in four HD brains the number of large neurones was about 25% that of controls. Although numbers were small, variation between the cases in numbers of large neurones did not correlate with either the clinical phenotype or loss of small cells. It is probably correct to say that there is loss of both large and small cells in the striatum but that small-cell loss predominates (Figure 5.10b).

Table 5.1 Light microscopic findings in Huntington's disease

Cerebral cortex	Neuronal loss and astrogliosis in 3rd to 5th layers. Increased lipofuscin and lipid laden macrophages in the perivascular spaces. (Tellez-Nagel *et al.*, 1973)
Striatum	Extensive (70%) loss, especially of small neurones. Normal ratio of large to small neurones increased (Dom *et al.*, 1973; Lange *et al.*, 1976). There is an absolute loss of astroglial cells but the ratio glial to nerve cells is increased (Lange *et al.*, 1976).
Globus pallidus	Neuronal loss and astrogliosis especially in the external part
Thalamus	Microneuronal loss especially in the ventrolateral nuclei
Subthalamus	Neuronal loss particularly in the lateral and medial parts (Lange *et al.*, 1976)
Substantia nigra	Neuronal loss especially in the pars reticulata
Cerebellum	Loss of Purkinje cells and cells of dentate nucleus most marked in juvenile cases (Jeste *et al.*, 1984)
Brain stem	Nerve cell loss in the olives, gracile and cuneate nuclei
Spinal cord	Gliosis of the posterior columns, pallor of the anterior and lateral columns
Lateral tuberal nucleus of the hypothalamus	Neuronal cell loss and gliosis similar to the pattern seen in the striatum (Kremer *et al.*, 1990)

Lange *et al.* also studied the globus pallidus and found a 50% reduction in volume; a reduction occurred in both the absolute number of nerve cells (44% and 43% for the lateral and medial parts respectively) and nerve cell density (42% and 27% for the lateral and medial parts respectively) with an increase in the glial index (60% and 53% for the lateral and medial parts respectively). Similarly the subthalamic nucleus showed a 23% reduction in volume as well as a reduction in the absolute number of nerve cells which did not reach statistical significance.

A quantitative study of the thalamus in seven cases of HD by Dom *et al.* (1976) demonstrated atrophy in the ventrolateral nuclei. Two populations of neurones were distinguished; macroneurones and microneurones. The degeneration was selective for the microneurone population. The percentage of microneurones fell from 23.5% in the normal brains to 11.4% in the HD brains.

In 1989, Oyanagi *et al.* quantified changes in the substantia nigra of four patients with HD compared with eight controls. There was a 57% reduction in the cross sectional area of the substantial nigra, with a 40% reduction in neuronal number; there was atrophy of both the pars compacta and pars reticulata. The loss of both pigmented (dopaminergic) and non-pigmented (GABAergic) neurones meant that the normal 2:1 ratio for these two groups was unchanged between HD and control brains. Pigmented neurones in the

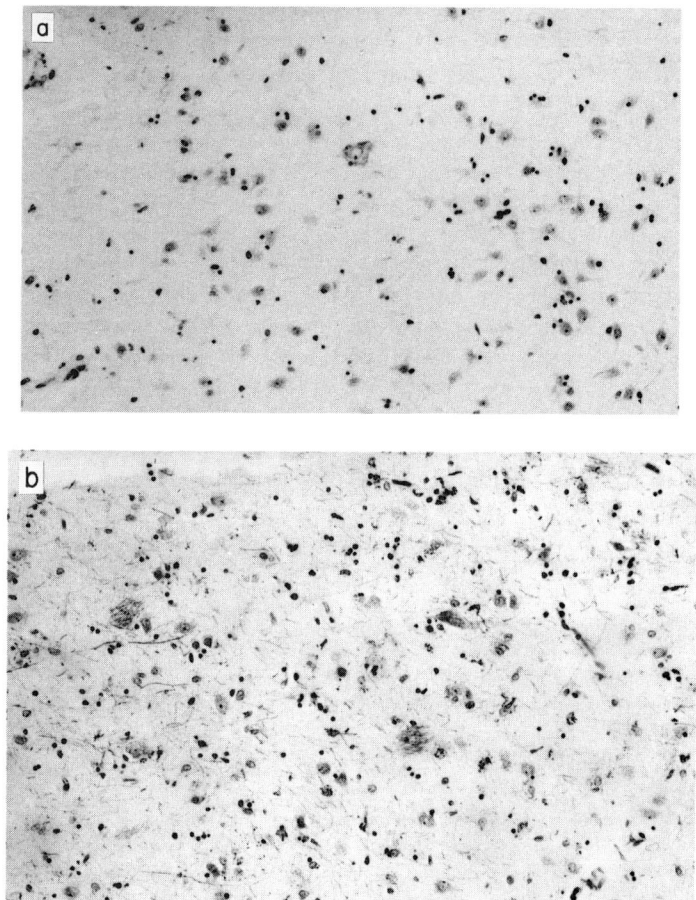

Figure 5.8 A high power photomicrograph of the normal caudate nucleus (a) as compared with the caudate nucleus in HD (b). In the caudate nucleus in (b) note the increase in the number of astrocytes as compared with (a). (Kluver–Barrera stain × 320.) Courtesy of Dr R Roos, Leiden.

central part of the substantia nigra were relatively spared, whereas there was a more uniform loss of the non-pigmented neurones. The authors suggested that the degeneration in the substantia nigra did not correspond to the topographical arrangement of fibres to and from the striatum. The authors concluded that degeneration within the substantia nigra was a primary process.

The loss of Purkinje cells within the cerebellum and loss of neurones in the dentate nucleus has already been mentioned in relation to cerebellar signs in Chapter 2. Certainly post-mortem studies of some juvenile cases have shown unequivocal abnormalities of the cerebellum; case 7 of Markham and Knox (1965) had loss of Purkinje cells and gliosis of the molecular layer. Case 4 of Jervis (1963) had considerable dysmetria and coarse tremors; pathologically there was cerebellar atrophy, absent Purkinje cells, and atrophy of the dentate

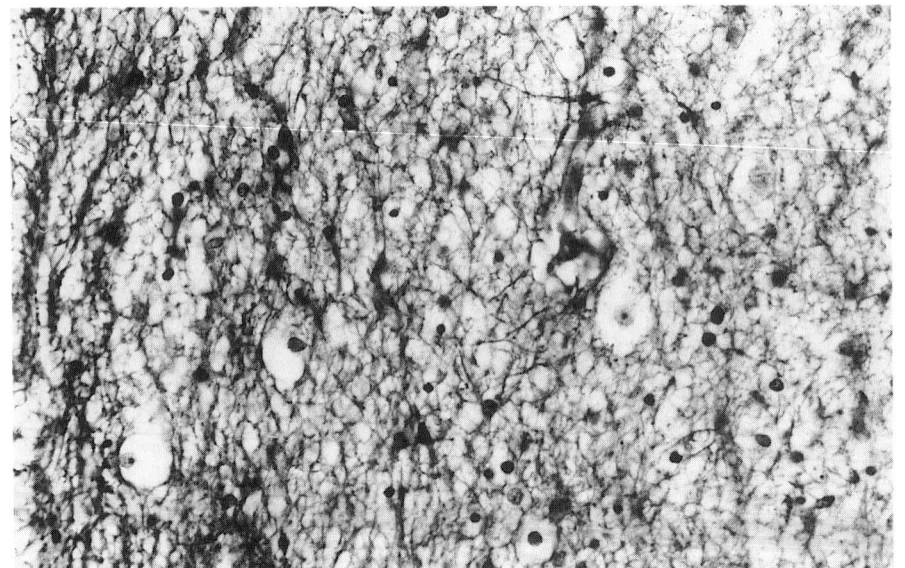

Figure 5.9 A high power photomicrograph of the caudate nucleus in HD, stained with PTAH, to show the increase in the number of dark glial fibres, astrocytes and vacuolation of the nucleus. (PTAH × 340.)

nucleus; his brother, case 3, had less marked cerebellar atrophy. Byers et al. (1973) also reported similar neuropathological findings in four children.

Rodda (1981) reviewed 300 neuropathological examinations of HD patients from one centre and found only three – two adults and a child, with cerebellar abnormalities. The child presented with falls, dysarthria, epilepsy and chorea. Both adults presented in their 50s; the first with incoordination, coarse tremor and dysarthria; the second had incoordination, ataxia, slow slurred speech, a fine nystagmus and marked dysdiadochokinesis. The disease progressed to give chorea and dementia in the first case and mixed chorea and hypertonicity in the second. More recently, Jeste et al. (1984) have undertaken detailed cell counts from HD and control brains and demonstrated a significant loss of Purkinje cells in 50% of the patients, suggesting that pathological involvement of the cerebellum may occur more frequently than had previously been supposed.

A pathological basis has been sought for the observed abnormalities of saccadic eye movements which occur in HD. As noted in Chapter 2 this abnormality occurs in other neurodegenerative disorders; focal lesions in the pons produce slowing of horizontal saccades and lesions in the midbrain produce slowing of vertical saccades. The recognition, neurophysiologically, of a group of cells, called burst cells, located in the rostral interstitial nucleus of the medial longitudinal fasiculus (riMLF) of the midbrain which discharge immediately prior to vertical saccades led Leigh et al. (1985) to undertake cell counts in this region in four HD patients who had abnormalities of both

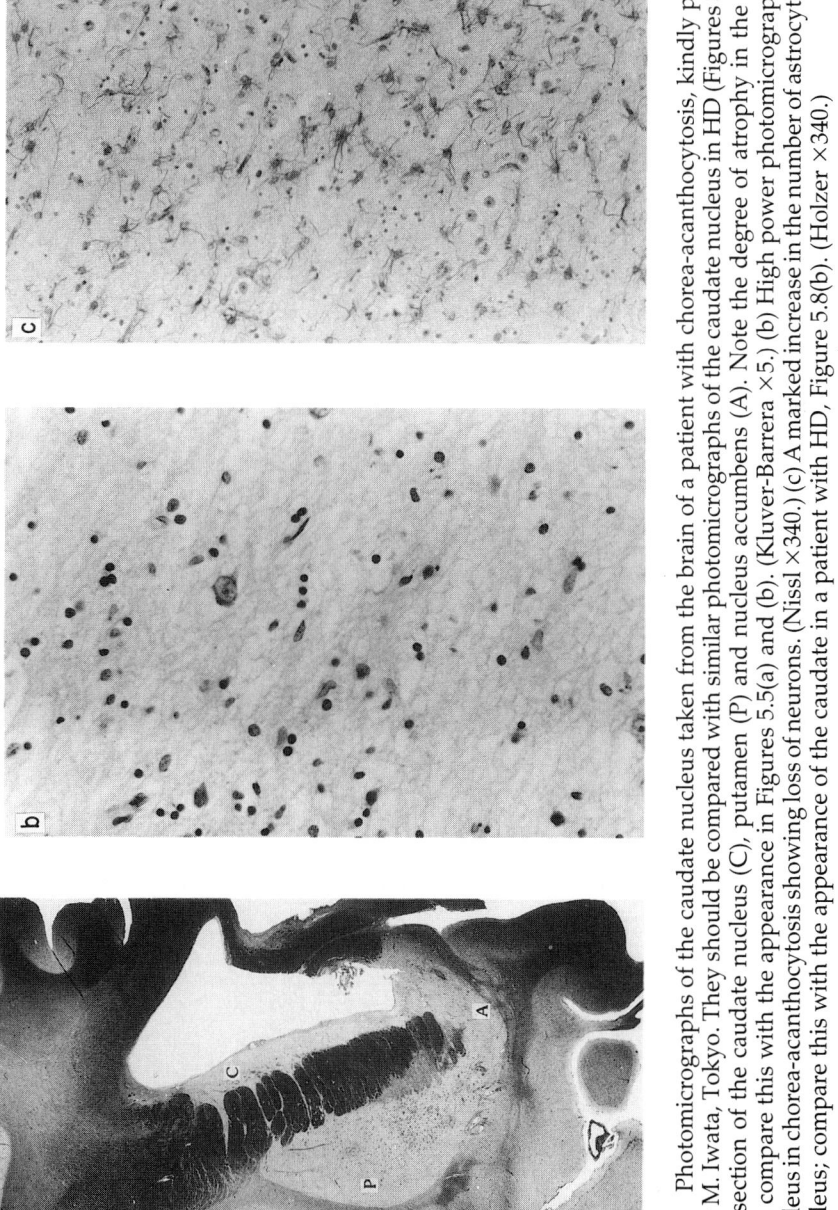

Figure 5.10. Photomicrographs of the caudate nucleus taken from the brain of a patient with chorea-acanthocytosis, kindly provided by Professor M. Iwata, Tokyo. They should be compared with similar photomicrographs of the caudate nucleus in HD (Figures 5.5–5.8). (a) Coronal section of the caudate nucleus (C), putamen (P) and nucleus accumbens (A). Note the degree of atrophy in the caudate nucleus and compare this with the appearance in Figures 5.5(a) and (b). (Kluver-Barrera ×5.) (b) High power photomicrograph of the caudate nucleus in chorea-acanthocytosis showing loss of neurons. (Nissl ×340.) (c) A marked increase in the number of astrocytes in the caudate nucleus; compare this with the appearance of the caudate in a patient with HD, Figure 5.8(b). (Holzer ×340.)

horizontal and vertical saccades. The HD patients had a reduced number of neurones but in only one patient was loss of larger neurones in the riMLF significantly different from the controls. The authors suggested that their results should be interpreted with caution because of general difficulties in obtaining cell counts in this region. They concluded that slow saccades may be due to altered metabolism in the burst cell, loss of burst cells and/or abnormal inputs to the burst cell population.

For very similar reasons Koeppen (1989) studied the nucleus pontis centralis caudalis in nine HD patients. Two of these showed abnormal saccadic eye movements, but of the others only two underwent detailed study, the remaining 5 being recorded as normal in the clinical notes. The area of the nucleus pontis centralis caudalis was reduced in 8/9 compared with controls. (The patient in whom this measurement was normal had developed HD only 2 years prior to death.) In addition there was loss of neurones, particularly large neurones, leading to the suggestion that this nucleus is regularly involved in the pathology of HD and may in part account for saccadic eye movement abnormalities seen in HD.

NEUROPATHOLOGICAL GRADING

There may be variation in the neuropathological features of HD depending on the time of death in relation to the course of the illness. Vonsattel et al. (1985) proposed a classification based on both the macroscopic and microscopic appearances of brains from 163 patients diagnosed clinically as HD. Five groups were defined and graded 0–4; the main characteristics for each group are summarized in Table 5.2 and the macroscopic appearances associated with each grade are illustrated in Figure 5.11. The five patients described as Grade 0 were diagnosed, in life, as having HD but no neuropathological findings were

Table 5.2 Neuropathological classification based on studies of the caudate putamen and globus pallidus

Grade	Macroscopic features	Microscopic features
0*	Normal	Normal
1	Normal	Some neuronal loss and astrocytosis in caudate and putamen
2	Caudate atrophy Medial surface concave	Neuronal loss and astrocytosis in caudate, putamen
3	Caudate and putamen atrophic Medial surface straight	As above but globus pallidus shows degenerative changes
4	Atrophy of caudate, putamen and internal capsule Medial surface concave	Severe neuronal loss and astrocytosis in caudate, putamen and globus pallidus

*Neurological signs and family history present before death.

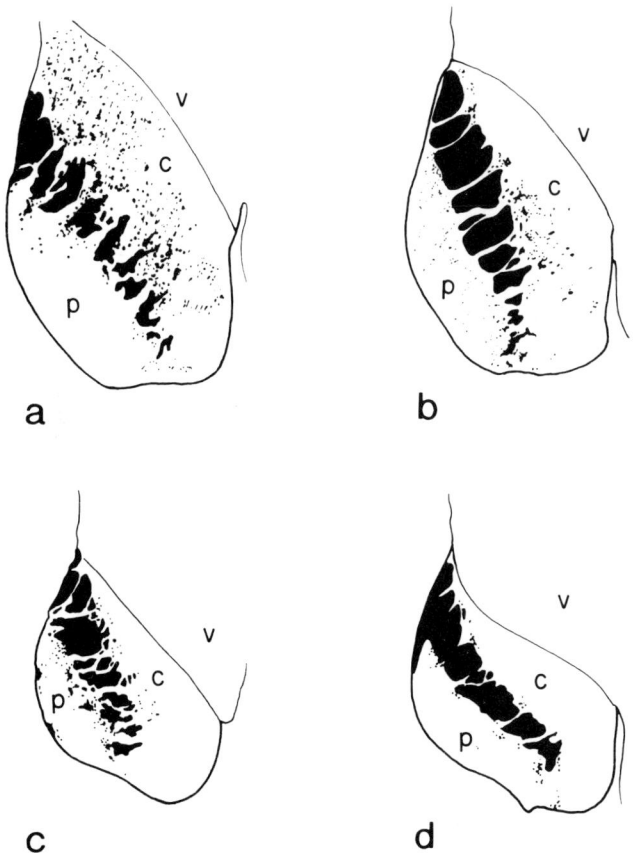

Figure 5.11 Line drawing of shape of the striatum in HD as the disease progresses. Based on the pathological grading of Vonsattel *et al.* (1985) (see text). C = caudate, P = putamen, V = ventricle. (a) Normal, Grade 0 or 1. (b) Grade 2, note slight atrophy. (c) Grade 3, note straight outline to the caudate nucleus. (d) Grade 4, note concave outline to the caudate and pyramidal fibres.

noted at the post-mortem examination. The distribution of the patients in Grades 1–4 was 8, 38, 90 and 18 respectively.

Neuronal cell counts of the head of the caudate nucleus were undertaken from only a few patients in each of the five groups: patients with Grade 1 and 2 HD had a 50% loss of neurones and those with Grade 4 HD had a 95% loss. It is interesting to note that routine examination of patients with Grade 1 HD would suggest only minimal neuronal loss; the real extent of the neuropathological damage was only revealed by cell counting. The pathological grading of brains in the earlier quantitative studies is unclear; Vonsattel *et al.* suggested that most were Grade 3–4.

In a later paper Myers *et al*. (1988) discussed the relationship between some of the clinical features and the proposed grading system. Not surprisingly, most of the patients in Grades 0–2 died as a result of either accidents or suicide. The neuropathological grading correlated only weakly with the duration of the illness; a greater correlation existed between the grading and the age of onset of symptoms, those with an earlier age of onset having the higher grades. Cell counts from the striatum were available for 23 HD patients and seven controls; as might be expected there was a significant relationship between the severity of the neuronal loss and the degree of clinical disability (the latter was assessed using a scale published in the same paper). The grading scale allowed an assessment of regional variation of neuropathological changes within the striatum. Changes were most noticeable in the tail of the caudate nucleus in Grade 1 cases; the paraventricular portion of the caudate and dorsal putamen showed early neuronal loss and gliosis; by the time Grade 4 had been reached there was widespread change throughout the striatum.

In a separate study of neuronal cell counts from 11 HD patients and nine controls Roos *et al*. (1985) were able to confirm the relative sparing of the ventral anterior part of the striatum termed the nucleus accumbens. This pattern of neuropathological damage is consistent with present concepts of topographic organization within the striatum; sensorimotor input is predominantly to the dorsal striatum, putamen and caudate nucleus; whereas inputs from the limbic system predominantly pass to the nucleus accumbens.

The clinical details of the five patients with Grade 0 HD were summarized by Myers *et al*. (1988). Their ages ranged from 36 to 69 years and the duration of signs and symptoms ranged between 2 and 13 years. In all five cases the family history was unequivocal and there were documented neurological abnormalities on reviewing medical records. Clearly, the existence of Grade 0 HD and need for cell counts to reveal the true extent of the neuronal loss in Grade 1 HD is suggestive that the gross pathological features lag behind the clinical features. This conclusion has considerable practical as well as theoretical implications but cannot be considered as an invariable rule in the light of two exceptional cases which have been reported.

In their original paper describing the grading system, Vonsattel *et al*. (1985) described a 73-year-old man, at risk for HD, who committed suicide. The family considered he had shown subtle signs of HD but he had never come to medical attention. General neuropathological examination was unremarkable but cell counts were in the HD range. A more worrying case has been reported by Carrasco and Mukherji (1986); a 48-year-old man, at risk for HD had an accidental death; his health, personality and intellect were said to be normal but it is unclear whether this statement was based on reports by the family or as a result of detailed medical examination. Given the very unusual nature of this case, it is most unfortunate that more clinical details were not provided. None the less, on sectioning the brain there was macroscopic evidence of striatal and pallidal atrophy and light microscopy showed neuronal loss and gliosis in these areas. As these changes are typical of HD, it is highly likely that he had inherited the HD gene. Interestingly, the GABA levels in the basal ganglia of this patient have been reported to be intermediate between those seen in normal controls and patients with established HD (Reynolds and Pearson, 1990).

Interpretation of these cases is difficult. In general, detailed neuropathological analysis of asymptomatic individuals, at risk for HD, dying from incidental causes is unavailable. Although Vonsattel et al. suggested the need for striatal cell counts in such cases it has to be admitted that firstly, detailed neuropathology is seldom undertaken in this group because the pathologist is unaware of the family history and secondly, cell counts are still considered to be of research value only. Confirmation of the presence of the HD gene in the above cases will have to await either transmission to the subsequent generation or specific mutation analysis when this becomes available. The need to document as many clinical, genetic and pathological data on those relatives who die and who could be either presymptomatic or in the early stages of HD is especially important in relation to predictive testing for relatives, as discussed in Chapter 12.

ULTRASTRUCTURAL CHANGES

The light and electron microscope (EM) findings in the frontal cortex from four patients who had brain biopsies were presented by Tellez-Nagel et al. (1973). Lipofuscin pigments were noted to have accumulated in both neuronal and glial cells. In those neurones which did not have prominent lipofuscin pigment there was an increased amount of smooth endoplasmic reticulum and vesicles associated with the Golgi complexes. Mitochondrial abnormalities (disrupted cristae, lamellar figures and peculiar tubular arrangements) were also noted in neuronal and glial cells. There was evidence of axonal degeneration and degeneration of pre-synaptic endings. Roizin et al. (1979) studied 18 cerebral cortex and caudate nucleus biopsies from HD patients; they drew attention to abnormalities of the nuclear membrane, disorganization of the nucleolus and irregularities of the rough endoplasmic reticulum with depletion of ribosomes. It is uncertain how these changes relate to the underlying pathology of HD.

Bots and Bruyn (1981) reported the presence of nuclear membrane indentations in neurones within the nucleus accumbens, the most rostral portion of the caudate nucleus, which is least affected by HD. This aspect was studied in greater detail by Roos (1984); the percentage of neurones from the caudate nucleus and nucleus accumbens with nuclear indentations was significantly greater in HD patients as compared with controls; the converse was true for neurones from the frontal cortex. The indentations were not thought to be artefacts of ageing or fixation and Roos considered them to represent a consequence of greater metabolism involving interaction between the nucleus and cytoplasm which, in the case of HD, represents a reponse by the neurone to the underlying pathological process.

Forno and Norville (1979) undertook observations of EM findings in the striatum of four HD patients; they stated that the most striking finding was of unmyelinated axons with synaptic vesicles without synaptic connections. They considered a likely explanation to be that afferent axons had lost their connections with striatal neurones.

SELECTIVE NEURONAL LOSS IN THE STRIATUM

Morphological studies

Although neuronal loss occurs in several areas of the HD brain, the loss may still be considered selective because not all neurones are equally susceptible to the pathological process, as demonstrated by the light microscopic finding of predominant small cell loss in the caudate nucleus. The hope of finding a specific neuronal pathway which degenerates, analogous to the situation seen in Parkinson's disease, has not been realized. A considerable research effort has focused on defining which neurones are most susceptible to the pathological process and which are spared.

Using the Golgi impregnation technique, which is not routinely used in neuropathological examinations, Graveland *et al.* (1985a) classified neurones from the normal human striatum (28 individuals) into five groups according to size, morphology and the presence or absence of dendritic spines. Spiny type I neurones were of medium size with spine-rich dendrites, while spiny type II were of medium and large size with sparsely spined dendrites. Aspiny type I were of medium size and aspiny type II were large neurones. The fifth group comprised small cells with variable morphology. This classification does not directly compare with the large cell/small cell distinction of earlier morphometric studies (Dom *et al.*, 1973; Lange *et al.*, 1976). Spiny neurones form the majority of the cells in the striatum and are preferentially lost in HD (Graveland *et al.*, 1985b). A representation of a spiny neurone is shown in Figure 5.12.

Neurotransmitter studies

Early work described the levels of the classical neurotransmitters GABA, acetylcholine and dopamine in the HD brain and later studies have considered additional neurotransmitters. Table 5.3 summarizes current thoughts regarding selective neurotransmitter changes within the striatum.

Following the observation of reduced amino acid levels in cerebrospinal fluid (CSF) from HD patients, Perry *et al.* (1973) investigated levels of various amino acids and related compounds from different brain regions at post mortem. The most striking finding was reduced levels of the inhibitory neurotransmitter GABA (gamma-aminobutyric acid) in the caudate, putamen, globus pallidus and substantia nigra. These authors recognized that the genetic defect need not necessarily be related to GABA metabolism; the results were interpreted as identifying a subset of neurones which was particularly susceptible to the basic pathological process. Further support for this observation came from the report

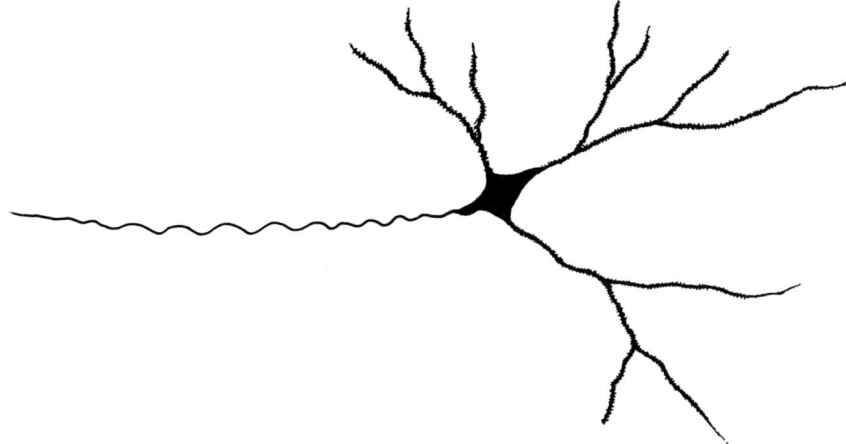

Figure 5.12 Diagram of a spiny neurone. The spines greatly increase the surface area of the dendrites.

of Bird et al. (1973) that glutamic acid decarboxylase, GAD, the enzyme involved in GABA biosynthesis, was significantly reduced in the caudate, putamen and globus pallidus.

GABA is one of the transmitters found in spiny neurones, so the finding of significantly reduced levels is consistent with the morphological observations. It is now known that two or more neurotransmitters may coexist in the same cell. Substance P may coexist with GABA; similarly met-enkephalin may coexist with GABA (Penny et al., 1986) and levels of both transmitters are reduced in the basal ganglia of HD patients (Kanazawa et al., 1977; Emson et al., 1980; Ferrante et al., 1986).

Table 5.3 Selective neuronal changes in HD

Neurones identified by transmitter	Morphological description	Effect of HD
GABA	Intrinsic spiny efferent	Decreased
GABA + met-enkephalin	Intrinsic spiny efferent	Decreased
GABA + substance P	Intrinsic spiny efferent	Decreased
SS/NP-Y/NADPH-d	Intrinsic aspiny interneurone	Relatively spared
Cholinergic	Intrinsic aspiny interneurone	Some loss probable, but may be relatively spared
Glutamatergic	Cortical afferent	Some loss likely
Dopaminergic	Midbrain afferent	Nigrostriatal pathway intact but inconsistent results for striatal dopamine levels

Acetylcholine cannot be measured directly but its biosynthetic enzyme, choline acetyltransferase is significantly reduced in striatal tissue; however, in some HD patients choline acetyltransferase levels are still within the normal range (Bird and Iverson, 1974). This reduction may represent the loss of some aspiny interneurones within the striatum although, as will be seen, this group of neurones may still be relatively spared during the pathological process. The pathological process appears to be relatively sparing of another group of neurones – those in which the transmitters somatostatin, neuropeptide Y and the enzyme NADPH-diaphorase (SS/NP-Y/NADPH-d) coexist, which morphologically also correspond to aspiny interneurones (Ferrante *et al.*, 1987). The percentage of these neurones changes from 2% up to as great as 30% in advanced cases of HD (Kowall *et al.*, 1987). As might be expected, this is consistent with the earlier observation of raised somatostatin levels (Aronin *et al.*, 1983; Beal *et al.*, 1984), raised neuropeptide Y levels (Dawbarn *et al.*, 1985) and increased activity of the enzyme NADPH-d (Ferrante *et al.*, 1985).

The above changes reflect neuronal degeneration occurring within the basal ganglia. Glutamate is the neurotransmitter found in the corticostriatal pathway. Striatal glutamate levels are reduced (Perry, 1982; Reynolds and Pearson, 1987; Ellison *et al.*, 1987). Perry (1982) reported normal glutamate levels for the frontal cortex; however, Reynolds and Pearson (1987) and Ellison *et al.* (1987) did observe significant reductions in the cortex of HD brains. Indirect evidence of abnormalities of glutamatergic neurones in HD includes reduced binding of ligand to presynaptic reuptake receptors (Cross *et al.*, 1986). In addition, Wong *et al.* (1982) noted reduced cortical activity of ornithine aminotransferase, a proposed biosynthetic enzyme marker for glutamate, in HD brains. These observations of glutamate abnormality are consistent with the light microscopic finding of cortical cell loss in layers III–V; corticostriatal fibres mainly arise from cells in layer V.

Another important striatal afferent is the dopaminergic nigrostriatal pathway. Given that the clinical sign of chorea may be considered as a mirror image of the rigidity and bradykinesia of Parkinson's disease, it is not surprising that interest should focus on dopamine levels within the striatum of HD brains. Table 5.4 summarizes the results from a number of studies; although there is some inconsistency between the various reports, it is still reasonable to consider that the striatonigral afferents are relatively spared and that given the

Table 5.4 Dopamine levels in HD

Author	Year	Caudate	Putamen
Bernheimer and Hornekiewicz	1973	Reduced	Normal
Bird and Iverson	1974	Normal	Normal
Spokes	1980	Raised	Raised
Melamed *et al.*	1982	Normal	Raised
Kish *et al.*	1987	Reduced	Normal

substantial loss of tissue in the striatum there may still be a relative excess of dopamine. Moreover, activity of tyrosine hydroxylase, the biosynthetic enzyme for dopamine, has been reported as normal (Bird and Iverson, 1974; McGeer and McGeer, 1976a). Melamed *et al.* (1982) suggested that HD was not associated with hyperactivity of nigrostriatal neurones because of the known loss of striatal dopamine receptors (see below) and normal or slightly reduced ratio of the main metabolite of dopamine, homovanillic acid, and dopamine in both the caudate and putamen. This aspect will be reconsidered in relation to animal models of HD on page 169.

Kish *et al.* (1987) observed a trend towards reduced striatal dopamine levels that was only significant in the caudal region of the caudate; they suggested that some of the discrepant findings could be explained by lack of regional assessment of the striatum. They argued that since the nigrostriatal pathway is relatively spared, the levels of dopamine are lower than might be expected; therefore, there has either been a down-regulation of dopamine turnover or a reduction in arborization of dopamine terminals.

Cross and Rosser (1983) studied ligand binding to dopamine receptors in the putamen and substantia nigra of post-mortem HD brains. D1 receptors use adenylate cyclase and within the striatum occur on neuronal cell bodies whereas D2 receptors occur on neuronal cell bodies and at the terminals of corticostriatal glutamatergic neurones. The reduction in binding of 45–50% in both D1 and D2 receptor binding is consistent with the striatal cell loss described in the previous sections. However, in the substantia nigra there was no loss of D2 receptors in either the pars compacta or the pars reticulata. There was a 50% reduction in D1 receptor binding in the pars reticulata. The authors concluded that this represented preservation of D2 receptors on the cell bodies but the loss of D1 receptors on striatonigral terminals and was consistent with the previously reported loss of striatal GABA neurones. Similarly, normal or increased levels of serotonin have been found in post mortem HD brains (Bernheimer and Hornekiewicz, 1973; Reynolds and Pearson, 1987).

Consideration of patch/matrix compartments

When viewed by light microscopy the striatum appears to be a homogeneous structure (see Figure 5.8); however, when stained to reveal the distribution of acetylcholinesterase a patch/matrix pattern is revealed (Graybiel and Ragsdale, 1978). The matrix represents an area of dense staining surrounding small acetylcholinesterase-poor patches which appear to have different functions (Figure 5.13). The patches receive afferent neurones from the prefrontal cortex and limbic system whereas the matrix receives afferents from other cortical areas (Gerfen, 1984). In both cases, cortical afferents use glutamate. Dopaminergic afferents are also distributed differently; those from the substantia nigra pars compacta go to the patches, whereas those from other areas of the midbrain are distributed to matrix areas (Gerfen, 1987). Similarly the distribution of efferent fibres differs: those from the patches go to the substantia nigra pars compacta, whereas those from the matrix are distributed to the

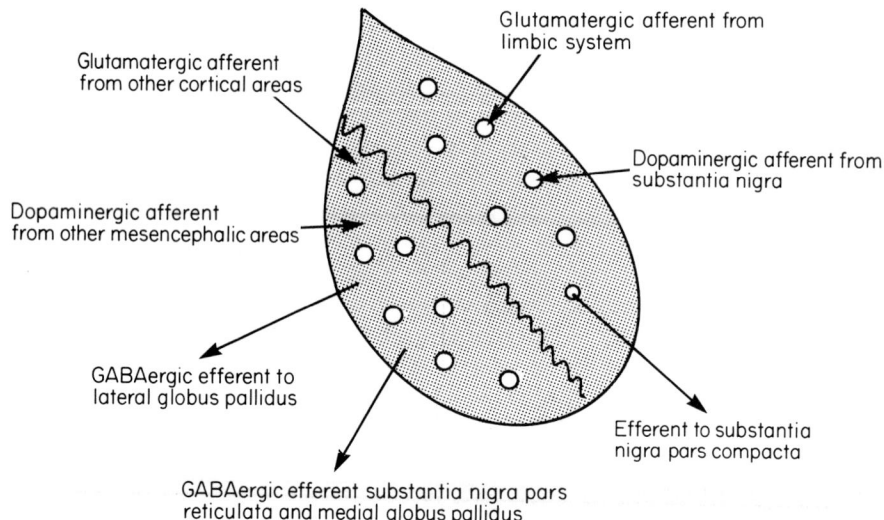

Figure 5.13 Diagram of the patch/matrix organization of the striatum (see text for explanation).

lateral and medial globus pallidus and the substantia nigra pars reticulata (Gerfen, 1987).

It has been demonstrated that most of the neurodegenerative process affects the matrix area (Ferrante *et al.*, 1987). Even in end-stage HD, areas of acetylcholinesterase-poor patches are surrounded by densely stained but much atrophied matrix, suggesting that the neurodegenerative process respects this histochemical boundary. Sources of acetylcholinesterase within the striatum are intrinsic and extrinsic; acetylcholinesterase is found in some of the large interneurones and it is also present in efferents from the substantia nigra and thalamus. It is therefore difficult to establish the extent to which this class of interneurone is involved in the pathological process. Interestingly, the aspiny interneurones, characterized by the presence of somatostatin, neuropeptide-Y and NADPH-diaphorase (SS/NP-Y/NADPH-d), that are selectively, spared are also found in the matrix compartment. Kiyama *et al.* (1990) have recently shown a marked loss in HD caudate nucleus of neurones reacting with the calcium binding protein calbindin 28K, the distribution of which corresponds to acetylcholinesterase in the matrix.

Alexander *et al.* (1986) considered that the basal ganglia consist of parallel neuronal circuits, the best characterized of which is the motor circuit. Other circuits include one involved in oculomotor control and at least two circuits involving the limbic system. It is unclear as to how much information passes between these circuits. In general terms, there is a division of afferents from the limbic system projecting to the patches and afferents from other cortical areas projecting to the matrix compartment, as has been noted earlier. It is also known that SS/NP-Y/NADPH-d neurones which have cell bodies in the matrix compartment also have dendrites which project to the patches, raising the possibility of some interconnection between the limbic and non-limbic systems (Gerfen, 1984). Whilst the complexity of the organization within the basal ganglia is increasing, in turn this should allow a more detailed understanding of the pathophysiology underlying the neurological signs and affective disturbances which occur in HD.

Reiner et al. (1988) suggested that in the early stages of the disease the spiny neurones containing GABA and met-enkephalin were most susceptible to the pathological process, neurones containing GABA and substance P being damaged later in the course of the illness. Matrix fibres containing met-enkephalin project to the lateral globus pallidus and those containing substance P project to the medial globus pallidus and substantia nigra pars reticulata (Graybiel, 1986). Moreover, substance P fibres projecting from the patch compartment to the substantia nigra pars compacta are least affected by the pathological process. These results were considered to be consistent with the earlier observation of Spokes (1980) of greater reduction of glutamic acid dehydrogenase (GAD) in the lateral globus pallidus than in the medial globus pallidus.

Albin et al. (1989) suggested that the observation of early loss of met-enkephalin fibres could explain the efficacy of dopamine antagonists in controlling choreic movements seen predominantly in the early stages of the disease; these drugs lead to stimulation of enkephalin production in certain cell subpopulations including striatal neurones.

Hemiballismus is noted in humans with lesions of the subthalamic nucleus; ablation of this nucleus in animal experiments also results in hemiballismus. Damage to the caudate nucleus alone does not cause a movement disorder. Albin et al. (1989) suggested that some of the signs seen in HD could be explained on the basis that loss of inhibitory GABA and met-enkephalin neurones to the lateral globus pallidus results in increased inhibition of the subthalamic nucleus. This suggestion is supported by the experimental observation that infusion of the GABA antagonist bicucilline into the lateral globus pallidus produces chorea (Crossman et al., 1988; Mitchell et al., 1989). Ultimately, the abnormalities in the striatum result in abnormal regulation of movement through fibres which pass from the ventrolateral thalamus to the supplementary motor cortex. Figure 5.14 summarizes one of the neural circuits in the basal ganglia and Figure 5.15 indicates the suggested pathophysiological basis of chorea. It is well established that the pattern of movement abnormalities changes as the disease progresses; these signs presumably reflect more widespread neurodegenerative processes within the basal ganglia and elsewhere, so it is likely that this explanation only offers a partial insight to the pathophysiology of the movement disorder.

The biochemical changes which occur in HD may not be specific; similar changes may be common to other disorders in which chorea is a prominent symptom. This point has recently received some attention; Kanazawa *et al.* (1985) demonstrated decreased GABA, substance P and choline acetyltransferase levels in the basal ganglia of three patients with the choreic form of dentatorubropallidoluysian atrophy (DRPLA). In a similar study of four patients with Pick's disease with atrophy of the basal ganglia there were similar biochemical changes; however the substantia nigra pars compacta was more severely affected than would be the case in HD and dopamine levels were reduced (Kanazawa *et al.*, 1988). The same authors (Kanazawa *et al.*, 1988) drew attention to one of their patients who had the least reduction in dopamine and in whom the clinical signs of grimacing and choreic movement had been noted. Clearly, the biochemical changes commented on here may not in themselves represent the primary pathological process in HD which produces neurodegeneration.

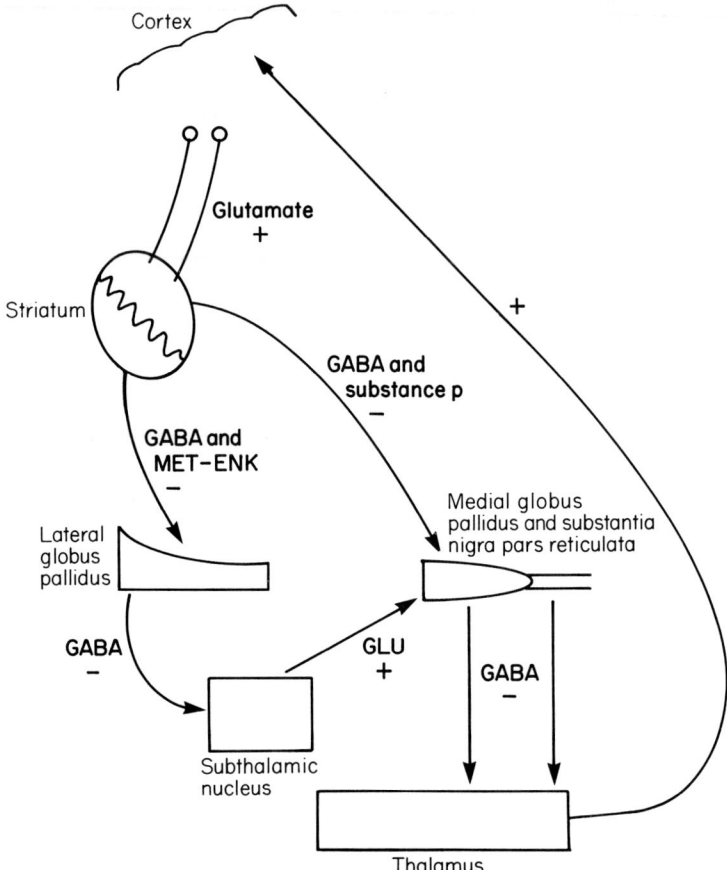

Figure 5.14 Diagram of one neural network within the basal ganglia. (After Albin *et al.*, 1989.)

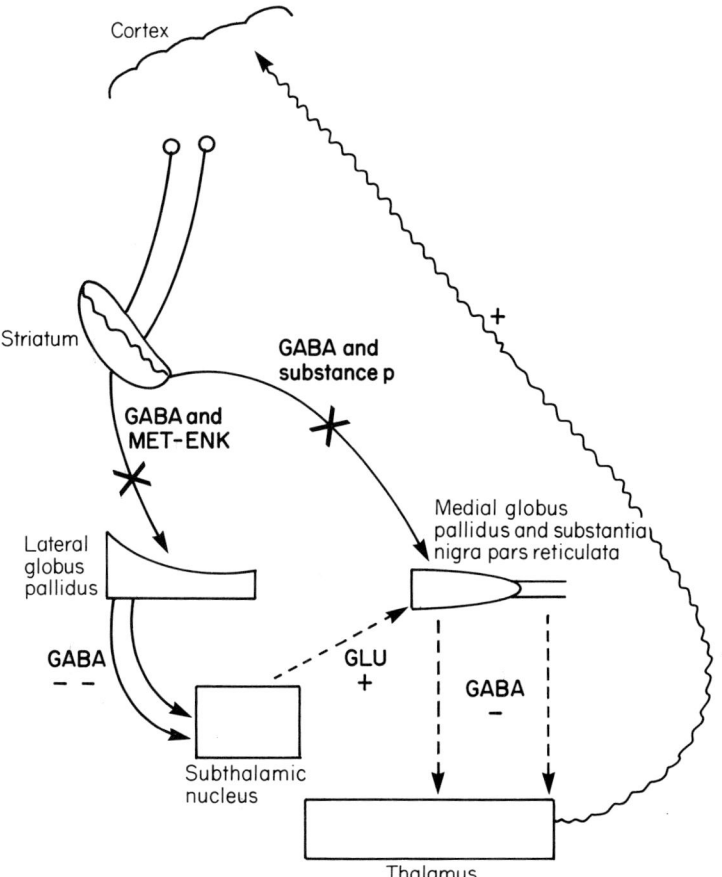

Figure 5.15 Effect of early HD, a time when the clinical sign of chorea is most prominent, on the neural network. The loss of GABA, met-enkephalin fibres to the lateral globus has the effect of increasing inhibition of the subthalamic nucleus, which in turn results in disinhibition of the thalamocortical fibres. (After Albin *et al.*, 1989.)

THE EXCITOTOXIC THEORY OF HD

The above description has not offered an explanation for the loss of GABAergic neurones within the striatum and elsewhere but knowledge that if the excitatory transmitter, glutamate, remains in the synapse then neuronal death will occur, has led to an interesting series of experiments. In this context glutamate is considered to be an excitatory neurotoxin as are its analogues.

Intrastriatal injection of the glutamate analogue, kainic acid, into rat striatum produced loss of neurones that had cell bodies within that structure but spared cells that had axons terminating there. Biochemical analysis revealed a decrease in glutamic acid dehydroxylase, choline acetyltransferase and increased levels of tyrosine hydroxylase (Coyle and Schwarcz, 1976; McGeer and

McGeer, 1976b). This pattern of neuronal cell death and biochemical profile, seen with kainic acid and other excitotoxins such as ibotenic acid, was strikingly similar to that which was being established for HD around that time (see above). Not surprisingly, theories were developed explaining loss of GABAergic cells based on the presence of abnormal production of excitotoxins. The kainic acid lesioned animal did not provide a perfect model for HD since selective sparing of SS/NP-Y/NADPH-d neurones was not observed (Beal et al., 1985). However, Beal et al. (1986) reported that intrastriatal injection of quinolinic acid into rat striatum produced a biochemical profile more closely resembling HD than that seen with other excitotoxins. In particular, high concentrations of quinolinic acid did not produce significant reduction in SS/NP-Y/NADPH-d neurones. In a later, more extensive, study Ellison et al. (1987) demonstrated that the only difference between biochemical changes in HD brains and quinolinic acid injected rats was a decrease in striatal taurine levels in the latter which was explained as a species difference in the distribution of this amino acid in striatal subpopulations.

Quinolinic acid has a further advantage over kainic acid as a model for an excitotoxin in HD in that it is a naturally occurring compound, being present as a metabolite in the kynurenine pathway of tryptophan metabolism (Figure 5.16). Quinolinic acid is a potent agonist of the N-methyl d-aspartate (NMDA) subgroup of glutamate receptors that are reduced in HD (Young et al., 1988). The quinolinic acid lesions in rats and primates may be blocked by pretreatment with the NMDA antagonist MK-801 (Beal et al., 1988, 1989). Further support for the hypothesis that the kynurenine pathway and NMDA-mediated excitotoxins are involved in the pathogenesis of HD is the observation that 3-

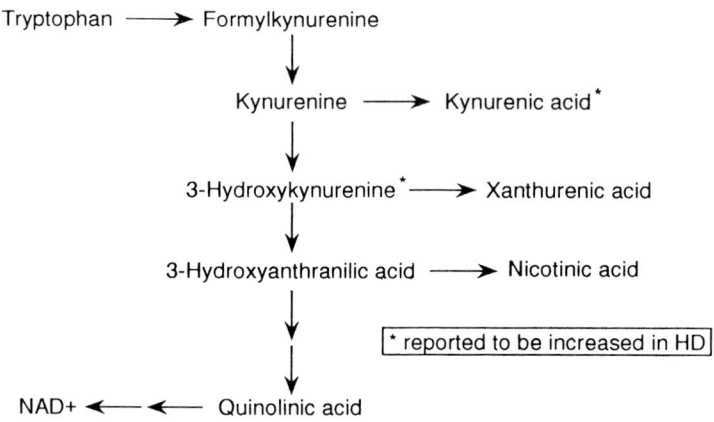

Tryptophan metabolism via the kynurenine pathway

Figure 5.16 Diagram to show main intermediates in the kynurenine pathway of tryptophan metabolism.

hydroxykynurenine, a precursor of quinolinic acid, and its parallel metabolite, kynurenic acid, are elevated in HD brains (Connick et al., 1989; Reynolds and Pearson, 1989). Moreover, activity of 3-hydroxyanthranilate oxygenase, the enzyme producing quinolinic acid from its precursor in the kynurenine pathway, was significantly elevated in HD brains. This enzyme is found in glial cells, so not surprisingly, greatest activity occurred in the striatum; however, increased activity was also noted in areas of the brain less affected by gliosis and there was no correlation with the severity of the neurodegeneration (Schwarcz et al., 1988).

Additional supporting evidence for the NMDA-receptor mediated excitotoxic hypothesis comes from the *in vitro* study of Koh and Choi (1988) in which striatal cultures from mouse embryos were exposed to NMDA; low concentrations (20–30 µmol) produced selective sparing of acetylcholinesterase and SS/NP-Y/NADPH-d containing neurones; higher concentrations (300 µmol) produced generalized neuronal damage. By comparison, exposure of the striatal cultures to 10 µmol kainic acid did not produce generalized neuronal damage but acetylcholinesterase and SS/NP-Y/NADPH-d neurones were selectively destroyed.

Cross et al. (1986) commented that glutamate was only weakly neurotoxic but toxicity could be increased if presynaptic uptake was reduced; significantly, binding of D-[^{3}H] aspartic acid, a marker for the glutamate re-uptake system, was reduced in HD brains. Whilst this result may or may not have a direct bearing on the quinolinic acid model it does support the concept that an excitotoxin is involved in the pathogenesis of HD.

Seto-Ohshima et al. (1988) noted in immunohistochemical studies that the vitamin D dependent calcium-binding protein was found in the matrix compartment of control brains and was decreased in HD brains. As previous reports had demonstrated that the matrix compartment was most susceptible to the pathological process this observation in itself was unsurprising but the authors speculated that loss of a calcium-buffering system may be relevant to a role in trying to ameliorate the effects of an excitotoxin.

Interestingly, Boegman and Parent (1988) found that rat cortical neurones containing NADPH-d were resistant to quinolinic acid toxicity, suggesting that this group of cortical neurones did not express the NMDA receptor. In this context it is interesting to note that Cudkowicz and Kowall (1990) observed relative sparing of neuropeptide Y immunoreactive neurones in the frontal cortex of HD brains even in those areas where there was significant loss of pyramidal projection neurones; the authors speculated that selective NMDA-mediated excitotoxic degeneration may not be limited to the striatum.

A further role for HD being related to an NMDA receptor abnormality is illustrated by a recent case report (Albin et al., 1990). A 32-year-old woman, from a family with well-documented HD, committed suicide; she had an 11-year history of psychiatric illness but relatives had noticed no movements or memory impairment to suggest that she had HD. Routine neuropathological analysis of the brain was entirely normal, a finding subsequently confirmed by three independent neuropathologists; however, analysis of a previously banked DNA sample gave a 95% probability that she had inherited the HD gene. Importantly, immunohistochemical studies of the basal ganglia identified loss of enkephalin and substance-P projection fibres and the density of NMDA receptors in the putamen was intermediate between values for controls and HD patients.

Problems with the quinolinic acid model

Although the findings described in the previous section are interesting, they by no means confirm that quinolinic acid or the kynurenine pathway is involved in the pathology of HD. An unresolved problem is lack of reproducibility of the selective sparing of SS/NP-Y/NADPH-d neurones when studied by other groups. Davies and Roberts (1987) reported substantial loss of both acetylcholinesterase and NADPH-d neurones following the administration of quinolinic acid (120 nmol) to rat striata. In a later study (Davies and Roberts, 1988), the same authors reported that a lower dose of quinolinic acid (30 nmol) produced an 82% reduction in SS/NP-Y/NADPH-d neurones but only a 26% reduction in acetylcholinesterase neurones, suggesting a relative sparing of the latter group. Boegman et al. (1987) and Boegman and Parent (1988) also reported that rat striatal SS/NP-Y/NADPH-d neurones were sensitive to both kainic and quinolinic acids (dose 60 nmol); acetylcholinesterase neurones were less severely affected. It is unfortunate that conflicting results have been reported with quinolinic acid injected rats. Until the conflict is resolved the model should be used cautiously.

There are further observations which do not support the quinolinic acid model: quinolinic acid levels in the putamen and frontal cortex from nine HD brains were not significantly elevated (Reynolds et al., 1988); normal or reduced urinary excretion of quinolinic acid has been found in HD patients (Heyes et al., 1985); and increased quinolinic acid levels have been reported in the CSF of patients with acquired immunodeficiency syndrome (AIDS) who did not have signs and symptoms suggestive of HD (Heyes et al., 1989). The decreased urinary excretion of quinolinic acid was explained as part of the general phenomenon of weight loss and malnourishment known to occur in HD. Moreover, the whole body reduction in quinolinic acid production may not reflect local cerebral metabolism. Whilst the above observations do not support a role for quinolinic acid (or the kynurenine pathway) in the aetiology of HD, neither do they exclude it as a possibility. A clear review of the current status of the excitotoxic theory in relation to HD has been provided by DiFiglia (1990).

OTHER THEORIES CONCERNING AETIOLOGY

In a recent review of the biochemical abnormalities of HD, Carter (1989) pointed out that particularly susceptible neurones were those which used

GABA, acetylcholine and glutamate as neurotransmitters. The biosynthesis of these neurotransmitters relies upon cerebral glucose metabolism and to a degree they are conserved and their metabolites reutilized. By contrast neurones which are selectively spared include those which use dopamine, serotonin and noradrenaline as neurotransmitters; although energy is required for their biosynthesis, it is not dependent upon cerebral glucose metabolism and attempts are not made to store these transmitters or reutilize their metabolites. It was suggested that those neurones which were most metabolically active were most susceptible to the pathological process. Moreover, this suggestion could explain the observation that some of the most susceptible neurones are small and have long axons.

Perry et al. (1987) described an interesting series of experiments in which neuronal cells from neonatal rats were cultured in medium containing various concentrations of sera from HD and control patients. The viability of the cultures was assessed by assaying GAD activity, this being an indirect assessment of GABA function. Culture in medium containing 30% by volume of serum from HD patients resulted in the suppression of GAD activity as compared with cultures containing 30% human serum from healthy controls. If the cultures were grown in medium containing 15% horse serum and 15% serum from HD patients results were more variable; in some cases GAD activity was increased when compared with controls. These effects were abolished if large molecular-weight molecules were removed from the serum by ultrafiltration but the putative toxin was not removed by dialysis nor if the serum was deproteinized using perchloric acid. The authors suggested that the serum of HD patients contained a neurotoxin of low molecular weight which was usually protein bound. Moreover, low concentrations of this toxin could stimulate GABA neurones whereas larger concentrations were toxic. The effect of serum from patients with other neurodegenerative disorders was not reported. The implication that HD results from a circulating neurotoxin must be considered as speculative.

ANIMAL MODELS OF HD

So far as is known, there is no natural disorder in animals which mimics HD. Previous sections have detailed the pathological changes that occur in the rat following striatal injection of kainic or quinolinic acid. Of necessity, this animal model of HD is imperfect for several reasons: firstly, the damage is acute whereas in HD there is a presymptomatic period lasting for decades and the pathological and clinical features evolve over a further two decades; secondly, there is dispute regarding the exact nature of the lesion in the rat; and thirdly, the rats do not develop chorea. Moreover, a rodent model is inadequate for studying intellectual decline and psychiatric disturbance, the other cardinal features of the disease.

Mention has already been made of non-human primate models of chorea, using Macaque monkeys, based on ablation of the subthalamic nucleus or injection of GABA antagonists in the lateral globus pallidus or the putamen, in the area immediately adjacent to the lateral globus pallidus (Crossman et al., 1988; Mitchell et al., 1989). Chorea affecting the contralateral limbs was noted soon after the infusion; less commonly myoclonus was noted, particularly when the injection had been in the putamen.

More recently, Kanazawa et al. (1990) have described a primate model of chorea, again using Macaque monkeys, but this time combining the intrastriatal injection of kainic acid with oral administration of L-dopa. Unilateral intrastriatal injection of kainic acid alone produced myoclonic-type seizures in the contralateral limbs which subsided after a few hours; contralateral bradykinesia was also noted but this subsided after a few days. No additional movement disorder was noted in those monkeys that had no further treatment for 14 days. L-Dopa was administered to seven monkeys 4–7 days after the excitotoxic lesion; both chorea and oral dyskinesia were noted in five monkeys which lasted for up to 6 h. Chorea did not occur in the two monkeys that had lost more than 60% of the striatum as a result of the kainic acid lesion. Administration of metamphetamine, which enhances presynaptic release of dopamine, produced a similar effect to administration of L-dopa; however, administration of apomorphine, which stimulates post-synaptic receptors, was not associated with chorea. Kanazawa et al. (1990) suggested that hyperactivity of dopaminergic fibres was presynaptic in origin; possibly arising from L-dopa released from damaged areas that had lost their partner cells. Even if this is not the correct interpretation of these results, it is not unreasonable to conclude that dopamine is involved in the genesis of chorea.

The animal models mentioned in this and previous sections have undoubtedly provided insights into the organization of the basal ganglia and have given rise to hypotheses regarding the nature of HD. Despite their limitations, the development of animal models should be pursued. Cloning the HD gene as described in Chapter 10 may have a complementary role in this respect since the significance or otherwise of the excitotoxic model should become more apparent. Moreover, more useful animal models may be developed using transgenic techniques for studying the action of the HD gene. In addition to providing an understanding of the underlying pathophysiology, animal models may also be valuable as a means of testing agents considered to be protective against the neurodegenerative process, and therefore potentially therapeutic for HD patients.

NEUROENDOCRINE ABNORMALITIES

Several studies of neuroendocrine function in HD were undertaken during the 1970s. The reason for this becoming a focus for research interest was threefold: firstly one of the abnormalities in HD could be characterized as relative dopamine excess; secondly there was evidence of glucose intolerance (Podolsky et al., 1972); and thirdly, there was no adequate explanation for the progressive weight loss observed in most HD patients. Basal levels of prolactin and growth hormone were measured, as were the responses of these hormone levels to stimulation tests. Some workers noted abnormalities which could not be replicated by others; consequently, no definite conclusions can be drawn. Good summaries and references to this body of work are provided by Bird (1979), Hayden (1981) and Kremer et al. (1989).

In 1972 Podolsky et al. described a form of glucose intolerance in six out of 10 randomly chosen HD patients; there was hyperglycaemia but no glycosuria or ketosis and no hypersecretion of insulin in response to administration of either glucose or arginine. In other studies on this topic such abnormalities could not be replicated (Davidson et al., 1974). A questionnaire method identified 65 cases of HD and diabetes out of 620 probands registered with the United States HD Roster (Farrer, 1985); this incidence, 10.5%, is significantly greater than that expected for the normal population. There was an increased incidence of diabetes in all age groups and a greater incidence in some families but not others. Farrer recognized the limitations of a questionnaire method, in particular the lack of a precise definition of diabetes, and called for a prospective study. Kremer et al. (1989) reported the results of oral glucose tolerance tests on 10 HD patients of whom six were reinvestigated 30 months later; only one patient had an abnormal glucose tolerance test. In considering their results and reviewing the literature the authors concluded: 'We think that the existing data do not suffice to state that impaired carbohydrate regulation is closely associated with HD, or has special clinical relevance to our patients'.

In a recent report Kremer et al. (1990) demonstrated, by cell counting techniques, atrophy of the lateral tuberal nucleus of the hypothalamus in five cases of HD; the authors called for a more systematic study of this area of the brain. The possibility that some neuroendocrine abnormalities result from structural changes in the hypothalamus must remain open.

EXPERIMENTAL ELECTROPHYSIOLOGICAL OBSERVATIONS

The electrophysiological data presented here are experimental only in that these techniques are not used in the clinical investigation of an HD patient or those at risk. Knowledge of the basic pathology of HD should allow an understanding of some of the abnormalities observed in evoked potentials, electromyographic recording of reflexes and event-related potentials. In general, abnormalities have been seen in those investigations which test polysynaptic responses. These abnormalities occur in aspects which are difficult to quantify and are not used in the routine evaluation of HD patients. A number of workers have detected electrophysiological abnormalities in asymptomatic individuals at risk for HD, the implication being that these abnormalities predate and predict the onset of HD. In our view such reports should be treated with caution until longer-term follow up has occurred, allowing validation of the preliminary results. This topic is given further discussion in Chapter 12.

The first report of abnormal evoked potentials in HD is attributed to Takahashi and Okada (1972); this study of synaptic evoked potentials (SEPs) and visual evoked potentials (VEPs) was written in Japanese with an English abstract, which may, in part, explain the delay between the earliest report and subsequent studies. Three modalities are commonly studied: somatosensory (SEP), visual (VEP) or brainstem auditory evoked potentials (BAEP). HD is not associated with clinically detectable sensory abnormalities but EPs test the integrity of complex systems, which in the case of SEPs include large diameter fast conducting group Ia muscle and group II cutaneous efferents, dorsal columns, brainstem, thalamus and cortex. SEPs are measured following electrical stimulation of a peripheral nerve (usually median or anterior tibial) by recording the amplitude and latency of potential changes from surface electrodes along the limb, over the spinal column and scalp. A number of studies have demonstrated reduced amplitude of cortical SEPs occurring within 100 ms of the stimulus (Oepen *et al.*, 1981; Noth *et al.*, 1984; Ehle *et al.*, 1984; Bollen *et al.*, 1985). One study found normal amplitudes of early SEPs but significant decrease after 100 ms (Josiassen *et al.*, 1982). The latencies of the wave forms were considered normal in three studies (Noth *et al.*, 1984; Ehle *et al.*, 1984; Bollen *et al.*, 1985); definitely increased in one (Josiassen *et al.*, 1982); and possibly increased in another (Oepen *et al.*, 1981). These findings reflect some lack of reproducibility of results but there is a reasonably consistent finding of decreased amplitude of cortical SEPs which is compatible with established pathological abnormalities in the thalamus and cortex.

VEPs may be recorded over the occipital cortex following a flashing light or patterned reversal (checkerboard) stimulus. Three studies have found normal latency and decreased amplitude in the majority of patients studied (Ellenberger *et al.*, 1978; Oepen *et al.*, 1981; Josiassen *et al.*, 1984). One study noted normal amplitudes and latencies but could not provide an explanation for the discrepancy with other results (Ehle *et al.*, 1984). Taken as a whole these studies indicate normal visual pathways but reflect diffuse cortical abnormalities.

BAEPs are recorded following auditory (click) stimuli and reflect activity in the auditory pathways and brainstem. This parameter was normal in two studies (Ehle *et al.*, 1984; Bollen *et al.*, 1985). In addition, lower brainstem function in HD patients was considered to be normal as respiration during sleep showed no difference from controls (Bollen *et al.*, 1986).

Blink reflexes are bilateral contractions of the orbicularis oculi muscles following electrical or mechanical stimulation of the supraorbital nerve. The reflex may be used to assess the function of the trigeminal nerve (afferent), pons and facial nerves (efferent). The electrical response has an early component (R1) which is unilateral and reflects the monosynaptic pontine reflex; it also has a late bilateral component (R2) which reflects a polysynaptic route involving the pons and lateral medulla. HD patients have a normal R1 (early) but delayed R2 (late) response (Esteban and Giménez-Roldán, 1975; Bollen *et* al., 1986; Agostino et al., 1988). The corneal reflex tests the same anatomical structures but an early component, equivalent to the R1 blink reflex, is not seen. Electrophysiological measurements of the corneal reflex has shown an increased latency but in two studies there was not a statistically significant difference between patients and controls (Bollen et al., 1986; Agostino et al., 1988). The explanation for the discrepancy between the corneal and blink reflexes is unclear; Agostino et al. (1988) suggested that the corneal reflex involves fewer interneurones than the late component of the blink reflex.

Additional evidence that polysynaptic responses are abnormal in HD patients comes from studies of long latency stretch reflexes. Stretching a contracting muscle produces a series of EMG responses; the short-latency responses are thought to be monosynaptic and are normal whereas the long-latency stretch reflexes are delayed and have a decreased amplitude. It has been hypothesized that the long-latency reflex involves a transcortical loop which is damaged in HD (Noth et al., 1983; Thompson et al., 1988; Abbruzzese et al., 1990).

Cognition is known to be abnormal in HD. This can be assessed electrophysiologically with long-latency event related potentials (ERPs). Subjects have to distinguish a target stimulus, auditory or visual from amongst a train of irrelevant stimuli. Of the many ERPs, the positive wave form at approximately 300 ms (P3) is consistently recorded in normal individuals. This component was abnormal in nine out of 13 patients following auditory stimuli and nine out of 12 patients following visual stimuli; 12 out of 13 patients had abnormalities with at least one modality (Rosenberg et al., 1985). A similar result was obtained in another study which tested the auditory modality; the P3 latency was 2 standard deviations above controls in 18 out of 27 patients (Hömberg et al., 1986). In both studies duration of symptoms at the time of testing had ranged between 1 and 9 and between 1 and 6 years respectively.

Finally, the integrity of the corticospinal tract has been assessed using an electromagnetic stimulation technique and recording surface EMG responses in either hand or leg muscles. Central conduction time can be measured by noting the difference in latencies between stimulation at the vertex and either the cervical or lumbar region. Thompson et al. (1986) reported normal central conduction times in seven HD patients and Hömberg and Lange (1990) reported similar results in 33 HD patients with varying degrees of severity of disease. These results imply that whilst there may be abnormalities of sensory input to the motor system and within the basal ganglia, the final efferent pathway is intact.

CHANGES OUTSIDE THE BRAIN

At the clinical level the major abnormalities found in HD can all be best explained by disordered neurological function. While advanced HD may result in more general disturbance, as seen in the marked weight loss, specific abnormalities outside the nervous system are relatively minor and probably secondary.

Despite this lack of systemic abnormalities it has been argued, quite reasonably, that the primary defect in HD might be a biochemical disturbance outside the nervous system resulting in a toxin, the effects of which are cerebral. Clearly, if this were to be the case the focus most relevant for prevention and therapy would be the source of the metabolic abnormality; the example of phenylketonuria, the clinical effects of which are almost entirely neurological, but whose primary defect is in the liver, is an obvious one.

This possibility has prompted numerous studies searching for metabolic defects outside the brain, in serum and in red and white blood cells, in the hope of identifying a generalized metabolic defect. The analysis of red and white blood-cell membranes has been a particular area of recent attention. None of these studies has been consistently abnormal, despite a large amount of effort. The 1979 volume on HD of *Advances in Neurology* gives details of many of these essentially negative investigations, while the reviews of Comings (1980) and Beverstock (1984) give critical assessments and detailed references.

The one field of work that has so far progressed almost entirely through studies on non-neural tissue is the mapping and isolation of the gene itself, discussed in detail in Chapter 10. Currently there is intense effort to identify a candidate gene which shows mutation in HD patients. Once this has been achieved it should become possible to identify a gene product using monoclonal antibodies raised to predicted peptides. At this point there will be the opportunity for molecular genetics to converge with neurochemistry and neuropharmacology and to begin to understand how the primary defect at the HD locus causes so many of the neurobiological effects described in this chapter.

REFERENCES

Abbruzzese G, Dall'Agata D, Morena M, Spadavecchia L, Ratto S and Favele E (1990) Impaired habituation of long-latency stretch reflexes of the wrist muscles in Huntington's disease. *Movement Disorders* **5**:32–35.

Agostino R, Bernardelli A, Crucco G, Pauletti G, Stocchi F and Manfredi M (1988) Correlation between facial involuntary movements and abnormalities in blink and corneal reflexes in Huntington's chorea. *Movement Disorders* **3**:281–289.

Albin RL, Young AB and Penney JB (1989) The functional anatomy of basal ganglia disorders. *Trends in Neurosciences* **12**:366–375.

Albin RL, Young AB, Penney JB et al. (1990) Abnormalities of striatal projection neurones and N-methyl-L-aspartate receptors in presymptomatic Huntington's disease. *New England Journal of Medicine* **322**:1293–1298.

Alexander GE, Delong MR and Strick PL (1986) Parallel organisation of functionally segregated circuits linking basal ganglia and cortex. *Annual Review of Neurosciences* **9**:357–381.

Aronin N, Cooper PE, Lorenz LJ et al. (1983) Somatostatin is increased in the basal ganglia in Huntington's Disease. *Annals of Neurology* **13**:519–526.

Beal MF, Bird ED, Langkus PJ and Martin JB (1984) Somatostatin is increased in the nucleus accumbens in Huntington disease. *Neurology* **34**:663–666.

Beal MF, Marshall PE, Burd GD, Landis DMD and Martin JB (1985) Excitotoxic lesions do not mimic the alteration of somatostatin in Huntington's disease. *Brain Research* **361**:135–145.

Beal MF, Kowall NW, Ellison DW, Mazurek MF, Swartz KJ and Martin JB (1986) Replication of neurochemical characteristics of Huntington's disease by quinolinic acid. *Nature* **321**:168–171.

Beal MF, Kowall NW, Swartz KJ, Ferrante RL and Martin JB (1988) Systemic approaches to modifying quinolinic acid striatal lesions in rats. *Journal of Neurosciences* **8**:3901–3908.

Beal MF, Kowall NW, Ferrante RJ and Cipolloni PB (1989) Quinolinic acid striatal lesions in primates as a model of Huntington's disease. *Annals of Neurology* **26**:137.

Beverstock G (1984) The current state of research with peripheral tissues in Huntington's disease. *Human Genetics* **66**:115–131.

Bird ED (1979) Neuroendocrine changes in Huntington's disease – an overview. *Advances in Neurology* **23**:291–298.

Bird ED and Iverson LL (1974) Huntington's chorea: post-mortem measurement of glutamic acid decarboxylase, choline acyltransferase and dopamine in basal ganglia. *Brain* **97**:457–472.

Bird ED, Mackay AVP, Rayner CN and Iverson LL (1973) Reduced glutamic acid decarboxylase activity of post-mortem brain in Huntington's chorea. *Lancet* **i**:1090–1092.

Boegman RJ, Smith Y and Parent A (1987) Quinolinic acid does not spare striatal neuropeptide Y-immunoreactive neurons. *Brain Research* **415**:178–182.

Boegman RH and Parent A (1988) Differential sensitivity of neuropeptide Y, somatostatin and NADPH-diaphorase containing neurons in rat cortex and striatum to quinolinic acid. *Brain Research* **445**:358–362.

Bollen EL, Arts RJ, Roose RA, Van Der Velde EA and Buruma OJ (1985) Somatosensory evoked potentials in Huntington's chorea. *Electroencephalography and Clinical Neurophysiology* **62**:235–240.

Bollen E, Arts RJHM, Roos RAC, Van Der Velde EA and Buruma OJS (1986) Brainstem reflexes and brainstem auditory evoked responses in Huntington's chorea. *Journal of Neurology, Neurosurgery and Psychiatry* **49**:313–315.

Bots GT and Bruyn GW (1981) Neuropathological changes of the nucleus accumbens in Huntington's chorea. *Acta Neuropathologica* **55**:21–22.

Bruyn GW (1973) Neuropathological changes in Huntington's chorea. *Advances in Neurology* **1**:399–403.

Byers RK, Gilles FH and Fung C (1973) Huntington's disease in children. Neuropathological study of four cases. *Neurology* **23**:561–569.

Carrasco LH and Mukherji CS (1986) Atrophy of corpus striatum in normal male at risk of Huntington's chorea. *Lancet* **i**:1388–1389.

Comings DE (1981) The ups and downs of Huntington's disease research. *American Journal of Human Genetics* **33**:314–317.

Connick JH, Carlà V, Moroni FV and Stone FW (1989) Increased kynurenic acid in Huntington's disease motor cortex. *Journal of Neurochemistry* **52**:985–987.

Coyle JT and Schwarcz R (1976) Lesion of striatal neurones with kainic acid provides a model for Huntington's chorea. *Nature* **263**:244–246.

Cross A and Rosser M (1983) Dopamine D-1 and D-2 receptors in Huntington's disease. *European Journal of Pharmacology* **88**:223–229.

Cross AJ, Slater P and Reynolds GP (1986) Reduced high-affinity glutamate uptake sites in the brains of patients with Huntington's disease. *Neuroscience Letters* **67**:198–202.

Crossman AR, Mitchell IJ, Sambrook MA and Jackson SJ (1988) Chorea and myoclonus in the monkey induced by gamma-aminobutyric acid antagonism in the lentiform complex. *Brain* **111**:1211–1233.

Cudkowicz M and Kowall NW (1990) Degeneration of pyramidal projection neurons in Huntington's disease cortex. *Annals in Neurology* **27**:200–204.

Davidson MP, Green S and Menkes JH (1974) Normal glucose, insulin and growth hormone responses to oral glucose in Huntington's disease. *Journal of Laboratory and Clinical Medicine* **84**:807–812.

Davies SW and Roberts PJ (1987) No evidence for preservation of somatostatin-containing neurons after intrastriatal injections of quinolinic acid. *Nature* **327**:326–329.

Davies, SW and Roberts PJ (1988) Sparing of cholinergic neurons following quinolinic acid lesions of the rat striatum. *Neurosciences* **26**:387–393.

Dawbarn D, DeQuidt ME and Emson PC (1985) Survival of basal ganglia neuropeptide 7 — somatostatin neurones in Huntington's disease. *Brain Research* **340**:251–260.

De la Monte SM, Vonsattel J-P and Richardson EP Jr (1988) Morphometric demonstration of atrophic changes in the cerebral cortex, white matter and neostriatum in Huntington's disease. *Journal of Neuropathology and Experimental Neurology* **47**:516–525.

DiFiglia M (1990) Excitotoxic injury of the neostriatum: a model for Huntington's disease. *Trends in Neuroscience* **13**:286–289.

Dom R, Baro F and Brucher JM (1973) A cytometric study of the putamen in different types of Huntington's chorea. *Advances in Neurology* **1**:369–385.

Dom R, Malfroid M and Baro F (1976) Neuropathology of Huntington's chorea. *Neurology* **26**:64–68.

Ehle AL, Stewart M, Lellelid NA and Leventhal, NA (1984) Evoked potentials in Huntington's disease. A comparative and longitudinal study. *Archives of Neurology* **41**:379–382.

Ellenberger C Jr, Petro DJ and Zeiger SB (1978) The visually evoked potential in Huntington disease. *Neurology* **28**:95–97.

Ellison DW, Beal MF, Mazurek MF, Malloy JR, Bird ED and Martin JB (1987) Amino acid neurotransmitter abnormalities in Huntington's disease and the quinolinic acid animal model of Huntington's disease. *Brain* **110**:1657–1673.

Emson PC, Arregui A, Clement-Jones V, Sandberg BEB and Rosser, M (1980) Regional distribution of methionine-enkephalin and substance P-like immunoreactivity in normal human brain and Huntington's disease. *Brain Research* **199**:147–160.

Esteban, A and Giménez-Roldán, S (1975) Blink reflex in Huntington's chorea and Parkinson's disease. *Acta Neurologica Scandinavica* **52**:145–157.

Farrer, LA (1985) Diabetes mellitus in Huntington's disease. *Clinical Genetics* **27**:62–67.

Ferrante RJ, Kowall NW, Beal MF, Richardson EP Jr, Bird ED and Martin JB (1985) Selective sparing of a class of striatal neurons in Huntington's disease. *Science* **230**:561–563.

Ferrante RJ, Kowall NW, Richardson EP Jr, Bird ED and Martin JB. (1986) Topography of enkephalin, substance P and acetylcholinesterase in Huntington's disease striatum. *Neuroscience Letters* **71**:283–288.

Ferrante RJ, Kowall NW, Beal MF, Martin JB, Bird ED and Richardson EP Jr (1987) Morphologic and histochemical characteristics of a spared subset of striatal neurons in Huntington's disease. *Journal of Neuropathology and Experimental Neurology* **46**:12–27.

Forno LS and Norville RL (1979) Ultrastructure of the neostriatum in Huntington's and Parkinson's disease. *Advances in Neurology* **23**:123–135.

Gerfen CR (1984) The neostriatal mosaic: compartmentalisation of corticostriatal input and striatonigral output systems. *Nature* **311**:461–464.

Gerfen CR (1987) The neostriatum mosaic: compartmental organisation of mesostriatal systems: In: *The Basal Ganglia II* (eds MB Carpenter and A Jayaram). New York: Plenum Press.

Graveland GA, Williams RS and DiFiglia M (1985a) A Golgi study of the human neostriatum: neurons and afferent fibres. *Journal of Comparative Neurology* **234**:317–333.

Graveland GA, Williams RS and DiFiglia M (1985b) Evidence for degenerative and regenerative changes in neostriatal spiny neurons in Huntington's disease. *Science* **227**:770–773.

Graybiel AM (1986) Neuropeptides in the basal ganglia. In: *Neuropeptides in Neurologic and Psychiatric Disease* (eds JB Martin and JD Barchas). New York: Raven Press.

Graybiel AM and Ragsdale CW Jr (1978) Histochemically distinct compartments in the striatum of human, monkey and cat demonstrated by acetylcholinesterase staining. *Proceedings of the National Academy of Science* **75**:5723–5736.

Hayden MR (1981) *Huntington's Chorea*. Berlin: Springer-Verlag.

Heyes MP, Garnett ES and Brown RR (1985) Normal excretion of quinolinic acid in Huntington's disease. *Life Science* **37**:1811–1816.

Heyes MP, Rubinow D, Lane C and Markey SP (1989) Cerebrospinal fluid quinolinic acid concentrations are increased in acquired immune deficiency syndrome. *Annals of Neurology* **26**:275–277.

Hömberg V, Hefter H, Granseyer G, Strauss W, Lange H and Hennerici M (1986) Event related potentials in patients with Huntington's disease and relatives at risk in relation to detailed psychometry. *Electroencephalography and Clinical Neurophysiology* **63**:552–569.

Hömberg V and Lange HW (1990) Central motor conduction to hand and leg muscles in Huntington's disease. *Movement Disorders* **5**:214–218.

Jervis GA (1963) Huntington's chorea in childhood. *Archives of Neurology* **9**:244–257.

Jeste DV, Barban L and Parisi J (1984) Reduced Purkinje cell density in Huntington's disease. *Experimental Neurology* **85**:78–86.

Josiassen RC, Shagass C, Mancall EL and Roemer RA (1982) Somatosensory evoked potentials in Huntington's disease. *Electroencephalography and Clinical Neurophysiology* **54**:483–493.

Josiassen RC, Shagass C, Mancall EL and Roemer RA (1984) Auditory and visual evoked potentials in Huntington's disease. *Electroencephalography and Clinical Neurophysiology* **57**:113–118.

Kanazawa I, Bird ED, O'Connell R and Powell D (1977) Evidence for a decrease in substance P content of substantia nigra in Huntington's chorea. *Brain Research* **120**:387–392.

Kanazawa I, Kwak S, Sasaki H, Muramoto O, Mizutani T, Hori A and Nukina N (1988) Studies on neurotransmitter markers of the basal ganglia in Pick's disease with special reference to dopamine reduction. *Journal of Neurological Sciences* **83**:63–74.

Kanazawa I, Sasaki H, Muramoto O, Masaaki M, Mizutani T, Iwabuchi K, Ikeda T and Takahata N (1985) Studies on neurotransmitter markers and striatal neuronal cell density in Huntington's disease and dentatorubropallidoluysian atrophy. *Journal of the Neurological Sciences* **70**:151–165.

Kanazawa I, Kimura M, Murata M, Tanaka Y and Fumiaki C (1990) Choreic movements in the macaque monkey induced by kainic acid lesions of the striatum combined with L-dopa. *Brain* **113**:509–535.

Kish SJ, Shannack K and Hornekiewicz O (1987) Elevated serotonin and reduced dopamine in subregionally divided Huntington's disease striatum. *Annals of Neurology* **22**:386–389.

Kiyama H, Seto-Ohshima A and Emson PC (1990) Calbindin D28K as a marker for the degeneration of the striatonigral pathway in Huntington's disease. *Brain Research* **525**:204–214.

Koeppen AH (1989) The nucleus pontis centralis caudalis in Huntington's disease. *Journal of Neurological Sciences* **91**:129–141.

Koh J-Y and Choi DW (1988) Cultured striatal neurons containing NADPH-diaphorase or acetylcholinesterase are selectively resistant to injury by NMDA receptor agonists. *Brain Research* **446**:374–378.

Kowall NW, Ferrante RJ and Martin JB (1987) Patterns of cell loss in Huntington's disease. *Trends in Neuroscience* **10**:24–29.

Kremer HPH, Roos RAC, Frölich M et al. (1989) Endocrine functions in Huntington's disease. A two-and-a-half years' follow up study. *Journal of Neurological Sciences* **90**:335–344.

Kremer HPH, Roos RAC, Dingjan G, Marani EU, Bots ThAM (1990) Atrophy of the hypothalamic lateral tuberal nucleus in Huntington's disease. *Journal of Neuropathology and Experimental Neurology* **49**:371–382.

Lange H, Thörner G, Hopf AV, Schröder KF (1976) Morphometric studies of the neuropathological process in choreatic diseases. *Journal of Neurological Sciences* **28**:401–425.

Leigh RJ, Parhad IM, Clark AW, Buettner-Ennever JA and Folstein SE (1985) Brainstem findings in Huntington's disease, possible mechanisms for slow vertical saccades. *Journal of Neurological Sciences* **71**:247–256.

Markham CH and Knox JW (1965) Observations of Huntington's chorea in childhood. *Journal of Pediatrics* **67**:46–57.

McCaughy WTE (1961) The pathologic spectrum of Huntington's chorea. *Journal of Nervous and Mental Diseases* **133**:91–103.

McGeer PL and McGeer EG (1976a) Enzymes associated with the metabolism of catecholamines, acetylocholine and GABA in human controls and patients with Parkinson's disease and Huntington's chorea. *European Journal of Pharmacology* **26**:65–76.

McGeer EG and McGeer PL (1976b) Duplication of biochemical changes of Huntington's chorea by intrastriatal injections of glutamic and kainic acids. *Nature* **263**:517–519.

Melamed E, Hefti F and Bird ED (1982) Huntington's chorea is not associated with hyperactivity of nigrostriatal dopaminergic neurons: studies in postmortem tissues and in rats with kainic acid lesions. *Neurology* **32**:604–644.

Mitchell IJ, Jackson A, Sambrook MA and Crossman AR (1989) The role of the subthalamic nucleus in experimental chorea. *Brain* **112**:1533–1548.

Myers RH, Vonsattel, JP, Stevens TJ et al. (1988) Clinical and neuropathologic assessment of severity in Huntington's disease. *Neurology* **38**:341–347.

Noth J, Freidemann H-H, Podoll K and Lange HW (1983) Absence of long latency reflexes to imposed finger displacements in patients with Huntington's disease. *Neuroscience Letters* **35**:97–100.

Noth J, Engel L, Freidemann H-H and Lange HW (1984) Evoked potentials in patients with Huntington's disease and their offspring. 1. Somatosensory evoked potentials. *Electroencephalography and Clinical Neurophysiology* **59**:134–141.

Oepen G, Doerr M and Thoden U (1981) Visual (VEP) and somatosensory evoked potentials in Huntington's chorea. *Electroencephalography and Clinical Neurophysiology* **51**:666–670.

Oyanagi K and Ikuta F (1987) A morphometric revaluation of Huntington's chorea with special reference to the large neurons in the neostriatum. *Clinical Neuropathology* **6**:71–79.

Oyanagi K, Takeda S, Takahashi H, Ohama E and Ikuta F (1989) A quantitative investigation of the substantia nigra in Huntington's disease. *Annals in Neurology* **26**:13–19.

Penny GR, Afsharpour S and Kitai ST (1986) The glutamate decarboxylase-, leucine enkephalin-, methionine enkephalin-, and substance P- immunoreactive neurons in the neostriatum of the rat and cat: evidence for a population overlap. *Neuroscience* **17**:1011–1145.

Perry TL (1982) Normal cerebrospinal fluid and brain glutamate levels in schizophrenia do not support the hypothesis of glutamatergic neuronal dysfunction. *Neuroscience Letters* **28**:81–85.

Perry TL, Hansen S and Kloster M (1973) Huntington's chorea: deficiency of gamma-aminobutyric acid in brain. *New England Journal of Medicine* **288**:337–342.

Perry TL, Yong VW, Hansen S et al. (1987) Tissue culture evidence for a circulating neurotoxin in Huntington's chorea. *Journal of Neurological Sciences* **78**:139–150.

Podolsky S, Leopold NA and Sax DS (1972) Increased frequency of diabetes mellitus in patients with Huntington's chorea. *Lancet* **i**:1356–1359.

Reiner A, Albin RL, Anderson KD, D'Amato CJ, Penney JB and Young AB (1988) Differential loss of striatal projection neurons in Huntington's disease. *Proceedings of National Academy of Sciences USA* **85**:5733–5737.

Reynolds GP and Pearson SJ (1987) Decreased glutamic acid and increased 5-hydroxytryptamine in Huntington's disease brain. *Neuroscience Letters* **78**:233–238.

Reynolds GP and Pearson SJ (1989) Increased brain 3-hydroxykynurenine in Huntington's disease. *Lancet* **ii**: 979–980.

Reynolds GP and Pearson SJ (1990) Brain GABA levels in asymptomatic Huntington's disease. *New England Journal of Medicine* **323**:682.

Reynolds GP, Pearson SJ, Halket J and Sandler M (1988) Brain quinolinic acid in Huntington's disease. *Journal of Neurochemistry* **50**:1959–1960.

Rodda RA (1981) Cerebellar atrophy in Huntington's disease. *Journal of Neurological Science* **50**:147–157.

Roisin L, Stellar S and Liu JC (1979) Neuronal nuclear-cytoplasmic inclusions in Huntington's chorea: electron microscope investigations. *Advances in Neurology* **23**:95–122.

Roos RAC (1984) Nuclear membrane indentations in Huntington's Disease. PhD Thesis. Leiden.

Roos RAC. (1986) Neuropathology of Huntington's disease. In: *Handbook of Clinical Neurology Vol 49. Extrapyramidal Disorders* (eds PJ Vinken, GW Bruyn and HL Klawans). Amsterdam: Elsevier.

Roos RAC, Pruyt JFM, De Vries J and Bots GTh (1985) Neuronal distribution in the putamen in Huntington's disease. *Journal of Neurology, Neurosurgery and Psychiatry* **48**:422–425.

Rosenberg C, Nudleman K and Starr A (1985) Cognitive evoked potentials (P300) in early Huntington's disease. *Archives in Neurology* **42**:984–987.

Schwarcz R, Okuno E, White RJ, Bird ED, Whetsell WO Jr (1988) 3-hydroxyanthranilate oxygenase activity is increased in the brains of Huntington's disease victims. *Proceedings of the National Academy of Science* **85**:4079–4081.

Seto-Ohshima A, Emson PC, Lawson E, Mountjoy CO and Carrasco LH (1988) Loss of matrix calcium-binding-protein-containing neurons in Huntington's disease. *Lancet* **i**:1252–1255.

Spokes EGS (1980) Neurochemical alterations in Huntington's chorea – a study of post mortem brain tissue. *Brain* **103**:179–210.

Stone TT and Falstein EI (1939) Pathology of Huntington's chorea. *Journal of Nervous and Mental Diseases* **89**:602–626 and **89**:773–797.

Takahashi K and Okada E (1972) Somatosensory and visual evoked potentials in Huntington's chorea. *Clinical Neurology (Tokyo)* **22**:381–385.

Tellez-Nagel I, Johnson AB and Terry RD (1973) Ultrastructural and histochemical study of cerebral biopsies in Huntington's chorea. *Advances in Neurology* **1**:387–397.

Thompson PD, Dick JPR, Day BL et al. (1986) Electrophysiology of the corticomotoneurone pathways in patients with movement disorders. *Movement Disorders* **1**:113–117.

Thompson PD, Berardelli A, Rothwell JC et al. (1988) The coexistence of bradykinesia and chorea in Huntington's disease and its implications for theories of basal ganglia control of movement. *Brain* **111**:223–244.

Vonsattel J-P, Myers RH, Stevens TJ et al. (1985) Neuropathological classification of Huntington's disease. *Journal of Neuropathology and Experimental Neurology* **44**:559–577.

Wong PT-H, McGeer PL and McGeer EG (1982) Ornithine aminotransferase in Huntington's disease. *Brain Research* **231**:466–471.

Young AB, Greenamyre JT, Hollingsworth Z et al. (1988) NMDA receptor losses in putamen from patients with Huntington's disease. *Science* **241**:981–983.

6

Social and Psychological Aspects of Huntington's Disease

INTRODUCTION

The social effects of Huntington's disease (HD) are frequently more severe than those occurring in other inherited disorders of the nervous system as a result of the severe mental and physical impairment that it causes and its pattern of inheritance which passes the disorder to generation after generation. Many early workers – Jelliffe *et al.* (1913), Davenport and Muncey (1916), Vessie (1932) in the United States, Spillane and Phillips (1937) in Wales, Pleydell (1954) in England – documented associated social effects such as alcoholism, delinquency, cruelty and neglect of children, broken homes and families living in degrading conditions. Later Oliver and Dewhurst (1969), Dewhurst (1970) and Bolt (1970) made notable contributions to our knowledge of the adverse environmental factors associated with HD. The US Commission report (Work Group, 1977) also presented a bleak and frightening picture of the suffering which this disease can cause.

None of these studies took account of cultural factors. It is not possible, therefore, to make a comparison with the frequency of these social problems in the general population and consequently the extent to which HD families were disadvantaged remains unclear. Also the true picture of the distress suffered by families overall is very hard to assess and quantify, since so much depends on the method of ascertainment which can 'select' for multi-problem families. There is also the natural desire of family members to conceal the condition and under-report 'stigmatizing' data. Much depends on who provides the information (Folstein, 1989) and it is always wise to use a variety of sources.

This chapter draws on the material obtained from a study in three counties of industrial South Wales (Gwent, south- and mid-Glamorgan) which attempted to measure the incidence of some psychosocial factors in an unselected population of HD families (Tyler, 1982; Tyler *et al.*, 1982, 1983; Tyler and Harper, 1983). The data have been updated 15 years later. On 1 January 1975 there were known to be 92 living patients in the study area; on 1 January 1990 the figure had increased to 137 patients, 15 of these being survivors from the original survey. The 122 new patients who have been identified do not represent the total number of new cases arising in the intervening 15 years; patients who were ascertained after January 1975 who had died before 1 January 1990 have not been included.

The social problems documented are seen in brain-impaired patients suffering from other chronic degenerative disorders. However, in HD families, these problems are magnified by the genetic implications which can distort a healthy adaptive response in a wide variety of family members, both those related through marriage and blood relatives.

Some of the characteristic features of the disease have a particular social impact; these include the slow progression of symptoms, often over 15–20 years, prominent dementing and psychiatric signs combined with loss of motor control, eventually resulting in total dependency. The care of such a patient can exact an immense toll from family members and great anguish can be caused to the patient who may already have suffered severely from living with and caring for an affected parent.

However, the manifestations and age of onset of the disease vary widely and influence the psychosocial effects so that some families experience relatively little disturbance. The age at which the patient develops symptoms and the type of symptomatology are important. The younger the patient and the more behaviour problems displayed, the greater the disruption caused, particularly if there are still young children at home. Family life is typically more disturbed if the affected parent is the mother; the economic effects are greater when the father is the ill parent.

Folstein *et al.* (1983) and Folstein (1989) have made a considerable contribution to our knowledge of the significance of psychosocial factors in the disorder by showing that conduct disorder in at-risk persons is not associated with the development of HD. This later finding was foreshadowed by Oliver and Dewhurst (1969) but it could not be confirmed since it was not known at that time whether the unaffected family members displaying anti-social behaviour, to which they referred, might later develop the disease.

Compensatory social networks in the family and in the community can be of great importance in supporting affected members but are often damaged by the fraught relationships which commonly exist in HD kindreds. The burden of guilt and anxiety carried by so many persons, springing from knowledge of the hereditary factor, diminishes their capacity to help and cooperate.

In the absence of effective treatment good social management is crucial in alleviating distress but this is often poorly understood and neglected. Also, the implications for health care personnel and social workers in a time of scarce resources, and ever-increasing demands, are likely to be considerable.

THE SOUTH WALES SAMPLE

Before discussing specific social aspects in detail, it is appropriate to describe the patient sample on which the South Wales study was based, since this has

Social and Psychological Aspects of Huntington's Disease

been the source of much of the data given here. More details of the original sample can be found in the publications already cited.

Sex
The original group divided into 57 females and 35 males and Tyler (1982) has speculated that the preponderance of females could have been due to the effects of World War II. The 122 new cases divided into 62 females and 60 males – the expected distribution; adding on the 11 females and four males who survived 15 years later, the total distribution by sex in 1990 was 73 females and 64 males.

Age range
The age range of the current 137 cases divided by sex and the numbers in each age group in 1975 and 1990 are contrasted in Table 6.1. It is noteworthy that the ages of the patients span over 60 years, from the twenties to the eighties, but there is no one under the age of 20 years. Also, in the whole of the survey period no case of the disease has been diagnosed in a child under the age of 10 years. There is a tendency for females to survive longer than males, most marked after the age of 65 years: five male patients compared with 15 female patients were aged over 65 years.

Civil (marital) state
The original sample is compared with the later sample in Table 6.2. Proportionately, fewer patients are single and a greater number divorced, in 1990. The trend for a greater proportion of males to be single and a higher proportion of females to be widowed at any one time continues.

Table 6.1 Social study of HD in South Wales. Age-range of patients by 5-year categories in 1990 (1975 totals also given for comparison)

Age-range	Females	Males	1990 Total	1975 Total
20–24	3	1	4	1
25–29	4	5	9	3
30–34	6	4	10	9
35–39	7	8	15	7
40–44	8	10	18	9
45–49	7	6	13	15
50–54	6	7	13	13
55–59	10	10	20	13
60–64	7	8	15	10
65–69	8	1	9	6
70–74	4	1	5	5
75–79	3	2	5	1
80–84	0	1	1	0
Total	73	64	137	92

Table 6.2 Social study of HD in South Wales. Civil state of patients in 1975 and 1990

	Original sample (1975)			New sample (1990)			Combined sample (all patients alive in 1990)		
	Male	Female	Total	Male	Female	Total	Male	Female	Total
Single	7	8	15	11	5	16	11	7	18
Married	22	36	58	35	38	73	37	44	81
Widowed	1	6	7	1	5	6	2	6	8
Divorced	5	7	12	13	14	27	14	16	30
Total	35	57	92	60	62	122	64	73	137

Following up the civil state of the 1975 sample, one patient married for the first time after showing symptoms; 42 marriages survived up to the patient's death. Six more marriages broke up in the intervening years and more may follow among the 15 survivors, eight of whom are still married. The total number of divorces in this group is actually 24 as five patients had married for the second time after divorce and one of these had been divorced for the second time.

Overall, in the total 1990 series, it is estimated that 10 patients had married for the second time while showing symptoms and four of these marriages had already broken up.

SOCIAL CLASS

Most studies have found that HD patients in general appear to belong to the lower socio-economic levels of society and that as a group they are very poorly represented in professional and semi-professional occupations (Panse, 1942; Pleydell, 1954; Reed et al., 1958; Bruyn, 1968; Mattsson, 1974). This finding has been confirmed in both of the Welsh series (see Table 6.3). A more recent Norwegian study (Saugstad and Odegard, 1986), however, found a normal occupational distribution. The reason for this disparity is not known but it was noted in this study that the majority were born in rural communities and that farmers, fishermen and workers, including farm labourers and seamen, accounted for 70% of the sample. None of the Welsh patients, living in an industrial community, were from these occupational groups.

EDUCATION

The original survey described the patients as generally of poor educational standard: only one affected man had achieved a higher educational (engineering) qualification. The patients identified in 1990 followed the same trend. Apart from two patients who are university graduates, and one patient who is a

Table 6.3 Social study of HD in South Wales. Social class of patients in 1975 and 1990

	1975 (n = 92)		1990 (n = 137)		Percentage of general population 1981 Census
	No.	%	No.	%	
SC1	1	1	1	1	6
2	8	9	18	13	23
3	49	53	49	36	49
4	12	13	19	14	17
5	18	20	39	28	5
Never worked	4	4	10	7	—
Student	—	—	1	1	—

General population figures taken from the National Report, Great Britain Part 11 1981 Census, London, HMSO. Office of Population Censuses and Surveys.

university student, only eight men and one woman had gained a technical qualification on leaving school. Their spouses, with one or two exceptions, were of similar background. Their educational expectations, generally, were low. This finding is perhaps not surprising when one considers that the home conditions often do not favour learning, e.g. the children's education can be frequently disrupted by the demands of the choreic parent. This seems particularly true of girls.

If the parent is affected when the children are of school age, the frequently tense atmosphere at home is not conducive to study. The children's sleep may be interrupted; their needs for food and clothing may not be met; their concentration may suffer and they may be inattentive at school. Often there is little money for books, school outings, etc. Although there may be no obvious neglect there may be very little encouragement to learn and to attend school. A minority persistently play truant. Furthermore, the children may be subjected to jeers and taunts from their classmates concerning the appearance and behaviour of their affected parent who may be accused of being drunk. The knowledge of being at risk can play a part (Tyler, 1989) but is insufficient on its own to lead to poor school performance. Rather, it can exacerbate the problems the youngster is already facing and may add to feelings of hopelessness about the future.

The effect of all this disturbance can be that children leave school as quickly as possible without educational qualifications in order to achieve some independence and a life of their own. A number of children were encountered in the course of the survey who had not been able to fulfil their potential at school.

EMPLOYMENT

None of the 15 survivors from the 1975 sample was working 15 years later but it is worth noting that two men and one woman had only ceased full-time work

2–3 years earlier. Four men in 1975 and three men in 1990 had been unable to work prior to the onset of HD by reason of other disability. Four self-employed men had been declared bankrupt – one in the earlier survey and three in the later one. Two persons in both surveys had not worked for many years because of the care needed by sick relatives.

Table 6.4 compares the employment state of patients in both series. The 1975 sample contained six patients who had never worked; three because of juvenile onset, one because of congenital disability and two who had stayed at home to look after sick relatives. The 1990 sample contained 10 patients who had never worked but none because of juvenile onset. One had been disabled from birth; the remaining nine had never been able to find work which reflects the trend towards greater unemployment in South Wales in the intervening 15 years. Five men were deemed to have worked on a casual basis only for many years prior to the onset of HD. They were heavy drinkers and one had 'slept rough' for a long time. Three similar patients had been identified in the earlier survey. In 1975 the heads of 75 households were manual workers; 18 of these were unskilled. In 1990, 92 were manual workers and 37 of these were unskilled; this reflects the trend away from heavy industries such as mining, in South Wales, to assembly line factory work.

The figures for the men 'signing on' for work reflect the difficulty that patients can experience in understanding and accepting the extent of their disability. In fact, none of them are/were fit for employment in the open market. The two identified in 1975 very quickly were persuaded to apply for sickness benefit (which increased their financial resources) as have been the two in the later survey. The reasons for manual workers becoming unemployable usually related to neurological problems, such as impairment of balance, which made the work place hazardous. Men in non-manual occupations where some decision-making was required, often left work earlier; in a few cases where their cognitive abilities were relatively unimpaired they could continue at work for as long as 10 years or more after diagnosis. Four out of the 13 men still in full-

Table 6.4 Social study of HD in South Wales. Employment status of patients

	1975			1990		
	Male	Female	Total	Male	Female	Total
Working full-time	8	1	9	13	1	14
Working part-time	1	4	5	2	5	7
Unable to work	21	52	73	43	67	110
Retired in normal way	3	0	3	3	0	3
Seeking employment	2	0	2	2	0	2
Student	0	0	0	1	0	1
Total	35	57	92	64	73	137

time work have been moved to lighter occupations, which present fewer hazards in the organizations in which they work and two others have taken part-time work. Four men in 1975 had been similarly treated. The situation with regard to women is not as clear cut. Most had families, were only working part-time, and giving up work could have been influenced by factors other than disability.

It is clear that HD symptoms of themselves are not incompatible with holding down employment and it is remarkable that very few significant accidents have been reported involving choreic individuals. The reason seems to be that other workers 'guard' them. One or two employers sustained serious losses due to their employee's clumsiness or 'negligence' but again, very few such losses have been reported. Most patients are able to continue working for the first 5 years after onset and a small minority for 10 years after onset. Much depends on the willingness and ability of workmates to 'carry' the affected person and the ability of the organization to redeploy. The most favourable conditions were found in large organizations among groups of workers, all employed on the same task. We know of no similarly collected figures with which to compare these findings. Dewhurst *et al.* (1970) analysed the work record of 55 male patients before hospital admission, and found that 29 were considered unemployable; 10 were known to be employed.

The loss of the patient's employment takes no account of the hidden costs. The economic impact of the disease is greater than the figures show, as discussed later; it is not possible to calculate, for example, how many women might have returned to work when their children were older, had the disease not become manifest. Spouses suffered also when they were able to remain in work. Opportunities for overtime or for promotion were foregone. Mobility was sacrificed. A number of men took early retirement to care for their disabled wives. Two self-employed men in the first series attributed their bankruptcies to their choreic wives' inappropriate behaviour.

MARITAL BREAKDOWN

In both the 1975 and 1990 surveys, factors associated with divorce (including separation) were found to be onset at under or around the age of 40 years and for the breakdown to occur within 5 years before or after the age of onset. Looking at first marriages only, 47 of the 214 patients have already been divorced and most likely more will follow; 27 occurred in patients under the age of 41 years and in the early years of the illness. Nine patients were divorced while they were asymptomatic; two of these were suffering from a major psychiatric illness.

Directly comparable figures from other sources are not available because of differing social conditions and methods of data collection but some are presented for interest. Mattsson (1974) found that 23 out of 162 patients in

Sweden were divorced, five of these within 5 years of onset; the figure obtained by Reed *et al.* (1958) was 20 out of 120 patients. The study of Oliver (1970) represents the nearest comparable group: of 82 married patients in Northamptonshire, 23 were subsequently divorced or separated. The finding that older couples are more likely to stay together is to be expected. Older couples are likely to be more stable and to have different attitudes towards marriage as an institution. Many spouses said that their partners had given them many years of happiness when well and that they wished to repay them by taking care of them when they became ill.

There is a widespread belief that the development of HD contributes to marital breakdown. This is confirmed in our own study by the reasons spouses gave for separating from their partners, mainly relating to their behaviour. Aggressive, violent, abusive behaviour was most commonly quoted by wives; slovenly, obsessional, neglectful behaviour by husbands. Examples of such behaviour were husbands kicking their pregnant wives; throwing plates of food over them; throwing them and their children into the street on a wet, cold, winter's night, poorly clad. Wives' behaviour included burying all their husband's clothing in the garden; sitting all day in front of an unmade-up fire, watching television, undressed, unwashed; persistently questioning their husbands about their activities outside the home and exhibiting morbid jealousy.

There has never been any evidence presented to the authors that at-risk persons were, or are, discriminated against in their choice of marriage partners. Saugstad and Odegard (1986) in Norway and Folstein (1989) in USA made the same observation.

Divorce occurring early on in the marriage can be the means of perpetuating ignorance of the hereditary implications. If ties with the affected parent are severed before diagnosis, children can grow up unaware of their genetic risk (Tyler *et al.*, 1983). This was the case in seven families overall in the series.

PATTERNS OF CARE

By 1990, 77 patients in the original South Wales series had died, 46 of whom were in institutional care and 31 cared for at home. Out of the 15 surviving patients, seven were occupying a long-stay bed and eight were still being cared for in the community. The type of residential care they have needed is set out in Table 6.5. Over half the patients had needed long-term residential care and it is very probable that some of the 8 patients still living in the community will require this. Tables 6.6 and 6.7 set out the way in which the 137 patients in the 1990 series were being cared for and type of residential care they occupied. Disregarding the seven long-term survivors, one in six of the 'new' patients was in long-term care, which is identical to the figure produced by the 1975 survey.

Table 6.5 Social study of HD in South Wales. Distribution of patients in long-stay care (1975 series; 46 deceased, seven alive in 1990)

	Male	Female	Total
Psychiatric hospital	14	18	32
Geriatric hospital	3	1	4
General practitioner hospital	0	7	7
Cheshire home	0	1	1
Old People's home	1	2	3
Nursing home	3	2	5
'Young' chronic sick unit	1	0	1
Total	22	31	53

Bolt (1970) found that just under one-third of the 154 patients ascertained in the west of Scotland in 1960 were in hospital. A similar proportion of hospitalized patients was found by Chiu and Teltscher (1978) in Australia. More recently, Barozak et al. (1987) in the Birmingham region and Peacock and Harris (1989) in the north western region of Britain found that 25% of HD patients at any one time are in residential care. The comparable figure for Maryland, USA, is 28% (Folstein, 1989).

It was originally estimated that approximately half the patients coming to the notice of the South Wales survey would die at home and half in long-term care (Tyler et al., 1982); an analysis of the deaths in the past 16 years has borne this out. Dewhurst (1970) found that about three-quarters of the patients in his series ultimately entered hospital and Bolt (1970) reported that two-thirds of

Table 6.6 Social study of HD in South Wales. Patterns of care for HD patients – 1 January 1990

	New patients (n = 122)			Patients from 1975 sample (n = 15)			Grand total
	Male	Female	Sub-total	Male	Female	Sub-total	
Long-stay beds	10	10	20	2	5	7	27
Living alone	13	14	27	0	2	2	29
Living in lodgings	1	0	1	0	0	0	1
Living with friend	1	0	1	0	1	1	2
Living with family	35	38	73	2	3	5	78
Total	60	62	122	4	11	15	137

Table 6.7 Social study of HD in South Wales. Type of residential care used by patients, 1 January 1990

	Patients from 1990 series			Patients from 1975 series, alive in 1990			Grand total
	Males	Females	Sub-total	Males	Females	Sub-total	
Psychiatric hospital	3	3	6	0	1	1	7
Geriatric hospital	1	4	5	1	0	1	6
Family practitioner hospital	0	0	0	0	2	2	2
Nursing home	4	1	5	1	2	3	8
Old People's home	2	2	4	0	0	0	4
Total	10	10	20	2	5	7	27

the patients surveyed died in hospital. The most likely reason for the apparently lower incidence and prevalence of hospitalization in South Wales compared with some other surveys is the different method of ascertainment. Of the 31 patients from the original series who died at home, 14 succumbed to another complaint before reaching the terminal stage: four persons suffered a cerebrovascular accident, two persons committed suicide, two died as the result of a fall, one from a gastrointestinal haemorrhage, one choked to death and four died from heart attacks.

The length of time on average that Welsh patients spend in an institution is estimated to be approximately 4 years. Bolt's (1970) figure for Scottish patients was 4.9 years and Hayden et al.'s (1980) figure for South African patients was 4 years. More recently, Simpson and Johnston (1989) found the average stay of hospitalized patients in the Grampian region of Scotland to be 7 years.

REASONS FOR HOSPITALIZATION

The main social reasons for admission to residential care are set out in Table 6.8. Many patients living alone were eventually admitted to institutional care when their home conditions became too degraded for them to continue. It is noteworthy that everyone in this group had a family member who attempted to provide some care for them, often at great cost to themselves. Similarly, elderly spouses and parents, themselves disabled, would struggle on for years caring for their affected relative until increasing dependency and their own increasing infirmity or death forced the provision of other care. All these patients had shown signs of HD for over 10 years. Six patients were unmanageable at home,

Table 6.8 Social study of HD in Wales. Main social reasons for admission of patients to an institution (1990 Series)

	Females	Males	Total
Disability of elderly spouse or parent	3	2	5
Death of spouse or parent	2	1	3
Unable to continue living alone	5	5	10
Unmanageable at home	5	4	9
Total	15	12	27

partly because of the distress caused to young children. Though no serious injury occurred they were very frightened by menacing behaviour and in two cases had been thrown across a room. For the three other patients either severe paranoid behaviour directed against the spouse (two) or manic behaviour (one) led to admission to a psychiatric hospital. These patients generally were younger and had shown signs of the disease for a shorter time. The earlier survey found that violent, destructive behaviour, the need for terminal care or living alone in degraded conditions were the usual reasons why residential care was required.

It is clear that families shoulder the main burden of caring for patients and often need considerable support, particularly if behavioural problems become severe or dependency needs become excessive. Those most in need have been found to be those patients living alone, or with carers who are themselves in poor health, or with young children at home.

Local authority services can be very important in enabling the affected person to remain in the community for as long as possible. Suitable housing is of paramount importance for patients whose balance is impaired; currently 15 in our study have been moved into ground floor accommodation or have had extensions built on to existing houses, or been provided with adaptations such as stair-rails and chair-lifts. Twenty-three patients are in receipt of health and social services such as home carers, home helps, district nurses, and meals on wheels. Though more have been offered the opportunity, only about six are attending a day centre regularly. Eleven patients are considered to be living in poor to appalling conditions, three with dependent children. Overall, 22 choreic patients with 69 children at home are considered to have 'multiple' problems but no children as yet have been taken into care, though 12 had been in the original series (eight from one family). Some are resisting all offers of help from the community-based services; others have accepted a minimum service; for a small group of about six families the normal social service provisions seem almost totally inadequate.

THE ECONOMIC COST OF HD

To estimate the direct cost of caring for an HD patient presents many difficulties. Looking purely at the take-up of social security benefits and the average cost of 2 years of institutional care (as half the patients die at home), we estimated in 1979/1980 that the minimum life-time cost was £21 500 (Tyler *et al.*, 1982). In practice, for many patients, the cost to the community is much greater since no account was taken of other medical costs and the use of local authority services. On the other hand, no attempt was made to estimate the costs which would have been incurred had HD not developed.

The Office of Health Economics (1980) estimated minimum costs of approximately £4 million per year for the UK's HD population in 1979, excluding direct financial payments. More recently Peacock and Harris (1989) analysed the cost and place of care of 158 patients on the north west of England regional register and found that 39 (25%) were in residential care. The cost of such care varied from £148.00 to £275.00 per week. Care at home could cost between £115 and £147 per week, looking only at the social security benefits to which the patient and carer would be entitled. Simpson and Johnston (1989) reported that the average length of stay of the HD patient in long-term care in the Grampian region of Scotland in 1987 was 7 years at a cost of £91 000.

The authors estimate that about 25% of patients who have died each year for the past 15 years in Wales have died from other causes before the need for institutionalization arose or the stage of considerable dependency was reached. For the remaining 75% of patients the disease imposes a considerable financial burden on the State and the family, as has been discussed.

EFFECTS ON THE PATIENT

The distress experienced by a patient obviously can arise from the disease state itself and from the implications for children; it is also influenced by any previous experience of HD sufferers that the patient has known (Wexler, 1979). In the first 5 years of the illness patients are usually able to perform most of their ordinary activities and to continue at work though neurological signs are evident (Tyler, 1982). However, short-term memory and activities demanding the exercise of higher mental functions such as the ability to organize, to calculate or to carry out complex processes, are usually impaired. Also, family members often notice behavioural changes of which the most prominent are emotional lability, confusion and social withdrawal (Tyler *et al.*, 1983).

In the next 5 years most patients are forced to give up their employment and become increasingly dependent on others for help in managing their affairs, though they are still able to maintain a considerable degree of personal independence. Also aggressive behaviour is more frequently reported.

Ten years after onset, neurological impairment is usually considerable; social disinhibition and demanding, bad-tempered behaviour are common, while irritability and bouts of rage are often present (Bolt, 1970; Folstein, 1989). It is at this stage that patients' behaviour is often perceived as dangerous; they are usually still ambulant and active but are liable to lose their balance and drop whatever they are trying to pick up. They suffer significant memory loss and are unable to foresee the consequences of their actions, e.g. if they smoke, they may set fire to household articles and themselves. If they feel thwarted they can become aggressive. However, serious injury to family members appears to be rare because of the patient's lack of coordination and relatives learn to take avoiding action.

After 15 years the behavioural problems are usually lessened, masked by the almost total dependency of the patient, who is often virtually mute, incontinent and profoundly demented by this stage, though some comprehension is usually maintained up to the end. Death usually supervenes within the next 5 years.

When patients first become aware of failing capacities (though some apparently never do) they can be very fearful of abandonment by spouses, of being 'put away', or 'losing control of body and mind'. They can become very depressed and self-conscious about going out in public, particularly when they are accused of being drunk or find themselves the objects of ridicule. Many will shun affected family members in a more advanced stage because they 'cannot bear to see themselves as they may become'. Fortunately, as the patients' affect becomes blunted, this phase usually passes though it may take several years.

Similarly, in the first stages of the illness, patients can worry about the implications for their children and sometimes this can lead to bizarre behaviour. In one large South Wales family the children of an affected mother continued to bear offspring in order to reassure her that their lives had not been ruined by the knowledge of their genetic risk. As the illness progresses, however, and patients become increasingly egocentric these worries seem to disappear and the patient often appears indifferent even if his or her own children are showing signs of the disease. Denial is a common reaction and most patients make use of it in varying degrees. Some patients adopt an attitude of total denial from the start and maintain it to the end of their days. The most usual pattern, however, is for insight to be intermittent, gradually lessening until it is totally lost with advancing dementia. Discussion of any fears and worries, support and encouragement offered with sensitivity and compassion, as well as drug therapy where appropriate, can alleviate the patient's anguish, particularly in the early and middle stages of the illness.

EFFECTS ON THE CARER

Given the patient's lack of insight, irritability, disorganized behaviour and memory loss, it is not surprising that carers become stressed. Dementia secondary to brain damage is one of the most disruptive of disorders (Brooks *et al.*, 1986) and creates adjustment problems for carers in a way that physical disability usually does not. In one study (Tyler *et al.*, 1983) 82% of HD carers reported feeling stressed, and a high rate of adverse behaviour symptoms in the patient correlated with a high rate of stress symptoms in the primary care agent.

As we have seen, the majority of carers are close family members – spouses, parents, siblings and children. Emotional and social bonds and expectations can make them particularly vulnerable to the patient's personality change and inability to function as before (Lezak, 1978). Role reversal is common and can lead to difficult problems of adjustment. Where there are dependent children the situation can become particularly stressful if the regressed patient competes with them for attention.

A common feature of the HD patient's behaviour is negativism which may make it very difficult for the carer to obtain the practical help needed. The authors have often found that community services which would have been of great benefit have been refused by the patient, the most frequent being attendance at a day centre. Aids and adaptations can be refused on the grounds that the patient does not consider himself or herself disabled. Similarly, the services of health professionals, e.g. district nurses, may be refused. It is perhaps not surprising, therefore, that in our first series six patients were reliably reported to have been battered by the carer and six men and two women carers were very worried that they would be driven to assault. Some family members responsible for an HD patient take refuge in heavy drinking and/or extra-marital relationships. Sometimes it takes a crisis, such as an attempted suicide on the part of the carer, to bring about a change in circumstances and provide the respite care they sorely need. A particular problem, which has faced a number of spouses with young children at home, is the conflict between the needs of the patient and the needs of the children. The decision to separate in order to protect them from destructive abuse and/or violence is often taken only after much heart-searching and accompanied by considerable feelings of guilt. However, most have felt that the interest of the children must come first (Folstein, 1989).

Although the picture painted above is a bleak one, not all patients present with multiple behaviour problems, nor do they necessarily exhibit them to a significant degree. The manifestations in older patients are generally milder; a number of older women in particular show a passivity or lethargy which is relatively easy to manage. Their care often demands considerable physical effort but the mental strain is tolerable and ailing spouses will often make heroic efforts to keep the patient at home. Good respite care and opportunities to share their burden are of immense importance.

EFFECTS ON FAMILY RELATIONSHIPS AND CHILDREN

The worst effects of the dementing process are seen in families where there are young children. Out of the 137 current patients in the South Wales series, 60 had children under the age of 20 years living with them at the time of known onset. The number of children involved was 139, 56 being under the age of 10 years, so the size of the problem is considerable.

Where the family has no knowledge of the patient's family history, the early signs of the disease, which are often subtle and insidious, can be overlooked or misinterpreted. Even where it is known about beforehand, the family may not be able to accept the full reality for several years. This mechanism of denial can serve a useful purpose (Kessler and Bloch, 1989) but may inadvertently contribute to the distress of young children who may experience a loved and idealized parent change into someone who is erratic, indifferent and bad-tempered. The child may perceive the rejection by the parent as being due to 'badness' in him or herself, not realizing that the parent is ill, and having been given no other explanation.

Adolescents may be bullied and belittled by the affected parent as their developing maturity threatens fragile self-esteem. Older children can display a high rate of 'acting-out' behaviour. Folstein *et al.* (1983) saw conduct disorder as a response to the breakdown of family structure that occurred primarily in large families where the HD parent had an early onset of symptoms and the spouse was inadequate; the experience of the South Wales survey is broadly in agreement.

Warriner (1989) vividly describes her experience as a young girl of witnessing the disease take hold of her mother: 'The loving wife becomes a jerking cripple, the caring mother an irrational tormenter, the intelligent teacher a babbling child'. Other damaging effects can occur. The spouse may be so overwhelmed and anguished that children are obliged to assume extra responsibilities too early, to grow up too soon. In effect they may take on part of the role that the affected parent has been obliged to relinquish and be expected to give emotional support to the unaffected parent. Many of these children suffer from a lack of nurturing; when adult, they may remark that they 'never had a childhood' and envied children who came from homes where their parents looked after them. The adverse effects of a mentally impaired parent can be reduced if the spouse is strong enough to shoulder the dual role or other environmental influences are favourable; grandparents can be an invaluable source of reassurance and comfort.

It is an unhappy fact, however, that although these families need a great deal of help, often it is not offered within the kindred network and their social support systems are poorly developed (Kessler, 1987). Unaffected not-at-risk family members often are carrying their own burdens of guilt and anxiety which may lead to fraught relationships. Branches as yet unaffected may shun those with an affected member. One of the reasons for this attitude may come from cumulative depression and anger which is a common reaction to the threat posed by HD. The disease brings with it several types of loss – not just of physical capacity but of a familiar and loved personality, perhaps many years before death supervenes. HD families generally have a great deal of grieving to do and repeated bereavements to endure. It is important to 'mourn well' (Boyd,

1989); if the process of mourning is not completed the grieving individual's relationships with others can suffer and avoidance can occur. Healthy parents can be particularly stricken as they may first nurse an ill spouse, then find themselves caring for their children, and finally be watching the disease appear in the third generation. Such persons (often mothers) will speak of 'the horror' of the disease and may say they live in a state of permanent sadness.

Another reason for tension-filled family relationships can be the desire of one set of parents to protect their children from knowledge of their genetic risks when children in other branches of the family are aware of them. This can result in a loosening or severing of family bonds. However, the guilt felt by these parents and the reactions engendered often disrupt already damaged lines of communication with their children even further. The problem does not go away by cutting links with other relatives. Children are often aware that there is a 'family secret' which cannot be discussed (Martindale and Bottomley, 1980). They may pick up clues from professionals, such as nurses and doctors who know the family history, or they may be given some information by relatives who believe they should be warned. A number have guessed the diagnosis in an affected relative and learned of the implications for themselves from the media. They may obtain information from the lay society or they may seek genetic counselling. A familiar situation often met with in the genetic clinic is that of parents worrying about telling the children of their risk, only to find that they already know but have not been able to tell the parents for fear of upsetting them.

In a significant proportion of cases the information gained by these children is distorted and incomplete, e.g. they may believe that they are bound to inherit the gene or that all boys are affected; in other words, their fantasies can be worse than the reality and can conduce to a feeling of hopelessness and a belief that they have no future. A comprehensive and on-going counselling service offered to these families can ameliorate some of the worst effects.

Children from stable families, where the disease is of late onset, who have left home before the symptoms were apparent, can escape relatively unscathed (Dewhurst et al., 1970; Folstein et al., 1983). In the authors' experience, also, where there is a strong and sustained relationship with the unaffected parent and support and protection offered by others, children can cope surprisingly well (Wexler, 1984).

DELINQUENCY, CRIME AND ANTISOCIAL BEHAVIOUR

Delinquency is a legal category not a pyschiatric syndrome, and the term refers to a person who has a criminal conviction registered against their name in the Criminal Record Office. Official juvenile delinquents are overrepresented by the lower social groups; in the South Wales series 15 at-risk children from 12 families in social classes 4 and 5 were known to have a history of juvenile delinquency with offences ranging from grievous bodily harm, possession of illicit drugs to 'joy riding' in cars. It is likely that this behaviour is underreported. However, 25% of boys from working-class backgrounds have a criminal record by the time they reach the age of 20 (West and Farrington, 1973), so it is possible that juvenile delinquency is no commoner in HD than in a general population group controlled for social class.

Adult HD patients have been considered to have a high rate of criminal activity. According to Lind (1927) 'the insanity of Huntington's chorea must be regarded as one of the dangerous type and sudden outbursts of violence requiring institutional control'. Dewhurst et al. (1970) found that assault was the most common violent crime in HD. Reed et al. (1958) reported violent crime in 2 HD patients out of 103 male choreics. In the South Wales survey no HD patient has been convicted of homicide and criminal convictions due to violent behaviour are uncommon. Similarly, none has been convicted for child abuse, though severe neglect has undoubtedly occurred both in South Wales and in other series (Pearlstein et al., 1982). Four children (girls) from two families in South Wales were considered to have been sexually abused but in one family the unaffected parent was the abuser. This figure may be an underestimate. For non-violent offences, petty theft was the most common. In Bolt's series (1970), non-violent crime such as indecent exposure and drunkenness were more common than violent crimes. Four Welsh patients are known to have been in prison but for relatively minor offences.

In an analysis of general antisocial behaviour, Folstein (1989) found that 25% of the offspring of a sample of choreic parents were considered to have adolescent conduct disorder or its adult equivalent, an anti-social personality. She noted that disorders of conduct often subsided when the child left the household.

ADOPTION AND FOSTERING

The problems of the child at risk who is placed for adoption and the implications for genetic counselling are discussed in Chapter 11. In this section the question of whether adoption and fostering are options for adults with a family history of HD and the problems faced by prospective adopters of at-risk children are further explored.

Many adults who are considering adoption as an alternative to having children of their own find that the family history of HD automatically disqualifies them. From the standpoint of the agency involved this is entirely understandable since the needs of the child are paramount and the best available family has to be chosen (Oxtoby, 1982). Prospective adopters in the direct line of descent from an affected relative may develop a chronic, degenerative and ultimately fatal disease, strongly associated with behaviour problems, in the vulnerable years of a child's development. Also, they may be unable to offer security and consistent parenting because of the unstable family life which they experienced by virtue of the mental disorder in the family.

Nevertheless, before adoption is refused, some aspects deserve careful consideration. The person wishing to adopt, if not the offspring of an affected parent, may be at low genetic risk; a Genetic Centre will give an accurate risk estimate. It is also important to consider the risk that exists over a specific period of time, rather than the life-time risk alone. A table of such risks is given in Chapter 11 and it can be seen that this may be relatively low in some situations. We know of about 30 adults and children, who were adopted, when babies, into HD families, before the disorder was identified and its hereditary nature known. In four cases the parent has developed HD. All these adoptees, except one, appear to have made affectionate attachments to their at-risk parents, and if they have become ill, have given them devoted care.

Adopting an older child or fostering on a short-term basis may be more feasible, but it must be acknowledged that such placements can present particularly difficult problems and generally need very experienced and mature parents. Nevertheless, we have known a few at-risk persons who have been very successful short-stay foster parents; one fostered over 50 babies on a short-term basis, in a stable and caring environment, until she developed HD at the age of 50 years.

Families who apply to adopt or foster children at risk should be given full information about the illness. Even if the child is born after the biological parent has developed symptoms (a frequent cause for seeking alternative care) there is no evidence that the child's general health is affected in any way. Thus any problems arising in childhood are likely to be due to the previous environment and upbringing rather than to genetic factors. A child who has been living with a severely disturbed parent may require much skilled care and attention before the harm can begin to be repaired. The family who undertakes the upbringing of such a child should understand that there are no personality or character traits which predict the development of HD in later life: in other words, there is no evidence that children of a particular personality or temperament are more or less likely to develop HD than those without them.

It is also important for these adoptive and long-term foster parents to be prepared to tell the children of their genetic risks at an appropriate age and to be offered skilled help in coming to terms with the implications themselves and in preparing the children beforehand. It is often helpful for the genetic clinic to be involved at an early stage. A more detailed discussion of this topic can be found in Harper and Tyler (1985), while the question of genetic testing in relation to adoption is discussed in Chapter 12.

THE EXPERIENCE OF BEING AT RISK

The experience of being at risk for HD has been until recently of living with ambiguity and uncertainty, often for many years; of not knowing whether or not one has been the victim of a random genetic accident which was decided at the moment of conception. We are currently in a time of change, which will undoubtedly accelerate, and we can anticipate that more and more persons at risk will be offered the opportunity of a molecular genetic test to alter their status. How many will take advantage of testing is still not fully known; for some, undoubtedly, the uncertainty will be preferable to the risk of being told with near total certainty that one carries the gene (Willday, 1989).

At-risk persons who know they are at risk can be perceived as living under threat; their experiences with their affected family member(s) can influence the amount of anxiety they feel and the type of responses they make (Wexler, 1979). Pearson (1973), on the basis of an intimate knowledge of a small self-selected group of 13 persons at high risk and a good reported knowledge of 21 others, considered that 40% of his subjects at one time or another manifested an anxiety neurosis which could be reactivated at crisis points and that spouses were similarly affected. In a random sample of 91 at-risk persons (Tyler et al., 1983), all under the age of 50 years, 20 (23%) rated themselves as suffering from chronic anxiety ranging from mild to severe in almost equal proportions. A high proportion of persons, suffering severely, attributed their anxieties primarily to their distressing experiences with a violent and/or psychotic parent. Persons with mild anxiety states considered that they led relatively normal lives but became easily depressed and symptom searched frequently. Supportive spouses were of immense importance to them. The majority of the sample experienced worry intermittently and only a few denied any worry at all. Most said that unless something happened to remind them of their risk status they could put the knowledge to the back of their minds and not think about it for lengthy periods.

The threat of the disease can exact a subtle toll and lead to the adoption of various behavioural devices to keep it at bay based on the individual's knowledge of the disease in the family. Some lead very active lives, taking on new challenges, in order to prove to themselves (and others) that they are not developing the disease. Some who observed temper tantrums to be the first sign of HD in a relative never allow themselves to lose control. Others are obsessional about their house and appearance since their affected mothers were slovenly and self-neglectful. One obese woman refused to diet as her affected mother was very thin. Many believe that stress can hasten onset and therefore they avoid contact with affected relatives, even to the extent of moving house.

The insecurity felt by a few at-risk adults (in conjunction with other personality factors) may cause them not to marry; they feel they cannot inflict themselves on anyone knowing the disastrous effect the disease had on their parents' marriage. (This is a fairly normal reaction for adolescents but one which in our experience, rarely lasts.) Another manifestation of a lack of self-confidence is deliberately to choose undemanding employment and refuse promotion for fear that the disease will strike and render them incapable of coping with greater responsibilities.

For some the threat of HD allows some gains; it can serve as a defence against self-knowledge. Willday (1989), herself at 50% risk, describes the mechanism thus: 'If we are brutally honest with ourselves HD can be a useful tool in our emotional armoury, to be wheeled out at intervals to explain away unpleasant truths about ourselves that may be totally unrelated to the master gene'.

Being at-risk can have implications in terms of family dynamics; unaffected persons can feel guilty that they have been spared and devote their whole lives to caring for their sick members, denying themselves marriage and independence. One single woman known to us has spent 50 years of her life caring for three generations – her mother, sister and niece.

Adolescents and children who are aware that they are at risk can have special problems. A minority may be singled out by family members or fallaciously 'pre-selected' as gene-carriers (Kessler and Bloch, 1989), perhaps because of some likeness to the affected parent, either in behaviour or appearance. This device can serve falsely to reassure other at-risk members that *they* have not inherited the gene. It is a myth which can exert a powerful effect even when the at-risk person is intellectually aware that there is no scientific basis for such a connection. A number of these persons (with others) go on to live their lives as if they are bound to develop the disease; believing they have no long-term future they may achieve very little or sometimes very much.

Commonly, at-risk persons only share the knowledge of their status with close family members and perhaps one or two close friends (Folstein, 1989). They usually prefer their employers, and even their general practitioners on occasion, not to be aware that they have an HD parent for fear of the consequences of lack of privacy. Confidentiality is of the utmost importance to them.

No obvious case of stigmatization in the field of employment has come to our knowledge though the Armed Forces and the Police have reservations about accepting applicants with a family history of HD. Insurance companies commonly charge higher premiums for certain types of policies or may refuse cover (see Chapter 7).

In summary, most persons at risk come to terms with their status and achieve a reasonable quality of life (Folstein, 1989). Counselling needs to be available to them at crisis points.

SOCIAL MANAGEMENT

In a dominantly inherited incurable disease, where the effect of medical treatment is limited and uncertain, social management assumes great importance. Though this chapter has identified numerous problems besetting the HD patient and family – some of which are probably insoluble – practical help, emotional support and counselling can help in alleviating their suffering. Obviously the amount and type of help needed depends on many factors, as previously discussed, but there are certain core problems which are common to a large spectrum of the patient population and which usually require a multidisciplinary approach. As HD is a relatively rare condition and is poorly understood by many agencies (Caro, 1977; Hayden et al., 1980; Simpson and Johnson, 1989), workers involved with the patients need to have access to specialized advice (Barette and Marsden, 1979).

Information on aspects of patient care, such as diet, nursing, special equipment and on eligibility for state benefits can be much valued and needs to be widely available. The lay organizations in America, Britain and some other countries (see Appendix 3) have produced a number of excellent free leaflets covering these and other areas.

HD families tend to be passive in their approach to problems and not to feel in control of their own lives. As discussed, their educational background is often limited and they may lack confidence in their ability to negotiate with outside bodies. Twenty years ago Hans and Gilmore (1968) in the US noted that 'few of the patients or their families were active in social, fraternal, political or civic affairs' and this is true of Welsh families today. Therefore a 'reaching out policy' from a specialist team, incorporating an educational and consultative function, is advocated, together with the assignment of a well-informed and experienced key worker at the time the patient and family come to notice. Needs can be assessed and resources coordinated on a long-term basis by this worker as they change with the progression of the illness.

Young children and adolescents living with an erratic, mentally impaired parent, perhaps in very deprived conditions, can have special problems, as has been described in earlier sections. Some recommendations have been made by Tyler (1989) entitled 'Helping children live with HD'. These include helping the patient to retain as much of the parental role as possible for as long as possible and making opportunities to give and receive affection from the children; even very small gestures towards children can soften the impact of the personality change in an affected parent.

These children desperately need the opportunity to develop a sense of achievement, to enhance their self-esteem and to obtain recognition in terms of societal goals. Actively looking for, and fostering, any interest and potential can sometimes result in introducing the child to another world of valuable new experiences. Mention has been made of the anti-social behaviour displayed by some children as a reaction to their disorganized home conditions. We believe that this behaviour can represent a perhaps unrecognized plea to be removed from the home and recommend a careful psychological and social assessment early on to determine whether placement elsewhere could be of benefit and prevent unnecessary distress.

Huntington families can feel, with some justification, that they tend to be a neglected group, shunned by doctors, in particular. Some of the reasons for this are easy to identify. Patients are often hard to help; they are notoriously poor attenders at the clinic, they can be quarrelsome and uncooperative; they may lack insight. There is no good treatment for their condition. They engender feelings of helplessness and despair in doctors and health professionals which can lead to avoidance (Martindale, 1987).

Two commonly neglected areas, in our experience, are the management of the dementia and the needs of the carers. As has been discussed previously, the worst problems for the carer and family are the behavioural ones. Management by psychologically based methods is crucial to minimize the disruption caused and to improve the patient's sense of wellbeing, but rarely is any attention paid to this aspect, except by community-based psychiatric nurses and psychiatric social workers. It is vital that carers understand what can and cannot be expected of the patient in the face of failing cognitive abilities. Strategies for promoting the optimum environment, such as maintenance of a familiar routine and the necessity of avoiding confrontation, have been described in a growing number of publications (Mace et al., 1985; Folstein, 1989; Marshall, 1990). The plan of management devised by family and workers together needs to address these issues.

The problems likely to be encountered by carers need to be recognized from the start of the illness. Opportunities for holidays, for regular 'time off' to pursue their own interests, to share the burden, the avoidance of undue fatigue, must be built in to any programme, even if it appears to go against the patient's wishes and causes initial distress. The authority of the health professional should be exercised, if necessary, to facilitate proper consideration of their needs.

COUNSELLING

This disorder is a family one and the 'ripples' caused by making the diagnosis can spread far and wide. The time of diagnosis can be the start of a counselling process that can encompass many persons within the total kindred. Martindale (1987) has defined the aim of counselling as being 'to foster emotional insight' so that family members can reach the decisions which are best for them. A related aim would be to enable family members (including the patient) to function at their optimum level.

Not all families can share their pain and some can only do so for brief periods and to a limited extent. Living with HD is essentially learning to live with loss and stress. Each family is a unique entity with its own set of individual experiences and circumstances. Their life history, the story of how their knowledge of HD was acquired, the reaction it caused and the impact of this knowledge on their lives is crucial to an understanding of how best they can be helped, and how they can help themselves. Some persons are almost immobilized by guilt and anxiety. Others feel overwhelmed by repeated grief. Bereavement counselling can help but some individuals will need specialized medical or psychiatric help. 'Magical' thinking is common (Wexler, 1979) and myths abound, particularly about personal guilt and responsibility which can add an unnecessary burden.

Family dynamics may also benefit from some explanation and interpretation; some parents react in understandable but potentially damaging ways;

they may deny their children nothing for fear that their span of health will be short. If communication is poor it has been found that holding family discussions in the presence of a counsellor may help. Other counselling opportunities often arise and are valued around the time of major life events such as the patient's admission to residential care or the separation of the spouse and children.

As has been described, the lives of at-risk persons can be hedged in by emotionally based ways of behaving which bear little relationship to reality. Also, there can be a tendency to project all problems on to the person's at-risk status which, again, leads to passivity and a feeling that 'nothing can be done'. This attitude seems particularly prevalent in relation to marital problems. Counselling can be beneficial by persuading individuals that they have some control over their lives and choices to make and that, where there is some goodwill, marital difficulties can be worked on.

From time to time the future can appear very bleak to at-risk persons, particularly if they are looking after an affected patient and are themselves approaching the most frequent age of onset. Counselling may help by compartmentalizing the problem and finding some aspects which can be improved (Wexler, 1979), though others are intractable. Another coping stratagem which some find valuable involves deliberately placing emphasis on, and finding joy in, the 'here and now' (as well as taking thought for the future) as this is all we are certain we have.

Some denial, which will fluctuate in intensity, is probably inevitable and necessary for most at-risk persons and family members coping with a disease which can strike at the root of an individual's identity, and continually show its presence through the generations. Counselling may help to free the family from some of the bonds of anxiety and guilt, improve their capacity to make decisions, and to make the best of their lives.

REFERENCES

Barette J and Marsden CD (1979) Attitudes of families to some aspects of Huntington's chorea. *Journal of Psychological Medicine* **9**:327–336.

Barozak P, Pedlar A, Hunter S and Betts T (1987) Institutional care for patients with Huntington's chorea: is there a better alternative? *Bulletin of the Royal College of Psychiatrists* **11**:187–188.

Bolt JMW (1970) Huntington's chorea in the West of Scotland. *British Journal of Psychiatry* **116**:259–270.

Boyd D (1989) Living through grief: the process, healthy and unhealthy. Horizon Report No 55, Huntington Society of Canada.

Brooks N, Kampsie L, Symington C, Beattie A and McKinlay W (1986) The five year outcome of severe blunt head injury: a relative's view. *Journal of Neurology, Neurosurgery and Psychiatry* **49**:764–770.

Bruyn GW (1968) Huntington's chorea: historical, clinical and laboratory synopsis. In: *Handbook of Clinical Neurology, Vol 6* (eds PJ Vinken and GW Bruyn), pp. 298–378. Amsterdam: North Holland.

Caro A (1977) Huntington's chorea: a genetic problem in East Anglia. University of East Anglia: PhD thesis.

Chiu E and Teltscher B (1978) Huntington's disease: the establishment of a National Register. *Medical Journal of Australia* **21**:394–396.
Davenport CB and Muncey EB (1916) Huntington's chorea in relation to heredity and eugenics. *American Journal of Insanity* **73**:195–222.
Dewhurst K (1970) Personality disorder in Huntington's disease. *Psychiatrica Clinica* **3**:221–229.
Dewhurst K, Oliver JE and McKnight AL (1970) Socio-psychiatric consequences of Huntington's chorea. *British Journal of Psychiatry* **116**:255–258.
Folstein SE (1989) *Huntington's Disease. A disorder of families*. Baltimore: Johns Hopkins Press.
Folstein SE, Franz ML, Jensen B, Chase GA and Folstein MF (1983) Conduct disorder and affective disorder among the offspring of patients with Huntington's disease. *Psychological Medicine* **13**:45–52.
Hans MB and Gilmore TH (1968) Social aspects of Huntington's chorea. *British Journal of Psychiatry* **114**:93–98.
Harper PS and Tyler A (1985) Huntington's chorea: problems in adoption and fostering. *Adoption and Fostering Journal* **9**(3):47–51.
Hayden MR, Ehrlich R, Parker H and Ferera SJ (1980) Social perspectives in Huntington's chorea. *South African Medical Journal* **58**:201–203.
Jelliffe SE, Muncey EB and Davenport CB (1913) Huntington's chorea. A study in heredity. *Journal of Nervous and Mental Diseases* **40**:796–799.
Kessler S (1987) Psychiatric implications of presymptomatic testing for Huntington's disease. *American Journal of Orthopsychiatry* **57**(2):212–219.
Kessler S and Bloch M (1989) Social system responses to Huntington's disease. *Family Process* **28**:59–68.
Lezak MD (1978) Living with the characterologically altered brain injured patient. *Journal of Clinical Psychiatry* **39**(7):592–598.
Mace NL, Rabins PV, Castleton BA, Cloke C and McEwen E (1985) *The 36-hour Day. Caring at Home for Confused Elderly People*. London: Age Concern and Hodder & Stoughton.
Marshall M (ed.) (1990) *Working with Dementia*. Birmingham: Venture Press.
Martindale B. (1987) Huntington's chorea: some psychodynamics seen in those at risk and in the responses of the helping professions. *British Journal of Psychiatry* **150**:319–323.
Martindale B and Bottomley V (1980) The management of families with Huntington's chorea: a case study to illustrate some recommendations. *Journal of Child Psychology and Psychiatry* **21**(4):343–351.
Mattsson B (1974) Huntington's chorea in Sweden. ii. Social and clinical data. *Acta Psychiatrica Scandinavica* (**suppl. 255**):221–235.
Office of Health Economics (1980) Scoones T. (ed.) *Huntington's Chorea*. Pamphlet No. 67. London.
Oliver JE (1970) Huntington's chorea in Northamptonshire. *British Journal of Psychiatry* **116**:241–253.
Oliver JE and Dewhurst K (1969) Six generations of ill-used children in a Huntington's pedigree. *Postgraduate Medical Journal* **45**:757–760.
Oxtoby M (ed.) (1982) *Genetics in Adoption and Fostering*. Practice Series 8. British Agencies for Adoption and Fostering.
Panse F (1942) *Die Erbchorea*. Leipzig:Thieme.
Peacock CE and Harris R (1989) Huntington's chorea: who cares? *Health Trends* **21**:15–17.
Pearlstein LS, Brill CB and Mancall EL (1982) Child abuse in Huntington's disease. *Paediatrics* **70**:630–632.
Pearson JS (1973) Behavioural aspects of Huntington's chorea. In: *Advances in Neurology, Vol I* (eds A Barbeau et al.). New York: Raven Press.
Pleydell MJ (1954) Huntington's chorea in Northamptonshire. *British Medical Journal* **2**:1121–1128.
Reed TE, Chandler JH, Hughes EM and Davison RT (1958) Huntington's chorea in Michigan, I. Demography and genetics. *American Journal of Human Genetics* **10**:201–225.
Saugstad L and Odegard O (1986) Huntington's chorea in Norway. *Psychological Medicine* **16**:39–48.
Simpson SA and Johnston AW (1989) The prevalence and patterns of care of Huntington's chorea in Grampian. *British Journal of Psychiatry* **155**:799–804.
Spillane J and Phillips R (1937) Huntington's chorea in South Wales. *Quarterly Journal of Medicine* **6**:403–423.
Teltscher B and Davies B (1972) Medical and social problems of Huntington's disease. *Medical Journal of Australia* **I**:307–310.
Tyler A (1982) The social, personal and economic burden of Huntington's chorea in South Wales. MSc Thesis, University of Wales.
Tyler A (1989) *Helping children live with Huntington's disease*. Special Report. Association to Combat Huntington's Disease, London.

Tyler A and Harper PS (1983) Attitudes of subjects at risk and their relatives towards genetic counselling in Huntington's chorea. *Journal of Medical Genetics* **20**:179–188.

Tyler A, Harper PS, Walker DA, Davies K and Newcombe RG (1982) The socio-economic burden of Huntington's chorea in South Wales. *Journal of Biosocial Science* **14**:379–389.

Tyler A, Harper PS, Davies K and Newcombe RG (1983) Family breakdown and stress in Huntington's chorea. *Journal of Biosocial Science* **15**:127–138.

Vessie PR (1932) On the transmission of Huntington's chorea for 300 years. The Bures Family Group. *Journal of Nervous and Mental Diseases* **40**:796–799.

Warriner S (1989) Huntington's disease: predictive testing. *Nursing Times* **86**:49–50.

Wexler NS (1979) Genetic Russian Roulette: the experience of being at risk for Huntington's disease. In: *Genetic Counselling, Psychological Dimensions* (ed. S Kessler). London: Academic Press.

Wexler NS (1984) Huntington's disease and other late onset genetic disorders. In: *Psychological Aspects of Genetic Counselling* (eds AEH Emery and I Pullen). London: Academic Press.

Willday P (1989) *The Difficulty of Choosing Not to Know*. Newsletter 36, Association to Combat Huntington's Chorea, London.

Work Group Report of the Commission for the Control of Huntington's Disease and its consequences (1977) Vol. I, US Dept of Health, Education and Welfare.

7

Management and Therapy

INTRODUCTION

In his original description, Huntington (1872) stated that 'treatment seems to be of no avail and indeed nowadays its end is so well known to the sufferer and his friends that medical advice is seldom sought'. More than a century later, there is still no effective or specific treatment for Huntington's disease (HD). Clinicians faced with the HD patient are often overcome by therapeutic nihilism: the disease will progress despite medical intervention and the prognosis may seem hopeless.

The successful management of the HD patient requires all of the clinician's skills and the efforts of a multidisciplinary team. Symptomatic pharmacological treatment is available but drugs are only part of the management of any patient, and in HD they play only a small part. Non-pharmacological management, which is often overlooked, can markedly improve the quality of life. Individual care-plans, based on a full assessment and identification of the patient's resources, should be constructed, preferably with the cooperation of the affected individual. Since the problems of the bedridden HD patient are much greater than those of the more ambulatory patient, the caring professions should make the preservation of mobility a primary goal. There are now numerous devices which can make the difference between dependence and independence but restitution of lost function should not be confined to mobility.

HD patients are vulnerable to concurrent medical illnesses, especially respiratory infections, so early and vigorous treatment of physical conditions is important. Attention to nutritional needs is essential; lack of concern for this in HD patients can result in weight loss, wasting and vitamin deficiencies. Social aspects of management including day care and residential accommodation will need consideration. In addition, the clinician will have to take into account the needs of the family and in particular the primary caregivers who are the single most important source of care for HD patients. Occasionally, the doctor will be called upon to deal with aspects of the civil law concerning matters such as the HD patient's capacity to care for his or her own property. The practitioner should also be familiar with legislation concerning compulsory admission to hospital.

The various forms of pharmacological treatment of HD are considered first here, followed by a review of neurosurgical and psychological forms of treatment. It is impossible to cover every aspect of treatment and rehabilitation and much helpful advice can be obtained from the voluntary organizations. Genetic counselling, a central part of management in HD, is considered in Chapter 11.

SPECIFIC APPROACHES TO TREATMENT

Pharmacotherapy

Numerous compounds have been assessed in the management of HD but most of these trials have had major methodological shortcomings. A substantial number of studies have been open trials which may offer evidence that a drug is beneficial but do not provide conclusive evidence of efficacy. The stringent test that a drug works is the double-blind controlled trial and there have been too few of these in the treatment of HD. In addition, many of the reported pharmacological studies have used small samples (less than 15–20 patients) so that it is difficult to draw clear-cut conclusions if a drug is stated to be beneficial. Many of these trials have been of short duration so that insufficient data have been collected on how long antichoreic drugs remain clinically useful. Some trials were performed when the patient was receiving multiple concurrent antichoreic agents. Another important consideration in reviewing the literature on pharmacotherapy in HD is which objective assessments (e.g. Abnormal Involuntary Movements Scale or functional assessment) were used to measure change.

Drug therapy will be described for chorea, rigidity, depression and psychoses. Details of agents used are given in Appendix 4.

Chorea
Among the earliest drug treatments for chorea were hyoscine and arsenic (Sinkler, 1889). Treatment by belladonna alkaloids continued until well into the present century (Tomlinson, 1947) when a number of reports showed that these drugs were unhelpful (Lazar, 1948; DeMeyer and Dyken, 1954). Other early drugs in treatment of HD were sedatives, in particular phenobarbitone. Nielsen and Butte (1955) advocated dimercaprol (BAL), a chelating agent, which had been found useful in Wilson's disease, another disorder of the basal ganglia. Its intramuscular mode of administration makes it unsuitable for prolonged use and in any case the value of dimercaprol has not been replicated. Similarly, the use of hydrolazine, an iron-chelating agent, has not been widely accepted (Hoerster *et al.*, 1961). Renewed interest in drug therapy was stimulated by Goldman (1952) who recommended procaine amide after the chance finding that a patient relaxed well after an injection for a dental extraction.

Table 7.1 World Health Organization classification of antipsychotic drugs

1. Phenothiazines, e.g. chlorpromazine
2. Thioxanthenes, e.g. flupenthixol
3. Butyrophenones, e.g. haloperidol
4. Diphenylbutylpiperidines, e.g. pimozide
5. Reserpine derivatives

Results of other trials with procaine amide were conflicting: Ganguili (1956) and Merskey (1958) found the drug beneficial for hyperkinesis whereas Lazarte et al. (1955), DeMeyer and Dyken (1954), Pleydell (1954) and Forrest (1957) failed to find procaine amide useful in the treatment of chorea.

The most useful drugs for the symptomatic relief of chorea are the neuroleptic or antipsychotic agents, which block dopamine neurotransmission. Conversely, dopamine and its agonists accentuate chorea. Antipsychotic compounds form a major category in the World Health Organization classification of psychotropic drugs (1967) (Table 7.1). They are subclassified into phenothiazines, butyrophenones and other groupings.

Some neuroleptic drugs can cause a number of adverse drug reactions, including anti-adrenergic, anticholinergic, cardiac and endocrine effects. The most common are extrapyramidal symptoms which occur in about 40% of patients treated with phenothiazines. These symptoms may be divided into two groups: in the first, the side-effects are relatively acute and dose-dependent whereas in the second, the side-effects are chronic, often irreversible and continue after cessation of neuroleptic treatment. Examples of the first group are the acute dystonias, which usually develop within 5 days of administration of the drug, pseudoparkinsonism, and akathisia which is characterized by uncontrollable physical restlessness. The most serious extrapyramidal effect is probably tardive dyskinesia which is a chronic syndrome of involuntary choreoathetoid movements affecting principally the face, tongue and lips ('bucco-linguo-masticatory syndrome') but also the arms, legs and trunk. It is estimated to occur in about 20–40% of patients treated with long-term antipsychotic medication (Marsden and Jenner, 1980). Unfortunately there is no effective way of ameliorating tardive dyskinesia, so prolonged treatment with these neuroleptic drugs should be limited to patients where this is essential.

The neuroleptic malignant hyperpyrexic syndrome, characterized by fever and rigidity, is another serious adverse reaction associated with this group of drugs. Burke et al. (1981) described its occurrence in a HD patient treated with dopamine-depleting agents.

Drug therapy for chorea should be held in reserve if the abnormal movements are slight, especially in the early stages of the disorder. The advantages of doing so are substantial. Unwanted effects, which increase with age, dosage and duration of treatment are avoided. In some patients, chorea fails to respond to medication and may even be worsened. Neuroleptic medication has a minimal effect on other motor manifestations such as gait and bradykinesia while dysarthria, dysphagia and dystonia may be aggravated. As the intensity of chorea lessens during the natural course of the disease, the apparently beneficial effects of anti-choreic agents may be artificial (Shoulson, 1983; Young et al., 1986). Finally, there is no evidence that drug therapy alters the progression of the disease and some authors have reported that antipsychotic agents may lead to a more rapid loss of functional capacity (Young et al., 1986; Shoulson, 1981; Shoulson et al., 1989a).

Pharmacotherapy will be indicated in patients with pronounced and disabling chorea and it is recommended that the clinician should use well-tried drugs whose actions and side effects are well known. Neuroleptic medication should be introduced cautiously and under close medical supervision, and the patient's age, weight, and other concurrent medications should be taken into account when dosage and dosage intervals are being determined. The clinician should explain to the patient or the relatives the likely therapeutic and unwanted effects. They should be asked to report any significant adverse reaction (e.g. sore throat on phenothiazines which may cause leucopenia) and warned that centrally-acting drugs may induce sedation. Time spent in discussing these issues should improve compliance.

Other points to be considered in treatment are whether the patient is taking any other drugs (e.g. sedatives, anti-hypertensive agents, anticonvulsants) that could be making the condition worse (Goldblatt and Bryer, 1987) and whether there are any concurrent disorders which could be improved by therapeutic intervention (e.g. thyroid or cardiovascular disease).

Choice of drug
A large number of antichoreic agents are available. All neuroleptic drugs block central dopamine receptors (Creese et al., 1976) but they differ in duration of action, unwanted effects and mode of administration. The phenothiazine, perphenazine, has been recommended as drug of first choice by the commission to control Huntington's Disease in the United States (1977). Haloperidol and reserpine were suggested as second-line therapies. A recent well-conducted trial involving haloperidol has shown that low doses of the drug (less than 10 mg/day) are the most effective (Barr et al., 1988). There is little justification for polypharmacy.

Replacement therapies
For the past 20 years, pharmacological investigators in HD, influenced by the success of the dopamine replacement model of Parkinson's disease, have based their studies on replacing or simulating neurotransmitter activity in HD. Replacement of various neurotransmitters has been advocated but the most extensively studied have been dopamine, γ-aminobutyric acid (GABA), and acetylcholine. Despite initial optimism, the limitations of the replacement strategy in HD are now clear. Firstly, several of the neurotransmitter systems,

not just dopamine, are affected, thus making full replacement difficult. Secondly, neuronal loss is progressive and receptors cannot be replaced even if neurotransmitter levels are raised. Thirdly, the drugs must cross the blood–brain barrier.

Anti-dopaminergic therapies
Presynaptic depletion of dopamine from neurones
RESERPINE. Reserpine is an alkaloid derived from *Rauwolfia serpentina*. It depletes neurones of the major neurotransmitters, including noradrenaline, adrenaline, dopamine, serotonin and histamine. It was advocated by several authors in the 1950s (Walther-Buel, 1955; Chandler, 1955; Lazarte *et al.*, 1955), but the drug has fallen into disfavour on account of its many adverse reactions, including depression.
TETRABENAZINE. Tetrabenazine, introduced in 1959, is a synthetic benzoquinolizine and is similar in pharmacologial action to reserpine in that it leads to neurotransmitter depletion of the pre-synaptic vesicles. Early trials of the drug in small numbers of HD patients found it had useful therapeutic effects (Sattes, 1960; Brandrup, 1960; Sattes and Hase, 1964). The suppressive effect of tetrabenazine on involuntary movement was confirmed in a series of 17 patients with choreiform, athetoid and ballistic involuntary movements (Swash *et al.*, 1972). These patients were filmed and assessed by 'blind' investigators. Among the adverse reactions were sedation, insomnia and depression, which may limit its clinical usefulness. Snaith and Warren (1974) warned of the risks of dysphagia and death from aspiration pneumonia with use of this drug.

Postsynaptic dopamine blockade. In HD, there have been trials involving drugs of all the subtypes of the antipsychotic agents. As stated previously, most of these have been anecdotal and for short periods. Some of the more systematic trials are discussed.

Barr *et al.* (1988) investigated the butyrophenone, haloperidol, in a study of 20 HD patients. They assayed serum haloperidol concentrations following administration of doses of 1–40 mg daily and used the Abnormal Involuntary Movement Scale (AIMS) to assess severity of chorea. They found that low doses of haloperidol (less than 10 mg/day; serum concentration between 2 to 5 ng/ml) produced significant improvement of abnormal movements and there was little clinical improvement on doses over this limit.

Sulpiride, which is a selective antagonist at the D-2 receptor site, and placebo were administered to 11 HD patients and nine with tardive dyskinesia in a randomized double-blind crossover trial (Quinn and Marsden, 1984). All patients had a 2-week 'wash-out' period before the trial began. Objective rating scales, including a functional disability instrument, were used to measure outcome. There was significant amelioration in chorea but functional improvement was not found.

GABAergic therapies

Gamma-aminobutyric acid (GABA) is an inhibitory neurotransmitter. Reductions in the concentration of glutamic acid decarboxylase (GAD) in postmortem brains of patients dying with HD have been demonstrated (Bird et al., 1973; Spokes, 1980). In HD, levels of γ-aminobutyric acid (GABA) are decreased in the substantia nigra, putamen, and globus pallidus (Perry et al., 1973) and in the cerebrospinal fluid (Glaeser et al., 1975). Three main groups of agents have been tried to enhance GABAergic neurotransmission in HD. They are the GABA precursors, such as glutamate, GABA-mimetic drugs, such as muscimol and GABA-transaminase inhibitors, such as isoniazid.

Barr et al. (1978) administered L-glutamate, the substrate for glutamic acid decarboxylase, and pyridoxine, its cofactor, to five HD patients. No significant improvement in chorea was found even after 2 years of medication. This is not surprising considering the diminished concentrations of GAD. Although, GABA itself was given to HD patients (Fisher et al., 1974) it is unlikely to cross the blood–brain barrier (Kuriyama and Sze, 1970).

Several drugs that exert GABA-mimetic activity have been evaluated. Muscimol, a GABA analogue that binds specifically to GABA receptors, was administered to 10 HD patients in a double-blind trial but failed to produce amelioration of chorea (Shoulson et al., 1978). Similarly, in clinical trials other GABA-mimetic drugs such as imidazole-4-acetic acid (Shoulson et al., 1975) and THIP (Foster et al., 1983) did not show improvement of chorea.

Baclofen [beta-(p-chlorophenyl)-γ-aminobutyric acid], another GABA agonist, was found to be beneficial as an anti-choreic agent in uncontrolled studies (Anden et al., 1973; Paulson, 1976). On this evidence and the evidence that baclofen retards corticostriatal release of glutamate and aspartate Shoulson et al. (1989b) conducted a placebo-controlled double-blind trial to investigate whether reduction in glutamergic transmission would result in a slowing of neuronal atrophy and hence of clinical progression. When 60 otherwise untreated HD patients who were randomized to either baclofen (60 mg/day) or placebo were followed up for 42 months, it was found that any short-term beneficial effects of baclofen were not sustained and that in the long term, the active drug may lead to a hastening of functional decline.

The third approach to elevating brain concentrations of GABA has been the inhibition of GABA transaminase, the first of two enzymes that degrades the neurotransmitter. Isoniazid is the most widely evaluated of these inhibitors. In animal studies, it was found to elevate brain GABA content (Perry and Hansen, 1973). Open trials have found that isoniazid is beneficial (in high doses) in ameliorating chorea (Perry et al., 1977, 1979). However isoniazid also inhibits glutamic acid decarboxylase, which synthesizes GABA and a double-blind, video-monitored crossover study found no objective improvement in chorea (McLean, 1982). Negative results with isoniazid were also reported by Neophytides et al. (1980). Other inhibitors of GABA transaminase, such as γ-acetylenic GABA (Tell et al., 1981) and aminooxyacetic acid (Perry et al., 1980) have not alleviated the motor symptoms of HD.

The finding that GABA receptors are reduced in HD brains (Bird, 1980) may help explain why increasing the concentration of GABA is not enough.

Benzodiazepines. The benzodiazepines are a group of widely prescribed drugs possessing anxiolytic, sedative, hypnotic, and anticonvulsant properties. Chlordiazepoxide in 1960 was the first to be introduced and was followed soon afterwards by diazepam. They act at the GABA/benzodiazepine receptor complex in the brain to enhance GABA action.

Specific receptors for benzodiazepines have been demonstrated and these may prove to be a target for an endogenous ligand (Möhler and Okada, 1978). Possible candidates for this ligand have included β carboline 3-carboxylic acid ethyl ester (which is now known not to occur naturally), diazepam binding inhibitor (a neuropeptide containing 105 amino acids) or a related peptide (Costa *et al.*, 1983; Guidotti *et al.*, 1983). Immunohistochemical studies show that diazepam binding inhibitor (DBI)-containing neurones are highest in certain hypothalamic nuclei, with decreasing frequency in the striatum and cortex (Alho *et al.*, 1985). In a post-mortem study of 11 HD brains, DBI-like immunoreactivity was more than 1.5-fold increased in the putamen, caudate, globus pallidus and nucleus accumbens compared with controls who had died of a non-neurological disorder (Ball *et al.*, 1988). DBI may be a modulator of GABAergic function and it may eventually explain how benzodiazepines increase GABA activity.

Reisine *et al.* (1979) have stated that the degeneration of striatal neurons and the associated loss of benzodiazepine receptors are central to the understanding of the neurochemistry of HD. It was therefore proposed that benzodiazepines would be effective in the treatment of chorea (Ban, 1969). However there have been few clinical pharmacological studies to test this hypothesis. Nilsen (1964) described the first HD patient in the literature to be treated with diazepam and identified a number of side-effects, including dizziness and ataxia. Clonazepam, which has the highest affinity for the benzodiazepine receptor, has been used in at least two studies of HD patients. Peiris *et al.* (1976) reported that in three HD sufferers, clonazepam was of value, both in subjective improvement and objectively, when patients were asked to copy geometric figures. Stewart (1988) supported this finding in a case study.

Cholinergic therapies
The finding of decreased striatal levels of choline acetyltransferase, which converts choline to acetylcholine (Spokes, 1980), in the brains of HD patients led to the 'hypocholinergic theory' of chorea (Aquilonius and Eckernas, 1977; Barbeau, 1978). This in turn led to the therapeutic strategy to increase cholinergic activity in the striatum. Aquilonius and Eckernas (1977) administered choline, the immediate precursor of acetylcholine, to five patients with HD, but reported that it was of little value. Similarly, 2-dimethylaminoethanol (Deanol), a possible precursor of brain acetylcholine, was found to be ineffective in

treating chorea (Laterre and Fortemps, 1975; Caraceni et al., 1978). Arecoline, a cholinomimetic, was studied in six HD patients and exacerbated rather than improved choreic movements (Nutt et al., 1978). Beneficial effects of anticholinesterases, such as physostigmine, on chorea have not been sustained and many patients had serious side-effects from these drugs (Aquilonius and Sjostrom, 1971; Klawans and Rubovits, 1972). Empirical evidence has therefore not supported the cholinergic therapies.

Miscellaneous drugs advocated for chorea
Lithium. Lithium salts were introduced as a treatment for hypomania and since then extensive clinical experience has been obtained in mood disorders. Lithium has been advocated in the treatment of choreic symptoms (Dalen, 1973; Mattsson, 1973; Bleiweiss, 1989). However, a number of double-blind trials have shown that there is no improvement in chorea or ability to perform everyday tasks (Aminoff and Marshall, 1974; Carman et al., 1974). Lithium has been suggested as a therapeutic measure for irritability and aggression in HD (Dalen, 1973; Leonard et al., 1974; Worrall, 1974) but more convincing trials are awaited. Lithium has a proven place in the treatment and prophylaxis of hypomania and so is indicated when this disorder occurs in HD.

Cysteamine. Reports of elevated somatostatin levels in the basal ganglia of HD patients led to a double-blind, placebo-controlled trial of cysteamine (2-aminoethanethiol), a somatostatin-depleting agent, in five affected patients (Shults et al., 1986). However, plasma and cerebrospinal fluid concentrations of somatostatin were not significantly reduced and clinically cysteamine failed to influence choreic movements even at maximum tolerated dosages.

Steroids. There have been conflicting studies on the immune response in HD. An anticaudate neuronal antibody has been reported (Husby et al., 1977) but others have not found specific or unique immunological abnormalities (Williams et al., 1978). Brown et al. (1979) gave prednisolone (30–45 mg/day) to two HD patients for 2–3 months and found no improvement of motor or other symptoms. However, Robinson and Thornett (1985) reported that prednisolone improved the frequency and amplitude of chorea in a 10-year-old boy suffering from benign familial chorea, as mentioned in Chapter 2.

Neurotoxins
Recently, experimental therapeutic attention has focused on the hypothesized excitotoxic mechanism underlying HD (see Chapter 5). The 'excitotoxin theory' suggests that neurodegeneration in HD is caused by a relative excess of excitatory neurotransmitters such as glutamate. Glutamate antagonists acting on the N-methyl-D-aspartate (NMDA) receptor such as the anticonvulsant MK-801 (Wong et al., 1986) may therefore have considerable therapeutic potential

(for reviews of anti-neurotoxic strategies, see Shoulson, 1983; Schwarcz and Shoulson, 1987; DiFiglia, 1990).

Rigidity
Most pharmacological studies in HD have involved attempts to ameliorate chorea and there is a dearth of clinical trials on other neurological features. Extrapyramidal rigidity, which is particularly seen in patients with juvenile onset (see Chapter 2), may respond to conventional antiparkinsonian drugs (levodopa, amantadine, bromocriptine, and anticholinergic agents). Many drugs used to control chorea, especially the antipsychotic drugs, produce extrapyramidal rigidity as a side-effect and they should be avoided in HD patients whose main neurological presentation is rigidity.

Depression
High rates (over 40%) of depression in HD have been reported (see Chapter 3) and it is probable that it is undertreated (Table 7.2). In a survey of 306 patients responding to a questionnaire requesting details on medication, it was found that only six patients (2%) were prescribed antidepressants (Ringel *et al.*, 1973). Similarly, only two out of 117 HD patients hospitalized in New York State institutions were on antidepressant medication (Korenyi and Whittier, 1967; Whittier, 1968).

There have been few studies on the pharmacological treatment of depression in HD. However, it is reasonable to prescribe conventional tricyclic antidepressants as first line drugs when significant depression occurs in HD (Whittier *et al.*, 1961; Folstein and Folstein, 1981; Caine and Shoulson, 1983). There is little to choose between the antidepressant effectiveness of the various antidepressants and the prescriber should be familiar with a small range of these compounds which has proven efficacy. Amitriptyline, which is sedating, has been recommended for the anxious depressed HD patient who is unable to sleep. Imipramine is another well-tried antidepressant but is less sedative. The

Table 7.2 Psychotropic medications prescribed for HD patients

	Korenyi and Whittier (1967) ($n = 117$)	Ringel *et al.* (1973) ($n = 306$)
No psychotropic medication	15 (13%)	29 (9%)
Neuroleptics	73 (62%)	204 (66%)
Antiparkinsonian drugs	13 (11%)	15 (5%)
Benzodiazepines	7 (6%)	57 (18%)*
Other tranquillizers	5 (4%)	— (—)
Anticonvulsants	2 (2%)	— (—)
Antidepressants	2 (2%)	6 (2%)

*Ringel *et al.* call this category minor tranquillizers.

drug, which can be given in a single dose (75–150 mg) at night, should begin to show a therapeutic effect after 2–3 weeks. Full dose should be continued for at least a further 6 weeks after which a maintenance dose (i.e. half of the therapeutic dose) of the drug is recommended for another 6 months (Mindham et al., 1973). If this is not done, patients may relapse because antidepressants do not cure depression but rather suppress symptoms.

HD patients are more vulnerable than the general population to the side-effects of antidepressants. The patient or carer should be told of the side-effects to be expected, such as dry mouth, constipation and blurred vision. A more beneficial side-effect may be weight gain.

Monoamine oxidase inhibitors, the other main group of antidepressant drugs, have been reported to be useful in the treatment of depression in three HD patients (Ford, 1986). However, these drugs should not be used as initial therapy for depression because of dangerous food and drug interactions and because comparative studies have shown that they are less effective in treating depression than tricyclic antidepressants (Medical Research Council, 1965).

Although double-blind controlled trials have shown that electroconvulsive therapy (ECT) is effective in treating severe depressive disorders (Freeman et al., 1978; West, 1981), no systematic studies exist on the indications, frequency of usage, problems (e.g. consent) and unwanted effects (e.g. memory disorder) of ECT in the treatment of depression in HD. The procedure requires a general anaesthetic but there are only anecdotal reports about good anaesthetic practice in HD. Affected individuals may be abnormally sensitive to the barbiturate sodium thiopental (Davies, 1966) and neuromuscular depolarizing drugs (Lamont, 1979).

Treatment of depression in HD is not confined to the prescription of antidepressants. Patients will need to be carefully assessed and some will need admission to hospital. All of them will require continuing support.

Psychoses
HD patients who become acutely psychotic should be treated in hospital where a full assessment can be made. It is preferable that the first few days are drug-free so that the mental symptoms can be elucidated and the patient's behaviour observed. Some disturbed patients will require early sedation with drugs such as chlorpromazine. The wide range of antipsychotic drugs has been mentioned but, as in the case of antidepressants, the clinician should become fully familiar with a small number of them. While the sedating effects of the antipsychotic drugs are immediate, the antipsychotic action of these drugs takes about 3 weeks. As mentioned previously, side-effects are common. Antiparkinsonian medication may be necessary, but should not be prescribed routinely. Depot preparations should be considered if patients will not take oral drugs reliably.

Cognitive impairment
There are no drugs which have been found to be effective for cognitive impairment in HD.

Neurosurgery
Stereotactic surgery
It is a century ago since Dr Gottlieb Burckhardt (1891) reported the first psychosurgical procedure in man but unfortunately his results were not encouraging. Nevertheless, 4 years later Dana (1895) described trepanation as a method of treatment of HD. Interest in neurosurgical treatment revived in the 1930s when Egaz Moniz, a Portuguese neurologist and Nobel Prize winner, cut the fibres between the frontal lobes and the subcortical structures in 20 patients with severe psychiatric disorders (Moniz, 1954). Early techniques, including the 'standard leucotomy', failed to take proper account of accumulating knowledge of neuroanatomy. More selective techniques were necessary and in the late 1940s, Ernest Spiegel and Henry Wycis of Temple University, Philadelphia, introduced stereotactic techniques. Patients with HD were among the first to be subjected to this form of treatment. Spiegel and Wycis (1950) operated on six patients with HD and selectively destroyed the medial globus pallidus because it was thought that increased pallidal activity caused by neuronal loss in the striatum was the neuropathological basis of chorea (Spiegel and Wycis, 1953). Three patients improved while two showed no change and the other suffered a hemiplegia. Several other workers performed psychosurgical interventions in HD but results were disappointing (Riechert, 1957; Nagao, 1962; for review, see Narabayashi *et al.*, 1973). Heimburger (1967) performed bilateral dentatotomies in one HD patient but there was no improvement in choreiform movements, although in patients with choreoathetoid quadriplegic cerebral palsy he found that rigidity and spasticity were relieved and that there was improved voluntary control of muscles. In the 1970s, scientific, and public opinion, turned against psychosurgery and legislation was later enacted in the states of Oregon, California and Alabama that effectively prohibited psychosurgery in these states.

Cell implantation
In the late 1980s, new interest has been stimulated in neurosurgical techniques for degenerative diseases. Researchers in Mexico described the clinical improvement with adrenal medullary grafts in two young patients (35 and 39 years) suffering from advanced Parkinson's disease (Madrazo *et al.*, 1987). However, patients who did not improve, got worse or died were not included in the original report. A substantial number of such operations (about 300) have now been performed, especially in the United States and China (Quinn *et al.*, 1989). The American results, reported at the 40th Annual Meeting of the American Academy of Neurology in April 1988 were not encouraging. Twelve-month postoperative mortality rate is about 7.5% and that improvement, if it occurs, tends to be transient. In the opinion of a *Lancet* editorial (1988), 'almost certainly, adrenal implants do more harm than good'. This conclusion was repeated in a later editorial in the British Medical Journal (Williams, 1990).

Adrenal autografts can increase the dopamine content of the striatum by up to 50%. While this would be of value in Parkinson's disease patients, it is more difficult to justify this procedure in HD patients because dopamine worsens choreiform movements. An adrenal graft has been performed in a single HD patient in Nashville, Tennessee. These workers justified it on the grounds that, in their animal experiments, adrenal transplants were protective against the toxic effects of quinolinic acid. This finding has not been replicated by other centres and at this stage it would be wise not to proceed with further adrenal transplants in HD until animal trials have yielded more conclusive results.

More recently, embryonic transplants have come to the forefront of investigation and debate. The first brain tissue transplant from one human being to another took place in Mexico in September 1987 when fetal brain cells were implanted in the brains of two patients with Parkinson's disease (Madrazo et al., 1988). Similar experiments have been performed in Britain (Hitchcock et al., 1988) and Sweden (Lindvall et al., 1990) and early results which were reported as promising have received wide media coverage. One of the largest series of embryonic grafts was presented in August 1989 at the Third International Symposium on Neural Transplantation by Molina (1990). These Cuban workers described neurotransplantation of embryonic cells into the caudate nucleus of 23 patients with advanced Parkinson's disease (Perry, 1990). In this disorder, the rationale is to implant fetal substantia nigra cells into the striatum so that dopamine levels may be permanently increased. Similarly, it is argued that fetal caudate cells would be beneficial in HD patients. However, this is unlikely because the pathological process in HD is considerably more extensive than that in Parkinson's disease (Quinn et al., 1989).

Such procedures raise major ethical issues concerning the use of tissue from the dead fetus. In Britain, a Government committee has addressed these controversial issues and recently published its report, 'Review of the Guidance on the Research Use of Fetuses and Fetal Material' (1989).

PSYCHOLOGICAL TREATMENT

Supportive psychotherapy

There are many different forms of psychotherapy but the most useful for the HD patient is supportive psychotherapy which may be defined as 'that form of psychological treatment given to a patient over an extended period, often many years, in order to sustain him, because he is unable to manage his life satisfactorily without such long-term help' (Bloch, 1986). Its aims include maintaining self-confidence yet accepting almost inevitable disability. The affected patient will have to adjust to role changes, such as giving up work, and loss events, such as loss of speech and coordination. Problems in relationships are also likely to arise. Reassurance, explanation and other aspects of supportive psychotherapy are considered by Bloch (1986). The clinician should also be alert to psychological defence mechanisms such as denial (Hans and Gilmore,

Table 7.3 Neurosurgical interventions in HD

Author	No. patients	Procedure	No. improved	Adverse effects
Spiegel and Wycis, 1952	6	Pallidotomy	3	Hemiplegia
Heimburger, 1967	1	Dentatectomy	—	Not described
Feinstein et al., 1965	5	?	—	Not described
South Wales 1950–1965 (unpublished)	5	Stereotactic surgery	—	One death soon after operation

1968; Martindale, 1987; see Chapter 3). Whittier (1963) reported on the particular problems of confidentiality when counselling HD patients.

Psychological intervention can be usefully extended from the affected individual to include the rest of the family (Pearson, 1973; Tibawi, 1978). Martindale and Bottomley (1980) described conjoint family therapy in HD and found that family members gained relief from open discussion of the disorder. Dynamic psychotherapy is probably inappropriate and may appear threatening to some HD families.

Behavioural psychotherapy

Problematic behaviours are well known in HD. Perhaps the most disconcerting are physical aggression and assault. These are often the result of disinhibited over-reactions to frustration, for example in response to an unexpected change in routine. Violent outbursts may, however, be inhibited in the presence of professional staff. Relatives also find verbal accusations and insults very distressing, especially if they are proclaimed in public.

In general the familiar daily routines of the HD sufferer should be maintained as far as possible. Reduction of environmental stimuli, avoidance of direct confrontation and clear, short, patient explanations should reduce aggressive behaviour. Patients' self-esteem should be preserved by inviting their views and not removing their independence and responsibilities prematurely. Measures such as day care, which other among things reduce constant direct contact between patient and carer should be considered.

Behavioural psychotherapy offers a more formal approach to the management of problem behaviours. This technique aims to change behaviour directly rather than by analysing childhood experiences. For the control of anger, a behavioural analysis may identify specific cues (e.g. being criticized), excessive behaviour (e.g. shouting and swearing) and deficit behaviour (e.g. listening). These principles and practical management advice are described in a useful Maudsley handbook (Marks, 1986).

Psychotropic medication may be necessary to help control irritable behaviour. This should be used sparingly, but particularly useful are the drugs thioridazine, promazine and haloperidol. Other drugs have been advocated for aggression, such as lithium (Worrall, 1974), propranolol (Stewart, 1988) and morphine but they have not been fully evaluated.

Memory aids

Memory impairment in HD can lead to difficulties in coping with everyday life (Oscar-Berman *et al.*, 1973). Memory aids can help those who have short-term memory problems. One of the most commonly used is the keeping of lists. Paper is often lost but a board in the patient's living room or kitchen may be more suitable. In the later stages of HD information such as the date and important tasks or events of the day can be written on the board and read out to the patient if necessary. A more formal technique to encourage correct verbal orientation is known as Reality Orientation Therapy, which involves group sessions of three to five patients meeting with staff for about 30 min twice weekly usually in an institutional setting. Visual and auditory aids are used to stimulate participation in for example naming common objects and discussing current events. New learning has been shown to occur in the elderly demented (Greene *et al.*, 1979; Woods and Britton, 1977) but the technique has not been specifically evaluated in HD.

SUPPORT SERVICES

Social services

The experience of most HD patients and their families, as well as those involved in their long-term care, has been that supportive management of a variety of types proves to be of considerably greater practical importance during the overall course of the illness than does the currently available drug therapy. Unfortunately, many patients still have difficulty in obtaining this support and in finding medical and other staff who have an interest and expertise in giving information. The following sections attempt to indicate some of the more important areas which are of help; the authors' experience is relevant mainly to the UK situation and it should be borne in mind that other countries will vary widely in how and to what extent these supportive mechanisms are available. Lay societies can play an especially important part in providing information and in attempting to raise standards of care for HD patients and their families.

In the UK, the most important of the publicly-funded services for the disabled and handicapped are those provided by local authorities through their departments of social services. The function of the social worker is to make an assessment of the disabled person's needs, to coordinate and develop appropriate services and to provide counselling and support to the patient and family. The most widely used services aimed at maintaining HD patients in their homes are described briefly.

Home helps

Home helps will do household duties like cooking, shopping, cleaning, washing and ironing (but not nursing). They also provide companionship for the client and relief for the carer. Home helps spend more time with the client than any of the other health and social services' workers (Levin *et al.*, 1989).

Local authorities provide this service on a means-tested basis usually only to households where the client lives alone or the carer is elderly or disabled. Home helps are the single most important personal social service for the elderly living at home and it is a pity that the service has not been offered to more elderly HD patients. Home helps are also provided by voluntary organizations such as the Women's Royal Volunteer Service (WRVS). They can be employed through the private sector though the cost of services may be much higher. There are about 130 000 home helps but they only cover the needs of about 10% of those over 65 (Audit Commission, 1985).

Care attendants
This group attends to the bodily needs (including washing, dressing and toileting) of the client and can be provided on a means-tested basis. The Care Sector Consortium has recently announced that better training and formal qualifications will be necessary for this group of workers (Murphy, 1990).

'Meals on wheels'
This service, consisting of a mid-day meal delivered to the patient's home, is provided by local authorities either directly (50%) or through voluntary organizations. A nominal charge may be made. Again, it is usually only available when the claimant lives alone or with a disabled or elderly carer. Only about 3% of the elderly receive such meals, often only once or twice weekly.

Laundry service
The laundry service, which is provided by both health and social services, is of great value, particularly for the bedridden and incontinent HD patient. Soiled linen is usually collected weekly. However, there is a great deal of underprovision. A general study found that only 7% of carers whose relatives were in residential homes received it during their relatives' last month at home (Levin et al., 1989). This may be explained by three factors: the service is rarely offered; relatives are often unaware of it; and even if they apply, only one in 20 families succeed in claiming it on first interview.

Day services
Sources of day services are day hospitals, day centres, luncheon clubs and sheltered workshops (see Table 7.4). Most insist that attenders live within a

Table 7.4 Day services

	Predominant provider	Main function
Day hospital	Health service	Occupational therapy
Day centre	Local authority	Occupational therapy
Day club	LAs/Voluntary	Social
Luncheon club	Voluntary agencies	Social/midday meal
Sheltered workshops	Dept. of Employment	Work

defined area or district. Transport, which is usually by ambulance or a special bus, can often create problems such as timing when patients can arrive and depart within an hour and the large distances (10–15 miles) some are expected to travel.

There have been definitional problems and demarcation disputes about the functions of various types of day service especially in the United States (Weiler et al., 1978) and France (Attias-Donfut, 1977). With some exceptions (e.g. Brocklehurst, 1970) this has been less of a problem in the UK, perhaps on account of joint funding arrangements between health authorities and local government.

A day hospital is a place in which patients spend a substantial portion of their waking time under a therapeutic regime and from which they return to their home or hostel to sleep at night. Day hospitals are a relatively recent phenomenon and they now tend to be specialized (e.g. psychiatric day hospitals, geriatric day hospitals). The officer in charge is a nurse who supervises medication and liaises with other professionals, community services and families. The programme often consists of occupational therapy, physiotherapy and group activities.

Local authorities provide most of the day centres (often in residential homes), although the voluntary agencies provide a smaller number. The range of occupation provided and the type of attender can vary widely. In areas of scattered population, the disabled may have access to only one day centre, regardless of age and type of disability. In areas of denser population, there is usually at least one separate day centre for the disabled individual under 65 years. It is often very difficult to occupy the HD sufferer in a day centre; the progressive neurological disability combined with memory impairment and short attention span make several of the activities provided unsuitable, but some attenders can benefit from social intercourse and simple musical and allied activities. Medical treatment is not provided.

Among HD patients, there are two main classes of day-care attender. The first is those who are awaiting long-term institutional care and the second those who are maintained in the community for a longer period. However, the uptake of day services by HD patients tends to be poor (see Chapter 6). Some have no insight into their disability and resent being placed among other disabled individuals. Others who are only too aware of their abnormal movements are too self-conscious to mix with others. Those with dysphagia may resent eating in company. In particular, HD patients with morbid jealousy may refuse to leave home so that they can maintain surveillance on the spouse. There can also be problems in placing HD patients together; some become distressed at seeing others similar to themselves. However, HD patients are more likely to accept day care as the disease progresses and cognitive decline supervenes.

Although day care may be valuable in the rehabilitation of the HD patient, the regular break (on average 2 days weekly) that it offers the supporter should not be underestimated. The amount of practical care (e.g. preparing food, feeding, changing incontinence pads, coping with demanding behaviour) that the supporter has to provide is reduced. Carers can go to work or pursue their own interests and so reduce social restriction. High expressed emotion (e.g. critical comments, hostility) has been reported in primary caregivers for a parent with dementia (Bledin *et al.*, 1990) and fewer hours spent together may reduce familial tensions. It is not surprising therefore that many supporters of HD patients not only welcome day care but want increased provision. In appropriate circumstances, every effort should be made to encourage the affected patient living at home to accept day services.

Health services
Family doctor
Many HD patients do not receive medical supervision over a lengthy period unless a regular home visiting service can be provided. It has been found in South Wales, with some notable exceptions, that general practitioners do not attend unless called and that families are reluctant to consult them unless they perceive a specific need. Prescriptions may have been collected for years without any enquiry being made about the patient's health. A periodic health check on the patient which gives an opportunity to discuss problems of care is considered very helpful and reassuring by almost all families. It can also enable the patient to remain in the community and delay or prevent admission to residential care by identifying problems at an early stage, which, if unattended, lead to the breakdown of existing arrangements. The family doctor may also detect other medical and psychiatric disorders. Chiu (1979) has outlined the role of the general practitioner in the management of HD families.

District nurses
In the UK, district nurses, who are usually attached to a group practice of family doctors, are the main providers of physical care in the home. Their duties include getting patients up and dressed in the mornings, administering medication and providing commodes and disposable bedding. They are occasionally accompanied by auxiliary nurses and bathing is sometimes done by special bathing attendants. Workloads are heavy and staffing levels are generally insufficient to provide a night-nursing service.

Health visitors
The profession of health visiting, which was established as long ago as 1862 in Britain, is concerned with the promotion of health and the prevention and detection of diseases that cause handicaps. Health visitors, who are usually attached to general practices, are qualified nurses who have spent a further year or more in specific training. They may help to support HD patients and families, especially where the welfare of young children is involved.

Community psychiatric nurses
Community psychiatric nurses may be employed by the health authority or the local authority. They can deal with psychiatric emergencies and enable many

patients to stay in the community. Their advice on the management of behavioural problems can be very helpful to carers of HD patients. They supervise depot medication but do not normally carry out physical care tasks such as dressing and feeding.

Dietitians
Dietitians, who mainly work in hospitals, although some also work in the community, can help in two main areas in relation to HD. Firstly, they can suggest the provision of a high calorie intake (up to 5000 cal/day) to minimise the weight loss which is a prominent feature of HD (Still, 1979; Morales *et al.*, 1989). Secondly, they can advise on swallowing difficulties (which compound the weight loss) and meal-time aids.

In advanced HD, the patient will be unable to feed him or herself and this may lead to feelings of frustration and anger and loss of self-respect. Patients should eat frequently and slowly with several pauses during the meal. More food should not be eaten until the previous mouthful is swallowed. Swallowing may be made easier by asking the patient to breathe out before placing food in the mouth and to breathe through the nose while chewing. After chewing, the patient should stop breathing and then swallow. Oral medication may be more easily administered in liquid or elixir form than in tablets.

Numerous meal-time aids are now available. Perhaps the simplest are durable plastic cups and plates. Handles of cutlery should be enlarged so that they are easier to grip; this can be easily done by putting sponge tubing or even rubber bicycle handlebar grips over them. If the HD patient is unable to grip at all, a leather strap that fits around the hand can be fitted to cutlery. HD patients often spill their food; a guard that fits around the edge of the plate can be useful. Non-slip mats or a damp cloth will prevent the plate from sliding on the table. A beaker with two handles (which should only be half filled to avoid spillage) is often useful to facilitate drinking.

Speech therapists
Speech therapists can be helpful in maintaining speech and communication for as long as possible. HD patients should have an initial assessment by the speech therapist who may refer them to a communication aids centre. However, assessed need often cannot be met because of staff shortages, so simple measures, such as allowing the patient ample time to say what he or she wishes, should not be ignored. Others include discussing one subject at a time, being aware of non-verbal communication and checking that dentures are properly fitted. As HD progresses, other communication aids should be considered (Wiltshire, 1987). Inexpensive methods include a series of cards with key words (e.g. 'hungry', 'thirsty') written on them. A 'letter board' consisting of letters (and numbers) allows the patient to form words by pointing to letters. A picture–word book may similarly be useful. In advanced HD, the patient should at least be able to indicate 'yes' or 'no'. The speech

therapist will be able to advise on these techniques as well as advising on the use of telephones and buzzers (Research Institute for Consumer Affairs, 1984).

Physiotherapists
After doctors and nurses, physiotherapists are the third largest direct patient-care profession in the National Health Service. They can make a significant contribution to the care of the HD patient, especially in an institutional setting, by working out programmes to improve or maintain balance, mobility and positioning. They can advise on a wide range of mobility aids and relaxation techniques. Increasingly, physiotherapists are working outside hospital settings and assess patients' physical capabilities in their homes. In addition, they can teach carers lifting and handling skills.

Occupational therapists
Occupational therapy became established as an important form of treatment at the turn of the century. Occupational therapists have considerable expertise in adapting the environment to suit the needs of disabled people. Most HD patients benefit from having their home surroundings assessed, and the occupational therapist can also advise on aids to daily living and how to make the home as safe as possible. In South Wales, provided consent can be obtained, patients are routinely referred to either the hospital or local authority occupational therapy department. The following sections describe the more commonly used aids but do not cover all the ones available (for further details, see Lavers, 1981).

Physical aids
The provision of rehabilitation aids for progressive disabling disorders such as HD has been relatively neglected in comparison with the facilities made available for diagnosis but many patients will benefit greatly from expert assessment and management in this area.

Wheelchairs
Wheelchairs can assist HD sufferers in retaining independence and enhancing mobility. They may however refuse to accept one because it would be to admit disability. A very wide range of wheelchairs exists. In Britain, the Department of Social Security (DSS), which provides wheelchairs on indefinite free loan, has a range of 66 models (Department of Health and Social Security). Other models can be bought privately. The three main types of wheelchair are the standard self-propelling wheelchair, the attendant (push) wheelchair and electric wheelchairs. The most suitable type for the HD patient is the attendant wheelchair; involuntary movements can often make self-propulsion an unrealistic goal. The wheelchair order form (AOF5G) must be completed by a

doctor. Patients should be referred to the nearest disablement service centre to define mobility need and for assessment concerning which model of wheelchair would be most useful (Male and Massie, 1990). This will avoid patient and family irritation that can arise if an unsuitable wheelchair is requested. The size of the wheelchair and the patient's home should also be considered as narrow doorways and steps will make wheelchair access difficult. A ramp and the widening of doorways may be a first requirement in these circumstances.

Owing to chorea, HD patients may need to be made more secure in the wheelchair. Truncal restraint straps and footstraps, which may prevent injury, are available on most models. HD patients who are unable to sit upright may need a reclining backrest and a headrest.

Bathing and toileting aids
Difficulty in bathing is common among HD sufferers and help with bathing is often provided by relatives or the district nurse. Useful bathing equipment includes a non-slip bath mat which should be placed on the bottom of the bath before the water is turned on. A board across the arms of the bath allows the patient to sit under a shower and to have the feet bathed. Bath rails are useful to help the patient get in and out of the bath or help to manoeuvre once there. Mechanical lifts are available to help the patient into the bath. Certain HD patients, especially those with muscular weakness or rigidity and those living alone may not benefit from the installation of the full complement of bathing aids. If the patient is living with a carer, it may be worthwhile to fix an alarm system in the bathroom. A raised toilet seat with integral grabrails is useful for the HD patient with rigidity and for the patient who cannot reach the toilet, a portable commode chair should be considered (Chamberlain *et al.*, 1978).

Anticipation of future needs is important (Turner-Stokes and Frank, 1990). Social services departments may take 6 months or more to provide a grant to build a downstairs toilet and another 6 months can elapse before planning permission is obtained and this facility is built.

Incontinence aids
Incontinence is not uncommon in HD (see Chapter 2) but unfortunately most HD patients with this problem do not seek professional help. Treatable causes of incontinence, such as urinary tract infections, should be sought. The most widely used of the incontinence aids are absorbent pads. They vary in design and patients should be offered a choice preferably by a continence adviser (in Britain, this is usually a specialist nurse) who can explain how each aid is used. A recent survey found the 'marsupial' pad and pants system can successfully control mild and moderate incontinence (Department of Health and Social Security, 1986). Other aids include adult diapers and incontinence sheets (Association of Continence Advisers, 1988).

ACCOMMODATION

Adaptations

The majority of HD patients live at home. Out of 137 living patients in South Wales, 110 (80%) are maintained in the community (see Chapter 6). Similarly in a series from the north-west of England, 119 patients (75%) were living at home (Peacock and Harris, 1989).

Home adaptations can help the patient lead a more independent life. Under the Chronically Sick and Disabled Persons Act, 1970, local authorities in Britain must find out the numbers of disabled people in their catchment areas and provide certain services including adaptations in the home. The main type of adaptation is the installation of special aids to daily living, for example rails on the stairway; safety measures, such as fixing loose floorboards, are also included. Aids are generally installed under the supervision of the social worker and the community occupational therapist.

Residential care

It will be necessary for a number of HD patients to transfer to accommodation more suited to their disabilities at some point during their illness. In a series of 77 deaths in South Wales, 46 HD patients died in institutional care and 31 at home (see Chapter 6). At any one time, one in five HD patients in South Wales is occupying a long-stay bed and recent figures from a north-west of England survey show that one in four HD patients there is in permanent care (Peacock and Harris, 1989). Residential care represents a mixture of public, private and voluntary provision.

Social services homes ('part III' homes)
Part III of the National Assistance Act 1948 provided the legal framework for the establishment of social services ('part III') homes. Many were built over the following 30 years, especially in the 1960s but this expansion has been markedly reduced over the past decade because of constraints on local authority resources. Nevertheless these homes provide the majority of the residential places available for the care of the elderly demented. One of their main advantages, especially for HD families, is that the clients remain in their locality with ready access to relatives and friends. On average, local authority homes have places for about 50 residents, over half of whom suffer from dementia. Some homes have also been provided for the care of the young chronically sick patient. As long-stay hospital beds are closed, patients who become progressively more disabled cannot be transferred to an environment with higher levels of nursing care. Another factor that accounts for the increasing disability of residents is that domiciliary services maintain the disabled in the community to a later stage.

Local authority homes are staffed by care assistants who have no nursing training and often little training in the problems of the demented and disabled.

However, in some homes specializing in the care of the more disabled patient, the officer-in-charge is a qualified nurse.

Problems not uncommonly arise in placing the HD patient in specialist local authority homes for the young chronic sick, especially if they exhibit disruptive behaviour which threatens the well-being of other residents. We are aware of one such home which will not accept any more HD clients. Nevertheless, some placements in these homes have been successful, particularly where a day centre is attached and beds are allocated for respite care which the patients have been able to use beforehand. Thus they are already known to the staff and the eventual transition to longstay care is much easier.

The legislation to 'provide care and attention which is not otherwise available' also permits local authorities to fund accommodation in residential homes organized by the voluntary and private sectors.

Private residential care (rest homes)

Private residential homes have burgeoned in the UK and the United States during the last decade. Fees vary enormously but do not always reflect quality of care. Under the Residential Homes Act (1984), residential homes must be registered with the local authority if they have more than four residents and the local authority can prosecute any home that has failed to register. The best residential homes are probably private but many others are unsatisfactory so that there is increasing demand for regulation of private rest homes. In the United States, Vladeck (1980) reported on the inadequate standards of private provision.

In South Wales, only one or two HD patients at any one time are in such homes. They can be suitable for patients who are still capable of maintaining a degree of independence but other provision has to be sought when they become more dependent. Problems can arise if the staff fail to understand the special needs of the HD patient.

Nursing homes

Nursing homes may be statutory or private. In both cases, they must be registered, regardless of the number of patients, with the local district health authority under the Nursing Homes Act (1975). If four or more of the residents are not patients, then the home must also be registered with the local authority. Regulations involving nursing homes are more strict than those concerning local authority homes. Nursing homes are reviewed at least twice a year by local authority medical and nursing officers. The National Association of Health Authorities has provided minimum standards, including the freedom of patients to go out whenever they wish and flexibility of visiting times. Size of rooms and design of the home must also satisfy regulations.

Patients in nursing homes are generally more disabled than those in residential homes and this is reflected in the higher fees. They often lack rehabilitative and recreational facilities, though some of the larger homes are making an effort to provide these. Medical input is usually weekly but there is little facility for investigations. These homes must provide 24-h qualified nursing cover. Pearson *et al.* (1990) identified problems of care which may arise in private nursing homes.

Voluntary and charitable homes
Religious and charitable bodies also provide residential accommodation. Certain criteria must normally be fulfilled such as membership of a certain church or occupation. Elderly Methodists for example, have priority admission to Methodist Homes for the Aged. There are homes for members of several occupations, including the police, nurses and miners. When trying to find a home for an affected patient, it is worth bearing his or her occupation and religious affiliation in mind. However, entry to such a home may mean moving a long distance.

Specialized centres for the HD patient
Specialized centres for HD patients have been established in a number of countries, including the United States, Canada, Australia and Britain. Some of those known to the authors are mentioned here.

The Arthur Preston Centre, Melbourne, Australia, cares for about 40 HD patients on a short and long term basis and offers advice to carers and community agencies on their management.

Stagenhoe, near Hitchin in Hertfordshire, UK, is a special centre for the care of patients with movement disorders. The estate, which is recorded in the Domesday Book, was bought by the Sue Ryder Foundation in 1969 and is now used for the care of 57 residents, about half of whom suffer from HD (Ryder, 1975, 1986). There is also provision for short-term stay to offer relatives a break.

Hamble ward at Knowle Hospital, Hants, UK is a ward where 16 brain-damaged patients, some of whom have HD, are managed. There is a multidisciplinary assessment profile on each patient to ascertain all areas of a patient's needs and to ensure that they are offered a wide range of activities (Shawcross *et al.*, 1990).

Oakwood Home, Stockport, UK, part of the Leonard Cheshire Foundation, opened in autumn 1990. It is a 16-bedded unit, with eight beds allocated for the care of HD patients and the remainder allocated to patients with severe accidental head injury. It offers regular short-term, but not long-term, care and a range of rehabilitation facilities.

Specialized homes which can develop expertise in the care of HD and offer a full range of facilities are obviously very desirable but it is perhaps unwise to organize any unit which admits only HD patients on a long-term basis. The strain on staff can be substantial and patients can be extremely upset at witnessing the inevitable decline in so many of their companions. In addition, relatives are often unhappy when affected individuals and indeed their whole families are 'labelled' when the patient is admitted to a HD unit. The pattern in the United Kingdom has been to set up mixed units.

Choosing residential care

A wide range of residential provision available for the HD patient in addition to hospital placement has been described. However, services vary widely in different parts of the UK; in particular, there is underprovision of specialized facilities for the care of the chronically ill patient under the age of 65 years.

What factors need to be taken into account when the decision to seek permanent residential care is made? Much depends on the age of the patient, the nature of symptoms, the degree of disability, social circumstances and the wishes of the patient and relatives. Other aspects influencing the choice of placement include distance, cost, ratio of staff to residents, expertise in the care of brain-damaged individuals and physical, medical and recreational facilities. It is unlikely that an ideal match will be found for all HD patients and each case should be assessed individually. The decision about placement will depend primarily on which of the patient's needs are paramount, but also on what is considered to be the best possible environment in the light of all the circumstances. Some HD patients, usually those exhibiting aggressive or very bizarre behaviour, are admitted to a long-stay bed in a psychiatric hospital. In the Maryland survey 33 patients (15%) needed hospitalization for irritability and aggression (Folstein, 1989). Comparable figures in South Wales are nine patients or 6.6% of a series of 137 patients (see Chapter 6).

In the terminal stage of the illness the overriding need is for good nursing care and to be within travelling distance of caring family members. In South Wales, small family practitioner hospitals provide a very high standard of care for women with HD. It can be very difficult to find a similar type of placement for men but some larger private nursing homes are organizing small units on their premises for male HD patients and will include those at an advanced stage of the disorder.

Elderly HD patients should, as far as possible, be admitted to local facilities, even if they are not entirely ideal, because these patients usually express a clear wish to remain in their familiar community and to be visited by relatives and friends who may be in the same age-group. There can be major difficulties in finding suitable placement for young HD patients. Families can complain with some justification that they are inappropriately admitted to psychogeriatric wards containing profoundly demented elderly people in surroundings which lack stimulation. Planned, phased care is the ideal (Peacock and Harris, 1989). However, so many patients refuse to attend day services or accept respite care that, in practice, this ideal is achieved in a very small minority of residential placements in South Wales.

SOCIAL WELFARE ENTITLEMENTS

It is necessary to include a section on this topic because many HD patients and their carers do not receive their rightful benefits. There are several reasons for this, including ignorance among professional staff. In Britain, large amounts of benefits are unclaimed; one authoritative body has estimated that $2000 million of benefits are not claimed annually (Association of County Councils, 1986). An estimated 33% of retirement pensioners fail to claim supplementary allowances to which they are entitled. It is likely that other countries will have similarly high rates of failure to claim state benefits.

Social security benefits may be divided into contributory (i.e. National Insurance contributions are required) and non-contributory benefits. There are a large number of benefits but only the ones most relevant to HD families are described. More details can be obtained from social services departments, self-help groups, Citizens Advice Bureaux and in social security leaflets and the handbook of the Family Welfare Association which is updated annually.

Contributory benefits
Statutory Sick Pay and Sickness Benefit
These benefits are applicable to HD patients who are unable to work. Statutory Sick Pay is paid by employers to employees who are incapable of work 'by reason of some specific disease or bodily or mental impairment' and have been so for four or more days. It is paid for up to 28 weeks during a spell of illness. Sickness Benefit is paid to someone who is not entitled (e.g. the self-employed) to Statutory Sick Pay because he or she has been excluded from the scheme. It is also paid for up to 28 weeks. Those who have never worked and never paid National Insurance contributions cannot get the flat-rate Sickness Benefit.

Invalidity Benefit
Invalidity Benefit is payable to individuals who remain unable to work and who have been getting Statutory Sick Pay or Sickness Benefit for 28 weeks. The requirements for Sickness Benefit contributions must be satisfied. There are two types of invalidity benefit: invalidity pension and invalidity allowance. The flat-rate Invalidity Pension is payable to all individuals who have been receiving Statutory Sick Pay or Sickness Benefit for 28 weeks. Invalidity allowance is paid in addition to invalidity pension and varies with the age at which incapacity began.

Non-contributory benefits
Attendance Allowance
Attendance Allowance is a tax-free social security benefit that is paid at two rates: first, if the claimant satisfies the prescribed conditions by day *and* night and second, if the conditions are satisfied by day *or* night.

To satisfy the day condition, the claimant must be so severely disabled that he or she requires from another person either:

(a) frequent attention throughout the day in connection with bodily functions, or
(b) continual supervision throughout the day in order to avoid substantial danger to himself/herself or others.

To satisfy the night condition, the claimant must be so severely disabled that he or she requires either:

(a) prolonged (over 20 min) or repeated (more than once) attention in connection with bodily functions from another person during the night, or
(b) another person to be awake for a prolonged period or at frequent intervals for the purpose of watching over him or her in order to avoid substantial damage to himself/herself or others.

The claimant must have satisfied one or both of the attendance conditions for a continuous period of 6 months before the claim. The allowance is not means-tested and the claimant must not be in certain kinds of accommodation provided for out of public funds (e.g. local authority home) or in hospital for 28 days or more.

Claimants should complete Department of Social Security (DSS) leaflet NI 205 and return it to their local DSS office. A doctor will conduct a medical examination at the claimant's home. If the allowance is refused, it is worthwhile seeking a review, which should be applied for within a period of 3 months. About two-thirds of individuals who seek a review are successful in being awarded the allowance. In South Wales, Attendance Allowance has been refused for certain HD patients because of the patient's lack of insight into the extent of disability, which may mislead the examining doctor regarding functional capacities. It is therefore recommended that a professional writes to the Medical Section of the DSS before the examination describing the patient's history, signs and extent of disability, stating also whether or not the patient is aware of the diagnosis.

Mobility Allowance
Mobility Allowance is a tax-free allowance for severely disabled individuals aged 5–80 years, although no new claims are permitted after age 65. The medical conditions are:

(a) the claimant must be suffering from a physical disablement to the extent that he or she is either unable or virtually unable to walk, or
(b) the claimant can walk but the exertion needed to walk would be a risk to life or would be likely to lead to a serious deterioration in health.

There is no fixed minimum distance laid down as a measure of walking ability although 100 metres is often used as a guide and the disability should be likely to last at least one year.

This flat-rate benefit may be claimed by completing form NI 211 which should be returned to the local Social Security office. Surprisingly, the money does not have to be spent on transport costs. The successful claimant is also exempt from road tax and receives an orange disabled badge to extend carparking rights.

Hospital travel costs
Travel costs to National Health Service hospitals may be claimed by two groups of patients: those who have automatic entitlement (e.g. in receipt of Income Support or Family Credit) and those who have low-income entitlement. The hospital pays the travel costs, usually public transport fares. If the hospital agrees that an attendant needs to travel with the patient, this person will also receive travel costs. Further details may be found on form H 11 available from the Department of Health.

Severe Disablement Allowance
Severe Disablement Allowance is payable to those who have been incapable of work for 28 weeks but are not entitled to Invalidity Benefit because not enough contributions have been paid. Claimants must be 80% disabled, which is a very difficult condition to satisfy. Medical assessments vary greatly and if a claimant has been refused it is worthwhile appealing to a medical appeal tribunal. Patients in receipt of Attendance or Mobility Allowance will be treated as 80% disabled.

Soon after the initial diagnosis of HD is a suitable time for a professional, usually a social worker (if the patient agrees to the referral), to discuss the range of benefits and services available and where further advice may be obtained.

Benefits for the carer
Invalid Care Allowance
Invalid Care Allowance is a taxable flat-rate payment to individuals who spend at least 35 h weekly caring for a severely disabled person who is in receipt of Attendance Allowance or Constant Attendance Allowance. It is payable to single or married people of both sexes but the carers must be under pension age when first claiming the allowance. It is not necessary to be a relative or to live with the disabled person. The claimant must not be earning more than a certain sum weekly; further details are available from the DSS (form NI 212).

Home responsibilities protection scheme
This scheme protects the pension rights of those who cannot work regularly as they are caring for an elderly or disabled person at home. For each tax year spent at home, the number of years of National Insurance contributions needed to qualify for a full pension is correspondingly reduced.

Tax allowances
Tax allowances are granted to people who care for elderly disabled people.

Dependent Relatives' Allowance
This allowance is available to people who support various relatives, including those aged over 65 years.

VOLUNTARY AGENCIES

Statutory organizations on their own cannot meet all the needs of the disabled and charities and voluntary organizations have a major contribution to make in the field of welfare. Many of these organizations depend on grants from local and central government.

There are two types of voluntary agencies: the self-help groups and the agencies which help other people rather than their own members. The self-help groups, which consist of affected people and their families, have the greater role to play in providing services for HD families. Examples of the second type of agency are the Samaritans, the Women's Royal Voluntary Service (WRVS) and the Citizen's Advice Bureau which offers advice on many legal matters and welfare entitlements.

In *Partnerships for Health* (1987), the National Council for Voluntary Organizations (NCVO) and the health authorities recognized the potential for closer collaboration between the health service and the voluntary organizations. The voluntary sector could enhance the health service in three main ways: by extending the range of flexible and informal provision; by representing the patient's needs; and by providing services which do not exist elsewhere. The HD societies not only attempt to fulfil these objectives but extend their responsibilities into supporting research, parliamentary lobbying and professional and public education.

Roles of the HD self-help groups
Service providers
The lay societies (see appendix 3) offer wide-ranging support services for the affected, the at-risk and carers. Many of them provide immediate practical advice by telephone. Individual counselling and support are also available together with advice on legal and insurance matters. Many family members find that group meetings and discussions provide mutual support and they are often relieved to discover that problems that they thought were unique to them have been experienced by other members of the group. A number of societies have established or helped to establish homes for respite or more permanent care.

Education and research
The lay organizations publish educational material on HD for professionals and relatives. Many aspects of HD are covered, including diagnosis, genetic counselling and the effect of HD on the families. Advice is given on other important topics, such as nutrition, sexual problems and incontinence. Most societies publish a regular newsletter to allow members, especially those who are unable to attend meetings, to share experiences and to keep them up to date on research activities. More recently, videos have been produced. National and local conferences for professionals are also arranged. The HD societies are one of the major sponsors of research.

Fund-raising
The HD societies are active in raising funds for education, research and patient and family services. Societies in several countries have now instituted an annual 'HD awareness' week or month with public appeals in the press and television. Certain governments have permitted charitable donations to be tax-free.

Pressure groups
The HD self-help groups can bring about changes in health services and they are uniquely placed to articulate consumer views. Perhaps the most influential result of lobbying of the self-help groups was the formation of the Commission to Control HD in the United States. The HD societies often join with other medical and paramedical organizations to respond to issues of common interest.

International Huntington Association
The International Huntington Association was established in 1974 by representatives from the United States, Canada, United Kingdom, Australia and Belgium. It now (1991) has 27 member societies, meets regularly with the World Federation of Neurology HD Research Group and has an important role in coordinating policies, including the guidelines on predictive testing (World Federation of Neurology, 1989, 1990). The European Confederation of Huntington Associations comprises the European members of the IHA and meets annually.

Information on HD societies was obtained by O'Shea (1989) who sent questionnaires to the societies in Britain, Ireland, Holland, USA and New Zealand. He collected data on the history of each group, the settings in which HD patients were cared for, attitudes towards genetic counselling and their relationship with the medical profession.

Further details are given here on the founder members of the International Huntington Association.

United States. Marjorie Guthrie, whose husband was the legendary folksinger Woody Guthrie, a sufferer from HD, founded the Committee to Combat Huntington's Disease (see Chapter 1). She was spurred to raise public awareness of the disorder, provide support for affected families and obtain research funds. She was highly successful and in 1977, she chaired the influential US Commission for the Control of Huntington's Disease and its Consequences.

A number of separate voluntary HD organizations in the United States have now been coordinated. Among these are the Hereditary Disease Foundation (formerly the California Committee to Combat Huntington's Disease) which was founded in 1974 by Dr Milton Wexler. Similarly, the National Huntington's Disease Association, which was formed in 1976, developed from branches of the Committee to Combat Huntington's Disease. In 1967, the Huntington's Disease Foundation was established by Mrs Alice Pratt in Houston, Texas. In 1986 their two bodies merged to form the Huntington's Disease Society of

America (HDSA). Local services are provided through its 100 chapters and affiliates and about 4000 inquiries are received and answered annually. Its quarterly newsletter, the *Marker*, is distributed to 36 000 people. Public service announcements are regularly broadcast on television. Together with patient services and research, the HDSA places considerable importance on advocacy; in 1989/1990, HDSA members visited 135 congressional offices and trustees of the society regularly offer testimony to congressional committees on issues such as health insurance and the need for more medical research funding. In collaboration with other health agencies, the HDSA has also been successful in havingا Medical Research Day declared by the governors of 39 states and in having the 1990s designated by Congress as the 'Decade of the Brain'.

United Kingdom. In Britain, the Huntington's Disease Association (formerly the Association to Combat Huntington's disease) was formed in 1971 by Mrs Mauveen Jones. It receives a grant from the Department of Health and, among other activities, it produces a twice-yearly newsletter, supports multidisciplinary workshops, sponsors a holiday home and a family counselling service; it also funds scientific research. Its membership (1990) is over 7000 of whom nearly 2000 are professionals. It is a founder member of the Genetic Interest Group, an influential organization of the self-help groups concerned with genetic services. The society has contributed to a number of television documentaries and often provides advice to the media.

Canada. The Huntington Society of Canada, founded by Mr Ralph Walker in 1973, was modelled on the Committee to Combat Huntington's Disease. At present, there are 45 chapters and area representatives in Canada. After the discovery of the G8 marker, the society opened the first DNA bank specifically for HD (at the University of British Columbia) which currently stores about 2000 samples. The society publishes a quarterly newsletter called *Horizon* and a French version of this periodical and other booklets are available from the Huntington Society of Quebec. A medical identification card for affected individuals is provided free on request.

Australia. The Huntington's Disease Association (New South Wales) was formed in 1975 and held its first national conference in 1979 at Melbourne. It provides a comprehensive service to families, supports the HD clinic and register at Lidcombe Hospital (since 1982) and has organized a holiday camp for HD patients since 1981. It has also established a successful 'Phone-a-Friend Service' in response to those who live too far away for direct participation. Its newsletter, published six times a year, is distributed to over 300 families, health professionals and various organizations.

CARERS

As has been seen, it is generally the immediate family who take on the responsibility of caring for an affected individual and more often than not, the primary caregiver is the spouse. Even if other family members are aware of the problems of caring it is not unusual for the whole responsibility to be left to one person. In such cases, it is often helpful for a professional to arrange a meeting with the family to discuss greater participation in the burden of caring. However, as more women join the work force the availability of carers is decreasing.

Informal carers face many problems and frustrations which should be recognized (Morris et al., 1988; Gilhooly and Whittick, 1989). These include loss of income and reduced social life. Support for carers is often lacking. There is insufficient provision of home helps, sitters-in, day and/or night care-attendant schemes and day services and this places a considerable burden on carers whose work is too often taken for granted. Stress, resentment and even violence can lead to collapse of informal care, at which point statutory services have to take over (see Chapter 6). This problem was recognized in the Griffiths Report on Community Care: 'The first task of the publicly funded services is to support and where possible strengthen these networks of carers. Public services can help by identifying such actual and potential carers, consulting them about their needs and those of the people they are caring for and tailoring the provision of services accordingly'.

The National Institute of Social Work has conducted a comprehensive study of the needs and strengths of carers of elderly, severely disabled patients (Levin et al., 1989). Most carers wanted to go on caring. Domiciliary and other services were effective in preventing the build-up of strain in carers and so postponing the need for admission to institutional care. The report proposed the following key requirements in domiciliary support:

(1) Early identification. Support services should be provided early before the carers are overly strained.
(2) Comprehensive medical and social assessments. Assessments should be multi-disciplinary, Carers welcomed the following conduct of assessors: early arrangement of interviews; identification of who they were, where they worked and why they were visiting; clear explanations; and the prompt implementation of a clear plan of management.
(3) Timely referrals by doctors and social workers to other agencies if they were unable to provide the necessary expertise themselves.
(4) Continuing back-up and reviews. Supporters valued the continued involvement of the relevant professionals, especially when patients had little prospect of improvement.
(5) Active treatment of concurrent medical conditions.
(6) Information, advice and counselling. Supporters wanted basic information on common care-giving activities like how to lift and how to manage incontinence. They also wanted to know about the course of the illness, the availability of local services and for which benefits the patients were eligible. In particular, carers appreciated professional recognition of the work that they did for their relatives. Counselling may also help deal with poor coping strategies, critical comments and hostility (Bledin et al., 1990).

(7) Regular help with household and personal care tasks. Practical help from the home help and laundry services improved the psychological well-being of the carers.

(8) Regular breaks. Although carers received some help from other relatives, friends and neighbours, few of them had been relieved for 24 h or more in the preceding year. Supporters indicated that there is a clear need for more short-stay admissions to homes and hospitals.

(9) Regular financial support. Heating bills, extra washing, new clothes and other expenses together with loss of earnings caused financial problems. HD patients are especially prone to household accidents, including the breakage of china and television sets. Social security payments had psychological value in that they were seen as official recognition of the heavy dependency brought about by illness.

Some services for carers

Care attendant schemes

Care attendant schemes, which are usually voluntary, can allow a carer some hours off. The 'Crossroads Care Attendant Scheme' is one of the best known of these schemes in Britain. Unfortunately, due to a shortage of volunteers, the service is not available in certain areas or there may be long waiting lists. Further local information can be found from the voluntary services officer at the social services office.

'Good neighbour' schemes

In 'good neighbour' schemes, such as that initiated by Kent Social Services Department, there is a special budget so that neighbours can be paid to provide domiciliary care (Ferlie, 1983; Tinker, 1984). The 'professional neighbour' may be contracted to cook, clean and shop for one or several disabled people.

Home care attendant schemes

These schemes have been developed by both Social Services and Community Nursing Departments with the aim of keeping the (younger) disabled person in his or her own home for as long as possible. There may be one attendant designated to a disabled person or a team who work on a shift system with several disabled individuals.

Short holiday breaks

Local authorities may be able to arrange the short-term admission of the patient to a designated 'Holiday Home' giving the carer an opportunity to have a holiday. The local authority should be informed of the level of disability of the affected person. Hospitals may also provide 'holiday relief' admissions. In some parts of the country, there are schemes whereby voluntary carers will look after a patient while the primary caregiver is away. Some families may agree to accept (or 'board-out') a disabled patient in their homes for short periods. In addition, private homes can offer more expensive short-stay residential facilities. In the UK the Huntington's Disease Association runs a small holiday home, as already mentioned.

LEGAL ASPECTS

This section is concerned with aspects of the law which may be relevant to HD patients. Firstly, consent to treatment and compulsory admission to hospital is considered. We then discuss aspects of the civil law concerning management of the patient's property. Although high rates of criminality have been described in HD, the issue of criminal responsibility and the preparation of court reports are not considered here; useful information may be found in other sources (Gibbens, 1974; Prins, 1980).

Consent to treatment and compulsory admission to hospital

Common law respects the right of people to decide what shall be done with their bodies. In medical practice, an individual can refuse treatment even if it is likely to benefit him. The precise meaning of the word consent has been much debated (Bean, 1986) but it implies awareness of information, choice and freedom from coercion.

Many HD patients are unable to give informed consent to necessary medical treatment, yet administration of treatment without consent constitutes assault. In practice, consent is taken to be implied if the patient cooperates or at least does not physically resist. It is usually wise to consult relatives on these matters although, in English law, relatives are unable to give consent on behalf of an affected individual (e.g. for an operation).

Compulsory powers of detention are necessary to protect the mentally ill and society from the effects of mental disorder. In Britain, there have been laws, beginning with the Poor Law and the Vagrancy and Madhouse Acts, for over 250 years concerning the admission and detention of psychiatric patients. In the last century, the Lunatics Act (1890) provided for the admission of patients to hospital after certification by a magistrate. Involuntary admission became a medical matter with the Mental Health Act (1959). Currently the Mental Health Act (1983) provides for the compulsory admission and detention of the mentally ill in England and Wales. No one can be defined as having a mental disorder by reason only of promiscuity, sexual deviancy or dependence on alcohol or drugs. In other countries, such as some states of the USA, compulsory admission is decided in court. Unhurried explanation and persuasion can often encourage the patient to accept admission voluntarily, thus avoiding compulsory admission. Family relationships may be irrevocably damaged if the nearest relative signs an application form for compulsory admission.

Section 2: admission for assessment (up to 28 days)
The grounds for admission under Section 2 of the Mental Health Act (1983) (England and Wales) are that the patient must suffer from a mental disorder and that admission is necessary for the patient's own health or safety or for the protection of others. Application is by the patient's nearest relative or an approved social worker and the order must be signed by two doctors.

Section 4: emergency admission (72 h)
The grounds for admission are the same as for Section 2. The application is by the nearest relative and only one medical recommendation (usually the general practitioner) is required. The patient must be seen and admitted within 24 h.

Section 47, National Assistance Act (1948)
This section is an older provision for the compulsory admission to hospital or residential home of individuals who are unable to look after themselves properly and who are a risk to themselves or other people. The order is issued by a magistrate on the recommendation of two doctors, one of whom is usually a community health physician. The patient or relative is given a week's notice of the order which only lasts 3 weeks but may be extended to a maximum of 3 months. (The National Assistance (Amendment) Act (1951) provided for the immediate admission to care on application to a magistrate.) In practice, this section is not commonly used, perhaps about 300 times a year in England and Wales, but there are no central statistics on its use (Bean, 1986). However, recent legislation (Chronically Sick and Disabled Persons Act (1970) which imposes a duty on local authorities to seek out those qualifying for help, and the Housing and Homeless Person Act (1977)) may have increased the usage of Section 47.

In the South Wales HD population, this legislation is known to have been used three times in the past 16 years. In one case, a 50-year-old patient who lived alone was found by neighbours lying wet and naked on top of her bed on a winter's night. The house was flooded and without electricity after the water pipes had burst. She refused to leave her house and compulsory admission to hospital was necessary.

Guardianship
In England and Wales, guardianship has been a part of mental health legislation since the Mental Deficiency Act (1913). Its powers were extended in the Mental Health Act (1959) but were rarely used (164 persons were under guardianship in 1982; Bean, 1983). Restrictions were introduced in the Mental Health Act (1983) to encourage greater use (Bean, 1983). The powers of guardians include the following (Bluglass, 1984):

(a) The power to require the patient to reside at a certain place specified by the person or authority named as a guardian.

(b) The power to require the patient to attend at places and times so specified for the purpose of medical treatment, occupation, education or training.
(c) The power to require access to the patient to be given, at any place where the patient is residing, to any registered medical practitioner, approved social worker or other person so specified.

Guardianship for patients in the community may be an alternative to compulsory detention but over the past 16 years there have been no guardianship applications for HD patients in South Wales.

Aspects of civil law relating to the patient's property and affairs

Power of attorney
If an individual is unable to manage their affairs because of physical disorder, advancing age and other reasons, they can, in countries such as Britain and Canada, sign a power of attorney authorizing a spouse, friend or another named person ('the attorney') to discharge financial and other affairs on his or her behalf. The patient must be fully aware of the consequences of signing a power of attorney and it can be revoked at any time. Therefore a person with advanced dementia cannot authorize these powers and in addition, an existing power of attorney is invalid if the patient becomes mentally incapable. The purpose of this is to protect the mentally infirm from exploitation by relatives or friends.

Enduring power of attorney
Enduring power of attorney, which was introduced in 1986, is a mechanism whereby the power of attorney continues if the individual becomes mentally incompetent. When it is drawn up initially, the person concerned must understand the nature of this legislation. If mental disability supervenes, the power must then be registered with the Court of Protection.

Court of Protection
The Court of Protection was established for the management of the property and affairs of the mentally incompetent. If a patient is so mentally disordered as to be unable to look after their own affairs, a relative must apply to this court (usually through a solicitor) so that a receiver may be appointed to administer the patient's affairs. Two medical certificates are required. If there is no relative to act as receiver, the judge or Master of the Court of Protection may appoint an official of the court. Legal and Medical Lord Chancellor's Visitors are appointed to ensure that the patient's affairs are being managed appropriately but because of the considerable numbers under the care of the Court (about 22 500 in 1985) visits to patients in hospital or residential homes are not given first priority (Gostin, 1983). Permission from the Court must be obtained before a wide variety of transactions, such as the sale of a home in joint names. Decisions of the Court can take several months.

Other authorization procedures
Agency. It is usually only necessary to apply to the Court of Protection when the mentally ill person has substantial savings or owns property. For patients of more modest means, procedures like agency and appointee may be more relevant. Pensions may be collected but not spent by a named individual ('the agent') after appropriate signature in the pension book and the agency authorization. The agent is usually the nearest relative or occasionally the local authority if the patient is in a residential home.

Appointee. A person, usually the next of kin, may be appointed by the Department of Social Security to collect and spend state benefits in the event of temporary or permanent incapacity. As with agency, the local authority may also be made the appointee.

Testamentary capacity. Testamentary capacity refers to the ability of a person to make a valid will. A will may be invalid if the testator is shown to have been mentally ill at the time the will was signed. The law uses the test of whether the person was 'of sound disposing mind' if there is doubt about testamentary capacity. This means that the testator is aware of what a will is; knows what assets exist; can make a fair and reasonable judgement about the claims of others to any property; and is not influenced by an abnormal state of mind (e.g. delusions) that might prejudice decisions.

Medical fitness to drive
In the UK, driving licences are issued by the Driver and Vehicle Licensing Centre (DVLC), Swansea. Licences extend until the driver is aged 70 or for a period of 3 years, whichever is the longer. Applicants must make a declaration of health and may be required to undergo a medical examination. The medical assessment of drivers at the Licensing Centre is the responsibility of the Medical Advisory Branch which is headed by the Medical Adviser. Applicants must declare any prescribed 'relevant disability' (e.g. epilepsy, impaired vision or 'any other disability likely to cause the driving of a vehicle to be a source of danger to the public'). The applicant must also declare any 'prospective disability' which may in future impair ability to drive and become a 'relevant disability'.

There is a duty imposed on licence-holders to inform the Licensing Centre as soon as they are aware of any relevant or prospective disability. This obligation is printed on every driving licence: 'You are required by law to inform the Drivers Medical Branch, DVLC, Swansea at once if you have any disability (includes physical and mental condition) which is or may become likely to affect your fitness as a driver unless you do not expect it to last for more than three months'. Patients often do not realize that the onus is on them to notify the licensing authority. Doctors and other professional staff have a responsibility to inform patients that they have a medical condition which may affect ability to drive and such advice should be recorded in the case-notes. Only in exceptional circumstances should the doctor inform the licensing authority without the consent of the patient.

HD is specifically mentioned by the Medical Commission on Accident Prevention (1985) as a condition likely to interfere with fitness to drive. Depending on the patient, an existing license may be replaced by one valid for 1, 2 or 3 years and this enables a regular medical review to take place. In the United States, there is a current public debate about the issue of 'limited licences', for example licences for journeys to the local shops but not for long freeway travelling; however, these may be difficult to monitor (Editorial, 1990).

Insurance and HD

Insurance companies normally charge higher premiums for certain types of policies, if the applicant is at risk of developing HD. However, at least one UK insurance company operates a sliding scale of charges, which reduces according to the age of the applicant, as long as the latter remains asymptomatic. After the age of 55 years, the charges are identical to those paid by the general population.

It does not appear to be the policy in the UK to refuse to insure any at-risk person, but the size of the premium requested may effectively put it out of their reach. Persons known to be suffering from HD at the time of application are probably uninsurable. The lay society, in Britain, provides a booklet and information about insurance brokers who are prepared to negotiate cover for at-risk applicants.

If a policy is taken out before the applicant is aware of any family history the contract will have been completed in good faith and should be honoured. But if the applicant knows of his or her at-risk status and does not declare it, although requested to do so on the application form, the policy could be rendered null and void and the premiums already paid may be lost. In Britain, the applicant's general practitioner is normally the doctor who provides the medical report for insurance purposes and is legally obliged to disclose all relevant details, which normally include a family history of HD in a parent. For these reasons persons at risk who seek advice on this topic should be advised to complete all contracts honestly.

Obviously practices regarding insurance vary between countries, and the financial consequences of being at risk for HD will vary accordingly. Some of the difficulties which may arise elsewhere, particularly in relation to private health care in the United States, have been discussed in detail by Brandt *et al.* (1989), Quarrell *et al.* (1989) and Quaid *et al.* (1989).

The arrangements for life insurance in Norway deserve special mention (Dr A. Heiburg, personal communication, 1990). Private life-insurance companies here do not face competition from foreign companies; they do not seek to identify and target low-risk groups; consequently individuals at risk for HD are not disadvantaged but may obtain cover at the usual rates. Individuals already diagnosed as having developed HD would be considered uninsurable.

With regard to presymptomatic testing, it is our understanding for the UK that a test result, whether the applicant's risk is raised or lowered, currently makes no difference at present to the amount of premium charged. This appears to be because insurance companies, as yet, do not have sufficient knowledge to fix accurate differential rates. The possibility exists however, that in the future, persons with a significantly raised risk will be considered uninsurable, a general issue which requires widespread public debate before the results of predictive testing are used in relation to insurance, for HD and for other genetic disorders.

REFERENCES

Alho H, Costa E, Ferrero P, Fujimoto M, Cosenza-Murphy D and Guidotti A (1985) Diazepam-binding inhibitor: a neuropeptide located in selected neuronal populations of rat brain. *Science* **229**:179–182.

Aminoff MJ and Marshall J (1974) Treatment of Huntington's chorea with lithium carbonate: a double-blind trial. *Lancet* **i**:107–109.

Anden NE, Dalen P and Johansson B (1973) Baclofen and lithium in Huntington's chorea. *Lancet* **ii**:93.

Aquilonius S-M and Eckernas S-A (1977) Choline therapy in Huntington chorea. *Neurology* **27**:887–889.

Aquilonius S-M and Sjostrom R (1971) Cholinergic and dopaminergic mechanisms in Huntington's chorea. *Life Sciences* **10**:405–414.

Ashbury JFP (1986) Confidentiality and third party interests. *Lancet* **ii**:1388.

Association of Continence Advisers (1988) *Directory of Continence Aids and Appliances*, 4th edn. London: Association of Continence Advisers.

Association of County Councils (1986) *Annual Report*. London: Association of County Councils.

Attias-Donfut A (1977) *Les Centres de Jour*. Paris: Le Ministère de la Santé et la Caisse Nationale de l'Assurance Vieillesse.

Audit Commission (1985) *Managing Social Services for the Elderly More Effectively*. London: HMSO.

Azima H and Ogle W (1954) Effects of largactil in mental syndromes. *Canadian Medical Association Journal* **71**:116–121.

Bachman DS, Butler IJ and McKhann GM (1977) Long-term treatment of juvenile Huntington's chorea with dipropylacetic acid. *Neurology* **27**:193–197.

Ball JA, Burnet PWJ, Bretherton-Watt, D and Bloom SR (1988) Increase in diazepam binding inhibitor-like immunoreactivity (51–70) in Huntington's disease. *Journal of the Neurological Sciences* **88**:177–184.

Ban TA (1969) *Psychopharmacology*. Baltimore: Williams & Wilkins.

Barbeau A (1978) Emerging treatments: replacement therapy with choline or lecithin in neurological diseases. *Canadian Journal of Neurological Science* **5**:157–160.

Barr AN, Heinze W, Mendoza JE and Perlik S (1978) Long term treatment of Huntington disease with L-glutamate and pyridoxine. *Neurology* **28**:1280–1282.

Barr AN, Fischer JH, Koller WC, Spunt AL and Singhal A (1988) Serum haloperidol concentration and choreiform movements in Huntington's disease. *Neurology* **38**:84–88.

Bean P (1986) *Mental Disorder and Legal Control*. Cambridge: University Press.

Bird ED (1980) Chemical pathology of Huntington's disease. *Annual Review of Pharmacology and Toxicology*. **20**:533–551.

Bird ED, MacKay AVP, Rayner CN and Iversen LL (1973) Reduced glutamic-acid decarboxylase activity of postmortem brain in Huntington's chorea. *Lancet* **i**:1090–1092.

Bledin KD, MacCarthy B, Kuipers L and Woods RT (1990) Daughters of people with dementia: expressed emotion, strain and coping. *British Journal of Psychiatry* **157**:221–227.

Bleiweiss H (1989) *Huntington's Disease and Lithium Carbonate*. First World Congress on Psychiatric Genetics, Cambridge, UK, 1986 (abs).

Bloch, S (1986) Supportive psychotherapy. In: *An Introduction to the Psychotherapies* (ed. S Bloch), pp. 252–277. Oxford: University Press.

Bluglass RS (1984) *A Guide to the Mental Health Act 1983*. Edinburgh: Churchill Livingstone.

Brandrup E (1960) Reserpin og tetrabenacin ved chorea Huntington. *Nordisk Medicin* **64**:968–969.

Brandt J, Quaid KA, Folstein SE et al. (1989) Presymptomatic diagnosis of delayed onset disease with linked DNA markers: The experience with Huntington's Disease. *Journal of the American Medical Association* **261**:3108–3114.

Brocklehurst JC (1970) *The Geriatric Day Hospital*. London: King Edward Hospital Fund.

Brown WT, Sanberg PR and McGeer PL (1979) Corticosteroids and chorea (letter). *Archives of Neurology* **36**:452–453.

Bruyn GW (1958) Some considerations on Huntington's chorea in connection with a case treated with procaine amide. *Folia Psychiatrica Neurologica et Neurochirurgica Neerlandica* **61**:375–378.

Burckhardt G (1891) Rindenexcisionen, als Bertrag zur operativen Therapie der Psychosen. *Allgemeine Zeitschrift für Psychiatrie* **47**:463–548.

Burke RE, Fahn S, Mayeux R, Weinberg H, Louis K and Willner JH (1981) Neuroleptic malignant syndrome caused by dopamine-depleting drugs in a patient with Huntington disease. *Neurology* **31**:1022–1026.

Caine ED and Shoulson I (1983) Psychiatric syndromes in Huntington's disease. *American Journal of Psychiatry* **140**:728–733.

Caine ED, Polinsky RJ, Kartzinel R and Ebert MH (1979) The trial of clozapine for abnormal involuntary movement disorders. *American Journal of Psychiatry* **136**:317–320.

Caraceni TA, Girotti F, Celano L, Parati E and Balboni L (1978) 2-Dimethylaminoethanol (Deanol) in Huntington's chorea. *Journal of Neurology, Neurosurgery and Psychiatry* **41**:1114–1118.

Carman JS, Shoulson I and Chase TN (1974) Huntington's chorea treated with lithium carbonate (letter). *Lancet* **i**:811.

Chamberlain MA, Thornley G and Wright V (1978) Evaluation of aids and equipment for bath and toilet. *Rheumatology and Rehabilitation* **17**:187.

Chandler JH (1955) Reserpine (*Rauwolfia serpentina* alkaloid) in the treatment of Huntington's chorea. *University of Michigan Medical Bulletin* **21**: 95–100.

Chhuttani PN and Singh S (1959) Reserpine in Huntington's chorea. *Journal of the Indian Medical Association* **32**:402–403.

Chiu E (1979) Notes on the management of Huntington's disease: a brief guide for family physicians. *Australian Family Physician* **8**:197–200.

Cohen C (1956) Huntington's chorea, with a report of a case treated with procaine amide. *East African Medical Journal* **33**:104–107.

Cohen NH (1962) The treatment of Huntington's chorea with trifluoperazine (Stelazine). *Journal of Nervous and Mental Disease* **134**:62–71.

Corsini GU, Onali P, Masala C, Cianchetti C, Mangoni A and Gessa G (1978) Apomorphine hydrochloride-induced improvement in Huntington's chorea: stimulation of dopamine receptor. *Archives of Neurology* **35**:27–30.

Costa E, Corda MG and Guidotti A (1983) On a brain polypeptide functioning as a putative effector for the recognition sites of benzodiazepine and beta-carboline derivatives. *Neuropharmacology* **22**:1481–1492.

Crawford JP (1969) DOPA in Parkinson's disease. *Lancet* **i**:984.

Creese I, Burt DR and Snyder SH (1976) Dopamine receptor binding predicts clinical and pharmacological potencies of antischizophrenic drugs. *Science* **192**:481–483.

Dalby MA (1969) Effect of tetrabenazine on extrapyramidal movement disorders. *British Medical Journal* **ii**:422–423.

Dalen P (1973) Lithium therapy in Huntington's chorea and tardive dyskinesia (letter). *Lancet* **i**:107–108.

Dana C (1895) The pathology of hereditary chorea; report of a case with autopsy; record of anomalies in degenerate brain. *Journal of Nervous and Mental Disease* **22**:565–583.

Davies DD (1966) Abnormal response to anaesthesia in a case of Huntington's chorea. *British Journal of Anaesthesia* **38**:490–491.

DeMeyer W and Dyken M (1954) Oral procaine amide for Huntington's chorea. *American Journal of Medical Science* **228**:70–72.

Department of Health (1989) *Caring for People: Community Care in the Next Decade and Beyond*. London: HMSO.

Department of Health and Social Security. *Handbook of Wheelchairs (MHM 408)*. London: HMSO.

Department of Health and Social Security (1986) *Incontinence Garments: Results of a DHSS Study*. HMSO: Health Equipment Information.

DiFiglia M (1990) Excitotoxic injury of the neostriatum: a model for Huntington's disease. *Trends in Neuroscience* **13**:286–289.

Editorial (1988) Embryos and Parkinson's disease. *Lancet* **i**:1087.
Editorial (1990) Driving and Parkinson's disease. *Lancet* **ii**:781.
Fahn S (1973) Treatment of choreic movements with perphenazine. *Advances in Neurology* **1**:755–764.
Feinstein B, Alberts WW, Levin G and Wright EW (1965) Some refinements of stereotaxic therapy for dyskinesias, and results of clinical evaluation. *Confinia Neurologica (Basel)* **26**:272–281.
Ferlie E (1983) *Sourcebook of Initiatives in the Community Care of the Elderly*. Headley: PSSRU.
Fisher R, Norris JW and Gilka L (1974) GABA in Huntington's chorea. *Lancet* **i**:506.
Folstein SE (1989) *Huntington's Disease: a Disorder of Families*. Baltimore: Johns Hopkins University Press.
Folstein S and Folstein M (1981) Diagnosis and treatment of Huntington's disease. *Comprehensive Therapy* **7**:60–66.
Ford MF (1986) Treatment of depression in Huntington's disease with monoamine oxidase inhibitors. *British Journal of Psychiatry* **149**:654–656.
Forrest AD (1957) Some observations on Huntington's chorea. *Journal of Mental Science* **103**:507–513.
Foy R and Pakkenberg H (1970) Combined Nitoman–Pimozide treatment of Huntington's chorea and other hyperkinetic syndromes. *Acta Neurologica Scandinavica* **46**:249–251.
Foster NL, Chase TN, Denaro A, Hare TA and Tamminga CA (1983). THIP treatment of Huntington's disease. *Neurology* **33**:637–639.
Freeman CPL, Basson JV and Crighton A (1978) Double blind controlled trial of electroconvulsive therapy (ECT) and simulated ECT in depressive illness. *Lancet* **i**:738–740.
Fulghum C, Buffaloe WJ, Smith D and Jernigan E (1960) Appraisal of Dartal in treatment of Huntington's chorea. *Diseases of the Nervous System* **21**:46.
Ganguili LK (1956) Procaine amide hydrochloride in Huntington's chorea. *Journal of the Indian Medical Association* **26**:274–275.
Gibbens TCN (1974) Preparing psychiatric court reports. *British Journal of Hospital Medicine* 278–284.
Gilhooly, MLM and Whittick JE (1989) Expressed emotion in caregivers of the dementing elderly. *British Journal of Medical Psychology* **62**:265–272.
Girotti F, Carella F, Scigliano G *et al*. (1984) Effect of neuroleptic treatment on involuntary movements and motor performances in Huntington's disease. *Journal of Neurology, Neurosurgery and Psychiatry* **47**:848–852.
Glaeser BS, Vogel WH, Oleweiler DB *et al*. (1975) GABA levels of cerebrospinal fluid of patients with Huntington's chorea: a preliminary report. *Biochemical Medicine* **12**:380–385.
Goldblatt, J and Bryer A (1987) Huntington's disease: deterioration in clinical state during treatment with angiotensin converting enzyme inhibitor. *British Medical Journal* **294**:1659–1660.
Goldman D (1952) New treatment for hereditary (Huntington) chorea. *American Journal of Medical Science* **224**:573–576.
Gostin L (1983) *The Court of Protection*. London: MIND.
Greene JG, Nicol R and Jamieson H (1979) Reality orientation with psychogeriatric patients. *Behaviour Research and Therapy* **17**:615–617.
Guidotti A, Forchetti CM, Corda MG, Konkel D, Bennett CD and Costa E (1983) Isolation, characterisation and purification to homogeneity of an endogenous polypeptide with agonistic action on benzodiazepine receptors. *Proceedings of the National Academy of Sciences* **80**:3531–3535.
Hans MB and Gilmore TH (1968) Social aspects of Huntington's chorea. *British Journal of Psychiatry* **114**:93–98.
Haslam MT (1967) Cellular magnesium levels and the use of penicillamine in the treatment of Huntington's chorea. *Journal of Neurology, Neurosurgery and Psychiatry* **30**:185–188.
Hawks DV and Silverstone JT (1962) Drug treatment of Huntington's chorea (letter). *British Medical Journal* **ii**:52.
Heimburger RF (1967) Dentatectomy in the treatment of dyskinetic disorders. *Confinia Neurologica (Basel)* **29**:101–106.
Hitchcock ER, Clough C, Hughes R and Kenny B (1988) Embryos and Parkinson's disease (letter). *Lancet* **i**:1274.
Hoerster S, Green JH, Gardner JH and Brown H (1961) The use of hydrolazine in the treatment of Huntington's chorea. *Austin State Hospital Medical Bulletin* **1**:1–2.
Huntington G (1872) On chorea. *Medical and Surgical Reporter* **26**:317–321.
Husby G, Li L, Davis LE *et al*. (1977) Antibodies to human caudate neurons in Huntington's chorea. *Journal of Clinical Investigation* **59**:922–932.

Kallail KJ, Godfrey NE, Suter G and Anthimides L (1989) A multidisciplinary approach to the management of Huntington's disease. *Kansas Medicine* **90**:309–311.
Kellam AMP (1990) The (frequently) neuroleptic (potentially) malignant syndrome. *British Journal of Psychiatry* **157**:169–173.
Kempinsky WH, Boniface WR, Morgan PP and Busch AK (1960) Reserpine in Huntington's chorea. *Neurology* **10**:38–42.
Kent DA (1965) Some clinical observations on haloperidol. *Clinical Trials Journal* **2**:166.
Kirkpatrick WL and Sanders P (1955) Clinical evaluation of reserpine in a state hospital. *Annals of the New York Academy of Sciences* **61**:123–143.
Klawans HL and Rubovits R (1972) Central cholinergic–anticholinergic antagonism in Huntington's chorea. *Neurology* **22**:107–116.
Klawans HL, Rubovits R, Ringel SP and Weiner WJ (1972) Observations on the use of methysergide in Huntington's chorea. *Neurology* **22**:929–933.
Kline NS and Stanley AM (1955) Use of reserpine in a neuropsychiatric hospital. *Ann NY Acad Sci* **229**:676–677.
Koller WC, Barr A and Biary N (1982) Estrogen treatment of dyskinetic disorders. *Neurology* **32**:547–549.
Korenyi C and Whittier JR (1967) Drug treatment in 117 cases of Huntington's disease with special reference to fluphenazine (Prolixin). *Psychiatric Quarterly* **41**:203–210.
Kuriyama K and Sze PY (1970) Aminooxyacetic acid and the blood–brain barrier to gamma-aminobutyric acid. *Pharmacologist* **12**:236.
Lamont AM (1979) Anaesthesia and Huntington's chorea. *Anaesthesiology and Intensive Care Medicine* **7**:189–190.
Larrichia R and Baseri A (1958) Osservazioni sull'impiego di un nuovo derivato della fenotiazina, la perfenazina, nella pratica neuropsichiatrica. *Minerva Medica* **49** (**suppl. 39**):1993–1995.
Laterre EC and Fortemps E (1975) Deanol in spontaneous and induced dyskinesias (letter). *Lancet* **i**:1301.
Lavers A (1981) *Remedial Involvement in the Management of Patients with Huntington's Chorea*, pp. 1–163. London: Association to Combat Huntington's Chorea.
Lazar, M (1948) The use of Bulgarian belladonna root in the treatment of Huntington's chorea. *Psychiatric Quarterly* **22**:136–140.
Lazarte JA, Peterson MC, Baars CW and Pearson JS (1955) Huntington's chorea: results of treatment with reserpine. *Proceedings of the Mayo Clinic* **30**:358–365.
Lehnhoff H (1973) Observation of the action of mesoridazine in Huntington's chorea. *Advances in Neurology* **1**:765–767.
Leonard DP, Kidson MA, Shannon PJ and Brown J (1974) Double blind trial of lithium carbonate and haloperidol in Huntington's chorea. *Lancet* **ii**:1208–1209.
Levin E, Sinclair I and Gorbach P (1989) *Families, Services and Confusion in Old Age*. Aldershot: Avebury.
Lindvall O, Brundin P, Widner H et al. (1990) Grafts of fetal dopamine neurons survive and improve motor function in Parkinson's disease. *Science* **247**:574–577.
Lyon RLL (1962) Drug treatment of Huntington's chorea: a trial with thiopropazate. *British Medical Journal* **i**:1308–1310.
Mackiewicz J and Reid AA (1965) Clinical and neuropathological investigations of four cases of Huntington's chorea treated with high doses of reserpine. *Medical Journal of Australia* 833–835.
Madrazo I, Drucker-Colin R, Diaz V, Martinez-Mata J, Torres C and Becerril JJ (1987) Open microsurgical autograft of adrenal medulla to the right caudate nucleus in two patients with intractable Parkinson's disease. *New England Journal of Medicine* **316**:831–834.
Madrazo I, Leon V, Torres C et al. (1988) Transplantation of fetal substantia nigra and adrenal medulla to the caudate nucleus in two patients with Parkinson's disease. *New England Journal of Medicine* **318**:51.
Male J and Massie B (1990) *Choosing a Wheelchair*. London: Royal Association for Disability and Rehabilitation.
Markham CH, Clark WG and Winters WD (1963) Effect of alpha-methyl dopa and reserpine in Huntington's chorea, Parkinson's disease and other movement disorders. *Life Sciences* **9**:697–705.
Marks IM (1986) *Behavioural Psychotherapy: Maudsley Pocket Book of Clinical Management*. Bristol: Wright.
Marsden CD and Jenner R (1980) Pathophysiology of extrapyramidal side-effects of neuroleptic drugs. *Psychological Medicine* **10**:55–72.

Martindale B (1987) Huntington's chorea: some psychodynamics seen in those at risk and in the responses of the helping professions. *British Journal of Psychiatry* **150**:319–323.
Martindale B and Bottomley V (1980) The management of families with Huntington's chorea: a case study to illustrate some recommendations. *Journal of Child Psychology and Psychiatry* **21**:343–351.
Mathews FP (1958) Dartal: a clinical appraisal. *American Journal of Psychiatry* **114**:1034–1035.
Mattsson B (1973) Huntington's chorea and lithium therapy. *Lancet* **i**:718–719.
Mattsson B and Boman K (1974) Buronil in Huntington's chorea (letter). *Lancet* **ii**:1323.
McLean DR (1982) Failure of isoniazid therapy in Huntington disease. *Neurology* **32**:1189–1191.
McLellan DL, Chalmers RJ and Johnson RH (1974) A double-blind trial of tetrabenazine, thiopropazate, and placebo in patients with chorea. *Lancet* **i**:104–107.
Medical Commission on Accident Prevention (1985) *Medical Aspects of Fitness to Drive*. London: HMSO.
Medical Research Council (1965) Clinical trial of the treatment of depressive illness. *British Medical Journal* **i**:881–886.
Merskey H (1958) A clinical and psychometric study of the effects of procaine amide in Huntington's chorea. *Journal of Mental Science* **104**:411–420.
Merskey H, Rice T and Troupe A (1961) An investigation of some therapeutic and physiological effects of perphenazine in Huntington's chorea. *Psychopharmacologia* **2**:436–445.
Mindham RHS, Howland C and Shepherd M (1973) An evaluation of continuation therapy with tricyclic antidepressants in depressive illness. *Psychological Medicine* **3**:5–17.
Molina H (1990) Neurotransplantation in Parkinson's disease: the Cuban experience. *Restorative Neurology Neuroscience* **1**:164 (abs).
Moniz E (1954) How I succeeded in performing the prefrontal leucotomy. *Journal of Clinical and Experimental Psychopathology* **15**:373–379.
Morales LM, Estevez J, Suarez H, Villalobos R, de Bonilla LC and Bonilla E (1989) Nutritional evaluation of Huntington disease patients. *American Journal of Clinical Nutrition* **50**:145–150.
Morgan DR (1956) Experimental use of reserpine in the treatment of tremor-rigidity. *Medical Journal of Australia* **i**:973–974.
Morris RG, Morris LW and Britton PG (1988) Factors affecting the emotional well-being of the caregivers of dementia sufferers. *British Journal of Psychiatry* **153**:147–156.
Murphy E (1990) Training for care assistants: new qualifications should raise their status and scope. *British Medical Journal* **301**:506.
Nagao T (1962) Electromyographical analysis of the choreic movements and effect of pallidotomy. *Psychiatria et Neurologia Japonica* **64**:559–578.
Narabayashi H (1962) Neurophysiological ideas on pallidotomy and ventrolateral thalamotomy for hyperkinesis. *Confinia Neurologica (Basel)* **22**:291–303.
Narabayashi H, Shimamura M and Ohye C (1973) Experience with stereotactic surgery on choreic movements with some physiological interpretations. *Advances in Neurology* **1**:789–793.
National Council for Voluntary Organizations/National Association of Health Authorities (1987) *Partnerships for Health*. London: National Association of Health Authorities.
Neophytides AN, Lieberman A, Foo SH et al. (1980) Treatment of Huntington's disease with isoniazid and pyridoxine. *Neurology* **30**:383.
Nielsen JM and Butte EM (1955) Treatment of Huntington's chorea with BAL (dimercaprol). *Bulletin of the Los Angeles Neurological Society* **20**:38–39.
Nilsen JA (1964) Valium therapy of Huntington's chorea. *American Journal of Psychiatry* **120**:1197–1198.
Nutt JG, Rosin A and Chase TN (1978) Treatment of Huntington disease with a cholinergic agonist. *Neurology* **28**:1061–1064.
Oscar-Berman M, Sax DS and Opoliner L (1973) Effects of memory aids on hypothesis behaviour and focusing in patients with Huntington's chorea. *Advances in Neurology* **1**:717–728.
O'Shea B (1989) Huntington's disease – the experience of voluntary organisations. *Psychiatric Bulletin (Royal College of Psychiatrists)* **13**:409–411.
Pakenham-Walsh R (1960) Perphenazine in Huntington's chorea. *Lancet* **ii**:767.
Pakkenberg H (1968) The effect of tetrabenazine in some hyperkinetic syndromes. *Acta Neurologica Scandinavica* **44**:391–393.
Paulson G (1976) Lioresal in Huntington's disease. *Diseases of the Nervous System* **37**:465–467.
Peacock CE and Harris R (1989) Huntington's chorea: who cares? *Health Trends* **21**:15–17.
Pearson JS (1973) Family support and counseling in Huntington's disease. *Psychiatric Forum* **4**:746–750.
Pearson J, Challis L and Bowman CE (1990) Problems of care in a private nursing home. *British Medical Journal* **301**:371–372.

Pearson SJ and Reynolds GP (1988) Depletion of monoamine transmitters by tetrabenazine in brain tissue in Huntington's disease. *Neuropharmacology* **27**:717–719.
Peiris JB, Boralessa H and Lionel NDW (1976) Clonazepam in the treatment of choreiform activity. *Medical Journal of Australia* **1**:225–227.
Perry, TL (1977) Isoniazid and Huntington's chorea. *New England Journal of Medicine* **298**:1092–1093.
Perry TL (1990) A brain transplant that works (letter). *Lancet* **i**:1042.
Perry TL and Hansen S (1973) Sustained drug-induced elevation of brain GABA in the rat. *Journal of Neurochemistry* **21**:1167–1175.
Perry TL, Hansen S and Kloster M (1973) Deficiency of gamma-aminobutyric acid in brain. *New England Journal of Medicine* **288**:337–342.
Perry, TL, MacLeod PM and Hansen S (1977) Treatment of Huntington's chorea with isoniazid (letter). *New England Journal of Medicine* **298**:840.
Perry TL, Wright JM, Hansen S *et al.* (1979) Isoniazid therapy of Huntington's disease. *Neurology* **29**:370–375.
Perry TL, Wright JM, Hansen S, Allan BM, Baird PA and MacLeod PM (1980) Failure of aminooxyacetic acid therapy in Huntington disease. *Neurology* **30**:772–775.
Pleydell MJ (1954) Huntington's chorea in Northamptonshire. *British Medical Journal* **ii**:1121–1128.
Prins HA (1980) *Offenders, Deviants or Patients?* London: Tavistock.
Quaid KA, Brandt J and Folstein SE (1989) Insurance and the presymptomatic diagnosis of Huntington's disease. *Journal of the American Medical Association* **262**:2384–2385.
Quarrell OWJ, Bloch M and Hayden MR (1989) Insurance and the presymptomatic diagnosis of delayed onset disease. *Journal of the American Medical Association* 2384–2385.
Quinn N and Marsden CD (1984) A double blind trial of sulpiride in Huntington's disease and tardive dyskinesia. *Journal of Neurology, Neurosurgery and Psychiatry* **47**:844–847.
Quinn NP, Dunnett SB and Oertel W (1989) Studies towards neural transplantation in Parkinson's disease. In: *Disorders of Movement: Clinical, Pharmacological and Physiological Aspects* (eds NP Quinn and PG Jenner), pp. 223–247. London: Academic Press.
Reisine TD, Wastek GL, Speth RC *et al.* (1979) Alterations in the benzodiazepine receptor of Huntington's diseased human brain. *Brain Research* **165**:183–187.
Research Institute for Consumer Affairs (1984) *Communication Aids: a Guide for People who have Difficulty in Speaking*. London: Research Institute for Consumer Affairs.
Review of the Guidance on the Research Use of Fetuses and Fetal Material (1989) London: HMSO.
Riechert T (1957) Die stereotaktischen Hirnoperationen in ihrer Anwendung bei Hyperkinesen (mit Ausnahme des Parkinsonismus) bei Schmerzzustanden und einigen weitern Indikationen. *Congress of International Neurologie und Neurochirurgie* **21**:7.
Ringel SP, Guthrie M and Klawans HL (1973) Current treatment of Huntington's chorea. *Advances in Neurology* **1**:797–801.
Robinson RO and Thornett CEE (1985) *Developmental Medicine and Child Neurology* **27**:814–821.
Ryder S (1975) *And the Morrow is Theirs*. Bristol: Burleigh.
Ryder S (1986) *Child of my Love: an Autobiography*. London: Collins Harvill.
Sattes H (1960) Die Behandlung der Chorea major mit dem Monoamin-Freisetzer Nitaman. *Psychiatria Neurologia Neurochirurgia* **140**:13–19.
Sattes H and Hase E (1964) Die Behandlung extrapyramidaler Hyperkinesien unter besonderer Berucksichtigung der Chorea Huntington. *Psychiatria, Neurologia Neurochirurgia* **67**:289–298.
Schwarcz R and Shoulson I (1987) Excitotoxins and Huntington's disease. In: *Animal Models of Dementia*, pp. 39–68. New York: Alan Liss.
Shawcross CR, Crowder J, Brownrigg K *et al.* (1990) Who cares for the adult brain damaged? Hamble Ward revisited. *Psychiatric Bulletin (Royal College of Psychiatrists)* **14**:526–527.
Shoulson I (1982) Care of patients and families with Huntington's disease. In: *Neurology 2. Movement Disorders* (eds CD Marsden and S Fahn), pp. 277–290. London: Butterworth.
Shoulson I (1983) Huntington's disease: anti-neurotoxic therapeutic strategies. In: *Excitotoxins* (eds K Fuxe, P Roberts and R Schwarcz), pp. 343–353. New York: MacMillan.
Shoulson I, Chase TN, Roberts E and van Balgooy JN (1975). Huntington's disease: treatment with imidazole-4-acetic acid. *New England Journal of Medicine* **293**:504–505.
Shoulson I, Goldblatt D, Charlton M and Joynt RJ (1978) Huntington's disease: treatment with muscimol, a GABA-mimetic drug. *Annals of Neurology* **4**:279–284.
Shoulson I, Kurlan R, Rubin AJ *et al.* (1989a) Assessment of functional capacity in neurodegenerative movement disorders: Huntington's disease as a prototype. In: *Quantification of Neurologic Deficit* (ed. JL Munsat), **20**:271–283. Boston: Butterworth.

Shoulson I, Odoroff C, Oakes D et al. (1989b) A controlled clinical trial of baclofen as protective therapy in early Huntington's disease. *Annals of Neurology* **25**:252–259.

Shults C, Steardo L, Barone P et al. (1986) Huntington's disease: effect of cysteamine, a somatostatin-depleting agent. *Neurology* **36**:1099–1102.

Sinkler W (1889) Two additional cases of Huntington's chorea. *Journal of Nervous and Mental Disease* **14**:69–91.

Snaith RP and Warren H de B (1974) Treatment of Huntington's chorea with tetrabenazine (letter). *Lancet* **i**:413–414.

Souder CL (1959) Treatment of Huntington's chorea with dihydrochloride: case histories. *Delaware State Medical Journal* **31**:249–250.

Soutar CA (1970) Tetrabenazine for Huntington's chorea. *British Medical Journal* **iv**:55.

Spiegel EA and Wycis HT (1952) Thalamotomy and pallidotomy for treatment of choreic movements. *Acta Neurochirurgica* **2**:417–422.

Spiegel EA and Wycis HT (1950) Pallidothalamotomy in chorea. *Archives of Neurology and Psychiatry* **64**:295–296.

Spiegel EA and Wycis HT (1953) Role of pallidum in genesis of some choreic movements. *Neurology* **4**:261–266.

Spokes EGS (1980) Neurochemical alterations in Huntington's chorea: a study of post-mortem brain tissue. *Brain* **103**:179–210.

Stewart JT (1988) Treatment of Huntington's disease with clonazepam. *Southern Medical Journal* **81**:102.

Still CN (1979) Nutritional therapy in Huntington's chorea: concepts based on the model of pellagra. *Psychiatric Forum*:74–78.

Straube W and Melliwa H (1962) Zur Dauerbehandlung extrapyramidaler Hyperkinesen mit Perphenazin. *Nervenarzt* **33**:549–553.

Swash M, Roberts AH, Zakko H and Heathfield KWG (1972) Treatment of involuntary movement disorders with tetrabenazine. *Journal of Neurology, Neurosurgery and Psychiatry* **35**:186–191.

Tarighati S and A'Brook MF (1968) Trifluoperidol in Huntington's chorea. *Lancet* **ii**:458–459.

Tarsy D, Sax DS, Leopold N and Feldman RG (1973) The effect of physostigmine on Huntington's chorea and *l*-dopa dyskinesia. In *Advances in Neurology*, pp. 777–788. New York: Raven Press.

Tell G, Bohlen P, Schechter PJ et al. (1981) Treatment of Huntington disease with gamma-acetylenic GABA, an irreversible inhibitor of GABA-transaminase: increased CSF GABA and homocarnosine without clinical amelioration. *Neurology* **31**:207–211.

Tibawi R (1978) A family with Huntington's chorea. *FSU Quarterly* **16**:11–19.

Tinker A (1984) *Staying at Home: Helping Elderly People*. London:HMSO.

Tomlinson PJ (1947) The treatment of Huntington's chorea with belladonna alkaloids. *Psychiatric Quarterly* **21**:447–452.

Turner-Stokes L and Frank AO (1990) Disability medicine. *British Journal of Hospital Medicine* **44**:190–193.

Vaughan GF, Leiberman DM and Cook LC (1955) Chlorpromazine in psychiatry. *Lancet* **i**:1083–1087.

Vladeck B (1980) *Unloving Care: the Nursing Home Tragedy*. New York: Basic Books.

Walker and Hunt (1989) *Clinical Neuropharmacology* **12**:322–330.

Walker FO, Winston-Salem NC, Young AB and Rodnitzky RL (1985) Diltiazem therapy in Huntington disease. *Neurology* **35** (**suppl. 1**):177.

Walker JE, Hoehn M, Sears E and Lewis J (1973) Dimethyl aminoethanol in Huntington's chorea. *Lancet* **i**:1512.

Walther-Buel H (1955) Das neuroplegische Prinzip in pharmakopsychiatrischer Betrachtung. *Monatsschrift für Psychiatrie und Neurologie* **129**:286–296.

Weiler PG, Rathbone E and McCurran I (1978) Adult day care. In *Community Work and the Elderly*. Berlin:Springer.

West ED (1981) Electric convulsion therapy in depression: a double-blind controlled trial. *British Medical Journal* **i**:355.

Whittier JR (1963) Research on Huntington's chorea: problems of privilege and confidentiality. *Journal of Forensic Science* **8**:568–575.

Whittier JR (1968) Treatment of Huntington's disease. *Modern Treatment* **5**:332–350.

Whittier J, Haydu G and Crawford J (1961) Effect of imipramine (Tofranil) on depression and hyperkinesia in Huntington's disease. *American Journal of Psychiatry* **118**:79.

Williams A (1990) Cell implantation in Parkinson's disease: more patients have probably been harmed than helped so far. *British Medical Journal* **301**:301–302.

Williams RC, Lewis M, Montano J *et al.* (1978) Immunological studies related to brain antigens in Huntington's disease. *Annals of Neurology* **3**:185–186.

Wiltshire ER (1987) *Equipment for Disabled People: Communication*. Oxford: Oxfordshire Health Authority.

Wong EHF, Kemp JA, Priestley T, Knight AR, Woodruff GN and Iverson LL (1986) The anticonvulsant MK-801 is a potent N-methyl-D-aspartate antagonist. *Proceedings of the National Academy of Sciences* **83**:7104–7108.

Woods RT and Britton RG (1977) Psychological approaches to the treatment of the elderly. *Age and Ageing* **6**:104–112.

World Federation of Neurology Research Group on Huntington's Chorea (1989) Ethical issues policy statement on Huntington's disease molecular genetics predictive test. *Journal of the Neurological Sciences* **94**:327–332.

World Federation of Neurology Research Group on Huntington's Chorea (1990) Ethical issues policy statement on Huntington's disease molecular genetics predictive test. *Journal of Medical Genetics* **27**:34–38.

Worrall EP (1974) Lithium in Huntington's disease (letter). *Lancet* **ii**:1323.

Young AB, Shoulson I, Penney JB *et al.* (1986) Huntington's disease in Venezuela: neurological features and functional decline. *Neurology* **36**:244–249.

8

The Epidemiology of Huntington's Disease

INTRODUCTION

Until now this book has concentrated on the effects of HD, whether clinical or pathological, on the individual. In Chapter 6 the social aspects have been considered as they affect the whole family and society in general, but HD needs also to be seen in its population context. Questions that arise are: how common is the disorder? Is it increasing or declining? What geographical variations are there and how can these be accounted for? Accurate population data are clearly essential for the planning of services for patients and families, as well as for the interpretation of preventive measures, and it is in the collection and analysis of these data that we look to the disciplines of epidemiology and genetics.

Epidemiology, evolving out of the study of infective diseases, has now assumed a major role in the study of chronic disorders whose causation is complex and usually poorly understood. Multiple sclerosis, motor neurone disease and stroke are examples of major neurological disorders where the epidemiological approach has played an important part in determining hypotheses about causation of these diseases as well as simply documenting their frequency and variation. For HD the situation is rather different. We know that, as a mendelian disorder, there must be a specific gene determining it; thus the route to isolating this gene, and hence understanding the pathogenesis of the disease, will principally be molecular, not epidemiological. Likewise the most satisfactory guide to frequency of the disorder will be a careful study of complete families rather than surveys restricted to primary cases; also those at risk for the disease are also mostly family members, rather than the population at large. For all these reasons most of the data discussed in this chapter have been obtained not by classical epidemiological methods but by means of family surveys. The methods of traditional epidemiology appear cumbersome and rather inadequate when applied to HD, as is clearly seen in the papers by epidemiologists in the special volume on HD (Schoenberg, 1979; Kurtzke, 1979; Hogg et al., 1979). Before the detailed situation is described, the different general approaches are outlined below.

FAMILY STUDIES

Systematic family studies have provided the great majority of existing population data for HD. Their success has usually resulted from the fact that they have been carried out by an experienced and dedicated worker or small team, based in the area under study and with considerable tenacity and determination. The sheer effort and detective work involved in tracing families from old and fragmentary records rarely receive a mention in the published results; nor too do the discomforts and even hazards of the work, not to mention the numerous fruitless visits that end in finding someone 'not at home'. The combined experiences and exploits of those involved in HD family surveys would make the basis of a series of fascinating novels – but sadly can be pursued no further here!

Returning to the scientific aspects, any family study will be meaningless in epidemiological terms unless it is systematic and complete. This essentially means that a point prevalence study is being aimed at, with the following criteria well defined from the outset:

(1) The geographical area must be well defined (usually corresponding to some administrative or political boundary). The area must be sufficiently large to avoid bias from a particularly large kindred, but sufficiently small to allow complete coverage of all of it by the investigator (500 000–5 000 000 probably represents outer limits for population size). All parts of the area should ideally be covered with equal thoroughness.

(2) As many separate methods of ascertainment should be used as possible, e.g. general and psychiatric hospital records, letters to family doctors, neurologists and other clinicians, existing genetics records, death certificates. If multiple ascertainment is frequent this is usually an indication that ascertainment is not going to be seriously deficient.

(3) Intensive search for secondary cases is essential. In HD, more than most genetic disorders, all investigators are struck by the number of new affected patients discovered through a detailed and systematic family survey, often as the result of a home visit, whose existence was totally unsuspected, and who would not have been included in any survey limited to primary cases. The thoroughness of this aspect will probably be the largest single factor in determining prevalence.

(4) Any study must be continued over a considerable period of time. Three years is probably a minimum and five preferable, but even then new families will be found in areas where one was confident one had achieved total ascertainment. It is even better to repeat the survey after an interval; this usually produces a significant increase in prevalence.

(5) A prevalence date should be chosen that is sufficiently remote (5–10 years) from the time of the study, to allow existing cases to be recognized and diagnosed but not to have died and been forgotten.

Table 8.1 Prevalence estimates of HD in South Wales

Study	Walker et al. (1981)	Quarrell et al. (1988)
Prevalence date	April 1971	April 1981
Census population	1 720 000	1 728 000
Number living affected	131	153
Prevalence (per 100 000)	7.61	8.85

When all these factors have been worked out, it remains to determine how many HD patients were alive and resident in the area on the prevalence date chosen, and to compare this with the total population of the area at the same date.

The authors' South Wales study (Walker et al., 1981) shows how this can be done: it was carried out over a period of almost 5 years (1973–1978), with a prevalence date of 25/26 April 1971, corresponding to the decennial census. The area chosen (industrial South Wales) was compact, had well-defined boundaries, and had a total population on the prevalence date of 1 720 901, giving a resulting prevalence estimate of 7.61 per 100 000. It should be noted that when re-analysed 10 years later, taking a prevalence date of 1981, the prevalence was found to be higher at 8.85 per 100 000 (Quarrell et al., 1988). Table 8.1 summarizes some of the epidemiological aspects of the South Wales study.

Once the prevalence rate is accurately determined, one can also work out the heterozygote frequency for the disorder and the estimated absolute number of gene carriers in the population, as described in Chapter 9.

GENERAL SURVEYS OF NEUROLOGICAL DISEASES

It is clearly important to be able to compare the frequency of HD with that of other neurological disorders, and a comparative epidemiological study of the major causes of neurological disability might be considered a suitable way to approach this, especially where medical diagnostic records are accurate. Unfortunately this does not work well in practice and such estimates are likely to be inaccurate. Thus Kurland (1958), studying a variety of neurological and neuromuscular disorders in Rochester, New York, estimated a prevalence of 6.7 per 100 000, but this is based on only two cases in a relatively small population. The reaction of any HD investigator would have been to visit these two personally, in the anticipation of finding several more unsuspected cases among relatives. As a result of a series of separate studies, data now exist for South Wales for a number of different inherited neurological disorders, based on extensive family surveys. While these estimates may well differ considerably from other areas, they are valuable for comparing the different disorders and are given in Table 8.2.

Table 8.2 Prevalence of inherited neurological disorders in South Wales*

Disorder	Number of living affected	Prevalence (per 100 000)
Huntington's disease	79	8.4
Neurofibromatosis (type 1)	125	13.3
Tuberous sclerosis	13	1.4
Hereditary spastic paraplegia	30	3.2
Charcot-Marie-Tooth disease	116	12.3
Myotonic dystrophy	65	6.9
Duchenne muscular dystrophy	40	8.8†
Becker muscular dystrophy	23	2.4†
Facioscapulonumeral muscular dystrophy	27	2.9

*Based on 1988 (June) total population, for Mid and South Glamorgan, of 939 300. Data by courtesy of Dr John MacMillan. (MacMillan J and Harper PS *Annals of Neurology*, 1991, in press.)
†Prevalence for males only.

ANALYSIS OF DEATH RATE

The use of statutory information on causes of death derived from death certificates is subject to a number of disadvantages, notably that HD may not appear on the certificate, either because the immediate cause of death was something else, or because the physician filling in the certificate may have wished to protect the family from adverse effects of HD being officially recognized as the cause. A further difficulty is that HD did not have a specific international Code until 1968 (ICD 331.0, hereditary chorea). For the subsequent 10 years this category reflects HD accurately, since other causes of hereditary chorea are very rare, but unfortunately a further change in 1979 relegated HD to a subset (333.4) so that the three-figure code (often the only one completed) now again contains other major diseases besides HD. Despite these problems, however, information on death rates is available from all areas without the need for special surveys or systems of collection. It is also likely to be comparable between regions in a way that individual surveys may not be. For this reason the study of Hogg *et al.* (1979) of death rate data for the United States is important. Based on the data of 1968–1974, the study showed a remarkably uniform pattern of mortality from HD, with an age-adjusted death rate ranging only between 0.96 and 1.35 per million. These data are considered further in relation to specific surveys in the United States, but while they are certainly an underestimate they provide a valuable baseline. A more recent study in the United States by Lanska *et al.* (1988) has analysed the mortality from HD when this is given as a subsidiary condition rather than the main cause of death. This considerably increased the recognized mortality rate by around 80% over the values previously found by Hogg *et al.* Comparable data

Table 8.3 Approximate average annual HD death rates per million population by sex and country. From Kurtzke (1979)

Country	Period	Male	Female
US total	1968–1974	1.1 (1.0–1.1)*	1.2 (1.1–1.3)
US white	1968–1974	1.2 (1.1–1.2)	1.3 (1.2–1.4)
US nonwhite	1968–1974	0.4 (0.3–0.6)	0.4 (0.3–0.5)
Sweden	1969–1974	1.7 (1.2–2.3)	1.7 (1.2–2.3)
Denmark	1951–1968	1.4 (1.1–1.9)	1.4 (1.1–1.9)
Denmark	1969–1975	1.8 (1.2–2.6)	1.9 (1.3–2.7)
England–Wales†	1960–1973	1.5 (1.4–1.6)	1.6 (1.5–1.7)
Japan	1969–1975	0.1 (0.1–0.1)	0.2 (0.1–0.2)

*Figures in parentheses are 95% confidence intervals.
†Code 331 total.

for the European countries are also available. Table 8.3 summarizes some of these.

Death certificates can also be used longitudinally to follow any possible increase or decrease in HD. Caro (1977) noted a marked increase between 1959 and 1974, and that some increase was still present when the data were corrected for inaccuracy of certification, something that he had found to improve over this period. The same trend was noted and discussed by the Office of Health Economics report on HD (1980). Figure 8.1 and Table 8.4 show these data extended up to the present time; it can be seen that the increase has

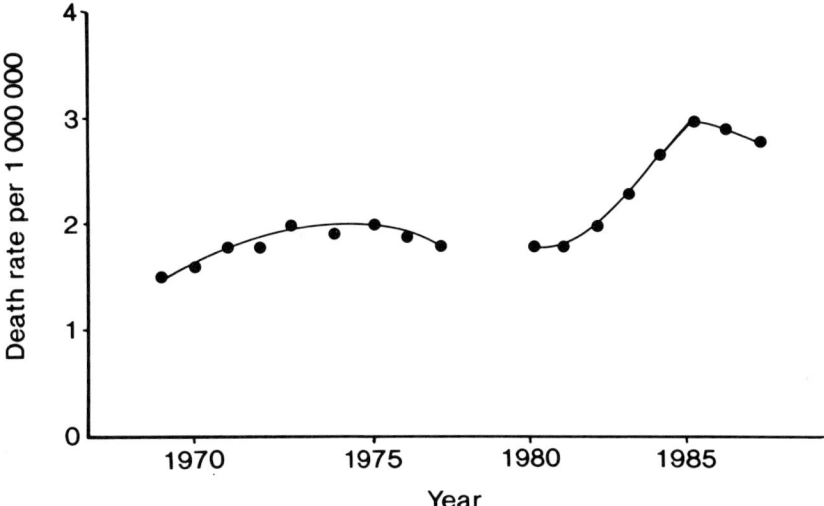

Figure 8.1 Death rate from HD in England and Wales, 1968–1988. Three year moving averages based on data of Table 8.4, kindly provided by the Office of Population Census and Surveys. Note that the discontinuity produced by the changes in ICD coding in 1978 does not appear to have altered the rate.

Table 8.4 Deaths from Huntington's disease in England and Wales in the years 1968–1988*

Year	No. of deaths	Population ($\times 10^3$)	Rate (per 1 000 000 population)
1968	71	48 593.0	1.5
1969	71	48 826.8	1.5
1970	78	48 891.1	1.6
1971	89	49 152.0	1.8
1972	99	49 327.1	2.0
1973	82	49 459.0	1.7
1974	111	49 467.9	2.2
1975	85	49 469.8	1.7
1976	96	49 459.2	1.9
1977	106	49 440.4	2.1
1978	71	49 442.5	1.4
1979	92	49 508.2	1.9
1980	82	49 603.0	1.7
1981	89	49 634.3	1.8
1982	96	49 601.4	1.9
1983	109	49 653.7	2.2
1984	142	49 763.6	2.9
1985	155	49 923.5	3.1
1986	151	50 075.4	3.0
1987	137	50 242.9	2.7
1988	127	50 393.3	2.5

*1968–1978 – Hereditary chorea (ICD 331.0). 1979 – Huntington's chorea (ICD 333.4).

continued, though whether the accuracy of certification has also changed in the recent period is unknown. The mortality rate is still considerably less than to be expected from the prevalence data shown in the detailed surveys. Broadly these data agree with the trend in different prevalence surveys carried out at different times in showing that both prevalence and mortality from recognized HD have increased. However, it would seem unwise to assume that any true increase in mortality can be inferred from these figures.

It is worth noting here that mortality data are of little use in monitoring changes resulting from genetic counselling and preventive programmes, since death rates reflect births occurring on average 50 years earlier. Hence the importance of birth-rate data based on those births known to be at risk, which will give a more sensitive indication of any recent changes. Any changes in death rate are likely to indicate factors operating half a century ago and may be very different to those that are occurring at present.

HD IN DIFFERENT COUNTRIES

Variation in the frequency of HD between countries has long been recognized, as outlined in Chapter 1. Early studies in North America stressed the common ancestry of most cases from a small number of founding immigrants, and the gene was thought to have come principally from England. It soon became clear, though, that this was an over-simplification, and a broader 'northern European' origin of the gene became accepted. It is only recently that surveys based on adequate ascertainment have been available from a wider range of countries and these are still far from complete. This section attempts to record the existing data as fully as possible and to plot them visually in the hope that more accurate comparative analysis will soon be possible. This should be of particular relevance now that specific DNA haplotypes associated with HD can be recognized; once it becomes possible to detect different HD mutations directly, all our epidemiological data will require reanalysis, so that as detailed a picture as possible of the relative frequency of HD in different parts of the world is of more than local interest.

United States of America

In looking at the detailed distribution of HD throughout the world it is appropriate to begin with the United States. Following George Huntington's description of the disorder, attention initially focused on its occurrence in New England and its British origin (Jelliffe, 1908). Much has been written of the supposed descent of most New England cases from three original migrants from East Anglia; as discussed in Chapter 1, most of this is likely to be inaccurate or erroneous. The extensive data of Davenport and Muncey (1916a) likewise deal mainly with the New England families and must be regarded more as a compilation than a survey, though one of considerable historical significance.

The first critical and detailed analysis was that of Reed and colleagues in Michigan (Reed *et al.*, 1958; Reed and Neel, 1959), discussed more fully in Chapter 9. Their prevalence rate of 4.2 per 100 000 is likely to be an underestimate, since ascertainment was based mainly on hospital records, but it represents a landmark in careful design of the study. Reed *et al.* studied the origin of their 124 kindreds and found that over half (73) could not be traced back to a progenitor outside the United States. Those that could were almost equally divided between three groups, Britain, Germany and other various European countries. An even greater predominance of German origin was found by Falstein and Stone (1939) in Iowa, German origin kindreds accounting for 27 of the 62 kindreds of known origin, while Britain and Ireland had contributed only nine.

The various prevalence estimates for different regional studies in North America are summarized in Table 8.5 and Figure 8.2. Much the most detailed of these is the Maryland study of Folstein *et al.* (1987), which used intensive ascertainment from multiple sources and which undertook a detailed family analysis and search for secondary cases. Folstein's monograph (1989) gives a full description of the methods used in this valuable study and gives practical information on guidelines to follow in undertaking work of this type. The overall prevalence estimate of 5.15 per 100 000 is likely to be the closest available

Table 8.5 Prevalence studies of HD: North America

Study	Publication year	Prevalence year	Region	No. affected	Population	Prevalence (per 100 000)
Reed et al.	1958	1940	Michigan	203	4 932 562	4.1 (overall) 4.2 (whites) 1.5 (blacks)
Shokeir	1975		Manitoba/ Saskatchewan	162	1 926 942	8.4
Pearson et al.	1955	1955	Minnesota	117	3 174 000	5.43
Folstein et al.	1987	1980	Maryland	217	4 217 000	5.15 (overall) 4.94 (whites) 6.37 (blacks)

to the true figure and the only one providing an accurate estimate for the black American population, a topic discussed below.

The US Huntington's roster, a record of HD families from all parts of the country (Conneally, 1984; Gersting et al., 1984), has already been described in Chapter 1. While it cannot be used for prevalence data and could be biased in some respects (e.g. towards the recording of large kindreds) its size and scope has made it an invaluable resource in providing large samples for quantitative analysis on a variety of genetic and other topics.

A number of other American studies for which prevalence estimates are frequently sited are too incomplete to be of serious use. Thus the survey of Korenyi and Whittier (1977) in New York City and Long Island was simply based on estimating the number of children of patients recorded in hospital notes and assuming that half would be affected. The estimate of Myriantho-poulos (1973) for HD in New York Jews was based on the 70 Jewish families who were members of the lay society. This certainly suggests that the disorder is not uncommon in Jews, but cannot be taken as a specific prevalence estimate. Kurland's (1958) estimate of 6.7 per 100 000 for Rochester, Minnesota was based on only two cases. Thus the United States remains remarkably under-studied in terms of detailed prevalence, when compared with Europe.

The population mobility seen in many parts of the United States makes it less easy to use regional genetic registers as a means of estimating and monitoring the prevalence of HD. It is thus difficult to draw conclusions as to whether the differences between the various prevalence estimates are real or simply reflect the methods of the different studies. An indication that differences are probably not great comes from the death certificate data which, as discussed earlier, have the advantages of being systematically and uniformly collected in all states, even though likely to be a considerable under estimate of the true incidence of HD. When states are grouped as major census regions, the highest age adjusted death rate (1.36 per 10^6) is in the north–central and western regions and the lowest in the southern region (0.96 per 10^6), a difference which is remarkably small (Hogg et al., 1979).

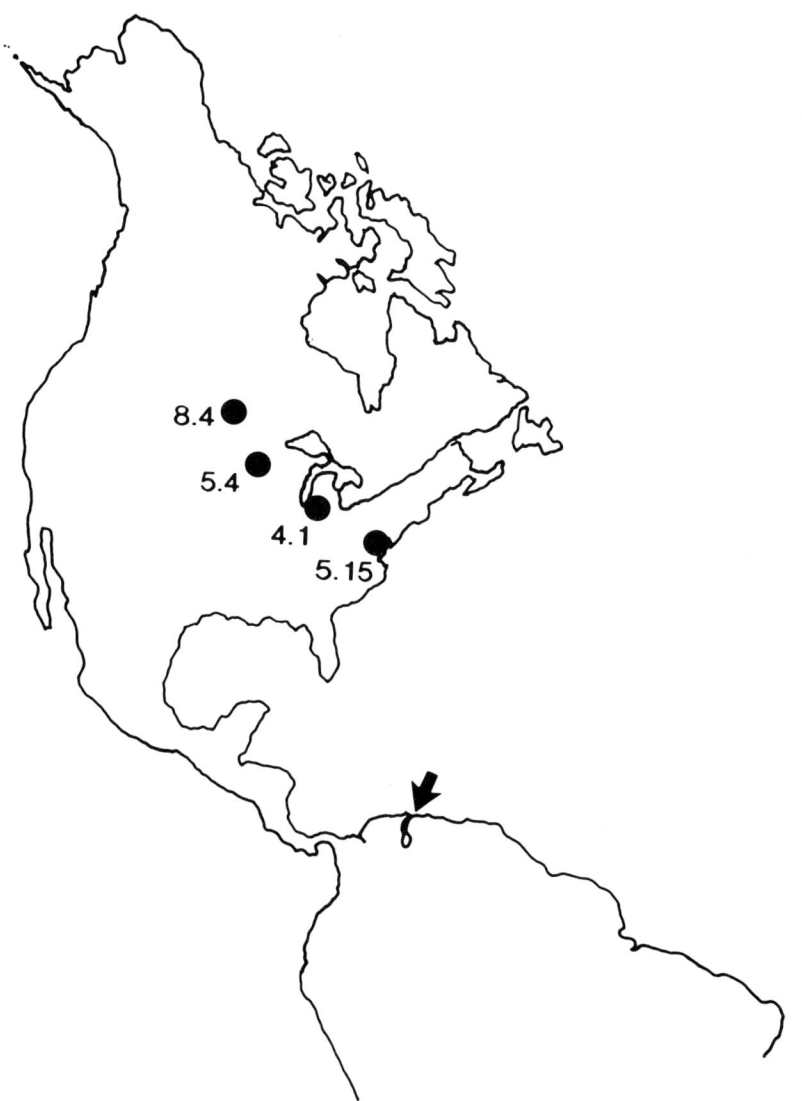

Figure 8.2 North American prevalences estimates for HD. (See Table 8.5 for details.) The number of detailed population studies has been few in comparison with Europe. The Venezuela focus of HD is arrowed.

Canada

Canada has produced some remarkable concentrations of genetic disorders, especially among its isolated French Canadian populations. Myotonic dystrophy, oculo-pharyngeal muscular dystrophy and tyrosinaemia are striking examples which have been well documented. HD, by contrast, appears to be principally of English origin and does not show a restricted geographical distribution, though the first Canadian report (MacKay, 1904) was of a French-Canadian family from Quebec.

Clarke and McArthur (1924) reported a large southern Ontario family with HD, of English origin, while further four-generation Ontario cases were reported subsequently and reviewed by Archibald (1938). Barbeau and Fullum (1962) undertook a major systematic study of HD in Canada, documenting no fewer than 820 cases from 104 families. Of these families 75 could be traced to an origin outside Canada, 55 coming from Britain and Ireland, the USA contributing only nine, while four were from Germany and only three from France. Barbeau *et al.* (1964) were able to trace 123 French Canadian cases to a single common ancestor who came from France to Montreal in 1645. The only other significant HD focus of probable French origin is that in Nova Scotia, originally reported by Hattie in 1909 and reassessed by Winsor and Welch (1973). This group originated from eastern France and came to Canada via the United States.

Barbeau and Fullum did not attempt a prevalence estimate, but Shokeir (1975) surveyed the Prairie provinces of Saskatchewan and Manitoba and found the remarkably high prevalence of 8.4 per 100 000; like several other investigators he found an increased fertility both compared with the general population and with unaffected sibs. There is no note as to the origin of the families in this study.

British Columbia was found by Hall and Te-Juceto (1983) as having 386 affected individuals from 121 kindreds; it has since become a major centre for HD research and for the development of predictive testing. Canada has also taken a major initiative in establishing a coordinated nationwide predictive testing programme (see Chapter 12), something which will allow valuable combined data to be obtained on this difficult topic, as well as ensuring equitable delivery of the service to the population as a whole.

South America

The remarkable Venezuela isolate and its contribution to HD research in general has already been mentioned in Chapter 1. Over 100 individuals are affected, with large families producing a particularly high number at risk among young people, some with both parents affected. The clinical aspects have been documented by Young *et al.* (1986) and Penney *et al.* (1990), and the probable homozygotes by Wexler *et al.* (1987).

The affected families are located in a number of villages at the edge of Lake Maracaibo, some very isolated (Figures 8.3 and 8.4). The combination of geographical and social isolation produced by fear of the disease has reduced

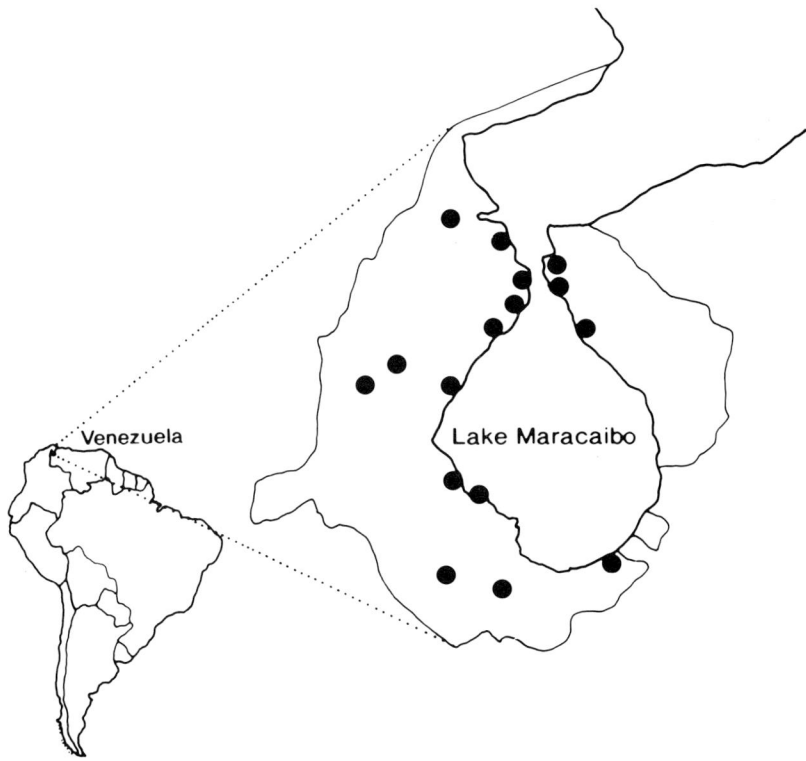

Figure 8.3 HD in Venezuela. Ocurrence of HD in communities around Lake Maracaibo, Venezuela (based on the US congressional report, 1977).

intermarriage with the surrounding communities and has resulted in extreme poverty and lack of medical facilities. While genealogical tracing has interconnected the different branches and identified a common ancestor living around 1830, it is still not certain whether the disorder has a European origin and, if so, from what country the person came.

In epidemiological terms there is much to be learned from this unique community, which should progressively emerge from longitudinal studies still in progress. It is already clear that the pattern of decline seen in the disease is comparable with that elsewhere despite the adverse socio-economic situation and absence of drug treatment (Young et al., 1986; Penney et al., 1990), while the likely homozygotes have so far shown no clinical differences from heterozygotes (Wexler et al., 1987). The community illustrates the rapid expansion of a dominantly inherited genetic disorder that can occur when the gene is introduced into a population with a high natural rate of increase. It also shows how social acceptance of a disorder such as HD may paradoxically be better achieved in a relatively simple community with a high frequency of the condition. It will be of great interest to see whether changes in community

Figure 8.4 The extensive pedigree of the Venezuela kindred, requiring several walls of a room for its display (courtesy of Dr Nancy Wexler).

attitudes will result in uptake of predictive and prenatal testing and what changes these developments in turn produce in the population.

In other parts of South and Latin America, a further focus of high prevalence has been reported from Peru and a survey carried out in Chile (Cruz-Coke, 1987) as well as cases from the Caribbean (Beaubrun, 1962). The long-standing recognition of the disorder is shown by reports from Cuba and Brazil that date back to 1890 and 1891 respectively (Arostegui, 1890; Couto, 1891).

Britain
The United Kingdom has been studied more intensively than any other part of the world, with systematic prevalence studies in no fewer than 15 different regions, in addition to less complete reports. In general, the more recent studies have shown a higher prevalence than the earlier ones and are likely to be closer to full ascertainment, especially those based outside major conurbations where population mobility is greater. From the data given in Table 8.6 and Figure 8.5 no obvious pattern emerges to suggest a single focus of the disease, with high prevalence seen in regions as far apart as Wales and north-east Scotland, while populations of Celtic and Saxon origin likewise do not differ appreciably. No accurate estimate is currently available for the Republic of Ireland, though the disorder is known to be frequent there. This relatively uniform pattern is against a single mutation being responsible for the disorder, or at least one of relatively recent origin.

Britain has been the origin of HD in numerous populations around the world, including the United States, Canada, Australia and South Africa – in fact

Table 8.6. Prevalence estimates for Huntington's disease in the United Kingdom

Study	Publication year	Prevalence year	Region	Map number	No. affected	Population	Prevalence $\times 10^{-5}$
Bickford and Ellison	1953	1950	Cornwall	1	19	340 941	5.57
Pleydell	1954	1954	Northamptonshire	2–5	13	263 000	5.0
Pleydell	1955	1954	Northamptonshire		17	263 000	6.5
Reid	1960	1954	Northamptonshire		19	263 000	7.2
Oliver	1970	1968	Northamptonshire		27	428 000	6.3
Heathfield	1967	1965	Essex	6	81	3 271 000	2.5
Cameron and Venters	1967		S.E. Scotland	7		1 163 877	7.2
Bolt	1970	1968	W. Scotland	8	154	2 959 600	5.2
Heathfield and Mackenzie	1971		Bedfordshire	9	30	427 970	7.5
Glendinning	1975	1965	Somerset	10	33	632 000	5.5
Stevens	1976	1966	Leeds and Yorkshire	11	133	3 190 000	4.17
Caro	1977	1971	E. Anglia	12	54	584 415	9.24
Harper et al.	1979	1971	S. Wales	13	131	1 720 901	7.61
Simpson and Johnston	1989	1984	Grampian, Scotland	14	47	462 891	9.95
Quarrell et al.	1988	1981	N. Wales	15	34	621 000	5.5
Quarrell et al.	1988	1981	S. Wales	16	153	1 728 000	8.85
Nevin and Morrison	1990*	1975	N. Ireland	17	93	1 536 000	6.05
Dennis	1990*	1987	Wessex	18	92	2 457 473	3.74
Garrett	1990*	1987	Devon	19	46	1 010 000	4.6
Garrett	1990*	1987	Cornwall	20	22	453 100	4.9

*Personal communication based on genetic registers; under-ascertainment of varying degree likely. The authors are grateful for permission to quote these unpublished data.

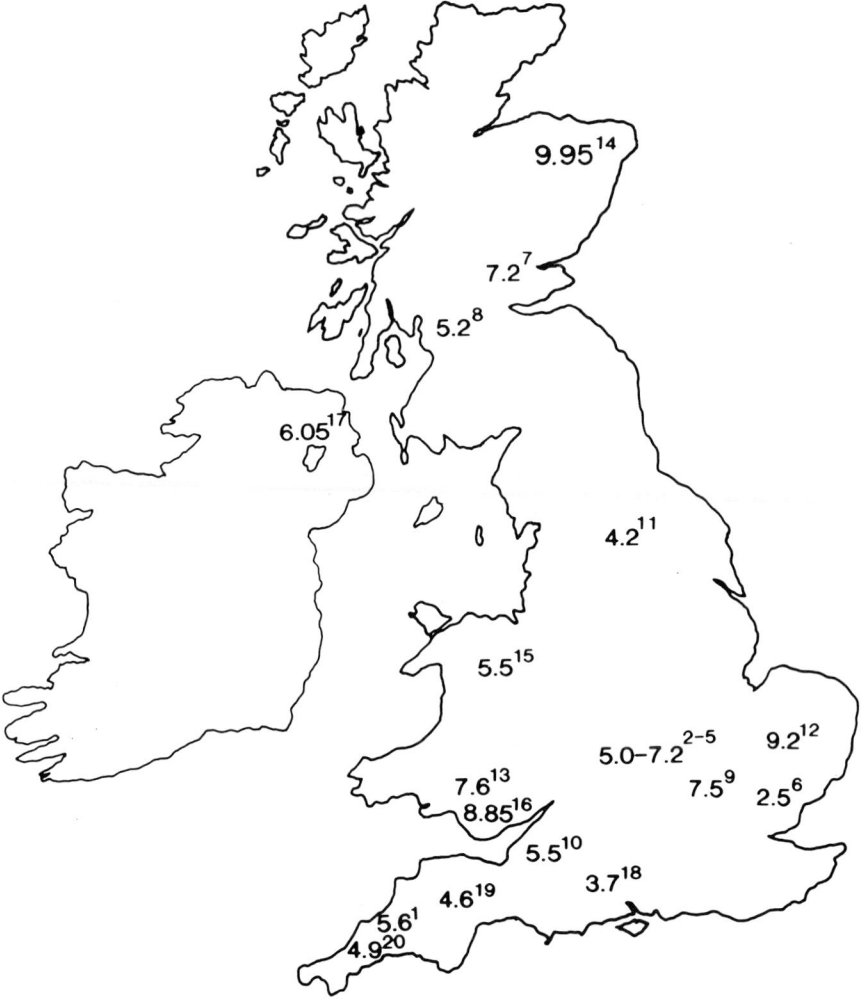

Figure 8.5 HD in Britain. The map shows the results of prevalence studies based on defined geographical regions (see Table 8.6 for details). The superscript numbers refer to the published sources given in the table.

everywhere that there was substantial British colonization. Almost all this migration has occurred during the past 300 years, much during the past century, so that it has often been possible to document the likely gene carriers who founded the new populations, and their area of origin in the UK. Not all of this research has been accurate, as witnessed by the over-elaboration of the history of the New England families derived from East Anglia; these were erroneously interpreted by Vessie (1932), Critchley (1934) and others as documented by Caro (1979) and Caro and Haines (1975) (see Chapter 1). So far it has not been possible to prove common descent between these distant HD populations and living HD patients in Britain, though this should become possible when the identification of specific haplotypes and mutations is more advanced.

Although population mobility continually militates against the complete ascertainment of HD, the well-developed and regionally based medical genetics services in the country are now resulting in a series of systematically maintained genetic registers for the disorder, mostly similar in nature to that operating in South Wales and described in Chapter 11. Thus a steady approach towards total ascertainment of patients and those at risk is likely that will allow not only more complete data on prevalence in different parts of the country, but also the monitoring of any trends resulting from genetic counselling and the application of predictive testing.

Germany

The shadow of the Nazi era still lies heavily on all aspects of work on HD in Germany. The study of Panse (1942) in the Rhineland region, following on the early work of Entres (1921) was one of the first and most thorough analyses of the genetic aspects, and his prevalence estimate of 3.1 per 100 000 is probably more accurate than later studies, agreeing closely with the estimate of Wendt and Drohm (1972) for the Kassel region in 1939. The association of Panse's work with the abuses of the Third Reich is discussed in Chapter 11. The principal later study has been that of Wendt (Wendt et al., 1959, 1960, 1961) brought together in a detailed monograph by Wendt and Drohm (1972). An attempt was made to ascertain all families in West Germany and detailed tables of prevalence are given for different regions; the overall prevalence for 1950 of 2 per 100 000 is likely to be an underestimate compared with the more intensive surveys in specific regions, particularly since the authors state that the work had to be curtailed owing to cessation of funding. Reluctance of families to cooperate could also have hindered ascertainment, though the Nazi policy of compulsory sterilization and killing of affected patients could have contributed to the lower post-war estimates. A recent study in Franconia (Przuntek and Steigerwald, 1987) has given a higher value of close to 5 per 100 000.

Not surprisingly, there remains extreme sensitivity in Germany to such topics as genetic registers and presymptomatic testing for HD, an attitude that applies generally to new developments in genetics, as discussed in Chapter 11.

A German origin is likely for a significant number of the HD families in North America, as already mentioned, and this also applies to other populations which received large-scale German immigration such as Australia.

Other European countries

Data available from specific studies are shown in Table 8.7 and Figure 8.6. Although it was thought in the past that the disorder is more frequent in

Table 8.7 Prevalence estimates for HD in Europe (excluding UK)

Country	Map No.	Author	Prevalence per 100 000
Belgium	1	Husquinet (1970)	1.63
Finland	2	Palo et al. (1987)	0.5
France			
(Northern)	3	Petit (1970)	4.0
(Haute Vienne)	4	Leger (1974)	4.8
Germany			
(Rhineland)	5	Panse (1942)	3.2
(Kassel)	6	Wendt and Drohm (1972)	3.2
(Federal Republic)	—	Wendt and Drohm (1972)	2.2
(Franconia)	7	Przuntek and Steigerwald (1987)	4.8
Iceland	8	Gudmundsson (1969)	2.7
Italy			
(Lazio)	9	Frontali et al. (1990)	2.56
(Emilia)	10	Mainini et al. (1982)	4.8
(Liguria)	11	Roccatagliata and Albano (1976)	4.5
(Toscana)	12	Arena et al. (1978)	2.34
(Florence)		Groppi et al. (1986)	4.1
Malta	13	Cassar (1967)	7.8
Norway	14	Saugstad and Odegard (1986)	6.7
Poland	15	Cendrowski (1964)	4.8
Sweden	16	Mattson (1974)	4.7
Switzerland	17	Zolliker (1949)	3.8–4.8
Yugoslavia			
(Rijeka district)	18	Sepcic et al. (1989)	4.46

northern than in southern Europe, it seems that the difference may well be slight, if it exists at all. Southern European countries such as Italy and Greece are well represented in the origins of migrant HD families in Australia and North America. Apart from a few older studies (e.g. Belgium) where ascertainment has almost certainly been incomplete, most specific prevalence estimates are in the 3–5 per 100 000 range, suggesting a relatively uniform frequency over a wide part of Europe. This distribution has considerable implications for the mutational origin of HD, supporting the occurrence of several common mutations. If the disorder were to be due to a single mutation, it would have to be of extremely ancient origin. A notable exception to the European pattern of HD (as for most genetic disorders) is Finland (Palo et al., 1987), whose exceptionally low prevalence of 0.5 per 100 000, comparable to that of Japan, reflects the unique genetic origin of this population and its unusual distribution of major disease genes, though even here recent molecular analysis has shown more than one DNA haplotype associated with the disease (Ikonen et al., 1990).

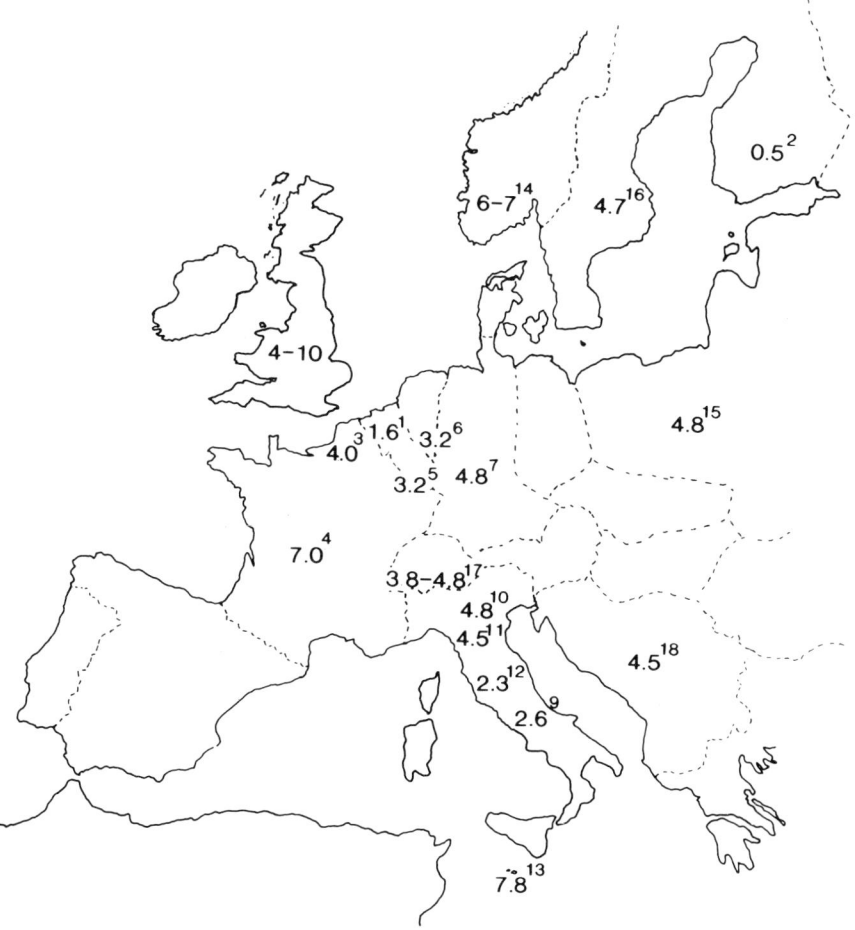

Figure 8.6 HD in various European countries (see Table 8.7 for details). The superscript numbers refer to the published sources given in the table. It can be seen that there is a high prevalence rate for HD in most European countries where it has been intensively studied; Finland is an exception.

Asia

There is a lack of detailed prevalence studies from any Asian country, but it cannot be assumed from this that HD is absent or even rare. In India there has been a series of case and family reports, principally from Punjab and other parts of north India (Chhutani, 1957; Singh et al., 1959; Khosla and Arora, 1973). One of the Caribbean families of Beaubrun (1962) originated from Madras. Further data have recently been provided by Shiwach and Lindenbaum (1990) who undertook a detailed survey of immigrants from the Indian sub-continent to Britain. They found 22 cases among an immigrant population of 1.26 million, giving an age-adjusted prevalence estimate of 1.75 per 100 000. Bearing in mind possible restrictions of immigration and likely under-ascertainment, this suggests that the true prevalence may well approach the European level.

HD in Chinese patients has been described principally outside China itself, from Hong Kong (Singer, 1962), Formosa (Tsuang, 1969) and Singapore (Tay, 1970). It is well recognized in mainland China (Lo, 1990, personal communication) but no prevalence estimate exists. In Soviet Central Asia a study by Kozlova et al. (1986) has reported a high frequency in the Shamkhor region of Azerbaijan, but again without a specific prevalence estimate.

On the western margin of Asia, Bayulkem and Turek (1961) recorded 35 cases of HD in Turkey between 1947 and 1959, and found them to be evenly distributed over the country.

Australia

Australia has been the site of a series of prevalence studies in different states, as well as of the more general HD research of Wallace, Brackenridge and others, referred to elsewhere. The surveys in Victoria (Brothers and Meadows, 1955; Brothers, 1964) and Queensland (Parker, 1958; Wallace and Parker, 1973) have confirmed HD as a major problem in these populations.

The origins and spread of HD in Australia provide a parallel to those that occurred in North America a century or more before. Rapid spread in an expanding population has occurred, with a mainly, but not exclusively, British origin (Gale and Bennett, 1969; Hahn, 1990, personal communication). The only documented cases in Aborigines have also had a European origin. Interestingly and perhaps relevantly in view of previous suggestions of criminality in association with the disorder, no association has been found between HD in Australian families and an ancestor having been 'transported' from Britain for criminal offences.

Table 8.8 and Figure 8.7 summarize the data for Australia, New Zealand and other Pacific regions; a central register was developed to record all cases in Australia (Chiu and Teltscher, 1978) but the specific prevalence studies in

Table 8.8 HD in Australasia and Asia

Country	Study	Prevalence Per 100 000
Tasmania	(Brothers, 1949; Conneally, 1984)	17.4
Tasmania	(Pridmore, 1990c)	12.1
Victoria	(Brothers, 1964)	(4.58)
Queensland	(Parker, 1958)	2.3
Queensland	(Wallace and Parker, 1973)	6.3
New Zealand	(Lintott, 1990, personal communication)	5.7
Japan	(Kishimoto et al., 1957)	0.38
	(Narabayashi, 1973)	0.45
	(Kanazawa, 1983)	0.11
Indian sub-continent (UK immigrants)	(Shiwach and Lindenbaum, 1990)	1.75

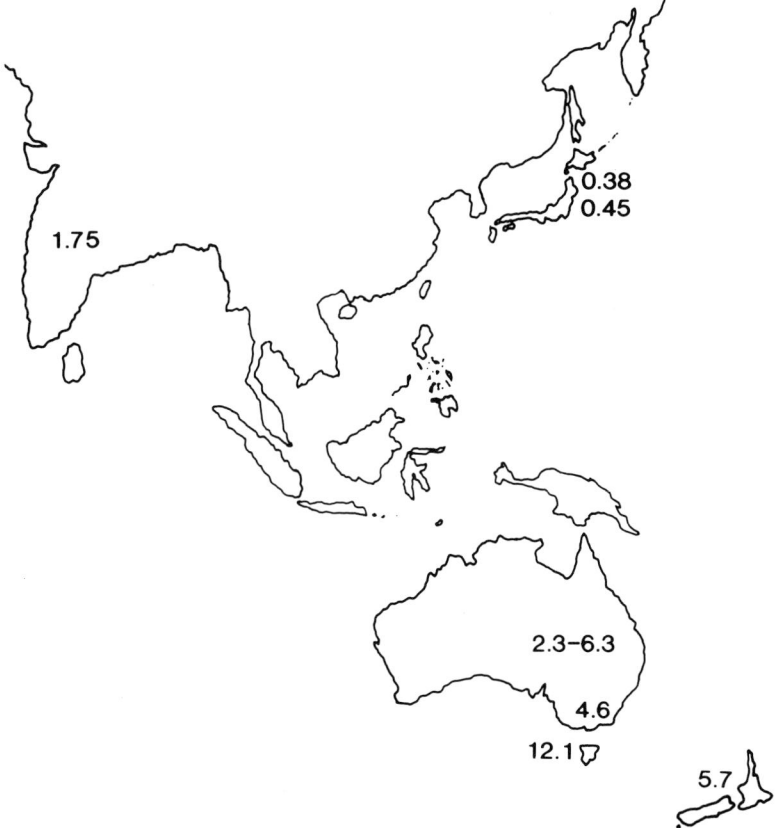

Figure 8.7 HD in Asia and Australasia (see Table 8.8 for details). Note the exceptionally low prevalence for Japan. The estimate for India is based on a study of immigrants in the UK.

Queensland and Victoria still probably represent the most accurate estimates for the Australian mainland; individual states are now maintaining their own registers.

The island of Tasmania has long been recognized as an unusual focus of HD. In 1949, Brothers reported an exceptionally large kindred on the island, derived from a single ancestor whose origin was Somerset, England. Forty years later a thorough study has been undertaken by Pridmore (1990a,b,c), who has established a genetic register and analysed a number of genetic aspects. The current situation can be summarized as follows: the kindred reported by Brothers is still responsible for most cases of HD on the island; as of January 1990 it had produced 198 affected individuals, 40 still living, together with 731 relatives with a risk greater than 10%; this kindred shows an unusually high mean age at onset (48.6 years) and death (62.9 years), though several juvenile cases have occurred. It is interesting that a study by Glendinning (1975) in Somerset, the country of origin of this family, has also documented a large late-onset kindred; it will be of great interest to see if they represent the same mutation.

While a small number of unrelated HD cases have been recognized in Tasmania, the overall data are still dominated by the one large kindred; Pridmore has shown that fertility is increased over the general population and that unaffected sibs show diminished fertility, as discussed in Chapter 9. The current prevalence estimate is 12.1 per 100 000, less than the earlier value of 17.4 per 100 000 based on Brothers' figures, but still much higher than most populations.

The Pacific Islands
HD has been found to occur in New Guinea and some of the smaller island groups of the Pacific and would appear to antedate any European settlement there (Hetherington and Wechsler, 1942; Scrimgeour, 1980, 1982). The interesting and plausible suggestion has been made that the gene could have been introduced by the crews of visiting whaling ships from North America which were active in the sperm-whale fishery in the first half of the nineteenth century. Scrimgeour (1983) searched the records of whaling voyages and found that a number of crew members had surnames present in known HD families from New England recorded in the Davenport and Muncey (1916) archives preserved at the University of Minnesota. While no direct link has been proved, the predominance of New Englanders in the whaling industry and the frequent sexual relations with islanders make this a reasonable hypothesis for these small populations, where genetic drift could have subsequently increased the gene frequency in some cases.

Japan
The remarkably low prevalence of HD in Japan is of great interest in relation to the possible origin of the disorder from a small number of mutations. Whereas for many parts of the world this could be attributed to lack of ascertainment, this is certainly not so for Japan, where well-developed neurological services and extensive neurological and neuropathological research would undoubtedly have identified most cases. Narabayashi (1973) reviewed the considerable number of publications on HD (over 40) from Japan, and it is of interest that some allied disorders such as chorea-acanthocytosis and DRPLA (see Chapter 2) appear to be unusually frequent.

Kishimoto *et al.* (1957) undertook a prevalence study of HD in Aichi prefecture and obtained a prevalence of 0.38 per 100 000; an updated estimate of 0.45 per 100 000 was given by Narabayashi (1973). These values are one-tenth the prevalence of most European origin populations. A more recent estimate has been made by Kanazawa (1983) in Ibaraki prefecture, which gives an even lower prevalence (0.11 per 100 000). Only three living cases (one doubtful) could be identified in a population of 2 638 280. Kanazawa points out that Kishimoto's original study was done in the part of Japan known to contain HD families, so it seems likely that HD is indeed exceptionally rare, though widely scattered.

There is no evidence for foreign admixture in the Japanese cases, but the reason for the lack of spread of the mutation or mutations raises points of great interest which will be resolvable only when the gene can be studied directly.

Africa
South Africa

The occurrence and origins of HD in South Africa have been documented in considerable detail, largely in the work of Hayden (Hayden, 1979, 1981; Hayden et al., 1980a,b). Following an earlier study of Klintworth (1962) Hayden set out to document all cases of HD in the Republic, work that provided the foundations of his subsequent monograph (1981) and for continued research on HD after leaving South Africa for Canada.

Hayden was able to document 481 cases of HD in South Africa and the most striking feature of his study was the relative rarity of the disorder among the black population, only 11 cases, three still living, being recorded. There were also no cases of Asian origin. By contrast the disorder was relatively common in both white South Africans and in those of mixed race (coloured), giving the prevalence estimates shown in Table 8.9. Since Hayden's study a report of eight further cases from four families of black South Africans has appeared (Joubert and Botha, 1988), but there is little doubt that the difference is a real one, even though exaggerated by differential ascertainment as discussed below.

Initially there were thought to be 134 separate kindreds of HD in South Africa, but careful genealogical study progressively reduced this to 74. All the main founding European groups were represented and of particular interest was the demonstration by Hayden (1979) that some of the largest family groups could be linked with ancestors from Holland who also have HD descendents living there today. One such group, traced over 14 generations to the seventeenth century, links together 50 present families containing over 200 individuals, representing three-quarters of all HD patients of Afrikaner origin.

A British origin was found for 15 kindreds, while smaller numbers have had their origin in Germany, Lithuania (a Jewish kindred) and France (via Mauritius). A definite origin could not be assigned to most of the mixed-race families,

Table 8.9 Prevalence of Huntington's disease in South Africa (based on Hayden, 1979)

	No.	Population	Prevalence (per 100 000)
White	97	4 367 000	2.22
Mixed race (coloured)	53	2 432 000	2.17
Black	3	16 647 000	0.01

but the similar frequency to the white population supports a largely European origin.

Detailed study of the South African patients showed essentially similar clinical and genetic features to HD elsewhere, except for a somewhat earlier onset and death especially in the mixed-race cases, a finding that could reflect the action of modifying genes as discussed in Chapter 9. Juvenile cases also showed a shorter duration of the illness, again differing from most other studies, but this estimate excluded living cases and is thus an underestimate.

Mauritius
This small island population, 2000 km east of South Africa, provides a good example of the 'founder effect'. A single kindred of French origin has produced six affected individuals and 60 at risk, giving a prevalence in the white population of 48 per 100 000 (Hayden, 1979). Some members of the kindred have since migrated to smaller islands and to mainland South Africa. While not of general significance, this illustrates the practical importance of 'micro-epidemiology' to the provision of health services in a community.

Other African populations and American blacks
Outside South Africa, there have been only isolated reports from other black African countries, including Uganda (Hutton, 1956), Kenya (Harries, 1973), Nigeria (Osuntoken, 1973) and Ghana (Haddock, 1973). This case material is summarized by Hayden (1979), Wright *et al.* (1981) and by Folstein *et al.* (1987); it does not allow any general conclusion to be drawn other than that HD exists in most African countries, but appears to be uncommon.

Information on American blacks is available from several studies, but the completeness of ascertainment can again be questioned in most of these, while the small numbers of blacks in each survey make accurate estimates impossible. Thus Reed *et al.* (1958) found a frequency in blacks of around one-third of that in the white population, but this was based on only three cases, a number not significantly different from that of eight expected on the basis of an equal prevalence in the races. Wright *et al.* (1981) in South Carolina, specifically studied HD in blacks and found a prevalence of 0.97 per 100 000, but again this was based on only nine patients.

The overall death rate data reported by Hogg *et al.* (1979) are of interest in comparing the frequency in the races in America, since these are standardized across the entire country. These show very small numbers of deaths recorded for blacks from HD in most states; the authors were reluctant to draw any firm conclusions on prevalence from the data.

Perhaps the most interesting study comparing HD in American blacks and whites is that of Folstein *et al.* (1987) in Maryland. They found, after intensive attempts to ascertain all cases in the state, that the prevalence in blacks was at least equal to that in whites (6.37 compared with 4.94 per 100 000). Interestingly this similar value is also seen in the death rates for the state given by Hogg *et al.* The Maryland study made the important point that most cases in blacks were not ascertained through the usual survey methods and would have been missed in many studies. Almost half the cases of HD in blacks were only ascertained by direct study of family members.

Folstein et al. made a detailed clinical comparison between HD in black and white patients and found an earlier onset in blacks, as had Hayden (1979); there was more severe bradykinesia, more frequent involvement of eye movements, and less severe psychiatric illness. Paternal transmission of early cases was found in both races, while the duration of disease was similar. Whether these differences reflect distinct mutations, or the different genetic background against which the same mutation is acting, remains to be determined.

It has previously been assumed that most HD in American blacks is the result of admixture with Caucasian genes, but Folstein et al. (1987) could trace such an origin in only one of 23 cases. It thus seems likely that HD in American blacks is longstanding in black populations, may well be more frequent than is generally recognized at present and could, at least in part, be of African origin. It will be of great interest to see whether any differences in DNA haplotype or specific mutations can be detected.

MIGRATION AND SPREAD OF THE HD GENE

The history of the spread of HD in different countries of the world is essentially that of the population movements of their inhabitants. Until now most attention has focused on the origins of the disorder in recently settled countries, but it is equally important to examine the migration patterns in countries where population is long established and, at first sight, more stable.

HD is found at comparable prevalence in all those countries derived from European settlement in the past three centuries, including the United States and Canada, Australia and South Africa. The situation for those countries colonized by Spain and Portugal, notably South and Central America, is less clear, partly because accurate prevalence data are not available, possibly also because the immigrants to these countries did not replace the indigenous population genetically to the extent of the first group.

A number of surveys have recorded the origins of the HD families studied, usually in terms of the earliest traceable ancestor. The early American surveys already mentioned in this chapter, are especially interesting in this respect. The studies of Falstein and Stone (1939) in Illinois, and of Marx (1973) in Minnesota, both show a predominantly German origin, this being three times more common than a British origin, while in the Michigan study of Reed et al. (1958) the two origins were equal. A Scandinavian origin was frequent in the Minnesota study, reflecting the origins of this population. In all these surveys, around half of the families could not be traced back to outside the United States. The Australian population studies show a similarly wide range of origins, showing that multiple introductions of the disorder have occurred in all those countries whose population has been derived from relatively recent immigration.

Equally important is the evidence these studies provide for internal migration. In both America and Australia an origin from other states in the same country is more prominent than origin from a foreign country and reflects, in America especially, the sucessive waves of settlement involving those already in the country.

There is much less evidence regarding migration patterns for those populations generally regarded as stable. The studies in Belgium (Husquinet, 1970) and Yorkshire, England (Stevens, 1976), show that while 'foreign' immigration is much less than for the recently settled countries, internal migration is still significant, accounting for around 40% of original ancestors. The South Wales study (Walker, 1979; Walker *et al.*, 1981) shows similar results, with only 4% of kindreds derived from outside Britain but 43% from outside Wales.

The South Wales study analysed the topic of migration in considerable detail and attempted to compare the findings for HD families with the general population in the area, which underwent considerable population expansion in the century 1820–1920 due to the industrial revolution. Of the 100 kindreds for which the earliest ancestor had a known origin, 46 came from the study area of industrial South Wales, 11 from other countries of Wales, while 18 were from the adjacent area of south west England. By contrast the general population of industrial South Wales received only half as many immigrants from south west England as from the rest of Wales at this time so that in terms relative to overall migration there has been a contribution of HD ancestors from south west England four times that from the rest of Wales.

While it is important to trace the origins of ancestors, and it will become still more so as unique mutations can be recognized, it is the migration in each generation that is more relevant to such activities as the maintaining of registers. When the South Wales data were analysed in this way only 43 of 412 affected individuals were born outside Wales.

Emigration as well as immigration can be studied. Seventy-seven per cent of South-Wales born patients and 79% of their first degree relatives at risk were found to have remained in the South Wales study area, while 8% and 13% had migrated out of Wales entirely. For a genetic register it is these relatives at risk that are of particular importance; a loss of around 20% per generation is considerable, but for many populations a much higher proportion could be expected.

The final level at which migration in HD can be analysed is the level of the individual kindred. Since such kindreds will often be extensive, and represent a selected one or few out of many in a region, it is unwise to generalize from them, but their study is of considerable significance to their immediate locality, as indicated earlier in this chapter. The Venezuela isolate, much the largest of such kindreds, has already been described, but one of the large South Wales families, descended from a single immigrant from outside Wales and already mentioned in Chapter 1, is an example of how static such a kindred may be,

The Epidemiology of Huntington's Disease 275

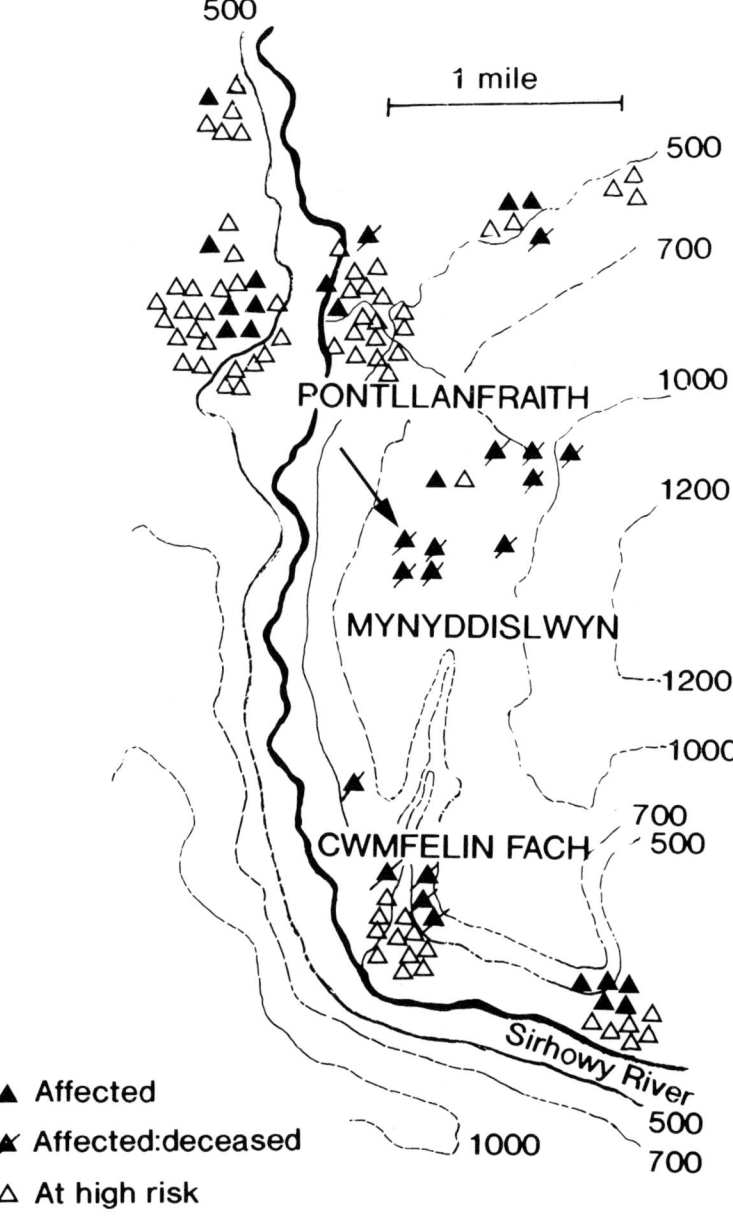

Figure 8.8 Persistent geographical clustering of HD in a large South Wales kindred (simplified pedigree shown in Chapter 1, Fig. 1.6). Note the small scale of the map; most descendants have remained close to the focus of origin of the family (arrowed).

Table 8.10 Local concentrations of HD due to large individual kindreds

Lake Maracaibo, Venezuela	Avila-Giron, 1979
Tasmania, Australia	Brothers, 1949; Pridmore, 1990c
Moray Firth, Scotland	Lyon, 1962
Gwent, South Wales	Walker et al., 1981
Northern Sweden	Sjögren, 1936

once it has established itself in a locality (Figure 8.8). Of the 55 known affected and 177 at risk descendants, only 24 had moved further than a 10 km distance from the original point of settlement 120 years ago.

Table 8.10 lists some of the well-documented concentrations of HD in particular populations that are either known or suspected to be derived from a single founding ancestor. As already mentioned, these local foci of HD can cause a considerable burden for the overall health of the community, especially when the population is isolated and without medical facilities. If in future it can be proved that each of these concentrations is indeed the result of a single mutation, this will make preventive measures such as presymptomatic or prenatal testing considerably easier to apply, though the social attitudes in the population will be the determining factor on whether such tests prove to be acceptable and effective. As has already proved to be the case for other genetic disorders (e.g. the haemoglobinopathies), the approaches to prevention will need to be adapted, in both social, clinical and even molecular terms, to the specific characteristics and needs of the particular HD population.

REFERENCES

Archibald CH (1938) Huntington's chorea in Canada. *National Health Review* **6**:19–21.
Arena R, Nuti M and Iudice A (1979) Rilievi epidemiologici della corea di Huntington nella Toscanna nord-occidentale. *Atti 5 Riun Limpe* **123**–133.
Arostegui G (1890) De la corea cronica progressiva. *Crónica Médico-quirúrgica Habana* **16**:74.
Avila-Giron R (1973) Medical and social aspects of Huntington's chorea in the State of Zulia, Venezuela. In: *Advances in Neurology, Vol. 1* (eds A Barbeau, TN Chase, GW Paulson), pp. 261–266. New York: Raven Press.
Barbeau A and Fullum G (1962) Origin and migration of Huntington's chorea in Canada: preliminary report. *Canadian Medical Association Journal* **87**:1242–1243.
Barbeau A, Witeux C, Trudeau JG and Fullum G (1964) La chorée de Huntington chez les canadiens francais: étude préliminaire. *Union Medicale du Canada* **93**:1178–1182.
Bayulkem F and Turek I (1961) Huntington's chorea in Turkey. *Psychiatric Quarterly* **35**:358–360.
Beaubrun MH (1962) Huntington's chorea in Trinidad. *Caribbean Medical Journal* **24**:45–50.
Bickford JAR and Ellison RM (1953) High incidence of Huntington's chorea in the Duchy of Cornwall. *Journal of Mental Science* **99**:291–294.
Bolt JMW (1970) Huntington's chorea in the west of Scotland. *British Journal of Psychiatry* **116**:259–270.
Brothers CRD (1949) The history and incidence of Huntington's chorea in Tasmania. *Proceedings of the Royal Australian College of Physicians* **4**:48–50.
Brothers CRD (1964) Huntington's chorea in Victoria and Tasmania. *Journal of the Neurological Sciences* **i**:405–420.
Brothers CRD and Meadows AW (1955) An investigation of Huntington's chorea. *Victoria Journal of Medical Science* **101**:548–563.

Cameron D and Venters GA (1967) Some problems in Huntington's chorea. *Scottish Medical Journal* **12**:152–156.
Caro AJ (1977) Huntington's chorea; a clinical problem in East Anglia. University of East Anglia: PhD thesis.
Caro A and Haines S (1975) The history of Huntington's chorea. *Update* **11**:91–95.
Cassar P (1967) Huntington's chorea with special reference to its incidence in Malta. *St Luke's Hospital Gazette* June, pp. 3–13.
Cendrowski W (1964) Some remarks on the geography of Huntington's chorea. *Neurology Minneapolis* **14**:839–843.
Chhutani PN (1957) Huntington's chorea in India. *Journal of the Indian Medical Association* **29**:156–157.
Chiu E and Teltscher B (1978) Huntington's disease: the establishment of a national register. *Medical Journal of Australia* **2**:394–396.
Clarke CK and MacArthur JW (1924) Four generations of hereditary chorea. *Journal of Heredity* **15**:303–306.
Conneally PM (1984) Huntington's disease: genetics and epidemiology. *American Journal of Human Genetics* **36**:506–526.
Couto M (1891) Da corea de Huntington. *Brasil-Médico* **5**:341–344.
Critchley M (1934) Huntington's chorea and East Anglia. *Journal of State Medicine* **42**:575–587.
Cruz-Coke R (1987) Epidemiologica genetica de Corea de Huntington en Chile. *Review of Medicine Chile* **115**:483–485.
Davenport CB and Muncey EB (1916) Huntington's chorea in relation to heredity and eugenics. *Eugenics Record Office Bulletin* **17**:195–222.
Entres JL (1921) *Zur Klinik und Vererbung der Huntington'schen Chorea. Monographien des Gesamtgebietes der Neurologie und Psychiatrie* **27**. Berlin: Springer.
Falstein EI and Stone H (1939) Huntington's chorea as a psychiatric and social problem in Illinois. *Illinois Medical Journal* **75**:164–168.
Folstein SE (1989) *Huntington's Disease: a Disorder of Families*. Baltimore: Johns Hopkins Press.
Folstein SE, Chase GA, Wahl WE, McDonnell AM and Folstein MF (1987) Huntington's disease in Maryland: clinical aspects of racial variation. *American Journal of Human Genetics* **41**:168–179.
Frontali M, Malaspina P, Rossi C et al. (1990) Epidemiological and linkage studies on Huntington's disease in Italy. *Human Genetics*, in press.
Gale F and Bennett JH (1969) Huntington's chorea in a South Australian community of aboriginal descent. *Medical Journal of Australia* **2**:482–484.
Gersting JM, Conneally PM and Yount EA (1984) Huntington's disease research roster data base support with MEGADATS-3m. *Journal of Medical Systems Supp.* **8**:163–172.
Glendinning N (1975) A study in Huntington's chorea. University of London: MD thesis.
Groppi C, Barontini F, Braco L, Sita D, Inzitari D, Amadulli L and Fratiglioni L (1986) Huntington's chorea: a prevalence study in the Florence area. *Acta Psychiatrica Scandinavica* **74**:266–268.
Gudmundsson KR (1969) Prevalence and occurrence of some rare neurological diseases in Iceland. *Acta Neurologica Scandinavica* **45**:114–118.
Haddock RDW (1973) Neurological disorders in Ghana. In: *Tropical Neurology* (ed JD Spillane), pp. 143–160. Oxford University Press.
Hall JG and Te-Juceto L (1983) Association between age of onset and paternal inheritance in Huntington's chorea. *American Journal of Medical Genetics* **16**:289–290.
Harper PS, Walker DA, Tyler A, Newcombe RG and Davies K (1979) Huntington's chorea. The basis for long-term prevention. *Lancet* **ii**:346–349.
Harries JR (1973) Neurological disorders in Kenya. In: *Tropical Neurology* (ed JD Spillane), pp. 207–222. Oxford University Press.
Hattie WH (1909) Huntington's chorea. *Proceedings of the American Medical–Psychological Association* **16**: 171–176.
Hayden MR (1979) Huntington's chorea in South Africa. University of Cape Town: PhD thesis.
Hayden MR (1981) *Huntington's Chorea*. Berlin: Springer-Verlag.
Hayden MR, Hopkins HC, Macrae M and Beighton PH (1980a) The origin of Huntington's chorea in the Afrikaner population of South Africa. *South Africa Medical Journal* **58**:197–200.
Hayden MR, MacGregor JM and Beighton PH (1980b) The prevalence of Huntington's chorea in South Africa. *South Africa Medical Journal* **58**:193–196.
Heathfield KWG (1967) Huntington's chorea investigation into the prevalence of disease in the area covered by the North East Metropolitan Board. *Brain* **90**:203–232.
Heathfield KWG and MacKenzie ICK (1971) Huntington's chorea in Bedfordshire. *Guys Hospital Reports* **120**:295–310.

Hetherington HB and Wechsler Z (1942) Huntington's chorea in a native Melanesian family of the British Solomon Islands. *Medical Journal of Australia* **1**:599–600.

Hogg JE, Massey EW and Schoenberg BS (1979) Mortality from Huntington's disease in the United States. *Advances in Neurology* **23**:27–33.

Husquinet H (1970) La chorée de Huntington dans les 4 provinces Belges. In: *CR 67E Congr Psychiat Neurol Langue Franc* (ed P Warot), pp. 1079–1118. Paris: Masson.

Hutton PW (1956) Neurological disease in Uganda. *East African Medical Journal* **33**:209–223.

Ikonen E, Palo J and Ott J (1990) Huntington's disease in Finland: Linkage disequilibrium of chromosome 4RFLP haplotypes and exclusion of a tight linkage between the disease and D4S43 locus. *American Journal of Human Genetics* **46**:5–11.

Jelliffe SE (1908) A contribution to the history of Huntington's chorea: a preliminary report. *Neurographs* **1**:116–124.

Joubert J and Botha MC (1988) Huntington's disease in South African blacks: a report of 8 cases. *South Africa Medical Journal* **73**:489–494.

Kanazawa I (1983) Prevalence rate of Huntington's disease in Ibaraki prefecture. Annual report of research committee of CNS degenerative diseases, Ministry of Health and Welfare of Japan, pp. 151–156.

Khosla SN and Arora BS (1973) Huntington's chorea – a clinical study. *Journal of the Association of Physicians of India* **21**:247–250.

Kishimoto K, Nakamura M and Sotokawa Y (1957) Population genetics study – Huntington's chorea in Japan. *Annual Report, Research Institute Environmental Medicine* **9**:195–211.

Klintworth GK (1962) Huntington's chorea in South Africa: a preliminary communication drawing attention to its frequent occurence. *South Africa Medical Journal* **36**:896–898.

Korenyi C and Whittier JR (1977) Huntington's disease (chorea) in New York State. *New York State Journal of Medicine* **77**:44–45.

Kozlova SI, Dadali EL, Prytkov AN, Bolshakova LP, Sibiryakova LG et al. (1986) Population, demographic and clinical-genetic studies of the Huntington disease in an Azerbaijan local region. *Genetika* **22**:2534–2539.

Kurland LT (1958) Descriptive epidemiology of selected neurologic and myopathic disorders with particular reference to a survey in Rochester, Minnesota. *Journal of Chronic Disorders* **8**:378–418.

Kurtzke JF (1979) Huntington's disease: mortality and morbidity data from outside the United States. *Advances in Neurology* **23**:13–25.

Lanska DJ, Levine L, Lanska MJ and Schoenberg BS (1988) Huntington's disease mortality in the United States. *Neurology* **38**:769–772.

Leger JM, Ranouil R and Vallat JN (1974) Huntington's chorea in Limousin: statistical and clinical study. *Revue Medicale de Limoges* **5**:147–153.

Lyon RL (1962) Huntington's chorea in the Moray Firth area. *British Medical Journal* **i**:1301–1306.

MacKay M (1904) Hereditary chorea in eighteen members of a family, with a report of three cases. *Medical News* **85**:496–499.

Mainini P, Lucci B, Guidetti D and Casoli C (1982) Prevalenzia della mallattia di Huntington nelle Provincie di Reggio Emilia e Parma.

Marx RN (1973) Huntington's chorea in Minnesota. *Advances in Neurology* **1**:237–243.

Mattsson B (1974) Huntington's chorea in Sweden. 1. Prevalence and genetic data. *Acta Psychiat Scandinavica Suppl* **255**:211–255.

Myrianthopoulos NC (1973) Huntington's chorea: the genetic problems five years later. *Advances in Neurology* 150–152.

Narabayashi H (1973) Huntington's chorea in Japan: review of the literature. *Advances in Neurology* **1**:253–259.

Office of Health Economics (1980) *Huntington's chorea*. Publication 67, pp. 1–35.

Oliver JE (1970) Huntington's chorea in Northamptonshire. *British Journal of Psychiatry* **116**:241–253.

Osuntoken BO (1973) Neurological disorders in Nigeria. In: *Tropical Neurology* (ed JD Spillane), pp. 161–190. Oxford University Press.

Palo J, Somer H, Ikonen E, Karila L and Peltonen L (1987) Low prevalence of Huntington's disease in Finland. *Lancet* **ii**:805–806.

Panse F (1942) *Die erbchorea: eine klinische-genetische studie*. Leipzig: Thieme.

Parker N (1958) Observation on Huntington's chorea based on a Queensland survey. *Medical Journal of Australia* **45**:351–359.

Pearson JS, Petersen MC, Lazarte JA, Blodgett HE and Kley IB (1955) An educational approach to the social problem of Huntington's chorea. *Proceedings Mayo Clinic* **30**:349–357.

Penney JB, Young AB and Shoulson I (1990) Huntington's disease in Venezuela: 7 years of follow-up on symptomatic and asymptomatic individuals. *Movement Disorders* 5:93–99.
Petit H (1970) La maladie de Huntington. In: *CR 67E Congr Psychiat Neurol Langue Franc* (ed P Warot), pp. 901–1058. Paris: Masson.
Pleydell MJ (1954) Huntington's chorea in Northamptonshire. *British Medical Journal* ii:1121–1128.
Pleydell MJ (1955) Huntington's chorea in Northamptonshire. *British Medical Journal* ii:889.
Pridmore SA (1990a) Age of onset of Huntington's disease in Tasmania. *Medical Journal of Australia* 153:135–137.
Pridmore SA (1990b) Age of death and duration in Huntington's disease in Tasmania. *Medical Journal of Australia* 153:137–139.
Pridmore SA (1990c) The prevalence of Huntington's disease in Tasmania. *Medical Journal of Australia* 153:133–134.
Przuntek H and Steigerwald A (1987) Epidemiologische untersuchung zur Huntington'schen Erkrankung in Einzugsgebiet der Würzburger Neurologischen Universitätesklinik Unterbesanderer Beruiksichtigung der Untefrankischen raumes. *Nervenarzt* 58:424–427.
Quarrell OWJ, Tyler A, Jones MP, Nordin M and Harper, PS (1988) Population studies of Huntington's disease in Wales. *Clinical Genetics* 33:189–195.
Reed TW, Chandler JH, Hughes EM and Davidson RT (1958) Huntington's chorea in Michigan. I. Demography and genetics. *American Journal of Human Genetics* 10:201–225.
Reed TE and Neel JV (1959) Huntington's chorea in Mitchigan. II. Selection and mutation. *American Journal of Human Genetics* 11:107–136.
Reid JJ (1960) Huntington's chorea in Northamptonshire. *British Medical Journal* ii:650.
Roccatagliata G and Albano C (1976) Storia naturale della corea di Huntington. *Rivista di Neurologia* 46:297–332.
Saugstad L and Odegard O (1986) Huntington's chorea in Norway. *Psychological Medicine* 16:39–48.
Schoenberg BS (1979) Epidemiologic approach to Huntington's disease. *Advances in Neurology* 23:1–11.
Scrimgeour EM (1980) Huntington's disease in two New Britain families. *Journal of Medical Genetics* 17:197–202.
Scrimgeour EM (1982) Huntington's chorea in Papua. *Papua New Guinea Medical Journal* 25:12–15.
Scrimgeour EM (1983) Possible introduction of Huntington's chorea into Pacific Islands by New England whalemen. *American Journal of Medical Genetics* 15:607–613.
Sepcic J, Antonelli L, Sepic-Grahovac D and Materljan E (1989) Epidemiology of Huntington's disease in Rijeka district Yugoslavia. *Neuroepidemiology* 8:105–108.
Shiwach RS and Lindenbaum RH (1990) Prevalence of Huntington's disease among UK immigrants from the Indian subcontinent. *British Journal of Psychiatry* 157:598–599.
Shokeir MHK (1975) Investigations on Huntington's disease in the Canadian prairies. I. Prevalence. *Clinical Genetics* 7:345–348.
Singer K (1962) Huntington's chorea in the Chinese. *British Medical Journal* ii:1311–1312.
Singh A, Singh S and Jolly SS (1959) Huntington's chorea: a report of 4 new pedigrees from Punjab. *Neurology (Bombay)* 7:7–8.
Simpson SA and Johnston AW (1989). The prevalence and patterns of care of Huntington's chorea in Grampian. *British Journal of Psychiatry* 155:799–804.
Sjögren T (1936) Vererbungsmedizinische Untersuchungen über Huntingtons chorea in einer schwedischen Bauernpopulation. *Vererb-Konstit-Lehre* 19:131–165.
Stevens DL (1976) Huntington's chorea: a demographic, genetic and clinical study. University of London: MD thesis.
Tay CH (1970) Huntington's chorea: report of a Chinese family in Singapore. *Journal of Medical Genetics* 7:41–43.
Tsuang M-T (1969) Case report. Huntington's chorea in a Chinese family. *Journal of Medical Genetics* 6:354–356.
Vessie PR (1932) On the transmission of Huntington's chorea for 300 years – the Bures family group. *Journal of Nervous and Mental Disease* 76:553–573.
Walker DA, Harper PS, Wells CEC, Tyler A, Davies K and Newcombe RG (1981) Huntington's chorea in South Wales: a genetic and epidemiological study. *Clinical Genetics* 19:213–221.
Walker DA (1979) Huntington's chorea in South Wales. University of Liverpool: MD thesis.
Wallace DC and Parker N (1973) Huntington's chorea in Queensland: the most recent story. *Advances in Neurology* 1:223–236.
Wendt GG and Drohm D (1972) *Die Huntingtonsche Chorea. Eine populations genetische studie.* Stuttgart: Thieme.

Wendt GG, Landzettel I and Unterreiner I (1959) Das Erkrankungsalter bei der Huntingtonschen Chorea. *Acta Genetica (Basel)* **9**:18–32.

Wendt GG, Landzettel I and Solth K (1960) Krankheitsdauer und Lebenserwartung bei der Huntingtonschen Chorea. *Archiv fur Psychiatrie und Nerven Krankheiten* **201**:298–312.

Wendt GG, Solth K and Landzettel I (1961) Kinderzahl, Erkrankungsalter und Sterbealter bei der Huntingtonschen Chorea. *Anthropologischer Anzeiger* **24**:299–309.

Wexler NS, Young AB and Tanzi RE (1987) Homozygotes for Huntington's disease. *Nature* **326**:194–197.

Winsor EJ and Welch JP (1973) Huntington's chorea in Nova Scotia. *Medical Bulletin* **52**:108–109.

Wright HM, Still CN and Abramson RK (1981) Huntington's disease in black kindreds in South Carolina. *Archives of Neurology* **38**:412–414.

Young AB, Shoulson I and Penney JB (1986) Huntington's disease in Venezuela: neurologic features and functional decline. *Neurology* **36**:244–249.

Zike V and Render ND (1949). Huntington's chorea. *Journal of the Iowa State Medical Society* **39**:386–388.

Zolliker A (1949) Die chorea Huntington in der Schweiz. *Schweiz Archiv für Neurologie und Psychiatrie* **64**:448–459.

9

Genetic Aspects of Huntington's Disease

INTRODUCTION

The hereditary nature of Huntington's disease (HD) has been clear since the earliest descriptions, as noted in the opening chapter of this book; indeed the term 'hereditary chorea' was commonly used in these descriptions to distinguish it from other forms of chorea. Mendel's work on the basic laws of inheritance had been published in 1865, but had to wait until 1900 before achieving recognition. Despite this George Huntington's original 1872 description shows clearly how the disorder follows the pattern of dominant inheritance: 'One or more of the offspring almost invariably suffer from the disease, if they live to adult age. But if by any chance these children go through life *without* it, the thread is broken and the grandchildren and great-grandchildren of the original shakers may rest assured that they are free from the disease'.

In the early years of this century, following the rediscovery of Mendel's work, scientists were eager to find examples of Mendelian inheritance in human diseases; HD soon became accepted as such an example and Jelliffe (1908) and Punnett (1908) both recognized this, as did Davenport in his early book (1911), which interestingly also suggested the possibility of Mendelian inheritance for other types of chorea. The specific and detailed study of New England HD families by Davenport and Muncey (1916) confirmed the view of HD as an autosomal dominant disorder. However, it was not until the systematic and quantitative analysis of Julia Bell (1934) that this conclusion was placed on firm and detailed foundations. A useful account of the early genetic studies on HD is that of Myrianthopoulos (1966), while Conneally (1984) has provided an excellent review of the subject, including a detailed assessment of possible modifying factors.

In recent years it has become clear that mendelian inheritance may be much more variable than previously supposed, and that even well-established Mendelian disorders can show significant departures from expected mendelian ratios. HD is one such condition, so before describing the discrepancies and unusual aspects it is worth examining closely the evidence on which autosomal dominant inheritance is based, rather than simply assuming this as an accepted fact.

HD AS A MENDELIAN DISORDER

The principal findings to be expected in a disorder showing autosomal dominant inheritance are summarized in Table 9.1, while Figure 9.1 shows diagramatically the pattern of segregation on which it is based. Equal sex incidence, equal transmission by both sexes, a 50% proportion of affected offspring born to an affected parent and lack of transmission by unaffected family members, are all features which appear evident from any large HD pedigree, but unless these are analysed quantitatively one can miss a significant deviation.

The sex incidence of HD found in the major studies has varied considerably but overall is close to the equality expected. Bell (1934) found a 53.5% male incidence in a total of 956 cases and Pearson *et al.* (1955) also found a slight male excess (66 out of 117) whilst Reed *et al.* (1958) found a deficiency of male cases (85 out of 203). Data from the Cardiff register (Sarfarazi *et al.*, 1987) for the South Wales area, where ascertainment has been close to complete for a considerable time, show 283 affected males and 284 affected females, as close to equality as it is possible to achieve.

Ascertainment and other factors can belie such overall estimates and a more accurate approach is to examine the sex ratio within sibships among the children of affected parents. Reed *et al.* found no significant difference from 50% regardless of the age cut-off used.

When parental transmission of HD is examined, most overall studies again agree closely with the expected 50% value. Bell (1934) found 55.2% of 744 cases to originate from the father, though there was an excess of father–son transmissions (60.3%), no such difference being seen when the disorder was maternally transmitted. The striking difference that emerges when the data are divided by age of onset, especially the paternal transmission of juvenile cases,

Table 9.1 Criteria for autosomal dominant inheritance, as applied to HD

Equal incidence in both sexes?
  Yes

Equal transmission by both sexes?
  Yes, but differential effect of sex of affected parent on age of onset

50% of offspring of an affected person also become affected?
  Yes, but only by age 60 years

No transmission of the disorder by those remaining free from HD?
  Yes, except for those parents dying young of another cause

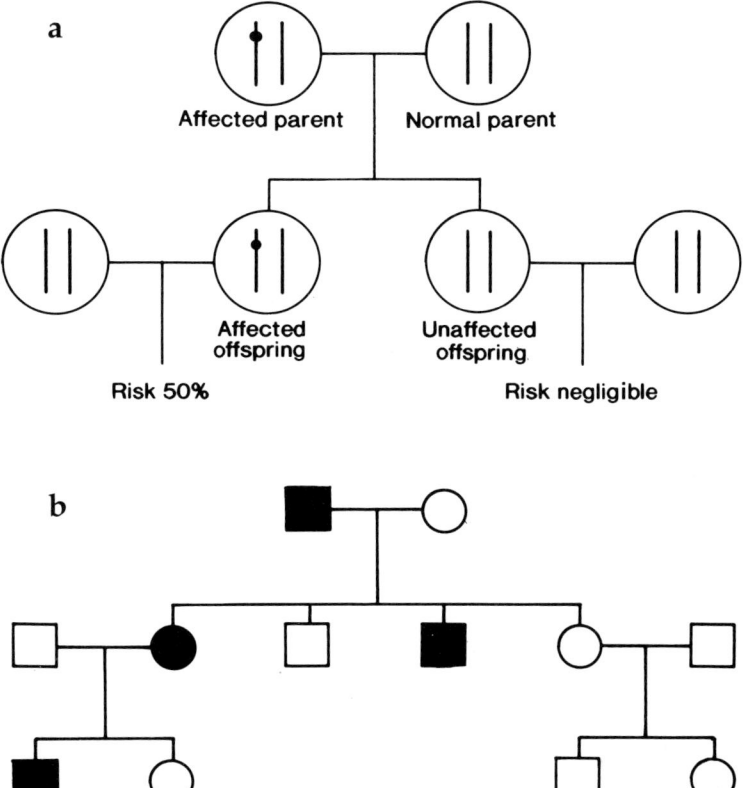

Figure 9.1 Mendelian autosomal dominant inheritance in HD. (a) To show segregation pattern in offspring of affected and unaffected individuals. Affected individuals are normally heterozygotes, with a 50% risk of transmitting the gene to offspring. (N.B. For a late-onset disorder such as HD, 'affected' implies ultimately affected, not necessarily at the time of reproduction.) Those not inheriting the gene have a negligible chance of having affected offspring. (b) A schematic pedigree to show autosomal dominant transmission in HD. Transmission is by and to each sex equally, with an average 50% of offspring of an affected person affected. Such an 'idealized' pedigree will rarely be seen in practice.

is discussed later in the chapter. The important point to be noted here is that there is no *overall* disturbance of the expected mendelian ratio.

Testing the expected ratio of 50% for the affected offspring of affected parents requires careful classification by age as well as the identification of propositi to minimize bias. Bell (1934) found a slight excess of affected offspring (55.6%) in her collected data on 1162 offspring from 240 sibships. Any sibship containing unaffected members under 40 years was omitted. She felt that ascertainment bias towards sibships with multiple affected members was likely and explained the deviation from 50%.

Table 9.2 Percentage of affected among offspring of affected parents. (Sibships containing unaffected members under 40 years omitted)

Author	% affected
Bell (1934)	55.6
Sjögren (1936)	43.0
Reed et al. (1958)	52.5 (48.2)*
Stevens (1976)	52.5

*Equivocal cases counted as normal

More detailed segregation analysis was done in the studies of Sjögren (1936), Panse (1942), Reed et al. (1958) and Stevens (1976) all of which gave values of close to 50%, though with some variation depending on the age criteria used. The data at the 40-year cut-off point, which allows comparison between most of those studies, are summarized in Table 9.2.

The fourth criterion for autosomal dominant inheritance, lack of transmission by an unaffected person, is also age related, and is bound up with the subject's of age at onset and penetrance of the gene, already discussed in Chapter 4 and dealt with further in relation to genetic counselling in Chapter 11. Penetrance, the proportion of individuals carrying the abnormal gene who show signs of the disease, is clearly minimal in childhood and low in early adult life, but the number of cases who have had an affected parent and child, but themselves have remained unaffected while living to old age, is essentially zero. As with new mutations, it is actually very difficult to document such a case satisfactorily, but in practice the only 'skipped generations' seen are those where the intervening parent has died relatively young or has not been documented medically (Figure 9.2).

Why should Huntington's disease follow dominant inheritance? It may appear strange to ask this question, but dominance and recessiveness are attributes of the phenotype, not the gene, and have long been recognized to be under genetic control themselves. Haldane (1941) discussed the role of modifying genes in affecting age at onset in a number of genetic disorders and concluded that selection was likely to delay the onset and reduce the severity. He expressed this in characteristically graphic terms: 'Modifiers are presumably being selected which delay the onset. Perhaps Huntington's chorea was a disease of infancy in Sinanthropus. And if eugenic measures are not taken against it, it may be confined to old age in our remote descendants, in which case the main gene will spread, and homozygotes appear, so that it will be, in effect, a recessive disease'.

As will be seen later in this chapter, there is now considerable evidence for the existence of modifying genes in relation to age at onset in HD. It is possible that future therapeutic approaches may be able to be exerted through these as well as directly through the function of the HD gene itself.

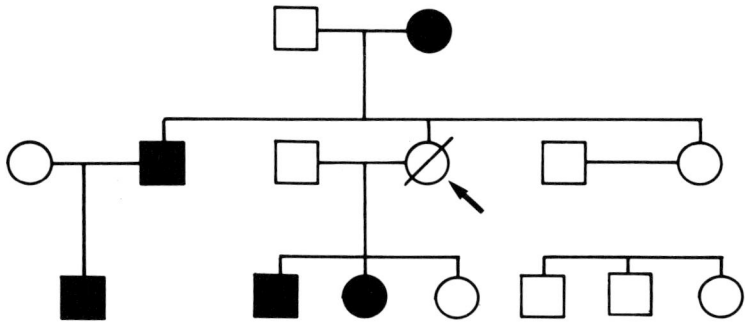

Figure 9.2 Apparent skipped generation in HD pedigree. The individual arrowed died of an unrelated cause aged 42 years. At this age her residual risk of having the HD gene but not having developed the disorder is around 40%, so the apparent 'skipped generation' is not surprising.

In summary, HD fulfils the basic criteria for mendelian autosomal dominant inheritance as well as can reasonably be expected for such a variable and late-onset disorder. We can now proceed to examine some of the other important genetic aspects of the disease before returning to the unexplained and puzzling features that do not fit with mendelian inheritance.

TWINS

The study of identical (monozygotic) twin pairs provides a special opportunity for analysing the interplay of genetic and environmental factors. It has proved of most value for those disorders (e.g. schizophrenia) where the mode of inheritance is not clear cut but where familial aggregation is marked. For a clearly mendelian disorder such as HD the value of twin studies is more to define the possible role of environmental factors in the range of clinical variation and in age at onset.

A number of monozygotic twin pairs with HD have been reported and the results are of considerable interest. Most of the early pairs were not adequately studied for zygosity. Table 9.3 summarizes the data on those where monozygosity seems likely; the remaining early pairs are reviewed by Myrianthopoulos and Rowley (1960).

The principal feature to emerge from these studies is the similarity between the co-twins, especially regarding age at onset. There are no reports of complete discordance, i.e. of the co-twin remaining healthy after a considerable interval. Although some pairs have varied in clinical presentation, such as that of Oepen (1973), where one had mainly rigidity, the other choreic movements, the overall impression is one of close similarity. This suggests that most of the considerable variation between patients of the same kindred is

Table 9.3 Monozygotic twin pairs with HD

Author	Sex	Evidence for zygosity	Age at onset	Clinical notes
Rosanoff and Handy (1935)	F	Appearance only	35/35	
Entres (1921)	F		41/41	Died at 54 and 56 years
	M		45/45	Died at 64 and 65 years
Parker (1958)	M		42/42	Died 3 months apart
Myrianthopoulos and Rowley (1960)	F	Blood groups, dermatoglyphs	22/22	Similar course; mental deterioration mild in one
Schiottz-Christensen (1969)	F		24/24	
Oepen (1973)	F	Blood groups; skin gråft	22/25	Similar; slight differences on detailed psychological tests
Husquinet et al. (1973a)	F	Blood groups and dermatoglyphs		Both severe psychiatric symptoms, slight chorea worsened by L DOPA
Bird and Omenn (1975)	F	Blood groups and RBC enzymes	Both mid 20s	Chorea marked in one minimal in another
Bachman et al. (1977)	F	HLA and blood groups	Both 5 years	Both with rigidity and dementia

likely to be the result of prenatal and developmental factors, including the action of other genetic loci, and that the post-natal environment plays a subsidiary role. This is not surprising in the light of studies on schizophrenia co-twins brought up apart, where remarkable similarity in detailed features of the illness has also been observed (Shields, 1962).

Identical twins also pose a special problem for presymptomatic testing, since an adverse result in one would automatically confirm that the co-twin also had the HD gene. In such a situation, testing would seem unwise unless both twins are requesting it.

NEW MUTATION

The diagnostic problems produced by the apparent isolated case of HD are considerable, and have already been discussed in Chapter 2. The implications for genetic counselling of whether an isolated case really is a new mutation are also of practical importance and are considered in that section. Here the questions that will be addressed are: how common (or rare) are true cases of new mutation? What proportion of isolated cases do they make up? Can we estimate an absolute value for the mutation rate in HD?

The first question can be easily answered in general terms – new mutation is undoubtedly rare and accounts only for a very small proportion of all cases of the disease. Problems arise in defining precisely how rare, both in absolute terms and in relation to the larger number of apparently isolated cases. Stevens and Parsonage (1969) were the first to address systematically the problems of defining a new mutation; they laid down four criteria, which should be met before accepting a case as a proven new mutation. The criteria, as now generally accepted, are as follows:

(1) The clinical features of the disorder should be typical for Huntington's disease, and other disorders mimicking HD excluded as far as is possible. The desirability of neuropathology in an affected person could well be added.
(2) Autosomal dominant inheritance should be demonstrated by transmission to the next generation by the individual representing the supposed mutation.
(3) The parents should be confirmed as healthy by examination or if dead should have been documented as unaffected beyond the age of likely onset of HD.
(4) Non paternity should be excluded.

These criteria are entirely logical, and it can be safely stated that any case fulfilling all of them will be a new mutation. However the criteria are so difficult to fulfil that one could predict that a mutation defined in this way would be impossible to observe even if new mutations existed at an appreciable frequency. The most difficult to fulfil is the transmission of the disorder to the next generation, for by the time this has happened the parents of the original case will almost certainly be dead, with likely doubt as to whether both were really free from the disease.

A series of possible mutations has been reported over the years; the fact that such cases still merit publication as case reports is in itself an indication of their rarity. Shaw and Caro (1982) provide a thorough and critical review of these cases and conclude that only one of the 10 possible candidates is likely to represent a true mutation. Several further cases have been described since then; the most likely to represent new mutations are the cases of Baraitser et al. (1983) and Wolff et al. (1989), as well as the earlier case of Wallace (1972). Non paternity is now readily testable by DNA fingerprinting, but lack of neuropathological confirmation remains a major limitation bearing in mind the overlap of clinical phenotype with other neurodegenerative disorders.

The most recent case of possible new mutation for HD to be reported, that of Wolff et al. (1989), is of particular interest on a number of counts. Not only were the parents unaffected and elderly (over 80 years) at the time of examination, but there were 15 unaffected sibs, 13 of them older than the patient; non paternity was excluded with a high degree of probability, while DNA typing with the G8 marker linked to HD showed a haplotype shared by several of the unaffected sibs. Advanced paternal age (55 years) also supported the occurrence of mutation, while the clinical features and investigations were typical for HD. It will be important for neuropathology to be studied when this patient dies.

All the major surveys of HD have reported isolated cases as accounting for only a small proportion of their total cases, and of these only very few can seriously be considered as possible, let alone likely, new mutations. Shaw and Caro (1982) combined data from a number of studies and estimated that isolated cases represented only 2% at most of all cases, or 5% of all families. These figures fell to 0.04% and 0.1% when applying more rigorous criteria. In the South Wales study (Walker et al., 1983) no case could seriously be considered as representing a new mutation among 418 patients from 101 kindreds; the same is true for several of the other major studies including Reed and Neel's study (1959) which found 36 'isolated' cases among 801 studied (196 kindreds) and that of Wendt and Drohm (1972) in Germany, which found no likely mutation among 1032 cases studied. As Shaw and Caro stated of their own cases: 'all peter out through lack of information rather than definite knowledge of the origin of the disease through mutation.'

In view of the difficulties in identifying new mutations is it possible to make a meaningful absolute estimate of the mutation rate at this locus? Direct estimates can easily be converted into a mutation rate (μ) by first dividing the number of cases representing new mutations by the total population number. The figure must then be halved since the mutation rate is per gamete (or chromosome locus), of which an individual will have two.

It is also possible to make an indirect estimate, based on the argument that if the disorder is remaining constant in frequency, there must be equilibrium between genes lost through diminished genetic fitness and those gained as a result of new mutation. For a disorder such as HD, with fitness reduced only slightly, if at all, it is clear that very few new mutations are needed to maintain equilibrium. Here the calculation is

$$\mu = 1/2\,(1-f)\,n/N$$

Table 9.4 Direct and indirect estimates of mutation rate in HD

Author	Direct	Indirect
Kishimoto et al. (1957)	–	0.67×10^{-6}
Reed and Neel (1959)	5.4×10^{-6}	9.6×10^{-6}
Wendt and Drohm (1972)	0	1.5×10^{-6}
Mattson (1974)	5.0×10^{-6}	0.8×10^{-6}
Stevens (1976)*	0.42–4.0×10^{-6}	0.07×10^{-6}
Hayden (1979)	0.13×10^{-6}	–
Walker et al. (1983)*	0–1.0×10^{-6}	–

*Higher estimate includes all possible mutations.

Table 9.5 Mutation rates in different inherited neurological disorders. Data from Vogel and Motusky (1986)

Disorder	Mutation rate (per 10^{-6} gametes)
Neurofibromatosis (type 1)	44–100
Tuberous sclerosis	6–10.5
Duchenne muscular dystrophy	43–105
Huntington's disease (direct estimates only)	0–5.4

(f = fitness; n = number of heterozygotes for HD; N = total population number)

It is clear that where genetic fitness is normal (1) this will give a mutation rate of 0, and if it is increased (as in some studies of HD such as the South Wales study of Walker *et al.*, 1983) one would have a negative mutation rate, which is meaningless.

Taking into account the difficulties in determining both what is a new mutation and the genetic fitness in HD any mutation rate calculated can only be very approximate. Table 9.4 gives some estimates, while in Table 9.5 HD is compared with some other inherited neurological disorders. It can be seen that the mutation rate for HD is extremely low compared with these other disorders. Even the inclusion of all possible mutations, as in the second of Stevens' (1976) estimates, still gives a very low value. The conclusion must be that mutation in HD is extremely rare, both in absolute terms and in relation to the total number of cases.

The problem of identifying a new mutation in HD will not be fully solved until the gene itself has been isolated, but meanwhile some independent evidence identifying specific mutations is beginning to come from the molecular analysis of marker haplotypes around the gene. These now show significant linkage disequilibrium (Snell *et al.*, 1989; Theilman *et al.*, 1989) with specific alleles and haplotypes preferentially associated with HD. Such a situation would not have been likely to be observed were there numerous different mutations at similiar frequency in a population, and suggests that one, or at most a few, common mutations will be found in each population, as discussed in Chapters 8 and 10.

HOMOZYGOSITY FOR THE HD GENE

Almost all the genes for Huntington's disease, as for other rare dominantly inherited disorders, are present in heterozygotes who will also possess a normal allele at this locus. For the gene to occur in a homozygous state would require one of several exceedingly rare events to occur. An individual inheriting the HD gene could undergo a new mutation at the normal allele (a vanishingly rare event given the rarity of mutation in HD). A person could

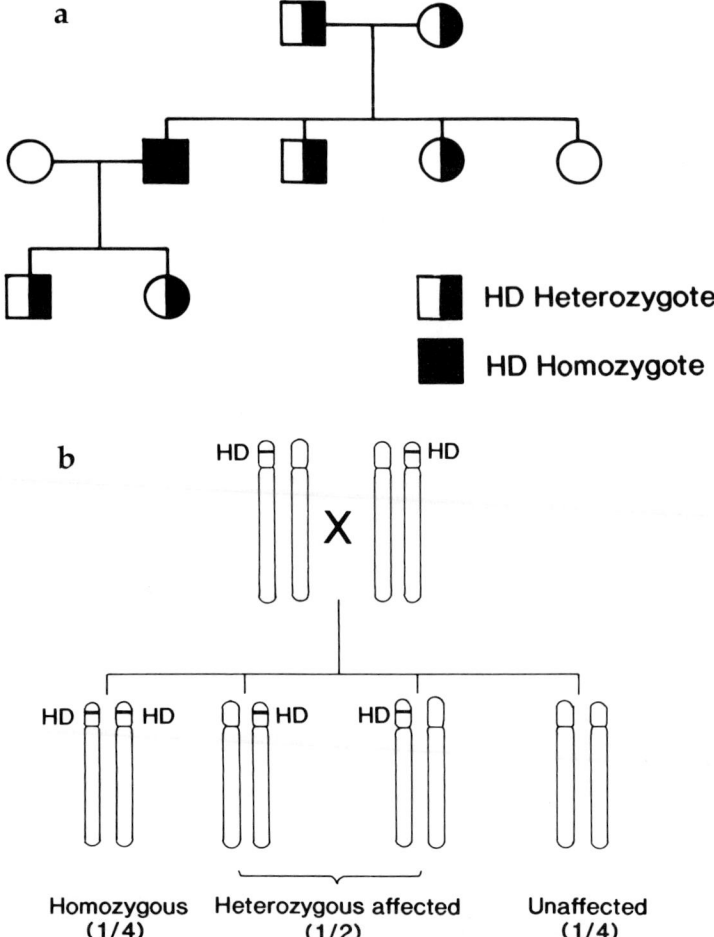

Figure 9.3 Homozygosity for HD. (a) Pedigree (b) diagram to show segregation of chromosomes. When both parents are affected, the chance of an offspring being homozygous for the HD gene is 1/4 (sex is not important). All offspring of a homozygous affected person would be heterozygous for HD (i.e. affected).

receive two copies of the HD gene from the affected parent and no normal gene from the healthy parent (this 'uniparental disomy' has been recorded for cystic fibrosis and may occur more commonly than previously recognized). Or, least unlikely, both parents might possess the HD gene, in which case there would be a one in four chance that any child would receive a 'double dose' (Figure 9.3). Until now all such homozygotes for dominantly inherited disorders have been much more severe in the homozygous state, often being fatal in early life. Achondroplasia, hereditary haemorrhagic telangiectasia and familial hypercholesterolaemia are examples of this. Any homozygote for HD might therefore have been expected to show a severe and distinctive neurological disorder, or alternatively the homozygous state might be so lethal during development as to appear as an excess of spontaneous abortions or stillbirths, or as a deficiency in the observed number of affected offspring.

Several instances of possible homozygosity for HD have been reported over the years; Hindringer (1935) reported a consanguineous marriage in a family from southern Germany where the two affected parents had five children, three with typical HD, one unaffected and one infant death. Eldridge et al. (1973) described an American family where both parents in a first cousin marriage were affected (clinical details on the father were scanty). The affected son had typical HD with onset aged 34; one of the two sibs had died in infancy, the other from a war injury. Both families were small and the information incomplete, so that no firm conclusions could be drawn other than to say that no unusual neurological features were observed among affected offspring. Most of our important information on homozygosity has come from the Venezuela isolate with HD already described, where the combination of high HD prevalence and extreme isolation has produced several instances where both parents have been affected.

A detailed report on the largest of these sibships has now appeared (Wexler et al., 1987) which promises to be of great value for our future understanding of the HD gene and its function. Figure 9.4 summarizes the pedigree; the investigators quite properly omitted details of sex and birth order that could breach confidentiality. The value of this study is that not only is the sibship extremely large (14 sibs) and both parents living, but that molecular studies of linked markers make it possible to identify with high probability which individuals are homozygous for the HD gene quite independently of their clinical status.

As can be seen from Figure 9.4, four of the 14 sibs are homozygous for the high-risk genotype, a proportion consistent with mendelian segregation and making prenatal lethality improbable. Nor did any members of the sibship show unusual clinical features that would not have been expected in heterozygotes in this population. At the time of the study (1986) the offspring were relatively young (16–42 years) and only minor neurological signs were present; subsequent follow up has confirmed definite HD in at least one of the four probable homozygotes, but has still not shown any distinguishing clinical features from heterozygotes. Analysis of closer markers has confirmed the original assessment of homozygosity.

In a separate study Myers et al. (1989) have analysed four possible homozygotes from three New England families where both parents were affected. One of these showed a 95% chance of being homozygous on molecular testing, but clinical features and age at onset were comparable with affected heterozygotes in the family.

The conclusion thus seems inescapable – the homozygote for HD is no more severely affected than the heterozygote, at least when the same mutation is inherited by common descent. This situation, quite unlike that so far encountered for other dominantly inherited disorders, has considerable implications

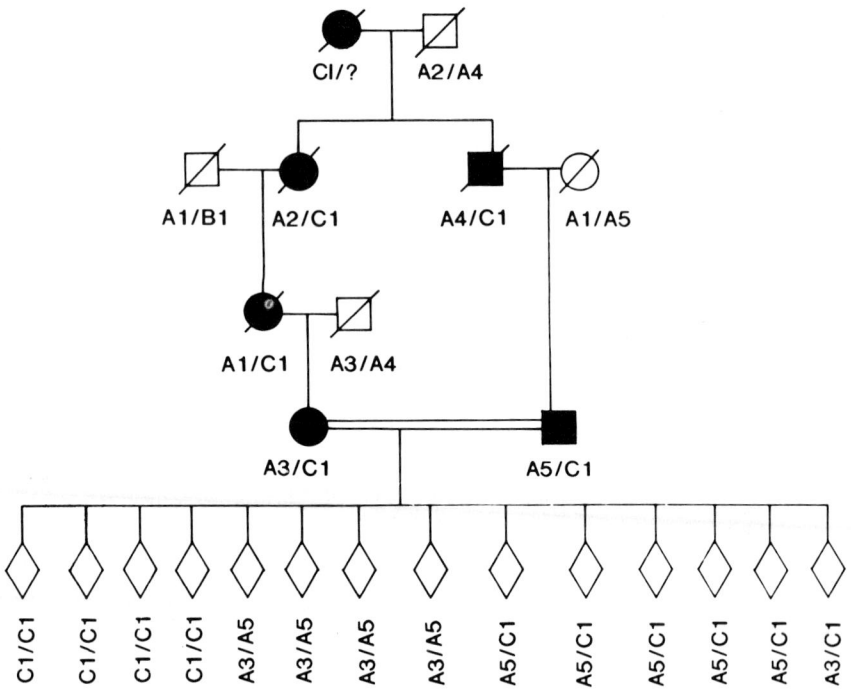

Figure 9.4 Probable homozygosity for HD in part of the large Venezuela kindred, demonstrated by typing of DNA markers. Those offspring typing as C1/C1 are expected to be homozygotes. After Wexler et al. (1987).

for the nature of the HD gene and its product. These aspects are discussed in more detail in Chapter 10, in particular the hypothesis of Laird (1990) but it is clear that these observations make it unlikely that the HD mutation is responsible for a simple loss of function in any enzyme or conventional structural protein, since in this situation one would expect a clear difference between a 50% loss of protein with one remaining normal gene and a total functional loss of both alleles.

The non-lethality of homozygous HD has a serious practical consequence for the offspring of such individuals. Regardless of the status of their spouse, all children will be affected, since the homozygous parent has no normal allele to pass on. This fortunately very rare situation raises ethical issues as to whether one should indeed test possible homozygotes if they already have children.

HETEROZYGOTE AND GENE FREQUENCY

The prevalence of HD and its variation in different populations have been discussed fully in the previous chapter, but a note is required here regarding the frequency of heterozygotes for HD and of the gene itself. For many genetic disorders these three factors will differ widely from each other; thus for a common autosomal recessive condition, such as cystic fibrosis, where most of the abnormal genes are present in heterozygotes, the heterozygote frequency is around 100-fold greater than the frequency of the disease (1 in 20–25 compared with 1 in 1600–2000). For an X-linked recessive disorder such as Duchenne muscular dystrophy the heterozygous state will be confined to females, themselves essentially unaffected. However for a rare fully penetrant autosomal dominant disorder, where the homozygotes are negligible in frequency and essentially all genes for the disorder are present in heterozygotes who will show the condition, the disease frequency and heterozygote frequency should be the same.

The reason that this is not the case in HD is the feature that underlies most of the problems in its genetics. Although essentially fully penetrant by old age, most heterozygotes will be symptomless for most (in some cases all) of their life. Thus only a proportion of heterozygotes for HD in a population will be recognized as being affected at any one time, even though all will develop the disease if they live long enough. The heterozygote frequency can thus be regarded as equivalent to the birth incidence of those expected to be affected in future – clearly an important figure to be able to measure when comparing different disorders, planning preventive programmes and monitoring trends.

At present the heterozygote frequency for HD cannot be measured directly; even when molecular developments make this feasible there would be strong ethical contraindications to any population survey in early life. We thus have to rely on an indirect estimate based on both the prevalence, which will be dependent on the particular population studied, and the population distribution of age at onset, which has proved to be remarkably similar across a wide range of populations.

Several methods are available for estimating heterozygote frequency using prevalence data; these have been well discussed by Stevens (1973, 1976), Walker et al. (1981) and Newcombe (1981); the formulae are given in these papers. Since the results closely parallel the prevalence estimates for a population, no comprehensive list is given here, but Table 9.6 shows as an example,

Table 9.6 Heterozygote frequency for HD. Estimates using different methods, based on the South Wales data of Walker et al. (1981)

Population studied		1 720 901
Number of affected living		131
Prevalence		7.61 per 100 000
Estimates of heterozygote frequency (see text for details)	1	34.9
	2	17.2
	3	24.9–26.8
	4	15.8–19.1
	5	20.2

the South Wales heterozygote frequency (Walker et al., 1981) estimated using a variety of methods. It can be seen that apart from method 1 based on crude prior risks and not recommended, the others give a reasonably comparable frequency. Method 2, originally used by Reed et al. (1958) and corresponding to method A of Stevens (1973), utilizes the age at onset distribution uncorrected, while method 3 (method B of Stevens) derives age-dependent risks for each individual and was used first by Reed and Neel (1959). Method 4 (Stevens method C) introduces an allowance for age at onset in a particular parent, while method 5 is based on the life table method of Newcombe (1981), in which the age at onset curve is adjusted to account for those individuals likely to be unascertained because they have yet to develop the disorder.

Regardless of the method used, the main question that those studying HD will wish to know is how much greater is the heterozygote frequency (representing birth incidence) than the prevalence? Fortunately there is close agreement on this between studies. The analysis of Walker et al. gave a value of 2.6, agreeing closely with those of 2.4–2.7 from other studies.

In summary it is reasonable to take as a rough guide of the heterozygote frequency a figure of 2.5 × the prevalence estimate; to put the figures for heterozygote frequency in a different way, there are likely to be 150 asymptomatic gene carriers for every 100 affected individuals. The gene frequency will simply be half the heterozygote frequency, since everyone has two genes at each locus.

Conneally (1984) has shown that the number of individuals at risk for HD in a population (only including first degree relatives) is around twice the heterozygote frequency. This is to be expected since for every symptomless or affected gene carrier there will be one who has not inherited the HD gene. The wider question of overall numbers at risk is considered in Chapter 11.

FERTILITY AND GENETIC FITNESS

All genetic disorders are subject to a balance between genes that enter a population by new mutation and migration, and those genes lost through early mortality or failure to reproduce. If we are to understand why a disorder such as HD is commoner in some populations than others, whether there has been in the past or still is an increase or decrease in the frequency, and whether genetic counselling or other newly introduced measures are altering the previous state, we must first understand as fully and accurately as possible the factors that are involved in this delicate balance.

Unfortunately for those undertaking such studies the balance is not a static one, but is likely to vary with time and with the population involved. We also need not only accurate and unbiased data from HD families, but equivalent information from the normal population, which may not be available in comparable form. It is thus not surprising that our information on this subject, while extensive, is variable and at times contradictory, but it is worth making a critical evaluation of it here.

Before proceding further it is important to clarify definitions, which may be confusing to the non-geneticist. By *fertility* is meant the number of offspring born to an individual with the condition under study. In making an estimate for a disorder such as HD, only livebirths are generally included, and data are often restricted to those considered likely to have completed their families (e.g. over 45 years). It is important to include those individuals who have never married and to avoid biases such as over-ascertainment of large families and combining data from widely different dates.

Genetic fitness differs from fertility in considering the reproduction of those possessing the HD gene rather than those who are, or later become, affected. Until now it has not been possible to estimate it directly, but it can be derived from estimates of fertility in the same way as can the heterozygote frequency from estimates of prevalence. Since almost all HD genes are present in heterozygotes, who will eventually develop the disorder if they live long enough, it matters little whether fitness or fertility is used as the measure, providing that like is compared with like. Finally, while fitness is a relative measure, using genetically normal relatives (usually sibs) as a comparison, fertility estimates may be relative or absolute, these latter using appropriate normal population data as the control group.

The results of the main studies on fertility and fitness in HD are summarized in Table 9.7. It can be seen that they vary considerably and that, taking a value of 1.0 as 'normal', some show a reduction, some are close to normal, while others are actually increased. However, all are agreed on one point: compared with most serious genetic disorders the values are near normal, supporting the basic clinical observation that most patients reproduce and that onset of the disorder is commonly after a family has been completed.

It is now worth examining in more detail some of the individual studies and asking why they should differ so considerably in some of their conclusions. The first major study was that of Reed and Neel (1959), based on a survey in Michigan, USA, and a model for thoroughness and for reporting detailed data. Their most striking finding was reduced fertility and fitness of male choreics when compared with females, and compared with either sex in the general population. This was partly due to a higher proportion of male choreics who never married (26% compared with 8% of females). When unaffected sibs were used as comparison, however, the reduction in male genetic fitness was much less, while female choreics were significantly more fit than their unaffected sibs. Taking the sexes together there was no difference between affected and unaffected. Their other major finding was that unaffected sibs also showed a reduced fertility in comparison with the general population. Reed and Neel suggested that this might reflect a general deterrence from reproduction by those at risk affecting equally those with and without the HD gene; however it could be that their general population data, though carefully chosen, were not

Table 9.7 Fertility and genetic fitness in HD

	Fertility			Fitness		
	Overall	Male	Female	Overall	Male	Female
Studies showing decrease						
Kishimoto et al. (1959)	–	–	–	0.65	0.61	0.68
Reed and Neel (1959)	0.82	0.66	0.98	1.03	0.82	1.25
Stine and Smith (1990)	(Coefficient of selection 0.36)					
Studies showing increase						
Marx (1973)	1.16	0.95	1.40	–	0.99	1.39
Shokeir (1975)	1.14	–	–	1.38	–	–
Stevens (1976)	1.36	1.09	1.61	1.39	1.03	1.52
Walker (1979); Walker et al. (1983)*	1.25–1.34	1.43–1.55	1.0–1.30	1.46	1.66	1.29
Studies showing close to normal level						
Wendt and Drohm (1972)	0.95	0.93	0.96	0.96	0.93	1.03
Wallace and Parker (1973)	1.0	0.88	1.13	1.28	1.11	1.52
Mattson (1974)	–	–	–	0.97	0.87	1.07
Mastromauro et al. (1989)	–	1.05	1.07	–	1.01	0.98

*General population estimates divided by decades of birth.

typical, since most later studies have shown no marked differences between unaffected sibs and population data. Thus it remains open whether their study should be interpreted as a reduced fertility (most marked in males) based on the absolute data, or as a near-normal fertility (somewhat increased in females) based on the comparisons with sibs.

One other early study, that of Kishimoto et al. (1957, 1959) also showed a reduced fertility; cousins were used for comparison but no detailed data were given. Given the much rarer prevalence of HD in Japan and the different social structures, a more detailed study is needed. A third and recent study which appears to show reduced genetic fitness in HD has used an entirely different approach. Stine and Smith (1990) analysed HD in the Afrikaner population of South Africa. Based on the proportion of current HD patients whose descent could be traced to original founders of the population, they estimated that this was less than would have been predicted from the known expansion of the Afrikaner population as a whole, suggesting some selective disadvantage to the HD gene. By contrast, no significant disadvantage was found for another autosomal dominant disorder common in the population, porphyria variegata. The assumptions underlying this study will be testable when it can be determined whether more than a single mutation for HD is responsible for the disease in Afrikaners.

The series of papers showing an apparent increase in fertility and fitness now requires examination. Two are from North America (Marx, 1973; Shokeir, 1975) and two from the UK (Stevens, 1976; Walker *et al.*, 1983). An increased fitness for the HD gene should imply some selective biological advantage, and would be highly unusual for a dominantly inherited genetic disorder. If this situation was maintained, a gradual replacement of the normal by the HD allele could be expected, a somewhat implausible situation. Criticisms of these findings have included bias from large families, where affected members are likely to have reproduced preferentially, the inappropriate use of unaffected sibs as controls, and inclusion of older data where non-reproducing affected patients may be under-represented. While these are valid points, it is difficult to explain the finding of increased fertility as entirely due to bias. Thus in the study of Walker *et al.*, the control values of normal sibs, normal second-degree relatives and general population data were all closely comparable, while the ascertainment was close to complete, at least for recent cases. This study did not show the sex differences for marriage and reproduction found by Reed and Neel (1959), nor was there any difference in the data for those patients whose main clinical features were psychiatric as compared with those whose presentation was principally neurological. There were no detectable differences in the pattern of family building and no evidence of clustering of births in the years preceding onset that might have suggested an effect of behavioural changes in this period.

Several studies have shown no clear difference in fertility and fitness estimates between HD and normal individuals; these include two European reports (Wendt and Drohm, 1972; Mattson, 1974) one from Australia (Wallace and Parker, 1973) and one from North America (Mastromauro *et al.*, 1989). The Australian study did, however, show a lowered male reproductive rate due to failure to marry. Wallace (1976) has discussed the various social reasons that could underlie this. The most recent study, that of Mastromauro *et al.*, found no differences between unaffected sibs and the general population. This may well be the last such study to be performed before population dynamics are altered by the widespread use of presymptomatic testing for the gene.

Taking all these studies together, a reasonable conclusion is that the HD gene has little, if any, biological disadvantage, but neither is there evidence of any specific factors on which to base a possible reproductive advantage. It seems quite possible that, during a century in which the normal family size has declined dramatically, and which has seen striking population increase associated with migration, the balance may have varied from time to time and place to place as to whether the HD gene has had a slight relative advantage or disadvantage. These same variations have made it impossible to provide entirely satisfactory matching normal data: but despite these drawbacks the various studies will be of long-term value in allowing comparison with future changes resulting from genetic testing. It can certainly be concluded that the

HD gene is unlikely to decline or die out spontaneously in the populations studied in the forseeable future and that very few new mutations are necessary to maintain the current population levels.

CHROMOSOMAL ABNORMALITIES

In a mendelian disorder such as HD one would not expect any visible chromosomal defect to occur, nor has any such change been reported. However it has become increasingly recognized that very rare instances of a small chromosomal deletion or translocation may be associated with a specific mendelian condition and may give the clue to its localization. Such changes gave the clue to the location of the Duchenne muscular dystrophy gene at band p21 on the X chromosome (see Harper, 1989a for review), while comparable translocations and deletions involving chromosome 17 in patients with neurofibromatosis have provided the basis for recent molecular work in the responsible gene.

Thus even a single HD patient showing a chromosome abnormality could be of great importance and there was considerable interest when a family was found in which HD appeared to be inherited together with a balanced translocation involving chromosomes 4 and 5 (Froster-Iskenius et al., 1986). Unfortunately detailed study showed that the part of chromosome 4 involved was the long arm, ruling out any causative association with the HD gene on the short arm of this chromosome.

The HD region of chromosome 4 is, however, involved in one well-defined chromosomal abnormality. Deletion of the distal part of the short arm of chromosome 4 provides a distinctive clinical phenotype known as Wolf–Hirschhorn syndrome (Figure 9.5). Affected children are severely mentally retarded and many die in infancy. All cases studied so far have had paternal origin of the deletion (Quarrell et al., 1991). So far there has been no detailed neuropathological study of the brain of such a patient, which is unfortunate. In an older patient one would expect to see changes of HD if the chromosomal deletion has involved the HD locus and if HD is really due to loss of gene function rather than some other mechanism. When such patients die in future it would be of great importance to search for the early changes of HD.

Wolf–Hirschhorn syndrome patients have proved relevant to HD research in a further way: the extent of deletions varies considerably, which has allowed a panel of cell lines to be constructed in which different portions of 4p are deleted. These, along with other cell lines showing rearrangement in this region, have proved a valuable tool in the localizing and ordering of newly isolated DNA probes thought to be close to HD (see Chapter 10). One patient with a particularly small distal chromosomal deletion (McKeown et al., 1988) has shown the G8 marker to be preserved, putting the chromosomal region responsible for this abnormal phenotype distal to it, as also is the HD locus.

Genetic Aspects of Huntington's Disease

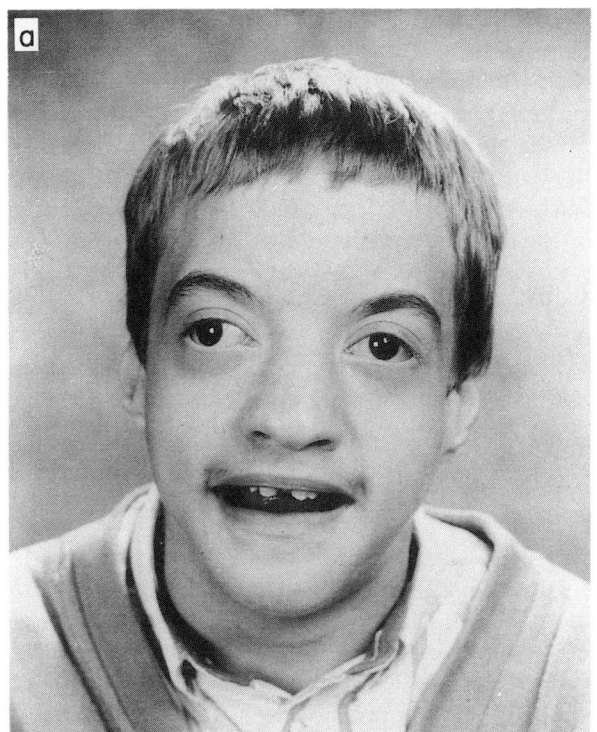

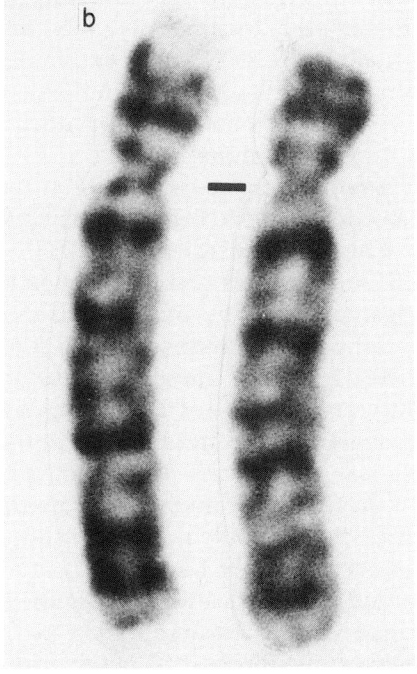

Figure 9.5 Wolf–Hirschhorn syndrome, a chromosomal disorder involving the HD region of chromosome 4. (a) Facial dysmorphic features (courtesy Dr Oliver Quarrell). (b) Karyotype to show loss of chromosomal material in the distal part of 4p (courtesy Dr Merryl Curtis).

Although no HD patient has been found with any chromosomally visible deletion, another clue might be gained from its occurrence with other, apparently coincidental disorders; such an occurrence might indicate the deletion of two or more contiguous genes, even though there was no visible chromosomal change. So far the few such situations reported are likely to be merely coincidence; Bruyn (1968) in his review mentions instances of neurofibromatosis, Charcot–Marie–Tooth disease and spastic paraplegia recorded together with HD, but all are relatively common conditions and two have a known separate chromosomal location. A patient with HD and congenital deafness was reported by Critchley and Secker-Walker (1966) but other family members had HD alone. Caro (1976) reported an HD family where affected members also had the bone abnormality Madelung deformity. This family deserves a fuller study as does any other association with another genetic disorder.

GENETIC HETEROGENEITY

The remarkable variability of both age at onset and clinical expression in HD has prompted investigators from the earliest period onwards to ask the question – is there more than one gene for the disease? There are two types of genetic heterogeneity that must be considered, however: the occurrence of separate mutations with differing clinical effects at the same genetic locus, and the existence of more than one distinct locus for the disorder. The difference between the two is fundamental, since in the latter situation one is postulating fundamentally different genes and gene products, with by implication an equally different pathogenesis for the separate forms of the disease. For separate mutations at a single locus, by contrast, one is dealing with a single disease process and gene product, though one altered in different ways by the different mutations.

Almost all the work done on this topic in HD was carried out at a time when it was impossible to distinguish between these two situations. Now, as discussed in Chapter 10, it is clear that all HD families so far studied, including some with unusual features, are determined by the same locus, or at least by loci so close together as to be indistinguishable. Thus it seems likely that the genetic variation to be discussed here is the result of different mutations at a single locus, together with the modifying effects of genes at both the same and different loci. One of the most important results of isolating the HD gene will be to see how far different mutational defects can be correlated with distinctive clinical phenotypes. Already it is becoming feasible to gain some information on these specific mutations from the existence of linkage disequilibrium of HD with closely linked markers, allowing the construction of specific marker haplotypes (see Chapter 10).

The simplest, and at present probably the most rewarding approach to identifying different mutations at the HD locus, is the careful documentation of phenotypic differences between large, multigeneration kindreds, particularly those occurring in the same geographical area, which allows major environmental influences to be excluded. The easiest features to document are age at onset and death; the latter can often be obtained on past generations for which detailed clinical assessment is lacking. The existence of genealogical registers that allow different branches of extended kindreds to be linked is helpful in ensuring that data are available on enough affected members to allow statistical comparison.

Went et al. (1975, 1983) analysed two large Dutch kindreds, extending over seven and eight generations, in this way and showed a clear difference in age at death between them (47.1 and 63.9 years). Such differences are unlikely to be due to the effects of genes at other loci, since these would (unless very closely linked on the same chromosome) be dispersed by recombination and show little or no correlation in distant branches of the kindred.

Similar examples have been found in other studies, including two of the largest families in the South Wales study (Walker, 1979). The significance of these reports is, however, reduced by the bias of their selection, their large size favouring late onset, and the existence of marked differences favouring their reporting.

A striking example of how a single kindred can dominate the pattern for an entire population is seen in Tasmania where the large kindred originally reported by Brothers (1949) and recently reassessed by Pridmore (1990) shows an extremely late age at onset and death which is reflected in the total picture of HD on the island. Interestingly, a similar late onset was found by Glendinning (1976) in families in Somerset, England, the area from which the Tasmanian kindred originated. That these large individual families do not represent an abrupt discontinuity in phenotype is clearly seen when they are placed in the context of an overall study.

Wallace and Hall (1972) examined the question of heterogeneity in their study of the Queensland, Australia, population. After noting the existence of individual families with unusual features, they analysed the variance for age at onset within and between families and found a higher correlation within families, regardless of whether the comparison was for sibs or more distant relatives. They concluded from this that the familial differences were likely to be due to different specific mutations at the HD locus, rather than to the effects of other loci.

Went et al. (1983) having compared age at death for the two large kindreds, already mentioned, collected similar data for 102 kindreds in which age of death was known for at least four members. The resulting range in different kindreds varied from 37.8 to 71.9 years, with a continuous and essentially normal distribution. Periczak-Vance et al. (1983) also examined the question of heterogeneity in age at onset and found clear differences between larger kindreds. However, these differences were greatly exceeded by the amount of variation within individual kindreds.

When possible heterogeneity is being considered, it is important to be clear as to which phenotypic features are primary ones and which secondary. The older studies considered heterogeneity mainly in terms of the form of clinical signs (e.g. the rigid form as opposed to the choreic), whereas later analyses concentrated on age at onset and death. Farrer and Conneally (1987) have undertaken a valuable analysis of clinical features in relation to the various familial correlations and have shown that there is essentially a continuum, rigidity in particular being strongly age-related, but with no clear correlation of this or other features within families independent of age at onset. Farrer and Conneally concluded that age at onset was influenced by a variety of genetic factors, as discussed later, and that the phenotypic differences in expression of the disease were related principally to age at onset, not being in themselves directly determined by these modifying factors or by heterogeneity. Van Dijk *et al.* (1986) have taken the alternative view that, at least for the juvenile form, age at onset reflects the occurrence of rigidity (see Chapter 2) but their data, while confirming the strong correlation between juvenile onset and rigidity, do not really conflict with those of Farrer and Conneally, especially when the importance of paternal transmission (see below) is taken into account.

HD pedigrees showing unusual but constant clinical features also provide evidence for different specific mutations, though until recently they have been open to the criticism that the diagnosis might be erroneous. Thus the demonstration by Zweig *et al.* (1989) that a black American family showing highly atypical clinical features and neuropathology was definitely linked to the D4S10 (G8) locus, allows this to be considered as a likely candidate for a different specific mutation, even though the racial background may have had some effect. The same group (Folstein *et al.*, 1985) also documented two other large kindreds showing constant characteristics over several generations; one of these, of German origin, showed a consistently late onset and high frequency of major psychiatric abnormalities. The other, a black Maryland (USA) family, showed early onset, with a high frequency of rigidity without severe psychiatric illness. Both showed linkage to D4S10.

That these differences do not just relate to a few atypical HD kindreds is illustrated by the further finding of Folstein *et al.* (1983) that there is a more general familial association for affective disorder within kindreds. For 23 kindreds where this was a major feature in the proband, it was present in a relative in 20, whereas in 23 kindreds where affective disorder was absent in the proband, only five showed it present in a relative. To what extent this more general correlation can be attributed to a series of differing mutations, or whether it is more related to other modifying genes, remains to be determined.

Another highly distinctive family that could have represented a specific mutation is the remarkable Australian family in which the presenting feature was aphonia, followed by involuntary movements and rigidity but with little mental deterioration (Tyrer, 1957; Wallace and Parker, 1973). This kindred has been reassessed clinically and is now considered to have a form of torsion dystonia, not HD (Dr Christine Oley, 1990, personal communication). DNA analysis is being undertaken.

The value of genetic linkage data in excluding other choreic disorders from being determined by the same locus as HD has already been shown for benign familial chorea (Quarrell et al., 1988) and for dentato-rubro-pallido-luysian atrophy (Kondo et al., 1990), both of which are unlinked to the HD region of chromosome 4.

MODIFYING GENES

From the work on possible genetic heterogeneity discussed above it is likely that some of the more striking differences between kindreds with HD will prove to result from different specific mutations at the HD locus, affecting the gene product in different ways and hence influencing phenotype. This has already proven to be the case for several neurological disorders where the genes have been isolated and characterized, such as Duchenne/Becker muscular dystrophy, Gaucher's disease and familial amyloid neuropathy. There remains, however, a large amount of variation both between and within HD families that cannot be clearly compartmentalized by studies of large kindreds, yet is still likely to be in part genetic. Much effort has gone into attempts to document and analyse this variation, and the work can be placed in three broad categories:

(1) Twin studies, particularly valuable in evaluating the extent of non-genetic variation, and discussed separately above.
(2) Sex-related factors involving differences of parental transmission, an unexpected feature for a mendelian disorder and dealt with fully in the next section.
(3) General analysis of variation in relation to possible modifying genes at the same or other loci. It is this general topic of modifying factors that will now be considered.

The early workers on HD were quick to recognize familial differences in the manifestations of the disease. The report of Davenport and Muncey (1916) provides the best example. In this study, Davenport claimed that he could divide his HD families into different 'biotypes', some showing a preponderance of mental disorder, some lacking this, and correspondingly for other features of the disease. Davenport's work (it seems unfair here to include Elizabeth Muncey, who was responsible only for the data collection) has to be treated with considerable reservation, for while he was able to draw on a large record base of almost 1000 affected individuals, his selection of data was arbitrary and his conclusions often idiosyncratic (at times frankly prejudiced). Also while his 'biotype' concept would seem to imply heterogeneity, he writes of 'crossing of biotypes' in a confused way and appears to have recognized a large number of types, even though his families were thought to have originated from only six ancestors.

The first systematic attempt to analyse data on variation in HD quantitatively was that of Bell (1934), whose thorough analysis of data in the literature was the basis not only for an accurate distribution of age at onset and death, but for detailed analysis of variance. Bell's finding of high correlations for age at onset between parent and child and between sibs pointed clearly to the involvement of genetic factors in this; she was also the first to find that duration of the disease was independent of age and the first to produce an accurate age at onset distribution curve, as mentioned in Chapter 4. Bell's data provided the basis for subsequent more detailed genetic analysis, notably that of Haldane (1941). Haldane was the first to suggest that most of the variation in age at onset of HD was the result of modifying genes rather than environmental factors or genetic heterogeneity. He pointed out that if environmental factors were principally responsible one would expect a greater correlation for sibs than between parent and child, whereas the reverse was if anything the case, Bell's value for the parent–child coefficient being 0.593 and the sib–sib coefficient 0.465. Likewise, if several distinct genes for HD existed, and were responsible for most of the observed variation, higher correlations (approaching 1) than those actually observed would be expected.

Bell's study was later criticized by Minski and Guttman (1938) as unsuitable because it was based on data from the literature. In fact Bell had close personal involvement in ascertaining and validating much of the data, and her study is a classical example of how a quantitative approach to genetic analysis can be based on pooled data sets, none of which alone could have been large enough to give useful information.

The finding of strong and similar correlations for age at onset and death between sibs and between parents and children has been confirmed by a series of specific studies, which are summarized in Table 9.8. They are comparable between American and European populations, and with the overall literature analysis of Brackenridge (1972). Two other consistent findings in numerous studies have been the relatively constant duration of survival irrespective of age at onset, discussed previously (Chapter 4), and the finding of a consistently earlier age at onset in children than in parents (Brackenridge, 1974; Myers *et al.*, 1982). This latter finding has led to considerable discussion as to whether any true 'anticipation' exists between generations; this indeed appears to be likely for the offspring of affected males, as discussed in the next section on sex-related effects; for female transmitted cases most analyses have found that the effect is no greater than to be expected from the biased ascertainment inevitable in a study, when later onset children will not yet have developed the disorder (Ridley *et al.*, 1988, 1991). Subsequent work has concentrated on determining what type of factor might be responsible for these various correlations and differences. The most detailed and convincing study in this respect has been that of Farrer and Conneally (1985) and Farrer *et al.* (1985), who used the extensive data of the American HD roster to examine the correlations for age at onset and death in these families. In addition to the expected correlations between affected members, they showed that there was an equally strong correlation with age at death in the normal (i.e. non-HD) parent, as well as between HD patients and their normal sibs and HD parents and their normal children. There was no correlation of age at death between HD patient and spouse, making environmental factors unlikely. This provides conclusive evidence that age at onset and death are not solely determined by the HD gene itself. Farrer et al. (1984), developing the previous suggestion of Finch (1980), proposed that 'normal' ageing genes affecting neuronal degeneration in the basal ganglia might be involved, and that these were probably not allelic to HD since, if they were, they could not be shared by sibs discordant for HD.

Other hypotheses have been more speculative, notably those of Brackenridge (1974, 1979) who suggested climate and stress as factors. The statistical basis of his suggestion that age at onset in a child could be altered favourably by the parent reproducing early was refuted by Burke (1976). Likewise the complex formulae of Burch (1969) who suggested that HD represented a clonal autoimmune disorder, are not borne out by any evidence of such a process acting in the disorder.

Until now the whole topic of modifying genes has been limited by the fact that, apart from the HD locus itself, there have been no specific genes that could be tested. This is now changing rapidly with the cloning of numerous genes of neurobiological importance, polymorphic variations in which can be tested in HD families to see if they are related to age at onset and other factors. Although these will not be likely to be 'candidate genes' for HD itself, their identification will help to resolve some of the theoretical points discussed here and bring the current rather unsatisfactory situation into a state where there is clearer evidence as to what are the factors that modify HD and how the influences act.

SEX-RELATED EFFECTS

The early reports of HD had soon shown a genetic pattern typical of autosomal dominant inheritance, as outlined at the beginning of this chapter. The sex incidence had been shown to be equal, with similar rates of transmission by the sexes and no marked differences in age of onset or clinical features between males and females. Thus the finding of a major sex-related effect in the transmission of HD came as a surprise, even a shock, partly because it was unexpected, but also because it conflicted with what was expected on the basis of mendelian inheritance. The sequence of events can be summarized as follows:

(1) The recognition that childhood HD was principally transmitted by the father.
(2) Recognition that onset in the children of paternally transmitted cases was generally reduced.
(3) Debate as to whether the effect was primarily paternal or maternal.
(4) Hypotheses on the possible mechanisms.

Paternal inheritance of juvenile HD

The first suggestion that childhood or juvenile HD was usually transmitted by an affected father came from Bruyn (1968) and from Merrit et al. (1969); Bruyn's collected data on transmission of the juvenile form were confused by his finding that there was also an excess of female juvenile cases, something not confirmed in later studies. It was the study of Merrit et al. (1969), reported at the International Neurogenetics Meeting the previous year, that showed clearly that this was a genuine biological phenomenon. They reported seven families containing 14 affected children and reviewed previous literature on a further 110 siblings with HD of onset under 21 years. The sex incidence of the affected children was approximately equal (70 male, 64 female) but there was a striking excess of affected fathers (84, compared with 22 affected mothers).

Confirmation of these data came from the pooled experience of the World Federation of Neurology research group on HD (Barbeau, 1970), which found 26 of 33 juvenile cases (under 20 years) to be paternally transmitted, only six maternally. Since then the finding has been observed in HD populations world wide, including Venezuela and the mixed race population of South Africa (Hayden, 1979).

A more detailed analysis of the situation showed two points of practical importance. First there is no absolute demarcation between the transmission of 'juvenile' and 'adult' cases. Myers et al. (1983) showed that the proportion of male-transmitted cases fell steadily when cases were grouped as juvenile, early, adult and late onset. Second, for the extremely rare instances of HD beginning before 10 years of age, there is virtually exclusive paternal transmission (Went et al., 1983), an observation of considerable significance when a young child in an HD family develops a neurological illness. As already discussed in Chapter 2, the clinical presentation in this age group is often most atypical, so that the parental transmission may be a major factor in the differential diagnosis or in making it less likely.

In view of the close correspondence between juvenile onset and the rigid form of HD, it is not surprising that the predominantly paternal transmission was found to apply to the latter group also.

The observation and confirmation of this striking sex-related effect made investigators reassess the earlier data to see if it had been overlooked, and also to look in more detail at the overall and unselected data for HD populations to see how far the findings might be relevant to adult and late-onset HD. The data of Myers et al. (1983) have already been mentioned; they showed a deficiency of paternal transmission in their late-onset group (onset 50–70 years) with 29% having an affected father and 71% an affected mother, contrasting markedly with values of 91% and 9% for the parental origin of their juvenile cases.

Bird et al. (1974) analysed age at death in parent–offspring pairs and showed a marked reduction in age at death in the offspring when the father was the affected parent (9.73 years) compared with little difference when the mother was affected (2.04 years). They obtained similar results from analysing previous data in the literature. Interestingly the overall lower age at death in offspring compared with parent had already been noted by Penrose (1948) in his reanalysis of Bell's (1934) data, but the significance of the difference between maternal and paternal transmission had not been appreciated. The use of age at death as opposed to onset can be queried in these studies, but it is certainly more definitive when older data are being considered, while no marked effect of age at onset on survival has been shown in any of the major studies when bias is allowed for, so that the conclusions regarding age at onset and age of death are probably comparable.

Stevens (1976) examined age at onset in relation to parental transmission in his UK data and found a reduction in onset with paternal transmission, though this only reached significance for male cases. Newcombe et al. (1981), using data from the South Wales study, analysed differences in age at onset by a log rank method as well as by previous methods and confirmed an overall correlation between paternal transmission and reduced age at onset. Both this study and that of Stevens also made the interesting observation that when data were available for three generations, the earliest onset group was that where transmission was grandfather/father/offspring, suggesting that any effect responsible might be cumulative over more than one generation. Myers et al. (1985) could not find evidence of such a multigenerational effect, a discrepancy which is important to resolve in view of the various hypotheses now being considered for the nature of the effect.

The largest data set available, the US Huntington's roster, has been analysed for effects of parental transmission by Conneally (1984) and Boehnke et al. (1983), with further detailed analyses by Farrer and Conneally (1985) and by Ridley et al. (1988, 1991). The age-at-onset curve in this material was clearly different for offspring of affected fathers and mothers across a wide age range (Figure 9.6). As in other studies this was principally due to an 8-year reduction in age at onset of paternally transmitted cases, offspring of affected mothers showing little reduction (Table 9.8). Even after excluding juvenile cases, there was still a 5.6-year reduction in the age at onset of paternally transmitted cases.

The independence of the effects of parental transmission from the racial genetic background of the population has been shown by the finding of Folstein et al. (1987) that the reduced age at onset in offspring of affected fathers was found equally among blacks and whites in the Maryland (USA) population, as well as in South African mixed-race families as already mentioned.

Table 9.8 Age at onset and death in HD. Differences between parent and child

Study	Father–child	Mother–child
Boehnke et al. (1983)	8.06 + 11.27 (n = 276)	1.41 + 7.62 (n = 281)
Bird et al. (1974)	9.73 + 1.92 (n = 70)	2.04 + 0.93 (n = 52)
Previous literature analysed by Bird et al. (1974)	9.67 + 1.14 (n = 114)	3.01 + 1.52 (n = 55)
Ridley et al. (1988)	6.73 (skewed)	1.35 + 8.49 (n = 899)

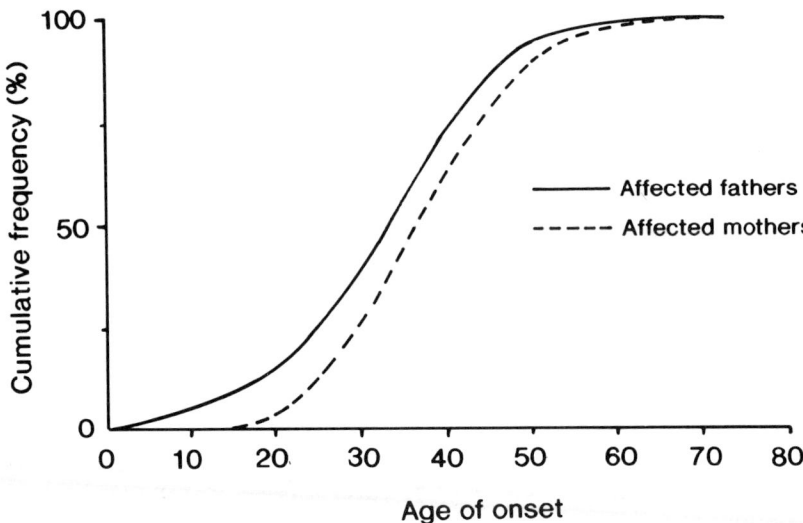

Figure 9.6 Cumulative frequency of age of onset in offspring of affected mothers compared with offspring of affected fathers. HD roster data; from Conneally (1984) with kind permission of the author.

Two early studies should be mentioned which failed to show an overall relationship between age at onset and parental transmission. Jones and Phillips (1970), in a relatively small study, showed no significant difference between 46 male-transmitted and 84 female-transmitted cases. The marked lack of male transmission in the study suggests they would probably have missed many juvenile cases; in fact their results do show a small difference (2.7 years) in the expected direction. Brackenridge (1971a, b), as part of his extensive literature survey, concluded that there was no clear overall relationship between onset and parental transmission, but did find a two to one paternal excess for juvenile cases and a significant difference in parental origin of juvenile and non-juvenile cases. Neither study would seem to invalidate the conclusion that sex of transmitting parent and age at onset are related over the entire range of onset in HD. The only large study failing to find an overall relationship has been that of Went *et al.* (1983, 1984), which used age at death as the criterion. Quite why their conclusions should differ from the others is not clear, but the overall weight of evidence strongly favours a reduced age at onset in male-transmitted compared with female-transmitted cases across the whole age range.

'Anticipation' of age at onset has already been mentioned in relation to paternal transmission in HD. While this phenomenon used to be considered an artefact resulting from ascertainment bias (Penrose, 1948), it is now clear that it can exist as a definite biological effect in certain situations, HD and myotonic dystrophy being the best-documented disorders. Ridley *et al.* (1988, 1991) have examined anticipation in HD in detail, using the US Huntington's roster data. Like previous workers they confirmed that anticipation existed for paternally transmitted cases, but not with maternal transmission. They also showed that this applied equally to those males with early age at onset, whose offspring had even younger (often juvenile) onset, whereas early-onset females showed the opposite, their offspring showing a later age at onset close to the population mean. Ridley *et al.* (1988, 1991) also showed that anticipation was preserved, even enhanced through two generations of male transmissions, but seemed to be reversed when the gene had passed through a female, no anticipation being observed regardless of the sex of the grandparent, in the limited number of cases when data was available on the generations of transmission.

Possible explanations
The existence of a marked effect of parental transmission in HD is undoubted; the possible cause is unknown, and probably will remain so until we can directly examine the HD gene and its action. Several hypotheses have been proposed, however, which are worth examining; it is also important to consider what is known of comparable situations involving other genetic disorders.

One explanation which can be discarded immediately is that the juvenile form might be the result of an unusual mutation in the HD gene, i.e. genetic heterogeneity. This is disproved by both its world-wide distribution and its occurrence within families where most cases are of typical adult onset, even those with an exceptionally late mean age at onset such as the large Tasmanian kindred (Pridmore, 1990). It is naturally likely that juvenile HD will occur more often in kindreds where mean age at onset is unusually low. Homozygosity for HD is likewise not a valid explanation, and it is of interest, as discussed on page 291, that the probable homozygotes documented in the Venezuela kindred have not shown a particular tendency to develop the disorder early.

The hypothesis that attracted most support until recently was that of a maternal factor which exerts a 'protective' action and delays age at onset in the offspring of affected women by comparison with those of affected men. Myers *et al.* (1983, 1985) and Boehnke *et al.* (1983) both proposed that this might result from a mitochondrial or comparable cytoplasmic factor and this would also fit with the observation in some, but not all, studies of a cumulative effect through both parental and grandparental generations. If a maternal factor is indeed operating one would expect to find a greater correlation for age at onset and death between mother–child than for father–child pairs. This was not observed by Myers *et al.* (1985) but was found in the larger data set of Boehnke *et al.* (1983), based on the US Huntington's roster.

Boehnke *et al.* (1983) are careful to point out that their model could be as well explained by the efforts of an autosomal or X-linked gene locus as by a mitochondrial factor and could not by its nature distinguish between these possibilities. The possibility of an X-linked modifying locus had previously been put forward by Stevens (1976), and has again been taken up in the position effect hypothesis of the HD gene put forward by Laird (1990) and discussed in the next chapter.

Erikson (1985), Ridley *et al.* (1988, 1991) and Reik (1988) have proposed a different mechanism and have suggested that the parental transmission effect may be an example of genetic imprinting. This phenomenon, well documented for a number of chromosomal regions in the mouse and other species (Cattenach and Kirk, 1985) and now demonstrated to be involved in several human disorders (Clarke, 1991; Hall, 1991), is the term given to the differential expression of a gene according to whether it is transmitted through sperm or ovum. The difference results principally from the degree of methylation of chromosomal regions in male and female gametes which affects the timing and degree of activation of the gene in the offspring. In HD the greater degree of methylation of an HD gene transmitted through the ovum is postulated to retard the expression of the gene.

At present there is no adequate evidence that directly favours either imprinting or mitochondrial inheritance as being responsible for the parental transmission effect in HD. Since the effects of imprinting are erased in each generation one should not see a multigeneration effect; it will thus be of particular importance to resolve whether one exists. The lack of any obvious difference between one and two generations of female transmission is against a mitochondrial basis and more compatible with imprinting as the cause. Now that the HD gene has being localized to a small region of chromosome 4p, it should also be possible to detect differential methylation effects in this region, and to look directly for evidence of imprinting in the homologous region of the mouse chromosome, which has been identified by the mapping of a sequence homologous to D4S10 in the mouse chromosome. Methylation effects have been noted in relation to the closely linked marker locus D4S95 (Pritchard *et al.*, 1989) but no clear differences in the HD region have yet emerged.

At this point it is worth observing that HD is proving to be far from unique among mendelian disorders in showing sex-related effects, though few of them have been adequately explained. Specific mitochondrial gene mutations have now been shown to be responsible for disorder such as Leber's optic atrophy and some maternally inherited myopathies (Wallace *et al.*, 1988a,b) as well as causing a variety of non-genetic disorders by somatic mutation. Imprinting has not yet been conclusively implicated in human mendelian disorders, though it is now well established in causing differences in disorders of specific chromosome regions such as the Angelman and Prader Willi syndromes involving chromosome 15q. Perhaps the most intriguing situation to compare is the maternal effect seen in myotonic dystrophy, where the severe childhood or congenital form is almost exclusively maternally inherited (Harper, 1989b), and where 'anticipation' is probably a true phenomenon in all age groups (Höweler *et al.*, 1989). Imprinting has been suggested as responsible for this also, but an intrauterine factor is equally likely.

In conclusion, while the basic mendelian inheritance of HD is not in doubt, the factors responsible for the striking variation in the disease phenotype remain highly speculative. The next few years will undoubtedly allow the various hypotheses to be tested fully when we understand the molecular basis of HD. This field of work, perhaps the most exciting currently and certainly the most rapidly developing, is the subject of the next chapter.

REFERENCES

Bachman D, Butler I and McKhann G (1977) The long term treatment of juvenile Huntington's chorea with dipropylacetic acid. *Neurology* **27**:193–197.

Baraitser M, Burn J and Fazzone TA (1983) Huntington's chorea arising as a fresh mutation. *Journal of Medical Genetics* **20**:459–460.

Barbeau A (1970) Parental ascent in the juvenile form of Huntington's chorea. *Lancet* **ii**:937.

Bell J (1934) Huntington's chorea. In: *Treasury of Human Inheritance* (ed RA Fisher). Cambridge: University Press.

Bird ED, Caro AJ and Pilling JB (1974) A sex related factor in the inheritance of Huntington's chorea. *Annals of Human Genetics* **37**:255–260.

Bird TD and Ommen GS (1975) Monozygotic twins with Huntington's Chorea in a family expressing the rigid variant. *Neurology* **25**:1126–1129.

Boehnke M, Conneally PM and Lange, K (1983) Two models for a maternal factor in the inheritance of Huntington's Disease. *American Journal of Human Genetics* **35**:845–860.

Brackenridge CJ (1971a) A genetic and statistical study of some sex-related factors in Huntington's disease. *Clinical Genetics* **2**:287–297.

Brackenridge CJ (1971b) The relation of type of initial symptoms and line of transmission to ages at onset and death in Huntington's disease. *Clinical Genetics* **2**:287–297.

Brackenridge CJ (1972) A statistical study of sibships born to parents affected with Huntington's disease. *Journal of Medical Genetics* **9**:17–22.

Brackenridge CJ (1974) Relation of parental age to rigidity in Huntington's disease. *Journal of Medical Genetics* **11**:136–140.

Brackenridge CJ (1979) Relation of occupational stress to the age of onset of Huntington's disease. *Acta Neurologica Scandinavica* **60**:272–276.

Brothers CRD (1949) The history and incidence of Huntington's chorea in Tasmania. *Proceedings of the Royal Australian College of Physicians* **4**:48–50.

Bruyn GW (1968) Huntington's chorea: historical, clinical and laboratory synopsis. In: *Handbook of Neurology* (eds PJ Vinken and GW Bruyn), pp. 298–378. Amsterdam: North Holland.

Burch PRJ (1969) The age-distribution of the onset of Huntington's chorea: an interpretation. In: *Progress in Neurogenetics* (eds A Barbeau and JR Brunette), pp. 674–681. Amsterdam: Excerpta Medica Foundation.

Burke W (1976) Age of onset in Huntington's disease: lack of prenatal age effect. *Journal of Medical Genetics* **13**:462–465.

Caro A (1976) Possible association of Madelung's deformity with Huntington's chorea. (Letter) *British Medical Journal* **ii**:1258.

Cattenach BM and Kirk M (1985) Differential activity of maternally and paternally derived chromosome regions in mice. *Nature* **315**:496–498.

Clarke A (1990) Genetic imprinting in clinical genetics. In: *Genomic Imprinting* (eds M Monk and A Surani), pp. 131–139. Cambridge (Supplement to Development).

Conneally PM (1984) Huntington's disease: genetics and epidemiology. *American Journal of Human Genetics* **36**:506–526.

Critchley EMR and Secker-Walker RH (1966) A deaf-mute with Huntington's chorea. *Journal of Neurology, Neurosurgery and Psychiatry* **29**:181–183.

Davenport CB (1911) *Heredity in Relation to Eugenics*. New York: Holt, re-issued by Arno Press, 1972.

Davenport CB and Muncey MD (1916) Huntington's chorea in relation to heredity and insanity. *American Journal of Insanity* **73**:195–220.

Eldridge R, O'Meara K, Chase T and Donnelly EF (1973) Offspring of consanguineous parents with Huntington's chorea. In: *Huntington's Chorea* (eds A Barbeau, TN Chase and GW Paulson), pp. 1872–1972. New York: Raven Press.

Entres JL (1921) *Zur Klinik und Vererbung der Huntingtonschen Chorea. Monographien des Gesamtgebietes der Neurologie und Psychiatrie* **27**. Berlin: Springer.

Erikson RP (1985) Chromosomal imprinting and the parent transmission specific variation in Huntington's Disease. *American Journal of Human Genetics* **37**:827–829.

Farrer LA and Conneally PM (1985) A genetic model for age at onset in Huntington disease. *American Journal for Human Genetics* **37**:350–357.
Farrer LA and Conneally PM (1987) Predictability of phenotype in Huntington's disease. *Archives of Neurology* **44**:109–113.
Farrer LA, Conneally PM and Yu P (1984) The natural history of Huntington's disease: possible role of ageing genes. *American Journal of Medical Genetics* **18**:115–123.
Finch CE (1980) The relationship of ageing changes in the basal ganglia to manifestations of Huntington's chorea. *Annals of Neurology* **7**:406–410.
Folstein SE, Phillips JA and Meyers DA (1985) Huntington's disease: two families with differing clinical features show linkage to the G8 probe. *Science* **229**:776–779.
Folstein SE, Chase GA, Wahl WE, McDonnell AM and Folstein MF (1987) Huntington's disease in Maryland: Clinical aspects of racial variation. *American Journal of Human Genetics* **41**:168–179.
Folstein SE, Abbott MH, Chase GA, Jensen BA and Folstein MF (1983) The association of affective disorder with Huntington's disease in a case series and in families. *Psychological Medicine* **13**:537–542.
Froster-Iskenius UG, Hayden MR, Wang HS et al. (1986) A family with Huntington disease and reciprocal translocation. *Americal Journal of Human Genetics* **38**:759–767.
Glendinning GM (1976) Genetic counselling with Huntington's chorea. (Letter) *British Medical Journal* **ii**, 46.
Haldane, JBS (1941) The relative importance of principal and modifying genes in determining some human diseases. *Journal of Genetics* **41**:149–157.
Hall JG (1990) How imprinting is relevant to human disease. In: *Genomic Imprinting* (eds M Monk and A Surani), pp. 141–148. Cambridge (Supplement to Development).
Harper PS (1989a) The muscular dystrophies. In: *The Metabolic Basis of Inherited Disease* (eds CS Scriver et al.). New York: McGraw-Hill.
Harper PS (1989b) *Myotonic Dystrophy*. 2nd Edn. Philadelphia: WB Saunders.
Hayden MR (1979) Huntington's chorea in South Africa. PhD Thesis, University of Cape Town.
Hindringer P (1935) Eine neue chorea-Huntington-sippe. Erlangen: Doctoral Thesis. (Quoted by Eldridge et al., 1973.)
Höweler CJ, Busch HFM and Geraedts JPM (1989) Anticipation in myotonic dystrophy: fact or fiction. *Brain* **112**:779–797.
Huntington G (1872) On chorea. *Medical and Surgical Reporter* **26**:320–321.
Husquinet H, Franck G and Vranckx C (1973a) Detection of future cases of Huntington's chorea by the L-dopa load test: experiment with two monozygotic twins. In: *Huntington's Chorea* (eds A Barbeau, TN Chase and GW Paulson), pp. 301–310. New York: Raven Press.
Husquinet H, Mackenzie-van der Noordaa MC, Myrianthopoulos NC, Petit H, Volkers W and Went LN (1973b) Analysis of Huntington's chorea in Northwestern Europe. In: *Huntington's Chorea* (eds A Barbeau, TN Chase and GW Paulson), pp. 171–177, New York: Raven Press.
Jelliffe SE (1908) A contribution to the history of Huntington's disease; a preliminary report. *Neurographs* **1**:116–124.
Jones MB and Phillips CR (1970) Affected parent and age of onset in Huntington's chorea. *Journal of Medical Genetics* **7**:20–21.
Kishimoto K, Nakamura M and Sotokawa Y (1957) Population genetics study – Huntington's chorea in Japan. *Annual Report, Research Institute Environmental Medicine* **9**:195–211.
Kishimoto K, Nakamura M and Sotokawa Y (1959) On population genetics of Huntington's chorea in Japan. In: *International Congress of Neurological Science* (eds L Van Bogaert and J Radermecker), Vol. 4, pp. 217–226. London: Pergamon Press.
Kondo I, Ohita H, Yazaki M, Ikeda JE, Gusella JF and Kanazawa T (1990) Exclusion mapping of the hereditary dentatorubropallidoluysian atrophy gene from the Huntington's disease locus. *Journal of Medical Genetics* **27**:105–106.
Laird C (1990) Huntington's disease: proposed mechanism of mutation, inheritance and expression. *Trends in Genetics* **6**:242–247.
Marx RN (1973) Huntington's chorea in Minnesota. *Advances in Neurology* **1**:237–243.
Mastromauro CA, Meissen GJ, Cupples LA, Berkman B and Myers RH (1989) Estimation of fertility and fitness in Huntington's disease in New England. *American Journal of Medical Genetics* **33**:248–254.
Mattsson B (1974) Huntington's chorea in Sweden. 1. Prevalence and genetic data. *Acta Psychiatrica Scandinavica Supplementum* **255**:211–255.
McKeown C, Read AP, Dodge A, Stecko O, Mercer A and Harris R (1987). Wolf–Hirschhorn locus is distal to D4S10 on short arm of chromosome 4. *Journal of Medical Genetics* **24**:410–412.

Mendel G (1865) Versuche über Pflanzenhybriden. *Proceedings of the Natural History Society of Brunn,* **4**:3–47. English translation reprinted 1965.
Merrit AD, Conneally PM, Rahman NF and Drew AL (1969) Juvenile Huntington's chorea. In: *Progress in Neurogenetics* (eds A Barbeau and JR Brunette), pp. 645–650. Amsterdam: Excerpta Medica Foundation.
Minski L and Guttmann E (1938) Huntington's chorea; a study of 34 families. *Journal of Mental Science* **84**:21–96.
Myers RH, Madden JJ, Teague JL and Falek A (1982) Factors related to onset of Huntington's disease. *American Journal of Human Genetics* **34**:481–488.
Myers RH, Goldman D and Bird ED (1983) Maternal transmission in Huntington's disease. *Lancet* **i**:208–210.
Myers RH, Sax DS and Schoenfeld M (1985) Late onset of Huntington's disease. *Journal of Neurology, Neurosurgery and Psychiatry* **48**:530–534.
Myers RH, Leavitt J, Farrer L *et al.* (1989) Homozygote for Huntington's disease. *American Journal of Human Genetics* **45**:615–618.
Myrianthopoulos NC (1966) Huntington's chorea. *Journal of Medical Genetics* **3**:298–314.
Myrianthopoulos NC and Rowley PT (1960) Monozygotic twins concordant for Huntington's chorea. *Neurology* **10**:506–511.
Newcombe RG (1981) A life table for onset of Huntington's chorea. *Annals of Human Genetics* **45**:375–385.
Newcombe RG, Walker DA and Harper PS (1981) Factors influencing age of onset and duration of survival in Huntington's chorea. *Annals of Human Genetics* **45**:387–396.
Oepen H (1973) Discordant features of monozygotic twin markers with Huntington's chorea. In: *Advances in Neurology* (eds A Barbeau, TN Chase and GW Paulson), Vol 1, pp. 199–201. New York: Raven Press.
Panse F (1942) Die erbchorea; eine klinische-genetische studie. Leipzig: Thieme.
Parker N (1958) Observations on Huntington's chorea based on a Queensland survey. *Medical Journal of Australia* **45**:351–359.
Pearson JS, Peterson MC, Lazarte JA, Blodgett HE and Kley IB (1955) An educational approach to the social problem of Huntington's chorea. *Proceedings Mayo Clinic* **30**:349–357.
Penrose LS (1948) The problem of anticipation in pedigrees of dystrophia myotonica. *Annals of Eugenics* **14**:125–132.
Periczak-Vance MA, Elston RC, Conneally PM and Dawson AV (1983) Age of onset heterogeneity in Huntington's disease families. *American Journal of Medical Genetics* **14**:49–59.
Pleydell MJ (1954) Huntington's chorea in Northamptonshire. *British Medical Journal* **i**:1121–1128.
Pridmore SA (1990) Age of onset of Huntington's disease in Tasmania. *Medical Journal of Australia* **153**:135–137.
Pritchard CA, Cox DR and Myers RM (1989) Methylation at the Huntington disease-linked D4S95 locus. *American Journal of Human Genetics* **45**:335–336.
Punnett RC (1908) Mendelian inheritance in man. *Proceedings of the Royal Society of Medicine* **1**:135–168.
Quarrell OWJ, Youngman S, Sarfarazi M and Harper PS (1988) Absence of close linkage between benign hereditary chorea and the locus D4S10 (probe G8). *Journal of Medical Genetics* **25**:191–194.
Quarrell OWJ, Snell RG, Curtis MA, Roberts SH, Harper PS and Shaw DJ (1991). Paternal origin of the chromosomal deletion resulting in Wolf–Hirschhorn syndrome. *Journal of Medical Genetics* **28**:256–259.
Reed TE and Neel JV (1959) Huntington's cnorea in Michigan. 2: Selection and mutation. *American Journal of Human Genetics* **11**:107–136.
Reed TW, Chandler JH, Hughes EM and Davidson RT (1958) Huntington's chorea in Michigan. I. Demography and genetics. *American Journal of Human Genetics* **10**:201–225.
Reik W (1988) Genomic imprinting: a possible mechanism for the prenatal origin effect in Huntington's chorea. *Journal of Medical Genetics* **25**:805–808.
Ridley RM, Frith CD, Crow TJ and Conneally PM (1988) Anticipation in Huntington's disease is inherited through the male line but may originate in the female. *Journal of Medical Genetics* **25**:589–595.
Ridley RM, Frith CD, Farrer LA and Conneally PM (1991) Patterns of inheritance of the symptoms of Huntington's disease suggestive of an effect of genomic imprinting. *Journal of Medical Genetics* **28**:224–231.
Rosanoff AJ and Handy LM (1935) Huntington's chorea in twins. *Archives of Neurology and Psychiatry* **33**:839–841.

Sarfarazi M, Quarrell OWJ, Wolak G and Harper PS (1987) An integrated micro-computer system to maintain a genetic register for Huntington's disease. *American Journal of Medical Genetics* **28**:999–1006.
Schiottz-Christensen E (1969) Chorea Huntington and epilepsy in monozygotic twins. *European Neurology* **2**:250–255.
Shaw M and Caro A (1982) The mutation rate to Huntington's chorea. *Journal of Medical Genetics* **19**:161–167.
Shields J (1962) *Monozygotic Twins Brought up Apart and Brought up Together*. London: Oxford University Press.
Shokeir MHK (1975) Investigations on Huntington's disease in the Canadian Prairies. I. Prevalence. *Clinical Genetics* **7**:345–348.
Sjögren T (1936) Verlungsmedizinische untersuchungen über Huntington's Chorea in einer schwedischen Bauernpopulation. *Zeitschrift für menschlische Vererbungs und Konstitutionslehre* **19**:131–165.
Snell RG, Lazarou LP, Youngman S et al. (1989) Linkage disequilibrium in Huntington's disease: an improved localisation for the gene. *Journal of Medical Genetics* **26**:673–675.
Stevens DL (1973) The classification of variants of Huntington's chorea. *Advances in Neurology* **1**:57–64.
Stevens DL (1976) Huntington's chorea: a demographic genetic and clinical study. University of London: MD thesis.
Stevens D and Parsonage M (1969) Mutation in Huntington's chorea. *Journal of Neurology, Neurosurgery and Psychiatry* **32**:140–143.
Stine OC and Smith KD (1990) The estimation of selection coefficients in Afrikaners: Huntington's disease, porphyria variegata, and lipoid proteinosis. *American Journal of Human Genetics* **46**:452–458.
Theilman J, Kanami S, Shiang R et al. (1989) Non-random association between alleles detailed at D4595 and D4598 and the Huntington's disease gene. *Journal of Medical Genetics* **26**:676–681.
Tyrer JH (1957) The differentiation of hysteria from organic neurological disease. *Medical Journal of Australia* **1**:566–571.
Van Dijk, van der Velde EA, Roos RAC and Bruyn GW (1986) Juvenile Huntington's disease. *Human Genetics* **73**:235–239.
Vogel F and Motulsky AG (1986) *Human Genetics. Problems and Approaches*. Berlin: Springer.
Walker DA (1979) Huntington's chorea in South Wales. University of Liverpool: MD thesis.
Walker DA, Harper PS, Wells CEC, Tyler A, Davies K and Newcombe RG (1981) Huntington's chorea in South Wales. A genetic and epidemiological study. *Clinical Genetics* **19**:213–221.
Walker DA, Harper PS, Newcombe RG and Davies K (1983) Huntington's chorea in South Wales: mutation, fertility and genetic fitness. *Journal of Medical Genetics* **20**:12–17.
Wallace DC (1972) Huntington's chorea in Queensland. A not uncommon disease. *Medical Journal of Australia* **59**:299–307.
Wallace DC (1976) The social effect of Huntington's chorea on reproductive fitness. *Annals of Human Genetics* **39**:375–379.
Wallace DC and Hall AC (1972) Evidence of genetic heterogeneity in Huntington's chorea. *Journal of Neurology, Neurosurgery and Psychiatry* **35**:789–800.
Wallace DC and Parker N (1973) Huntington's chorea in Queensland: the most recent story. *Advances in Neurology* **1**:223–236.
Wallace DC, Singh G, Lott MT et al. (1988a) Mitochondrial DNA mutation associated with Leber's hereditary optic neuropathy. *Science* **242**:1427–1430.
Wallace DC, Zheng X, Lott MT, Shoffner JM, Hodge JA, Kelley RI, Epstein CM and Hopkins LC (1988b) Familial mitochondrial encephalomyopathy (MERRF): Genetic, pathophysiological and biochemical characterization of a mitochondrial DNA disease. *Cell* **55**:601–610.
Wendt GG and Drohm D (1972) Die Huntingtonsche Chorea. Eine populationsgenetische Studie. Stuttgart: Thieme.
Went LN, Vegter-Van der Vlis M, Volkers W and Collewijn H (1975) Huntington's Chorea. In: *Early Diagnosis and Prevention of Genetic Disease* (eds LN Went et al.), pp. 13–25. Leiden: University Press.
Went LN, Vegter-van der Vlis M and Bruyn GW (1984) Parental transmission in Huntington's disease. *Lancet* **i**:1100–1102.
Went LN, Vegter-van der Vlis M, Bruyn GW and Volkers WS (1983) Huntington's chorea in the Netherlands: the problem of genetic heterogeneity. *Annals of Human Genetics* **47**:205–214.
Wexler NS, Young AB, Tanzi RE et al. (1987) Homozygotes for Huntington's disease. *Nature* **326**:194–197.

Wolff G, Deuschl G, Wienker TF, Hummel K, Bender K, Lucking CH, Schumachers M, Hammer J and Oepen G (1989) New mutation to Huntington's disease. *Journal of Medical Genetics* **26**:18–27.

Zweig RM, Koven SJ, Hedreen JC, Maestri NE, Kazazian HH and Folstein SE (1989) Linkage to the Huntington's disease locus in a family with unusual clinical and pathological features. *Annals of Neurology* **26**:78–84.

10

Molecular Genetic Approaches to Huntington's Disease

WHY A GENETIC APPROACH?

As is apparent from some of the other chapters in this volume, the function and nature of the Huntington's disease (HD) gene are not understood, nor are there any reliable indications from biochemical or pathological studies (Chapter 5). On the other hand, the genetic basis and mode of inheritance of the disease are not in question. Recent developments in molecular genetics have made it feasible to approach a disease gene of unknown structure and function by purely genetic means, and eventually to isolate it, deduce its structure, and use this information to study the types of mutation responsible for the disease and develop assays for the gene product itself. This approach, which has become known as 'reverse genetics' (Ruddle, 1984), has recently had spectacular successes in the isolation of the genes causing Duchenne muscular dystrophy (Monaco and Kunkel, 1987) and cystic fibrosis (Rommens et al., 1989). In this chapter we will first describe the classical approach to gene mapping, and then look at the way in which reverse genetics is being applied to try to isolate the HD gene.

GENETIC LINKAGE

When two genes are located on different chromosomes, it will be randomly determined whether or not the copies received from a particular parent are also passed on together to a child. By contrast, if the two are placed close to each other on the same chromosome, such co-inheritance will occur more often than the random 50%. This non-random co-inheritance of genes is termed genetic linkage. Thus the clue to the location of disease genes like that for HD may be given by studying inherited markers and testing whether any of them shows departure from random values of segregation. The greater the departure from random, the closer must be the linkage between the marker and the disease gene, or in other words, the smaller the physical distance between them on the chromosome.

This simple but fundamental principle formed the basis of our understanding that genes were carried on specific chromosomes. Experimental studies on species that could easily be bred in large numbers (such as the fruit fly, *Drosophila*) soon showed that linked genes could not only be organized into specific linkage groups corresponding to each chromosome, but also that they could be ordered along the chromosome.

The application of gene mapping to humans was slow to develop for two reasons: the impossibility of planned breeding to test linkage, and the scarcity of inherited variations that could be used as genetic markers. Until recently only a handful of blood groups and other variable proteins were available in which genetic polymorphism was sufficiently frequent to allow testing for linkage in families with rare disorders such as HD. None the less the possibility was not overlooked, nor was the important implication that a linked marker might allow prediction of the disease status of a person at risk for HD. In their classic paper over 50 years ago reporting the first genetic linkage for a human disease (haemophilia and colour-blindness on the X chromosome) Bell and Haldane (1937) wrote: 'If, however, to take a possible example, an equally close linkage were found between the genes determining blood group membership and that determining Huntington's chorea, we should be able, in many cases, to predict which children of an affected person would develop the disease, and to advise on the desirability or otherwise of their marriage'.

That it took almost half a century for such a linkage to be detected in HD was partly due to the lack of suitable markers, as mentioned above, and partly because of the difficulty in finding HD families of a suitable structure to analyse for linkage. In trying to follow a disease gene through successive generations of a family it is much easier when the disease gene shows itself early, yet is not fatal, allowing three or more generations to be studied. In HD, individuals rarely develop the disease until their affected parent is already dead, while apparently healthy members contribute little useful information until they are elderly because of uncertainty as to whether they might still possess the gene. Dependence on this fragmentary family material meant that linkage would not be obvious simply from examining individual pedigrees, but that pooled data would be needed, necessitating complex mathematical analysis to determine the likelihood of linkage; the 'lod score', representing the logarithm of the ratio of the odds for and against linkage, is the most widely used measure of the strength of evidence for linkage. A lod score of $+3$ or greater is considered to be reasonable evidence in support of linkage, whereas a score of -2 or less indicates that the loci are not linked.

Apart from some fragmentary data from early studies (Leese *et al.*, 1952; Beckman *et al.*, 1974) the first systematic search for genetic linkage in HD was that of Lindstrom *et al.* (1973), followed by a series of others in America, Australia, Britain and Holland (Pericak-Vance *et al.*, 1978; Brackenridge *et al.*,

1978; Hodge et al., 1980; Volkers et al., 1980). All of these studies were negative, which was not surprising since the markers used, mostly blood groups, serum proteins and red-cell enzymes, covered less than 30% of the total genome. Thus by 1980 the situation was not a hopeful one, since while the pooled data had excluded around 20% of the genome as the site of the HD gene, available markers had almost been exhausted and without the development of new approaches, it appeared that little more could be done.

RESTRICTION FRAGMENT LENGTH POLYMORPHISMS

Two factors were soon to improve dramatically the prospects for mapping the HD gene by linkage. The first was the recognition that for linkage to be detected, it was essential to combine material from the few very large complete families in the world; this prompted the Hereditary Disease Foundation (HDF) and others to organize a collaborative study to ascertain these families and set up immortalized lymphoblastoid cell lines as a renewable resource for future studies. These included large families such as those from the USA and Wales, that had already been studied for linkage, but by far the largest was the extended Venezuelan kindred (see Chapter 8) that became the focus of a major study of which genetic linkage formed the core (Wexler et al., 1985). The large sibships (up to 15 individuals) and numerous branches of this kindred meant that there was at least a reasonable chance of a marker linked to HD being recognized as such.

The second and crucial factor was a series of technical developments in molecular biology (Southern blotting, DNA hybridization, and restriction enzymes) that allowed individual DNA sequences in human genomes to be analysed directly, together with the recognition that human DNA carried in its structure inherited variation that was far more abundant and accessible than that which could be obtained from blood groups and protein polymorphisms. This variation is based on sequence differences that can affect whether or not the DNA strands are cut by particular enzymes (restriction enzymes), the presence or absence of cleavage at a particular point determining the length of the resulting DNA fragments. These DNA sequence polymorphisms (Figure 10.1) are known as 'restriction fragment length polymorphisms' (RFLPs) and can be detected by electrophoresis, which separates DNA fragments into bands according to their lengths. The bands can be detected by hybridization with a complementary, radioactively-labelled version of the DNA sequence (a 'probe'). A second class of DNA polymorphism results from the presence of variable numbers of short, repeated sequence elements; these are known as VNTRs (variable number of tandem repeats). Figure 10.2 shows the basis of techniques for the detection of DNA polymorphisms, which have since been improved by further technical advances.

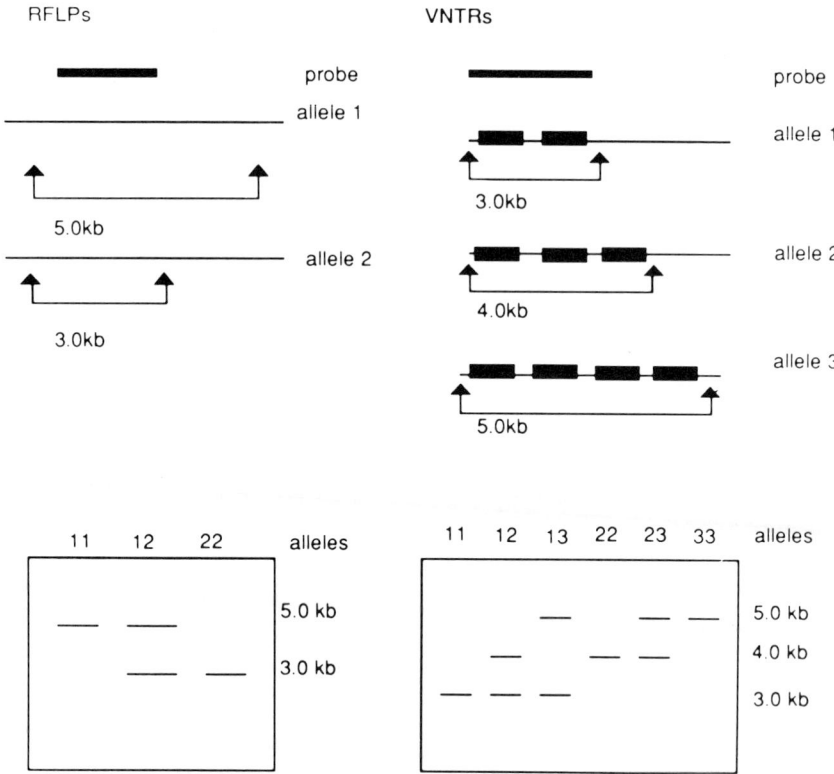

Figure 10.1 DNA polymorphisms: illustration of the structural basis of restriction fragment length polymorphisms (RFLPs) resulting from a single base mutation, and variable number of tandem repeat polymorphisms (VNTRs).

LINKAGE TO HUNTINGTON'S DISEASE

Botstein *et al.* (1980) proposed that a complete human linkage map could be constructed using RFLPs, and that the gene causing any human trait segregating in a mendelian manner could be located. Initially there was scepticism from many quarters, but the methods started to be applied around 1982 to X-linked and other diseases where the chromosome on which the disease gene was located was known, and then also to those where it was not, including HD, cystic fibrosis, and neurofibromatosis. This was in spite of the fact that a complete human linkage map was still a long way off, and the approach had therefore to be the use of RFLPs one by one until linkage was found.

A significant advantage of RFLPs over classical polymorphisms is that since the structure of the DNA is essentially the same in all types of cell, any nucleated cell may be used for the analysis. The blood samples and lymphoblastoid cell lines from the Venezuelan and other large kindreds had been

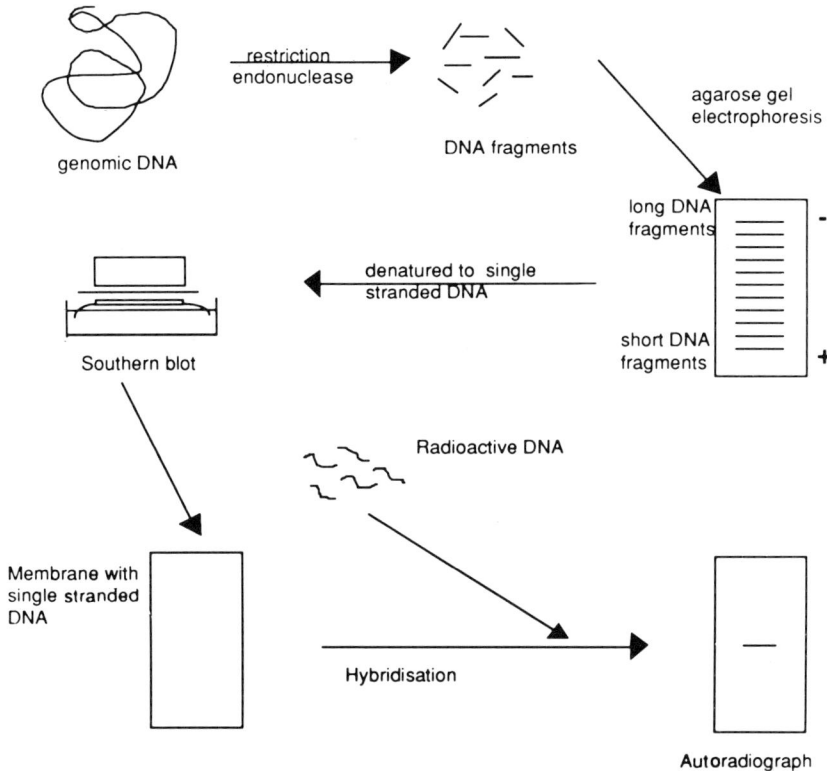

Figure 10.2 The detection of DNA polymorphisms.

collected with the RFLP approach in mind, and Dr Jim Gusella, initially with Dr D. Housman at the Massachusetts Institute of Technology, began to collect RFLP-detecting probes with which to begin the linkage analysis. At this time, (1982) many people were discussing how long such a study would take before it found a linkage. Estimates ranged from 10 years up to at least 50 years. All agreed that it would probably be necessary to test hundreds of markers, but Gusella and the HDF were not deterred. By an incredible piece of good fortune Gusella found linkage in his first dozen tries. The marker G8 is a functionless piece of DNA detecting several RFLPs (Figure 10.3), that Gusella showed to be linked to HD initially in American families; positive lod scores began to appear in the early summer of 1983, but it is probably fair to say that at this stage Gusella and his collaborators feared that the linkage might disappear when more families were analysed! By August 1983, however, data from the much larger Venezuelan kindred, already described in Chapters 1 and 8, had given further positive results, so that when the HD investigators met in Rochester, New York at the end of the month, there could be no doubt that the gene had been firmly localized. The resulting paper in *Nature* (Gusella *et al.*, 1983) was a

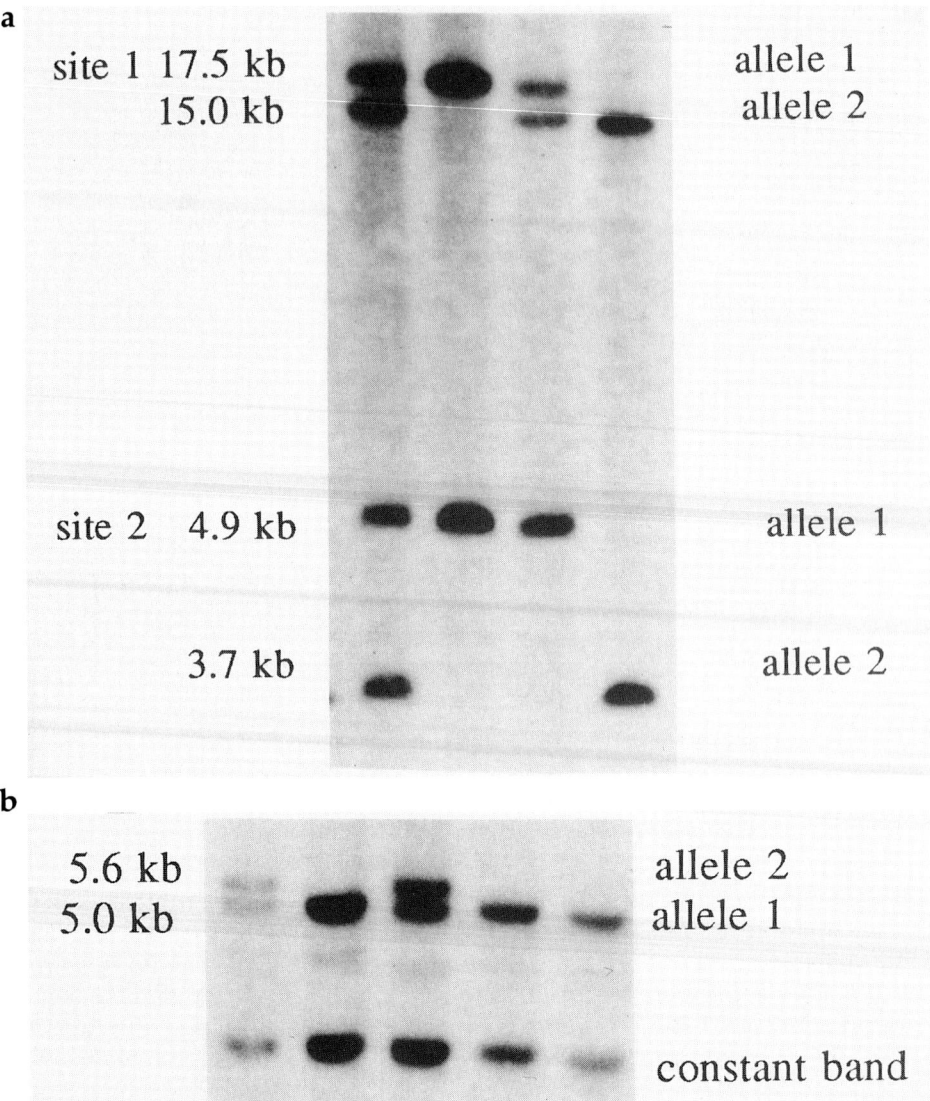

Figure 10.3 RFLPs detected by two markers from chromosome 4. (A) Two polymorphisms for the restriction enzyme *Hin*dIII, at the D4S10 locus (probe G8). (B) Polymorphism for the enzyme *Pvu*II, at the D4S90 locus (probe D5).

vindication of the new approach to gene mapping and of the effort that had gone into the ascertainment and sampling of the HD family panel.

Gusella *et al.* (1983) showed that D4S10 (the chromosomal locus recognized by the G8 probe) was located on chromosome 4, by the use of somatic cell hybrids. (These are cultured cell lines of mouse or hamster origin, which have been fused with human cells such that they contain one or a small number of human chromosomes, in addition to the rodent genome.) Furthermore, they obtained DNA samples from patients with the congenital anomaly Wolf–Hirschhorn syndrome, which results from a deletion of the terminal cytogenetic band of the short arm of chromosome 4 (4p16), and showed that the D4S10 locus was deleted in these chromosomes (Gusella et al., 1985). Hence the HD locus was localized with D4S10 to the region 4p16 (Figures 10.4 and 10.6).

The discovery of linkage made it possible to investigate the possibility of genetic heterogeneity, which would result from the existence of more than one genetic locus for HD. Studies of other large kindreds from America and Europe, and a collaborative study involving families from 14 different locations including Japan, showed no evidence for the existence of other loci (Youngman

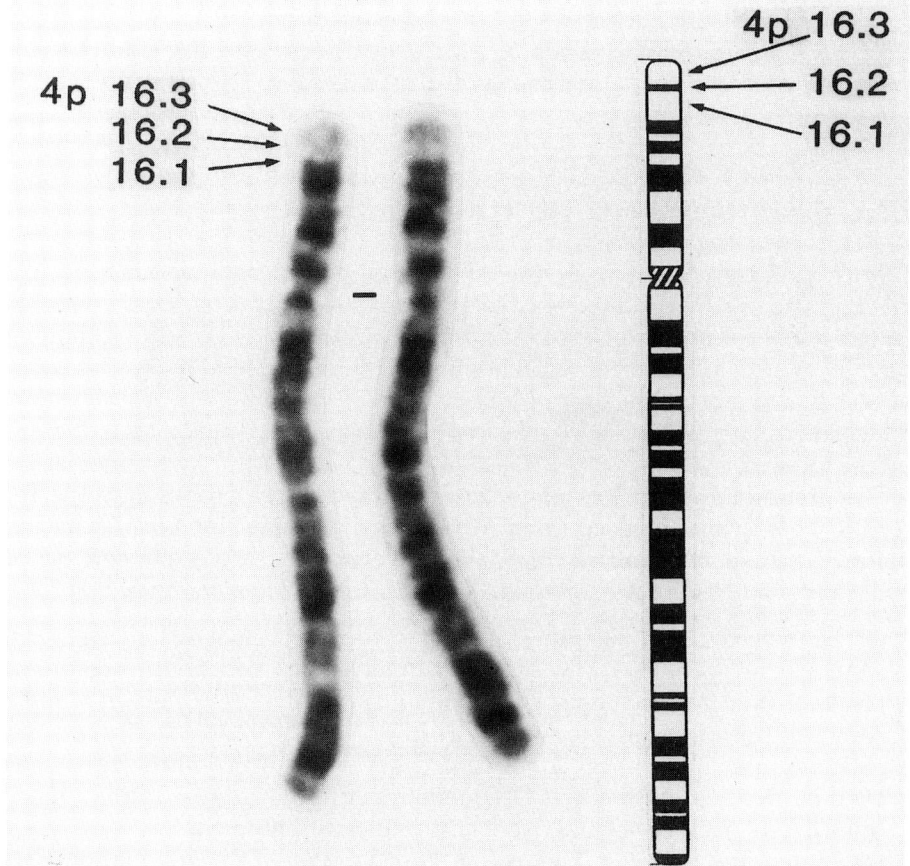

Figure 10.4 Photomicrograph and ideogram of human chromosome 4, with the 4p16 region indicated.

et al., 1986; Conneally *et al.*, 1989), nor was there evidence of heterogeneity in several clinically atypical families (Folstein *et al.*, 1985; Zweig *et al.*, 1989). The combined genetic data gave a recombination fraction of 4% between HD and D4S10, and with no evidence of linkage disequilibrium (see below).

The discovery of a close linkage to HD immediately had three major implications:

(1) It would allow the possibility of presymptomatic testing of individuals at risk for HD, and for prenatal diagnosis of at-risk pregnancies. These issues are discussed in Chapter 12.

(2) It demonstrated that RFLP analysis could indeed be used to find a human disease gene of completely unknown function, and caused those who had said it was an impractical exercise to eat their words. It also captured the attention of some of the best minds in molecular biology and medical genetics, thus ensuring rapid future progress.

(3) By localizing the HD gene to a small region of the genome, the way was opened for an as yet completely untried procedure now known as 'reverse genetics' to be applied, in order to isolate the disease gene itself, and hence to deduce its role in the pathogenesis of HD. This enterprise is the subject of the remainder of this chapter.

FINE MAPPING OF THE HD LOCUS

Following Gusella's discovery, the HDF decided to set up a working group on HD to try to ensure productive collaboration between the various groups that were now becoming interested, and hasten progress towards the isolation of the HD gene. This has been on the whole a highly successful exercise, which now includes most of the groups working on HD around the world. The many television documentaries and press reports have also ensured a high degree of public interest and support.

The first requirement was a good source of new DNA probes from the 4p16 region. These could be used to refine further the localization of the gene, and define the smallest region in which it should be located. Molecular genetics has generally made use of the following sources for chromosome-specific probes:

(1) Flow-sorted chromosomes. Using a fluorescence-activated cell sorter (FACS), chromosomes may be individually separated according to their physical size. Within a reasonable time it is possible to sort about a million copies of chromosome 4, which are then used to construct a complete collection (or 'library') of cloned DNA sequences. This approach has been used by the Los Alamos and Lawrence Livermore National Laboratories in the USA to produce several useful chromosome 4 libraries.

(2) Somatic cell hybrids. The use of these in relation to gene mapping was discussed earlier. If a hybrid cell line can be produced containing a single human chromosome, or better still only the relevant region as a sub-

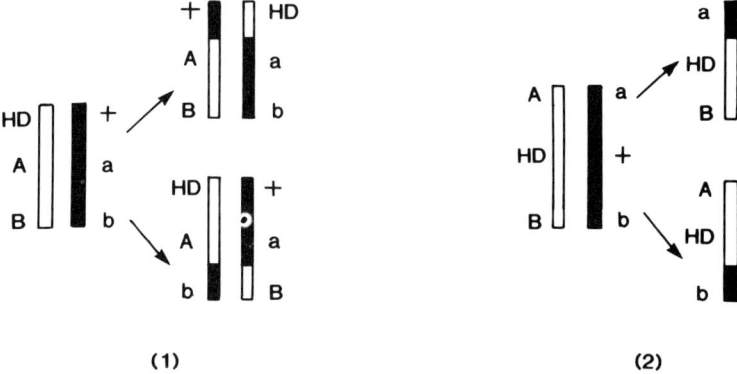

Figure 10.5 The genotypes of offspring resulting from genetic recombination between the HD locus and closely linked marker loci A and B. The + indicates the normal (healthy) allele of the HD gene. In (1), the marker loci are both on the same side of HD, whereas in (2) they are flanking the HD gene.

chromosomal fragment, then the DNA can be used to produce a library. Initially this will contain both human and rodent DNA clones, but the human ones can easily be recognized by hybridization with radioactively labelled human DNA. Hybrid cell lines that have been extensively used for this were constructed in several laboratories (Wasmuth *et al.*, 1986; MacDonald *et al.*, 1987; Smith *et al.*, 1988; Cox *et al.*, 1989).

A great deal of effort was by now going into the search for new probes. At this time the 'Holy Grail' was the so-called flanking marker, a DNA probe detecting an RFLP on the opposite side of the disease gene from D4S10. The flanking marker, together with D4S10, would define the interval on the chromosome in which the HD gene must be, and lead the attempts to isolate the gene in the right direction. To show that a marker flanked the HD gene, it would be necessary to observe crossovers (instances of genetic recombination) between the marker and HD, in individuals where there had been no crossover between HD and D4S10. The converse should also be true: in individuals where there had been a crossover between HD and D4S10, there should be no crossover between HD and the flanking marker (Figure 10.5). Because the markers used are close to HD and therefore by definition seldom cross over with it, this can be a difficult requirement to satisfy.

At the same time as the above, other groups were putting much effort into obtaining the most exact localization possible for the D4S10 locus. This was done by two means:

(1) *In situ* hybridization (Zabel *et al.*, 1985; Wang *et al.*, 1986; Magenis *et al.*, 1986). A radioactive probe DNA sequence is hybridized to a spread of metaphase chromosomes on a microscope slide, prepared as for routine cytogenetic analysis. After hybridization the excess probe is washed away

and a photographic emulsion applied to the slide, which reveals the site of hybridization of the labelled probe.

(2) Further use was made of cells containing Wolf–Hirschhorn chromosomes (see above) and other re-arranged derivatives of chromosome 4 (Wasmuth *et al.*, 1986; MacDonald *et al.*, 1987; Smith *et al.*, 1988), to construct a panel of hybrid cell lines that could be used to localize DNA markers accurately (Figure 10.6).

At first there was a slight disagreement between the two types of approach, with *in situ* hybridization favouring a localization to the chromosomal region 4p16.1–4p14 (Magenis *et al.*, 1986) or 4p16.1–4p16.3 (Wang *et al.*, 1986), whereas somatic cell hybrid analysis put D4S10 into the most terminal region, 4p16.3 (Figure 10.6). Subsequently, the breakpoint in the translocation used in the study of Magenis *et al.* (1986) was reassessed and assigned to 4p14–4p16.3 (MacDonald *et al.*, 1987). The consensus now favours the 4p16.3 localization for D4S10, and a more recent study using non-radioactive *in situ* hybridization provided further evidence that the correct localization was in fact 4p16.3 (Landegent *et al.*, 1986). The implication of this finding was that the distance between D4S10 and the telomere (end of the chromosome) was probably not more than 6 million bases of DNA (6 Mb), a relatively short distance in terms of the total length of the human genome (3000 Mb).

Many new DNA sequences had now been obtained from chromosome 4. These were being localized to specific sub-chromosomal regions, largely by the use of somatic cell hybrids, in the hope of finding some in the 4p16.3 area. This

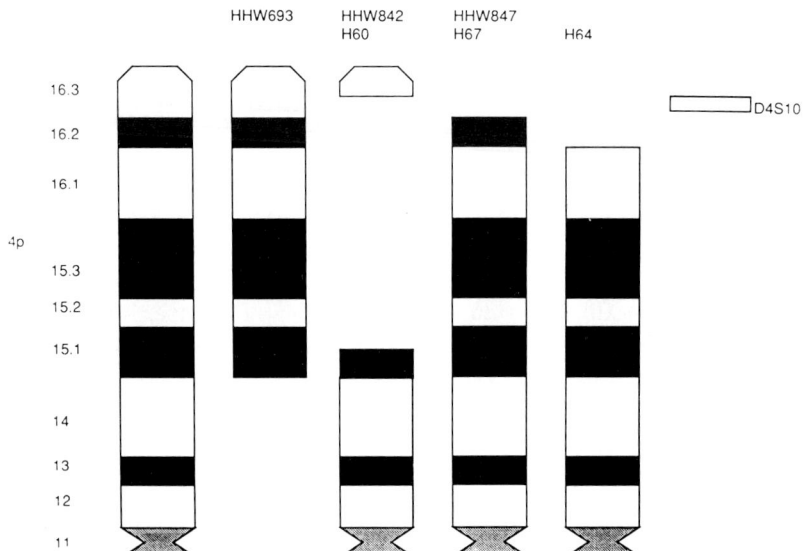

Figure 10.6 The chromosome 4 content of the somatic cell hybrids that have been used for localization of DNA sequences close to HD.

proved to be rather difficult, probably because 4p16.3 represents a very small fraction of chromosome 4, and also perhaps because of the biologically unusual nature of much of the DNA in this region (see below, in the discussion of telomeres). Two new markers proved to be particularly useful: D4S43 (probe C4H; Gilliam *et al.*, 1987) and D4S95 (probe 674; Wasmuth *et al.*, 1988). By looking at how these new loci were inherited in individuals where crossovers had been observed between HD and D4S10, and by making use of another marker (D4S62) that had been shown using somatic cell hybrids to lie proximal to D4S10, the order of markers was deduced to be as follows:

Centromere – D4S62 – D4S10 – (D4S95, D4S43,HD) – telomere

The brackets indicate that because no crossovers between these markers were observed, the relative order of HD, D4S43 and D4S95 could not be established. However it was clear that the HD gene was flanked by D4S10 and the telomere, and that the hunt for the gene could therefore be restricted to this region.

NEW TECHNOLOGY

Recently a number of strategies for the large-scale cloning and physical mapping of DNA have been developed. These include pulsed-field gel electrophoresis, chromosome hopping (or jumping) and yeast artificial chromosomes (YACs). Pulsed-field gel electrophoresis (PFGE) is a method that overcomes the limitation of normal agarose gel electrophoresis and allows very large DNA fragments, of up to at least 5 Mb length, to be separated (Schwartz and Cantor, 1984; Carle and Olson, 1984). Such large fragments can be generated using restriction enzymes whose recognition sequences are very infrequent in human DNA ('rare cutters'). Using more than 20 probes that were by then available, a long-range restriction map of the 4p16.3 region was constructed, largely the work of Dr Maja Bucan in the laboratory of Dr Hans Lehrach (Bucan *et al.*, 1990). A region of about 5 Mb was mapped, and the positions of the various probe loci defined. The map (see Figure 10.7) extends from the D4S10

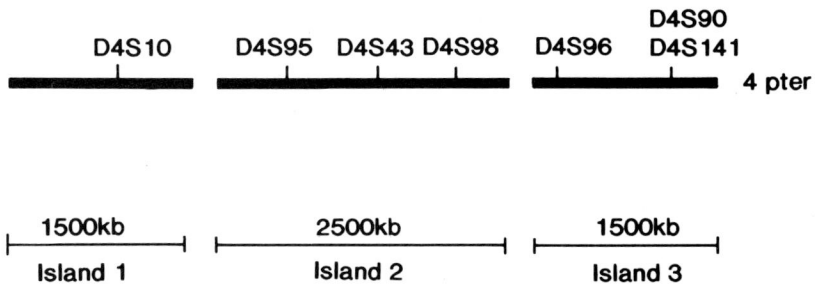

Figure 10.7 Long-range restriction map, constructed by pulsed-field gel electrophoresis, of the 4p16.3 region. The positions of a limited number of DNA markers are shown.

locus to the telomere, and has two gaps (unmapped regions) within it. The existence of a gap in such a map is due to the lack of a probe for the DNA in the gap. However, it is generally accepted that the gaps are unlikely to account for more than about 25% of the length of the entire region, since the distance from D4S10 to the telomere has been estimated at not more than 6 Mb. The three separate mapped segments, called 'islands', were oriented relative to each other by the use of genetic linkage analysis (MacDonald et al., 1989; Youngman et al., 1989) and somatic cell hybrids (MacDonald et al., 1987; Smith et al., 1988).

Related to PFGE is the group of techniques known as chromosome linking, jumping and hopping. These were pioneered independently by Drs Francis Collins and Hans Lehrach (Collins and Weissman, 1984; Poustka and Lehrach, 1986). What they all have in common is the use of specialized types of DNA library that contain sequences either from around rare cutter sites, or that are derived from both ends of a large rare-cutter DNA fragment. DNA probes of this type are of immense value in constructing a long-range restriction map, and in being able to proceed along a chromosomal region in a series of defined 'jumps' (or 'hops'). For example, a linking clone contains DNA from both sides of a rare-cutter restriction site, and can be used to link together the two adjacent fragments generated from the chromosome by the restriction enzyme and separated by PFGE. A clone from a jumping library contains sequences from both ends of a large DNA fragment generated by a rare-cutter restriction enzyme, and can be used together with a linking library to 'jump' along the chromosome.

A linking library was constructed using the rare-cutter enzyme NotI, and DNA from the hybrid cell line HHW693, which contains the HD region of chromosome 4 translocated onto part of human chromosome 5 (Pohl et al., 1988). The results of this study indicated a higher than expected number of NotI sites in the HD region, which may indicate a high density of genes, since the sites for enzymes such as NotI are often associated with the ends of genes (Lindsay and Bird, 1987). Further markers, isolated by chromosome jumping and by cloning the ends of rare-cutter restriction fragments, have been isolated and have contributed greatly to the physical mapping of the HD region (Whaley et al., 1988; Pohl et al., 1988; Richards et al., 1988; Bucan et al., 1990).

CONFLICTING RESULTS

It was felt that since it had now been shown that HD was between D4S10 and the telomere, and a physical map of the region had been constructed, it should be possible to jump to a point beyond the HD gene and hence narrow down the region of DNA in which it must be located. This would represent an attractive alternative to the attempts to isolate a flanking marker by screening large numbers of random DNA probes, a procedure that had by then only resulted in one likely flanking marker (the probe D5, locus D4S90; Youngman et al., 1989). Several jumping and linking clones were isolated, resulting in new RFLPs located along with D4S90 in the most distal part of the chromosome (Figure 10.7). Linkage studies using D4S90 and other markers in this distal region showed a recombination fraction with the disease of about 5%, and a slightly greater recombination fraction with D4S10. Because it was well established that HD was distal to D4S10 but with a recombination rate of only 4%, this result suggested that the new markers might be distal to HD. However when the markers were tested in the small number of HD crossover families a surprising result emerged. Instead of indicating a single location for the HD gene, the analysis of some crossovers favoured a location of HD distal to all the RFLPs (in the 300 kb of DNA between D4S90 and the telomere) whereas others suggested that HD should be somewhere within a larger proximal region of DNA, between D4S10 and markers close to D4S96 (Figure 10.7) (MacDonald et al., 1989; Robbins et al., 1989; Snell et al. (Cardiff), unpublished data). There are a number of possible explanations for this result, none of which is generally accepted. However, when the result emerged, the consensus of opinion was in favour of a telomeric (distal) location for the gene, since that was supported by a majority of the crossovers.

Several groups decided to try to clone the 4p telomere itself, because this would provide not only a definitive flanking marker but also a further point from which to jump. Since for technical reasons it is impossible to isolate the end of a chromosome using conventional cloning techniques, an alternative strategy had to be devised. The way in which this was achieved was extremely ingenious, and made use of a recently developed system for cloning mammalian DNA fragments in the common yeast, *Saccharomyces cerevisiae* (the yeast artificial chromosome or YAC system; Burke et al., 1987). The telomeres of a number of lower organisms had recently been isolated, and been shown to have a characteristic repeated DNA sequence that appeared also to be conserved in higher organisms. This suggested that a human telomere might be functional in a yeast cell, and that a hybrid YAC including the human 4p telomere at one end should be maintained in yeast cells. Such a hybrid molecule was recently constructed by Dr G. Bates in H. Lehrach's Laboratory; the YAC contains the terminal 115 kb of chromosome 4, and includes the region in which the HD gene is predicted to lie by one of the recombinants described above (Bates et al., 1990). The use of an RFLP right at the end of the chromosome should allow the question of the conflicting recombinants to be resolved, and the telomere YAC itself will allow the role of the telomeric sequences to be studied in relation to HD.

LINKAGE DISEQUILIBRIUM

Other workers were still not convinced that the HD gene was within 300 kb of the telomere, and started to try other approaches for fine genetic mapping. One such is the use of linkage disequilibrium, which is a property of a pair (or a small number) of genetic markers that are extremely close together on a chromosome, usually with a recombination fraction of less than 1%. If the distribution of alleles at each of these markers is studied in a large population of chromosomes, and found to be random, then the markers are said to be in genetic equilibrium. If on the other hand the pattern is not random, with a tendency for one particular allele at one locus to be associated more often than would be expected with a particular allele at the other locus, then the two loci are said to be in disequilibrium.

As well as close proximity, the two loci must have low mutation rates to show disequilibrium. This appears to be the case in HD, where new mutations are seldom reported. In the Cardiff laboratory it was decided to look for disequilibrium between HD and a selection of RFLPs from throughout the 4p16.3 region, using a large panel of HD families collected both for research purposes, and (with informed consent) for presymptomatic testing. In summary, disequilibrium was discovered between HD and two of the markers situated within the central region of the map (D4S95 and D4S98) but not with markers located at either end of the map (Snell et al., 1989). Essentially similar results were reported from Dr M. Hayden's laboratory in Vancouver, Canada, based on a sample of families of mainly British origin (Theilman et al., 1989). These results favour the hypothesis that the HD gene is located centrally within the mapped region, and not very close to the telomere. Linkage disequilibrium has now been studied using a further series of DNA markers and HD families from areas other than Great Britain; unfortunately this has not so far provided a more exact gene localization, possibly because of the existence of more than one common HD mutation (MacDonald et al., Snell et al., unpublished data).

CURRENT RESEARCH EFFORTS

As we have already seen, the use of recombination events in family studies has not resolved the issue of the exact location of the HD gene. If the terminal location is correct, then the gene should be present in the 115 kb of DNA contained in the 4p telomere YAC clone (Bates et al., 1990) and this possibility is being actively pursued. If on the other hand the more proximal location is correct, then a much larger region of DNA, region of 2.5 MB of DNA (Bates et al., 1991), remains to be studied (Figure 10.7). For several of the research groups the emphasis in the search for the HD gene has now moved away from the telomeric region and toward the area between and around the markers D4S95 and D4S98. It is hoped that further disequilibrium studies will help to narrow its location. It is not understood why the evidence from crossovers between HD and the marker loci was confusing, and all those involved in the search for the gene are in agreement that there is 'something unusual going on'. The eventual isolation of the gene should certainly throw some light on this matter, and may well tell us something new about the nature of human genetics, as well as about the disease.

It is fair to say that it is proving more difficult to proceed from the original discovery of linkage, to the isolation of the gene itself, than was generally expected. The reason for this is the apparently unusual genetic behaviour of the end of chromosome 4p in which the gene is situated, and the great difficulty encountered in trying to define a flanking marker. The results of recent studies have narrowed down the location of the HD gene to an area which is still big enough to contain 40 or more genes. Indeed it appears that genes may be up to four times as densely packed in the HD region as they are in an average stretch of chromosome, if the density of rare-cutter restriction sites is taken into account (see above). As well as continuation of the studies described above to try to reduce further the area to be searched, many workers are collaborating to assemble a complete set of cloned DNA sequences, using phage, cosmid and YAC systems, representative of the chromosomal region in which the gene is now thought to be. This will enable a systematic search for candidate genes to begin. All of the coding sequences within the region will have to be identified, both by the use of rare-cutter restriction sites (see above, 'New Technology') and also by looking for sequence conservation in animals other than humans. One of the most distinctive features of genes, as opposed to non-coding ('junk') DNA, is the fact that their base sequences are usually very similar in different species, especially those that are closely related. Sequence conservation is readily studied by the use of 'zoo blots', whereby a DNA sequence from one species (in this case human) is hybridized to DNA samples from a number of other species. A significant degree of hybridization indicates similarity between the DNA sequences of the different species.

CANDIDATE GENES AND THE NATURE OF THE HD MUTATION

Although it is quite obvious that HD is a disease of the brain, it is not necessarily the case that the gene causing it must be expressed in brain cells. The characteristic pathology of HD is the death of certain types of neuron (medium-sized spiny neurons) in the caudate and putamen (Chapter 5). This could be the result of a defect in a gene expressed within the neurons, or could be caused by a toxic effect of a gene expressed inappropriately elsewhere. Candidate genes for HD are isolated from cDNA libraries, which are collections of cloned DNA sequences that have been derived from messenger RNA, and therefore represent only expressed genes. This approach requires choices to be made concerning not only which tissue is used to make the mRNA, but also at what stage of development, and whether from a normal person or a HD patient (if available). In practice the choice can be no better than an educated guess, because of our total ignorance of the nature of the HD gene product.

At present, normal adult or fetal brain is the tissue being generally used for the isolation of HD candidate cDNA clones. This approach makes the assumption that the gene is expressed in normal brain, and that HD is the result of a mutation in a brain gene leading to its dysfunction. It is not known what the nature of this mutation might be, either at the molecular level (i.e. whether it is a single base change in a coding sequence or regulatory region, or a deletion of a stretch of DNA) or at the functional level. It is unlikely to be loss of an enzyme because HD is a true dominant condition, in which homozygotes for the disease gene are no more severely affected than heterozygotes (Wexler et al., 1987; Myers et al., 1989). More likely is a 'gain of function' mechanism, whereby a gene product acquires a new and deleterious function by mutation, possibly because it becomes deregulated and is expressed in an inappropriate manner. Laird (1990) has proposed a model for the nature of the HD mutation that seeks to explain the true dominance of the disease allele. The model, which is called 'position-effect variegation', is based on observations of a particular class of mutation in the fruit-fly, *Drosophila melanogaster*. Inactivation of the wild-type allele is caused by the spreading of proteins normally associated with heterochromatin (genetically 'silent' regions), and this effect can occur in *trans*, i.e. a mutant allele can affect the activity of the wild-type allele on the homologous chromosome. It is proposed that the HD mutation is not a single base change but an alteration in the DNA structure that places the gene close to a heterochromatic region, for example, the telomere. Pairing of the mutant and normal alleles would allow the inhibitory effect of the heterochromatic proteins now associated with the mutant allele to spread to the normal homologue. Thus, the phenotype of the heterozygote would be no different from that of the HD homozygote.

As well as being a truly dominant condition, HD has another unusual genetic feature, namely the parental origin effect. The affected offspring of affected fathers tend to have an earlier age-at-onset than those with affected mothers, in some cases showing a juvenile-onset form of the disease (Merrit et al., 1969). Two explanations for this phenomenon have been proposed. The first is genomic imprinting, whereby the paternal and maternal genetic contributions to the offspring are not functionally equivalent. The experimental evidence for this theory comes largely from studies of the mouse, in which it is possible to construct strains that have received both copies of a certain chromosomal region from one parent, rather than one copy from each parent as is normal. These strains show a variety of phenotypic effects, depending on the chromosomal region involved, ranging from nothing through growth abnormalities to lethality. Reik (1988) has proposed that germline modification of the HD gene, leading possibly to an earlier or higher level of expression of the gene product when paternally derived, could account for the parental origin effect. Because not all HD fathers give rise to juvenile-onset offspring, it is necessary to assume that the extent or degree of imprinting is variable. However, the imprinting hypothesis does not account for the occasional observations of mothers with juvenile-onset offspring. In order to test the hypothesis, it is necessary to propose a molecular mechanism for the imprinting effect, but it is not clear at present what this mechanism might be. DNA methylation has been proposed, and has the advantage of being easy to analyse using methylation-sensitive restriction enzymes. These kinds of studies will be more accessible once the gene has been isolated.

The second possible explanation for the parental-origin effect is to propose the existence of an X-linked, recessive modifier gene (Laird, 1990). The wild-type allele of this gene would have no effect on the age at onset of the HD offspring, whereas the proposed mutant allele would cause the disease to have the juvenile-onset form. If the frequency of the mutant allele is f, then the model predicts that only a proportion (f) of HD fathers and also a much smaller (f^2) proportion of HD mothers will give rise to juvenile-onset offspring. Laird (1990) claimed that this prediction fits well with the observed data.

Since it has now been shown that there is linkage disequilibrium in HD, it may be possible to use this approach to help test candidate genes. Any RFLP detected by an HD candidate should show disequilibrium with the HD phenotype at least as strong as that shown by the markers D4S95 and D4S98. The ultimate test would be the discovery of a polymorphism that appears only in HD chromosomes and never in normal ones. This could be regarded as '100% disequilibrium'. At the present time it is not possible to say whether there is more than one mutation resulting in HD; further disequilibrium studies might help to resolve this, but the evidence is compatible with a number of different scenarios. The fact that disequilibrium is observed at all suggests that there are only one or two common mutations, with the possibility of a number of additional rare ones. The recent isolation of the cystic fibrosis gene was greatly aided by the fact that there is one common mutation, a three base-pair deletion, that accounts for 60–70% of all CF chromosomes (Kerem et al., 1989). Similarly, in the autosomal dominant neuromuscular disease myotonic dystrophy there appears to be one common mutation, with a frequency of around 60%, and possibly other rarer mutations also (Harley et al., 1991).

The finding of a mutation in a gene sequence that is exclusively associated with the HD phenotype would constitute the strongest genetic evidence for that gene being HD itself. How to prove it from the functional point of view is not so obvious. As we have already seen there is no direct test for the HD gene product, nor any animal model. It has been reported that injection of kainic acid, which is an analogue of the neurotransmitter glutamic acid, into rat brains caused lesions and degeneration very similar to that observed in HD, although other workers have disputed the significance of these findings (Davies and Roberts, 1987) (see Chapter 5). This suggests that an animal model for HD could be created. A candidate gene could be introduced into a rat or mouse embryo, creating a transgenic animal. If the gene were expressed appropriately, then the effects of creating different mutations in it on the phenotype of the animal could be studied. The HD mutation is dominant and would be expected to show its effects in the presence of normal copies of the equivalent host-animal gene. An animal system such as this could be used to test various potential therapies for HD, as well as being invaluable for investigating the molecular pathology of the disease.

REFERENCES

Bates GP, MacDonald ME, Baxendale S et al. (1990) A yeast artificial chromosome telomere clone spanning a possible location of the Huntington disease gene. *American Journal of Human Genetics* **46**:762–775.

Bates GP, MacDonald ME, Baxendale S et al. (1991) Defined physical limits of the Huntington's Disease gene candidate region. *American Journal of Human Genetics* (in press).

Beckman L, Cedergren B, Mattsson B and Ottosson JO (1974) Association and linkage studies of Huntington's chorea in relation to 15 genetic markers. *Hereditas* **77**:203–211.

Bell J and Haldane JBS (1937) The linkage between the genes for colour-blindness and haemophilia in man. *Proceedings of the Royal Society* **123B**:119–150.

Botstein D, White RL, Skolnick M and Davis RW (1980) Construction of a genetic linkage map in man using restriction fragment length polymorphisms. *American Journal of Human Genetics* **32**:314–331.

Brackenridge CJ, Case J, Chin E, Prospect DN, Teltscher B and Wallace DC (1978) A linkage study of the loci for Huntington's disease and some common polymorphic markers. *Annals of Human Genetics* **48**:203–211.

Bucan M, Zimmer M, Whaley WL et al. (1990) Physical maps of 4p16.3, the area expected to contain the Huntington disease mutation. *Genomics* **6**:1–15.

Burke DT, Carle GF and Olson MV (1987) Cloning of large segments of exogenous DNA into yeast by means of artificial chromosome vectors. *Science* **236**:806–812.

Carle GF and Olson MV (1984) Separation of chromosomal DNA molecules from yeast by orthogonal-field alternation gel electrophoresis. *Nucleic Acids Research* **12**:5647–5664.

Collins FS and Weissman SM (1984) Directional cloning of DNA fragments at a large distance from an initial probe: A circularisation method. *Proceedings of the National Academy of Science USA* **81**:6812–6816.

Conneally PM, Haines J, Tanzi R et al. (1989) No evidence of linkage heterogeneity between Huntington disease (HD) and G8 (D4S10). *Genomics* **5**:304–308.

Cox DR, Pritchard CA, Uglum E, Casher D, Kobori J and Myers RM (1989) Segregation of the Huntington disease region of human chromosome 4 in a somatic cell hybrid. *Genomics* **4**:397–407.

Davies, SW and Roberts PJ (1987) No evidence for preservation of somatostatin-containing neurons after intrastriatal injections of quinolinic acid. *Nature* **327**:326–329.

Folstein SE, Phillips JA, Meyers DA et al. (1985) Huntington's disease: Two families with differing clinical features show linkage to the G8 probe. *Science* **229**:776–779.

Gilliam TC, Bucan M, MacDonald ME et al. (1987) DNA segment encoding two genes very tightly linked to Huntington's disease. *Science* **238**:950–952.

Gusella JF, Wexler NS, Conneally PM et al. (1983) A polymorphic DNA marker genetically linked to Huntington's disease. *Nature* **306**:234–238.

Gusella JF, Tanzi RE, Bader PI et al. (1985) Deletion of Huntington's disease-linked G8 (D4S10) locus in Wolf–Hirschhorn syndrome. *Nature* **318**:75–78.

Harley HG, Brook JD, Floyd J et al. (1991) Detection of linkage disequilibrium between the myotonic dystrophy locus and a new polymorphic DNA marker. *American Journal of Human Genetics* (in press).

Hodge SE, Spence MA, Crandall BF et al. (1980) Huntington disease: linkage analysis with age-of-onset corrections. *American Journal of Medical Genetics* **5**:247–254.

Kerem B, Rommens JM, Buchanan JA et al. (1989) Identification of the cystic fibrosis gene: genetic analysis. *Science* **245**:1073–1080.

Laird CD (1990) Proposed genetic basis of Huntington's disease. *Trends in Genetics* **6**:242–247.

Landegent JE, in de Wal NJ, Fisser-Groen YM, Bakker E, van der Ploeg M and Pearson PL (1986) Fine mapping of the Huntington disease linked D4S10 locus by non-radioactive *in situ* hybridisation. *Human Genetics* **73**:354–357.

Leese SM, Pond DA and Shields J (1952) A pedigree of Huntington's chorea with a note on linkage by RR Race. *Annals of Eugenics* **17**:92–112.

Lindsay S and Bird AP (1987) Use of restriction enzymes to detect potential gene sequences in mammalian DNA. *Nature* **327**:336–338.
Lindstrom JA, Bias WB, Schimke RN et al. (1973) Genetic linkage in Huntington's chorea. *Advances in Neurology* **1**:203–208.
MacDonald ME, Anderson MA, Gilliam TC et al. (1987). A somatic cell hybrid panel for localising DNA segments near the Huntington's disease gene. *Genomics* **1**:29–34.
MacDonald ME, Haines JL, Zimmer M et al. (1989) Recombination events suggest potential sites for the Huntington's disease gene. *Neuron* **3**:183–190.
Magenis RE, Gusella J, Weliky K et al. (1986) Huntington disease-linked RFLP localised within band p16.1 of chromosome 4 by *in situ* hybridisation. *American Journal of Human Genetics* **39**:383–391.
Merrit AD, Conneally PM, Rahnan NF and Drew AL (1969) Juvenile Huntington's chorea. In: *Progress in Neurogenetics* (eds A Barbeau and TR Brunette), pp. 645–650. Amsterdam: Excerpta Medica.
Monaco AP and Kunkel LM (1987) A giant locus for the Duchenne and Becker muscular dystrophy gene. *Trends in Genetics* **3**:33–37.
Myers RH, Leavitt J, Farrer LA et al. (1989) Homozygote for Huntington disease. *American Journal of Human Genetics* **45**:615–618.
Pericak-Vance MA, Conneally PM, Merritt AD, Roos R, Norton JA and Vance JM (1978). Genetic linkage studies in Huntington's disease. *Cytogenetics and Cell Genetics* **22**:640–645.
Pohl TM, Zimmer M, MacDonald ME et al. (1988) Construction of a NotI linking library and isolation of new markers close to the Huntington's disease gene. *Nucleic Acids Research* **16**:9185–9198.
Poustka A and Lehrach H (1986) Jumping libraries and linking libraries: the next generation of molecular tools in mammalian genetics. *Trends in Genetics* **2**:174–179.
Reik W (1988) Genomic imprinting: a possible mechanism for the parental origin effect in Huntington's chorea. *Journal of Medical Genetics* **25**:805–808.
Richards JE, Gilliam TC, Cole JL et al. (1988) Chromosome jumping from D4S10 (G8) toward the Huntington disease gene. *Proceedings of the National Academy of Science USA* **85**:6437–6441.
Robbins C, Theilmann J, Youngman S et al. (1989) Evidence from family studies that the gene causing Huntington disease is telomeric to D4S95 and D4S90. *American Journal of Human Genetics* **44**:422–425.
Rommens JM, Iannuzzi MC, Kerem B et al. (1989) Identification of the cystic fibrosis gene: chromosome walking and jumping. *Science* **245**:1059–1065.
Ruddle FH (1984) Reverse genetics and beyond. *American Journal of Human Genetics* **36**:944–953.
Schwartz DC and Cantor CR (1984) Separation of yeast chromosome-sized DNAs by pulsed field gradient gel electrophoresis. *Cell* **37**:67–75.
Smith B, Skarecky D, Bengtsson U, Magenis RE, Carpenter N and Wasmuth JJ (1988) Isolation of DNA markers in the direction of the Huntington disease gene from the G8 locus. *American Journal of Human Genetics* **42**:335–344.
Snell RG, Lazarou LP, Youngman S et al. (1989) Linkage disequilibrium in Huntington's disease: an improved localisation for the gene. *Journal of Medical Genetics* **26**:673–675.
Theilman J, Kanani S, Shiang R et al. (1989) Non-random association between alleles detected at D4S95 and D4S98 and the Huntington's disease gene. *Journal of Medical Genetics* **26**:676–681.
Volkers WS, Went LN, Vegter van der Vlis M, Harper PS and Caro A (1980) Genetic linkage studies in Huntington's chorea. *Annals of Human Genetics* **44**:75–79.
Wang HS, Greenberg CR, Hewitt J, Kalousek D and Hayden MR (1986) Subregional assignment of the linked marker G8 (D4S10) for Huntington disease to chromosome 4p16.1–16.3. *American Journal of Human Genetics* **39**:392–396.
Wasmuth JJ, Carlock LR, Smith B and Immken LL (1986) A cell hybrid and recombinant DNA library that facilitate identification of polymorphic loci in the vicinity of the Huntington disease gene. *American Journal of Human Genetics* **39**:397–403.
Wasmuth JJ, Hewitt J, Smith B et al. (1988) A highly polymorphic locus very tightly linked to the Huntington's disease gene. *Nature* **332**:734–736.
Wexler NS, Conneally PM, Housman D and Gusella JF (1985) A DNA polymorphism for Huntington's disease marks the future. *Archives in Neurology* **42**:20–24.
Wexler NS, Young AB, Tanzi RE et al. (1987) Homozygotes for Huntington's disease. *Nature* **326**:194–197.
Whaley WL, Michiels F, MacDonald ME et al. (1988) Mapping of D4S98/S114/S113 confines the Huntington's defect to a reduced physical region at the telomere of chromosome 4. *Nucleic Acids Research* **16**:11769–11780.

Youngman S, Sarfarazi M, Quarrell OWJ et al. (1986) Studies of a DNA marker (G8) genetically linked to Huntington disease in British families. *Human Genetics* **73**:333–339.

Youngman S, Sarfarazi M, Bucan M et al. (1989) A new DNA marker (D4S90) is located terminally on the short arm of chromosome 4, close to the Huntington disease gene. *Genomics* **5**:802–809.

Zabel BU, Naylor SL, Sakaguchi AY and Gusella JF (1985) Regional localisation of a DNA polymorphism (D4S10) linked to Huntington's disease at 4p16–p15. *Cytogenetics and Cell Genetics* **40**:787.

Zweig RM, Koven SJ, Hedreen JC, Maestri NE, Kazazian HH and Folstein SE (1989) Linkage to the Huntington's disease locus in a family with unusual clinical and pathological features. *Annals of Neurology* **26**:78–84.

11

Genetic Counselling in Huntington's Disease

INTRODUCTION

By devoting a specific chapter to this topic, the authors hope to emphasize their view that genetic counselling is probably the most important single factor in the management of a family with Huntington's Disease (HD). Until recently one might have added – and the most neglected – but this has changed rapidly in recent years with the increased awareness of the genetic aspects of the disorder and of genetics in general, together with the more widespread availability of specialist genetic counselling centres. For those involved regularly in genetic counselling, HD often represents a combination of most of the difficulties that are encountered individually in other disorders. It is a challenging and stressful area of work, with no two families ever the same and new problems continually arising.

Genetic counselling is so central to the management of HD that all clinicians involved with patients and families need to be fully aware of its scope, its limitations and pitfalls. Even if they prefer to refer families to a clinical geneticist, or are fortunate enough to have such a person as part of a team involved with HD management, genetic questions will arise frequently and often unexpectedly when patients and relatives are being seen for other reasons. The way in which such questions are handled, even in a preliminary manner, can often deeply influence the attitude of family members to fuller genetic counselling later.

This chapter looks first at some of the basic aims and features of genetic counselling in general, most of which are as relevant to HD as to other disorders. Some of the particular problems encountered in genetic counselling for HD are then examined, and finally the rather small number of objective studies relating to the subject analysed.

THE PRINCIPLES OF GENETIC COUNSELLING

It is worth beginning with a definition of genetic counselling before proceeding further: 'Genetic Counselling is the process by which patients or relatives at risk of a disorder that may be hereditary are advised of the consequences of this disorder, the probability of developing or transmitting it, and of the ways in which this may be prevented or ameliorated' (Harper, 1988). From this definition it is clear that we are dealing with a complex process, containing a number of different elements which must be carefully integrated if genetic counselling is to be satisfactory. These elements can be summarized as:

(1) Diagnosis. Without a firm and accurate diagnosis genetic counselling will inevitably be less than definitive, often provisional and at worst erroneous.
(2) Pedigree documentation and determination of the pattern of inheritance that is operating in the family.
(3) Genetic risk estimation for relatives.
(4) Communication of the information to those seeking advice.
(5) Information and support regarding available preventive measures, treatment and other aspects.

It is also worth emphasizing at this stage what genetic counselling is *not*, since families and some clinicians may be confused on what are its aims. First, it is not synonymous with the prevention of genetic disease, although genetic counselling may be an important element in any programme of prevention, as discussed later; it is an individual activity, related primarily to the problems of particular people and families. Its success or failure should thus not be judged in terms of how these individuals have been helped with their problems, rather than in terms of births prevented or a reduction in disease prevalence. Genetic counselling is not, or at least should not be, directive but aims to help individuals and families reach decisions that are best for themselves. Finally, genetic counselling is not simply a mathematical calculation of risks; any numerical risk has to be set in the context of the severity of the disorder, how it is perceived by the family, and what measures are available for treatment and prevention.

In none of these aspects does genetic counselling for HD differ significantly from genetic counselling for other disorders. For this reason it is helpful for those whose genetic counselling activities are largely confined to HD to gain broader experience of the process as it applies to other diseases. The books of Kelly (1986) and Harper (1988) give general information on the topic. The special problems of genetic counselling for HD now need consideration.

DIAGNOSTIC DIFFICULTIES IN RELATION TO COUNSELLING

In many (probably most) of the families seen for genetic counselling in relation to HD, the diagnosis is all too certain; one frequently has to disappoint relatives who have been hoping against the odds that the true diagnosis is something else; occasionally it can be difficult to convince such individuals that there really is no significant doubt. Often, however, the diagnosis of HD is far from clear cut; this may create serious, sometimes insuperable problems for those involved in counselling. Some of the diagnostic difficulties encountered in HD have already been discussed in Chapter 2, but common problems include the following:

(1) All the affected individuals are dead and no neuropathological or other definitive details are available.
(2) The person seeking advice has migrated or has lost touch with their family and the whereabouts of any affected members are unknown.
(3) The affected individual appears to represent an isolated case.
(4) The disorder in the family may appear to be dominantly inherited, but the clinical features are not typical for HD.
(5) The family disorder is clearly HD but the individual concerned (or the affected parent) appears to have a different condition. Could this still be HD?

The first two problems can often be resolved with persistence, but the work involved can be considerable. Death certificates, the tracing of old hospital records (those of psychiatric hospitals are especially likely to be kept), and the linking of a new family into a well-established HD pedigree may all be possible solutions. Some countries, notably Scandinavia, have well-established civic and church records extending over several centuries. Likewise it is usually possible to trace individuals or records relating to HD anywhere in the world through the close network of those professionals involved with the disorder, especially when a region has a well-maintained genetic register. The special problems in genetic counselling for the isolated case of HD are considered later in more detail, but here it must be noted that no case should be accepted as isolated until all avenues have been explored to exclude HD in a relative.

The family where genuine uncertainty exists regarding the diagnosis represents a major problem, especially if no neuropathology has been obtained. In the past it was possible to minimize the implications for genetic counselling by taking the view that the genetic risks were likely to be the same even though the disorder might prove to be something else, but the advent of predictive testing has now made it crucial to obtain firm evidence that the disorder really is HD. As discussed in Chapter 9, some atypical families have already been shown to be linked to the HD locus. Unfortunately many families will be too small or too fragmented for this to be achieved, but the least one can do is to store DNA on any affected members and try to ensure a detailed neuropathological study when one of them dies.

Families with HD are not immune to the occurrence of other disorders, but where an apparently distinct neurological or psychiatric disorder occurs in a person at 50% genetic risk, the possibility of the condition representing atypical HD must be weighed carefully against the supposition that two distinct diseases are present. Many clinicians are still unfamiliar with the range of variation to be found in HD, especially in juvenile cases. Again neuropathology where possible and DNA storage for the future are important to bear in mind.

Most people seen for genetic counselling regarding HD are the symptomless children of an affected patient, seeking reassurance for themselves and information regarding the risks for their children, present or future. Since the genetic risks given will depend on the assumption that the individual concerned is clinically unaffected at the time seen, it is essential to be sure that this is indeed the case. A careful neurological examination is important and a normal result gives both clinician and person examined a confident foundation on which to base further discussions. However, it is rare to find clear-cut formal neurological abnormalities when the general interview has given no suspicion; much more often it becomes clear during general conversation and the taking of family details that the person is actually affected, but is unaware of it.

This difficult situation is encountered more than occasionally by all those involved in genetic counselling for HD. It has proved to be frequent in those requesting presymptomatic testing, as discussed in Chapter 12. It poses a particularly traumatic situation for all concerned, for the individual has come asking one question – what is the risk of developing HD in the future? – but is faced with an adverse answer to a question that was probably not even raised i.e. the disease is present now.

No easy solution can be given as to how to cope best with this situation, but a few suggestions are given, based on our own experience, which may be helpful to others, though they may not agree with them:

(1) Avoid communicating the view that the person is affected at the initial interview, even if asked directly; it would seem unreasonable to give such important information as a complete stranger and after a single assessment, especially if circumstances have dictated a less than thorough examination.
(2) Attempt to find out from the patient and family whether they have noticed any symptoms. It may become apparent that requesting genetic counselling has been the only way by which medical attention could be sought for symptoms that are recognized but not yet accepted. Again this information may only be obtained at a second interview.
(3) Suggest more detailed neurological assessment before disclosing the diagnosis. This may well not show anything that is not already apparent clinically but it will help in exclusion of other (possibly treatable) disorders and again will give time for accommodation to the situation.

Colleagues may sometimes suggest that it is preferable not to tell a person that he or she is affected until a later stage. The broader aspects of disclosing the diagnosis are discussed elsewhere, but our own policy has been to be open and frank about it once it is definite and once there has been adequate time to adjust to the possibility of being affected. Sometimes, when the patient has clearly indicated that they do not wish to know the results of any examination or investigation, only close family members are informed.

A related and even more difficult problem encountered in genetic counselling is the at-risk individual who is not clearly affected but nevertheless shows intangible suspicious features. Most people working with HD families (not just medical staff) develop remarkably sensitive 'antennae' for detecting the individual who is 'not quite right'. It is common for this not to be accompanied by any firm neurological abnormality, though equivocal abnormalities may be found and the Quantitative Neurological Examination may show changes; the best one can do is to note one's concern, to emphasize that the genetic risks for offspring will be dependent on a person's own status, and try to maintain contact with the family.

PEDIGREE ANALYSIS AND RISK ESTIMATION

For HD, even more than for other genetic disorders, there is no substitute for the construction of a full and detailed pedigree as part of a genetic counselling referral. Given the amount of time and effort involved, it may well be wise to undertake this in a preliminary session, especially if one is fortunate enough to have a trained fieldworker who can make a home visit beforehand. Such a visit not only allows family details to be collected by talking to elderly relatives who may not be able or wish to attend a formal genetic counselling session, but it also allows any fears or worries about what genetic counselling may entail to be dispelled and can alert one to any particular problems that exist in the family or which they may wish to bring up.

The value of a systematic and accurately drawn pedigree is not only that it will most likely confirm the autosomal dominant inheritance expected for HD, but also that it will provide information on other branches of the family that are at significant risk and who may be seen for genetic counselling subsequently. Often individuals will be identified by drawing up a pedigree who are suspected by relatives to be affected, but who have not yet been diagnosed. Even though medical information may not be accurate on older generations, their very existence may be the clue that allows two different HD kindreds to be linked together through a common ancestor. Even if detailed information on the kindred as a whole is of no immediate importance to that branch seeking counselling, it may well prove of great value in the future to other family members. In short, a full and accurate pedigree is the cornerstone of good medical practice in genetic counselling, in HD as for other inherited disorders.

With care and practice it should not be difficult to construct a 'working pedigree' (Figure 11.1) in a reasonably short time, which does not require to be redrawn later to look respectable. Such good resolutions rarely happen in practice, so it is preferable to draw the pedigree well in the first place. A rule, pen and paper have not yet been superseded by the computer in this respect! Once such a pedigree has been constructed, it should be remembered that it represents a valuable, confidential and highly sensitive document. It should not be left lying around, photocopied unnecessarily, or sent to colleagues other than professionals who will respect it similarly. Likewise it is best not placed in the patient's hospital file, which relates primarily to the person's own medical situation. Our policy is to keep the pedigree in the family genetic file, with a

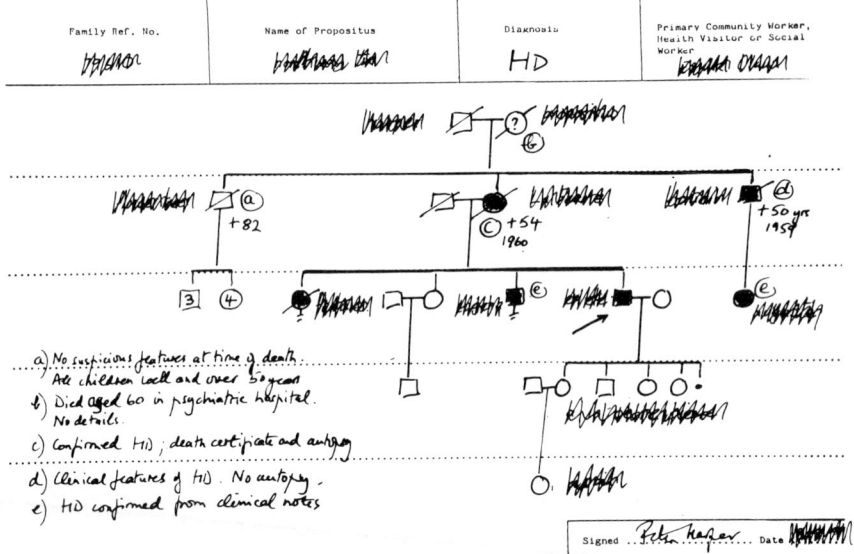

Figure 11.1 A 'working pedigree' drawn for a family with HD (identifying details deleted). With practice an accurate pedigree of this type can be drawn rapidly at a clinic or home visit; brief clinical details can be added in the margin. The pedigree should be signed and dated for future reference.

copy of relevant correspondence in the hospital notes. If the person involved in genetic counselling is also involved in HD research, then a clear method of separating service from research data must be decided upon, so that research workers do not have inappropriate access to counselling data. Such a separation may not always be necessary, but it must be carefully considered, especially if research data are likely to be made available to other centres. The general issue of confidentiality is discussed further below when the topic of registers is considered.

Once a clear pattern of autosomal dominant inheritance has been established for an HD pedigree, the estimation of basic risk for the children of affected persons is simple – since essentially all HD patients are heterozygous for the gene, having one normal as well as one abnormal copy, the chance of a child receiving either will be equal, giving a risk of one half or 50%, regardless of the sex of parent or child (see Figure 9.1). However, in reality the actual risk given may differ considerably from this and will be critically dependent on the age of the person concerned.

The shape of the age-at-onset curve for HD has already been discussed in Chapter 4, as has the need to correct the directly observed curve by a life table or other method to take into account those individuals not likely to have been ascertained and thus obtain the probability of an HD heterozygote showing the disorder at a particular age. By combining this age-at-onset curve (Figure 4.3) with the prior risk of 50%, a comparable curve can be constructed (Figure 11.2)

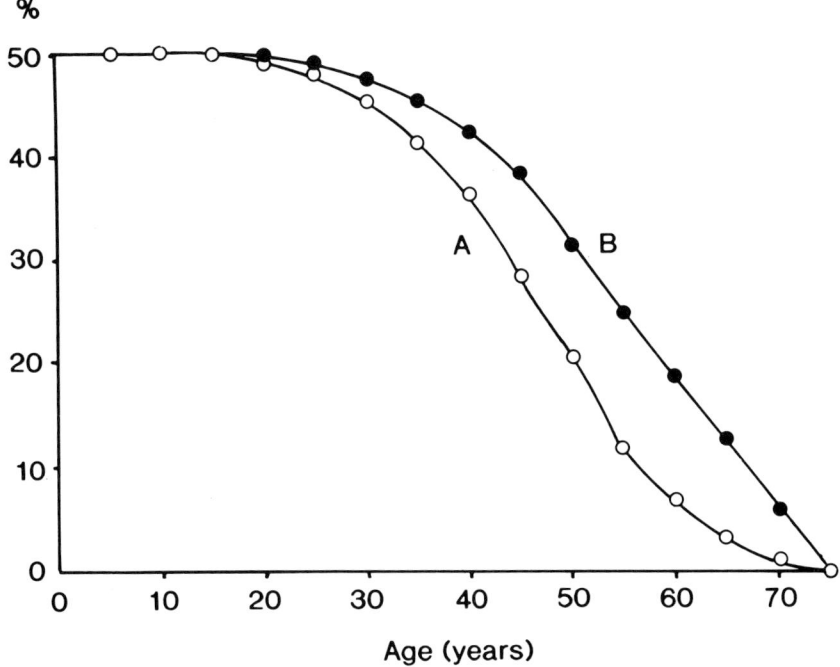

Figure 11.2 Risk of an unaffected individual at 50% prior risk for HD, carrying the HD gene at a particular age (see also Table 11.1). Curve A is based on the uncorrected age-at-onset curve given in Figure 4.3, while curve B is based on life-table analysis, and is more appropriate for genetic counselling. Note the considerable difference between the two curves particularly for older individuals. Data from the South Wales study kindly re-analysed by Dr Robert Newcombe.

which gives the risk for the offspring of an HD parent carrying the HD gene according to age. This risk can be read easily from the curve or from a table (Table 11.1) for any age required. It can be seen that the curves differ considerably from the uncorrected age-at-onset curve, which should not be used for genetic risk estimates.

In practice the risk does not decline significantly from the initial 50% until after 30 years of age (except for sibs of a juvenile case), so that those children of an HD parent wishing to make reproductive decisions will have to make these before there is any useful alteration. The risk curve is perhaps most useful in older age groups, where the question often is: how old do I have to be before I can be reasonably sure of remaining free from HD? It can be seen that the risk actually remains significant longer than many relatives expect, reflecting the higher than recognized frequency of the disorder in old age. It is possible that the curve could be unduly cautious in this respect, since it is based on age at definite onset, something that will certainly be several years away at least in an individual who is found to be totally normal on coming for genetic counselling. As discussed in Chapter 4, a recent study (Penney *et al.*, 1990) suggests that an

Table 11.1 Risk for a healthy individual at 50% prior risk of HD carrying the HD gene at different ages. Based on the life-table analysis (Newcombe, 1981) of South Wales data

Age (years)*	Risk (%)	Age (years)*	Risk (%)
20	49.6	47.5	34.8
22.5	49.3	50	31.5
25	49.0	52.5	27.8
27.5	48.4	55	24.8
30	47.6	57.5	22.1
32.5	46.6	60	18.7
35	45.5	62.5	15.2
37.5	44.2	65	12.8
40	42.5	67.5	10.8
42.5	40.3	70	6.2
45	37.8	72.5	4.6

* Figures are for the mid point of each 5-year grouping shown in Figure 11.2.

individual at risk with an entirely normal quantitative neurological examination has only a 3% chance of developing established HD over the next 3 years.

Should age at onset in a particular family be taken into account in risk estimation? The definite correlation of age at onset between parent and child has already been mentioned, and Stevens (1973) has drawn up a series of curves taking this into account. However the range of variation remains so great within kindreds that it seems unwise to alter the risk unless the kindred is so large that a specific age-at-onset curve can be based on it alone. For most families seen, this is impossible, or has such wide confidence limits as to be useless. The one situation where adequate evidence is available to depart from the general age curve is where a sib has already developed juvenile HD. Hayden et al. (1985), using data from the US Huntington roster, showed that no sibs of juvenile cases developed HD after the age of 45 years; at the ages of 30 and 40 years the risks were 33% and 5%, compared with 43% and 30% for families in general. This curve is shown in Figure 11.3. Clarke and Bundey (1990) have recently produced comparable data for sibships where onset in the proband was under 10 years of age.

Second-degree relatives, with an affected grandparent, are often seen for genetic counselling, since they may be planning marriage or childbearing when their healthy parent is still at an age of significant risk. Such individuals are still often given the simpler risk figure of 25%, based on half their healthy parent's prior risk. In fact their risk will invariably be much less than this, the actual figure depending not so much on their own age but on that of their parent. Figure 11.4 shows the curves for such individuals where the child is between 20 and 40 years younger than the parent at risk; Table 11.2 summarizes the data. In practice the risk for any grandchild under 30 years is close to half the parent's age-adjusted risk.

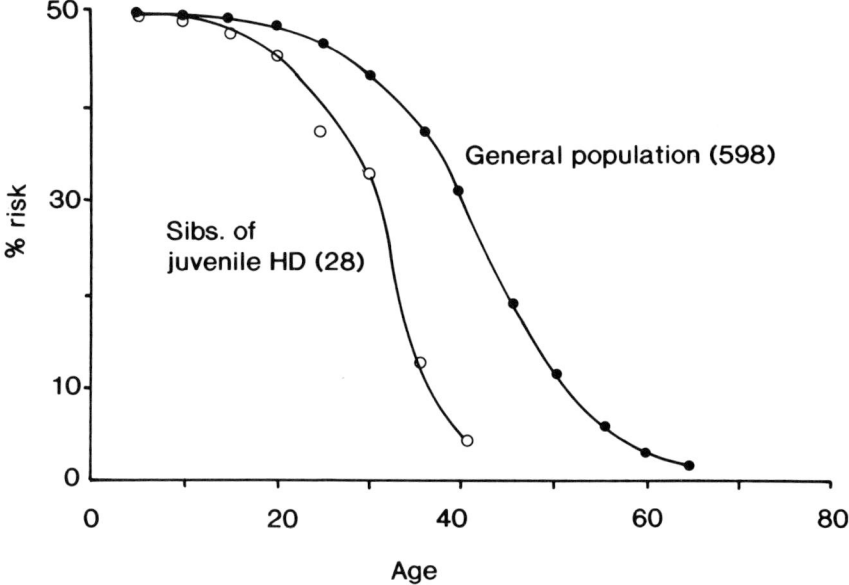

Figure 11.3 Risk curves for individuals at 50% risk of HD based on the US Huntington disease roster (redrawn with permission from Hayden *et al.*, 1985). Note the more rapid decline of risk in sibs of juvenile cases, based on the observed age of onset in this group. NB. These curves are based on uncorrected age at onset, not a life-table approach.

Not infrequently one sees individuals for counselling whose at-risk parent has died relatively young from a different cause. Wherever such a situation can be anticipated, it is vital that a detailed neurological examination is undertaken and that the brain should have detailed neuropathological study for minor changes of HD. Where this has not been done Table 11.2 can again be used for risk estimation in this situation for a second degree relative, the age of the parent being the age at death.

Once the third-degree relatives are reached, risk estimates have usually become negligible. Put simply, this means that by the time grandchildren are grown up and wanting to know their own risks, the grandparent will be past the age of significant risk for developing HD.

COMMUNICATION AND COUNSELLING

However skilled one may be in estimating the genetic risks, this is of little help if one cannot communicate them satisfactorily to the family one is seeing. Indeed it sometimes seems as if those most expert in thinking mathematically have particular difficulties in this respect. Conversely, it is probably even worse to be a good communicator of inaccurate information! The fact remains that ability to communicate with people of all levels of education and background is an essential element in genetic counselling. Sadly, communication skills remain greatly undervalued in medical education, and students can hardly be blamed for not acquiring them when they see their seniors pursuing their medical careers successfully without such skills. Even clinical geneticists, who like to consider themselves good communicators, often have received little or no formal training in counselling or communication; while much can be learned through experience this would be of greater value if based initially on formal teaching.

It is difficult to define precisely the characteristics that make for good communication in genetic counselling for HD: an unhurried approach, sensitive manner, rapport with the person counselled, willingness to allow the topics the person wishes to raise be discussed rather than dictate the flow of events, obvious interest in and concern for the person and their problems and, not least, experience in seeing HD patients and families, are but a few that come to mind.

Genetic counselling can be considerably helped (or hindered) by its setting. A room that is quiet, informal and comfortable – in other words quite unlike most hospital out-patient clinics – will help to dispel the very real fears that most families have when coming for genetic counselling. Absence of students (at least more than one) and the presence of the fieldworker who may have had previous contact are of particular help when seeing a family for the first time.

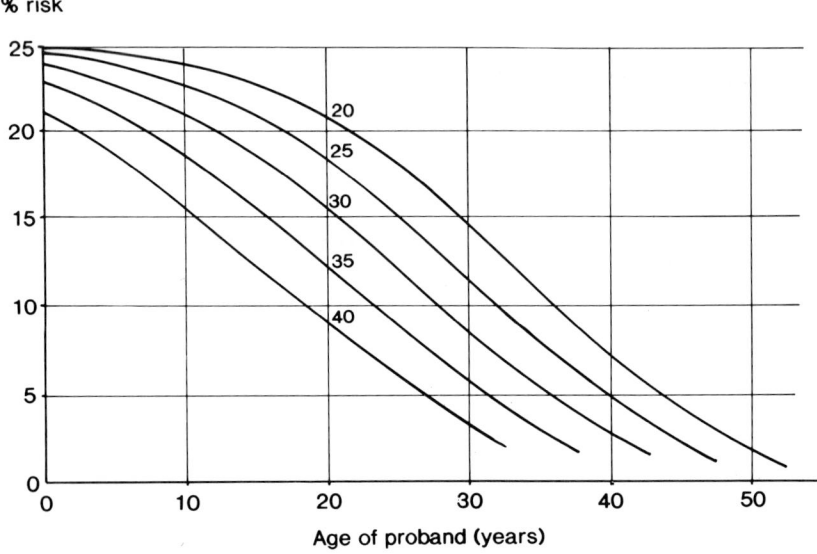

Figure 11.4 Risk curves for second-degree relatives. This series of curves shows the decline in risk of carrying the HD gene with age for a proband who is the grandchild of a patient with HD and who has an unaffected living parent. The different curves correspond to the age of the intervening parent at risk at the time the proband was born. A more detailed table is given in Table 11.2. Data are from the South Wales study, kindly re-analysed by Dr Robert Newcombe.

Table 11.2. Risk estimates (%) for second-degree relatives of a patient with HD

Age of child (years)	Age of parent (years)										
	20–	25–	30–	35–	40–	45–	50–	55–	60–	65–	70–
70–	1.6	1.5	1.5	1.4	1.2	1.0	0.8	0.6	0.4	0.3	0.1
65–	3.8	3.7	3.5	3.3	3.0	2.5	1.9	1.5	1.0	0.7	0.3
60–	5.6	5.4	5.2	4.9	4.3	3.7	2.8	2.2	1.5	1.0	0.4
55–	8.5	8.3	7.9	7.4	6.7	5.6	4.4	3.4	2.3	1.6	0.7
50–	11.2	10.9	10.5	9.8	8.9	7.5	5.8	4.6	3.1	2.1	0.9
45–	14.9	14.6	14.0	13.2	11.9	10.1	7.9	6.2	4.2	2.9	1.3
40–	18.1	17.8	17.0	16.1	14.6	12.5	9.8	7.7	5.3	3.7	1.5
35–	20.6	20.2	19.4	18.4	16.7	14.3	11.3	9.0	6.1	4.3	1.8
30–	22.2	21.8	20.9	19.9	18.1	15.5	12.3	9.8	6.7	4.7	2.0
25–	23.5	23.1	22.2	21.0	19.2	16.5	13.1	10.4	7.1	5.1	2.2
20–	24.2	23.7	22.8	21.7	19.7	17.0	13.6	10.8	7.4	5.2	2.3

The table shows the residual risk of a healthy second-degree relative carrying the HD gene at various combinations of age for the child and the intervening unaffected parent. Data from the South Wales life-table analysis (Newcombe, 1981), kindly recalculated by Dr Robert Newcombe.

Communication can often be hindered by specific factors which can sometimes be avoided. A family seen for counselling shortly after the diagnosis of HD has been made may well be too preoccupied with the immediate implication of the disease to take in details of inheritance. Likewise pregnancy is an unfortunate time to have to break the first news of risk for a genetic disorder such as HD; it is particularly difficult for a couple to weigh up the desirability of prenatal testing, or even of continuing the pregnancy, when they are faced with this situation. In our experience some couples coming for genetic counselling for the first time during pregnancy may have been urged to do so by well-meaning professionals or relatives; left to themselves they might not have sought further information or advice until the pregnancy had been completed. It is important for the counsellor to be aware of this possibility, so that having satisfied the couple's initial wish for information, more detailed discussion can be postponed until a more suitable time.

Most of these factors represent nothing more than common sense and consideration, yet it is surprising how often families are seen where these basic aspects have been ignored.

How detailed should one attempt to be in communicating risk estimates? Families vary considerably as to whether a precise figure is best given, or whether this is 'rounded off' to an approximation, or even given in general terms such as 'high' or 'low'. Regardless of this it is preferable to give an accurate figure in any letter summarizing the consultation, otherwise later misunderstanding may occur. Sometimes families are encountered who had apparently been told that 'it could not possibly happen' or that they have 'a chance of one in a million'; such certainty of reassurance is sadly rarely possible in genetic counselling.

SUPPORT FOLLOWING GENETIC COUNSELLING

Few families are able to understand fully the genetic implications of HD as a result of a single consultation. Questions almost always arise later that had not been thought of or which were considered less pressing at the time – or even too serious to be asked. The opportunity for a follow-up visit is thus strongly advisable; this will often be required anyway if records and other information have to be checked. A home visit by a fieldworker after a genetic counselling visit is often especially helpful in ensuring that the information given has actually been understood, while the opportunity to talk to a non-medical person who is already known may allow personal and sensitive family matters to be discussed more freely, such as possible illegitimacy, impending divorce, or fear of early symptoms in a spouse.

A written letter summarizing the genetic counselling information has already been mentioned; at the very least it will provide an address and telephone number for future contact. Valuable written information on HD has been produced by the various lay societies. This is often especially helpful in outlining in simple terms the range of problems that can be expected and what can be done to help. Whether a family seen for genetic counselling will actually wish to join such a society is a different matter. While some families find the support of these groups of the greatest help, others prefer not to join; the choice and timing should be left to the individuals concerned, without any pressure from professionals or other family members.

A genetic counselling visit frequently uncovers numerous unmet needs of the affected family members and those caring for them, which are quite unrelated to the genetic aspects. Many of these have been mentioned in Chapter 5, but it often falls to the genetic clinic, or whoever is undertaking counselling, to initiate contacts with social services departments or other bodies whose help may be needed to obtain benefit entitlements, house adaptations or other aids, and in handling the multiple social problems that the disorder often poses. The need for psychological and practical support in relation to genetic counselling in HD has been well discussed by Wexler (1984).

GENETIC COUNSELLING AND THE EXTENDED FAMILY

The difficulties involved in genetic counselling for HD are real enough when an individual or family have requested it; they can become even more difficult when the process is initiated by others. With large families it is frequently found that one branch actively seeks counselling, leading to the recognition that there are other parts of the family at equal risk, but who have not so far asked to be seen. What is one's duty to such individuals? Should one actively seek them out and offer information, or should one wait until a request is received? Frequently one may not know whether these family members are aware of their risks, or even of the disease itself. Sometimes there may be major rifts between different parts of the family; advice given by family members may not be well received, which is not surprising considering how tactless some relatives can be. Young couples at risk deciding to have children may come in for particular criticism from older relatives for not having had an abortion or sterilization. It is thus particularly important that those involved in genetic counselling avoid being drawn inadvertently into such family problems and approach other members in the most sensitive way possible.

Before any approach is made to other family members it is usually wise to attempt to find out how much they know already, and what their general attitude is likely to be. The family doctor is likely to be the person of greatest help in this, especially in countries such as Britain, where almost everybody is registered with one. Equally a failure to contact the family doctor may cause real problems if the discovery of a previously unknown genetic risk precipitates an emotional crisis or of there is already an unrelated serious illness present. A helpful family doctor is also in the best position to be a bridge between the HD extended family and the unit providing counselling, either by direct referral or by preliminary explanation. Two valuable extended studies of

HD have been made by family doctors in Britain (Glendinning, 1975; Caro, 1977), and it is worth remembering that George Huntington was himself a family doctor. In the absence of such a link the role commonly falls to a particular individual within a family who may be able to communicate with the other branches. Where such a person exists this can be extremely helpful, but often family members quite understandably feel that they should not have to carry this burden.

Precisely how to approach a previously unascertained family branch is difficult to lay down and, as indicated, will depend considerably on what one has previously found out from family doctors and relatives. In our own experience the most acceptable approach has usually been an initial letter, offering a home visit by a nurse or social worker, at which questions can be answered and information provided in the light of how much or little the family prove actually to know. Any hurried or pressured attempt to suggest hospital referral or formal genetic counselling is usually neither helpful nor acceptable until these preliminary steps have been taken and it is clearer what are the principal needs and what is the existing level of understanding.

The question remains to be answered – should one actually take the initiative in contacting such families or parts of families at all? Our own view is one that has been arrived at gradually and may well differ from that of others. We consider that the overall trauma and distress resulting from lack of information on HD almost always outweighs that resulting from knowledge, providing that the information has been given sensitively by people knowledgeable about HD and its problems. The bitterness that can be caused by the disorder being concealed in the family and the distress resulting from feeling that there is no-one that can be turned to for information and emotional support were perhaps the most striking features of our own original family survey, and have been commented on by many workers over the years. The great majority of individuals have expressed relief rather than negative feelings when contacted, as is borne out by the results of the few formal studies on the topic discussed below.

A further reason for taking the initiative, less recognized previously, is that almost all family members do eventually find out that they are at risk for HD – and often in a very traumatic and unsuitable way. In the more open society of today it is actually very difficult for such a risk to remain hidden permanently; we have encountered numerous examples of people being informed totally unexpectedly as the result of a chance remark by a relative, an unrelated visit to the doctor, an insurance assessment, or even from neighbours. The shock caused by such unexpected, even casual disclosures of devastating information is understandably profound and often causes great resentment at why information was not given before, as well as the way in which it was given. For all these reasons our view remains that when an individual or part of an HD family is recognized, but is thought to be unaware of the disorder, they should be given at least the clear opportunity to be fully informed, rather than no action being taken because it has not been actively requested.

A special problem that occasionally arises in counselling of the extended family is refusal of permission by a family member to allow other relatives, most often their children, to be contacted. While a sensitive and cautious approach will help to avoid this situation, it can nevertheless occur however careful one is and however clearly one explains the need for the family members to be given information. Such a person is usually acting from the highest motives, feeling strongly that they wish to protect their children from the trauma HD has caused to themselves. The 'children' in question will usually be adult, planning marriage or even already married, and the dilemma arises whether one should leave them in ignorance, with the possible later realization of risk when they are told by others, or develop the disorder themselves, or whether one should go against the wishes of the parent and inform them.

Our own view is that the second course is preferable provided that one has used every means to get around the problem first. The legal situation in both Britain and North America would seem to support this (Shaw, 1987), though it does not seem to have been tested for HD in an actual case.

COUNSELLING THE ADOLESCENT

There has been considerable debate about the age at which children should be offered genetic counselling but very little systematic research has been undertaken. Folstein (1989) recommends that parents should tell their children about their risk for HD between the ages of 14–16 years, depending on the child's maturity. She doubts whether children can understand it earlier. In a survey carried out by Tyler and Harper (1983), which is still ongoing, most subjects at risk wanted to be told when they were young and not beyond the age of 30 years, but opinion was divided as to whether the teens or twenties or even younger was the best time. A significant number of children felt that 'the earlier you know the better, as it doesn't come as such a shock'. Others felt that they 'could not have handled such information in their teens', and many of these recalled being given some information about the 'family complaint' which they blocked out totally, as having no relevance to their lives at that time.

However, while there is not a consistent viewpoint about age, there is a widespread belief among young at-risk subjects that it is the family's responsibility to give them some knowledge of their possible genetic risks; they may feel extremely resentful if ignored by a genetic clinic or general practitioner. There is also no doubt that this is a very difficult and painful subject for most parents; as has already been noted, the information they give is likely to be inadequate, or inaccurate. Barette and Marsden (1979) also came to the conclusion that the burden of telling the children the risks is too great for parents and that professional help is usually needed. A partnership between the parents and the genetic clinic would seem the most appropriate method.

In our experience the right time for the genetic clinic to initiate contact with a young at-risk person depends on a number of factors, which include the following:

(1) The parent should have had time to come to terms with the diagnosis and feel able to communicate information to the adolescent in a supportive way. The adolescent should also have a trusting relationship with some family members, or carer, who can offer ongoing support.

(2) The teenager should not have significant instability or excessive stress in his or her life. The threat posed by the knowledge of being at risk can strike at a possibly fragile sense of identity. If he or she is already very insecure, it can exacerbate existing problems and lead to anti-social behaviour. This is particularly true of adolescents who have been taken into care. If one or more of these circumstances is present it may be wise to postpone counselling if possible.

As has already been discussed a sensitive approach to the young person's needs and fears is necessary for effective counselling. Often the first expressed needs of the adolescent are for information about their relative's illness and help in understanding their reactions to it. It is only later that they can talk about their own risk and the implications for their lives. Commonly, their immediate worries can be around their sense of self-worth and attractiveness. They may tend to feel diminished, and that no one will wish to marry them. They may worry about becoming like their affected relative and wonder how they can 'inflict' themselves on any partner with this prospect ahead. Also, a surprising number believe that they are inevitably bound to develop the disease.

It is obviously important to tell the truth but also to take a cautiously optimistic standpoint. Adolescents who have had advice which, although well-meaning, has stressed the more disturbing aspects of HD, can take this as a personal affront and refuse further professional intervention for many years. 'The doctors made me feel worse; I thought they were just trying to frighten me', has been a complaint not infrequently heard by the authors.

The role of the general practitioner in the counselling of the at-risk adolescent can present problems, particularly where there is a different family doctor from the affected parent. The young person at risk may be very unwilling for their status to be divulged to their doctor, on the grounds that this is unnecessary as long as they are perfectly well. They fear unnecessary stigmatization for themselves and any future children if this knowledge is held within the surgery, but not kept confidential; unhappily the authors have found that, on occasion, these fears are justified. They may also doubt the capacity of the family doctor to offer support as they often 'have never seen him' (except briefly for a minor disorder). They may distrust doctors anyway, having perceived the illness in the family members as being mishandled. Our policy in these cases has been to respect their wishes, if they cannot be persuaded to change their minds.

Finally, the information and support needs of some adolescents may differ from those of their parents. Some ask very wide-ranging questions. Particularly if they are from small families it seems to break down their feelings of isolation to learn about other young persons at risk meeting the same problems and to know that HD is being studied in many different countries.

OPTIONS FOR HIGH-RISK FAMILY MEMBERS

Genetic counselling does not simply consist of giving risk information but involves helping individuals to avoid the risk so far as this is possible. How can those at high risk for HD be helped in this way?

While some individuals at high genetic risk for HD will decide to have children even though knowing of this risk for themselves and their future children, many will not be prepared to do so. What are the options available for such couples, who would like to have children, but who are deterred by the genetic risk of HD? Sadly, these remain limited at present, but the prospect is now becoming more optimistic for some of them. Table 11.3 summarizes the current situation; most of these aspects are dealt with more fully in other sections, but they deserve mention here.

Sterilization

Sterilization is a clear solution for those who have already had children and are clear that they do not wish for more. It is also a valid option for those childless couples or individuals who are certain that they do not intend to have children, but here there are factors which must be carefully considered before an essentially irreversible measure is taken. How certain is the decision? With a young person especially, could there be a change of mind in later years? Is the decision really a free one, or has pressure come from family or other sources? Is the person aware of current (and likely future) genetic advances that may allow them to have a child at low risk of HD? All these factors need careful consideration, and it is not surprising that gynaecologists, who see people later requesting reversal of the procedure, are cautious about undertaking it. In some countries there is strict legislation governing its use. The study of Tyler and Harper (1983) showed that sterilization had been performed in 27 out of 100 individuals at high risk for HD; in 23 cases HD was the indication for this,

Table 11.3 Reproductive options for couples at high risk for HD

Sterilization (male or female)
Adoption/fostering
Artificial insemination (donor) or *in vitro* fertilization.
Presymptomatic testing
Prenatal diagnosis
Prenatal exclusion testing

---

though some had already had children. Occasionally the partner not at risk was sterilized.

Adoption

Adoption is a topic that raises many issues in relation to HD (Tyler and Harper, 1983; Harper and Tyler, 1985) and is dealt with in Chapter 6. As an option for couples at genetic risk for HD it is currently rarely a realistic option, since adoption policy is based more on finding suitable adoptive families for children needing placement than on the needs of a couple at risk for a genetic disorder. The combination of a decline in the number of children available for adoption and the potential severity of HD means that those at risk will rarely be accepted. A partial alternative which may be acceptable to some couples and which deserves more serious consideration is long-term fostering or adoption of an older child whose needs are principally for the next few years. In this respect, it is important to ask the question – what is the chance of HD developing over a specified period? This risk is often quite small and may prove acceptably low given the inevitable difficulties of this type of placement. Table 11.4, based on the life-table curve of age-related risks already shown (Newcombe, 1981) gives figures which could be useful for those making decisions of this type.

A separate problem posed by adoption is keeping in touch with, or tracing, children at risk for HD who have been adopted before the onset of HD in the at-risk parent or even before the family history was known. There has been much debate on the ethics of actively seeking out such children to inform them of their possible genetic risk. Our experience again indicates that, provided the natural parent(s) has given consent and the disclosure is handled sensitively and sympathetically, adoptive parents generally agree that their child should have access to this information. This approach is recommended by the British Agencies for Adoption and Fostering (Carter, 1982).

How should contact be maintained with the at-risk parent? We suggest discussion between the genetics centre, the adoption agency and the natural

Table 11.4 Risk (%) of onset of HD over a defined period in a person at 50% prior risk; calculated by Dr Robert Newcombe from the original South Wales life-table analysis (Newcombe, 1981)

Age of person at risk (years)					
50–	7.3	14.8	19.1	24.3	24.3
45–	9.8	16.4	23.1	27.0	31.7
40–	8.5	17.4	23.5	29.6	33.2
35–	6.5	14.4	22.8	28.4	34.2
30–	4.2	10.5	18.0	26.1	31.5
25–	3.4	7.5	13.5	20.8	28.6
20–	1.8	5.1	9.1	15.1	22.2
	5	10	15	20	25

Time period from present (years)

parent at the time the child is placed for adoption. Our experience has been that the natural parent may agree to follow-up; if he or she is then diagnosed as affected in later life the genetic centre can (with the parent's consent) inform the adoption agency who can put them in touch with the adoptive parents.

The problems of tracing the adopted child where the family history is unknown at the time of adoption are much more difficult to solve, particularly if, as often happens, the child is living outside the region. The local family practitioner committee may be able to provide the name of the child's general practitioner.

Artificial insemination by donor

Artificial insemination by donor (AID) is a logical approach to removing the genetic risk of HD where the risk is in the male line. However, the number of couples taking up this option is low. Tyler (1982) found that it had not been used as an option in any of the 100 relatives at risk studied, though several had considered it. McCormack *et al.* (1983) undertook a survey of attitudes to AID in America among HD families and controls; the two groups were similar in their attitudes, with more women than men (around two-thirds) regarding AID as the best option for reproduction in families at risk from HD, while the proportions were reversed in response to the question of whether they would personally use it. *In vitro* fertilization (IVF) has likewise not yet been used to avoid the risk in the female line on our own, or in other, published experience.

Abortion

In most countries termination of pregnancy is now legal where there is a significant risk of serious genetic disorder in the child. Although most such terminations are for disorders causing serious handicap in early life, it seems unlikely that a request from a person carrying a pregnancy at risk for HD would be refused. Pleydell (1968) and Oliver (1968), both writing before abortion was generally performed in Britain, both stressed the need for its availability.

In practice the number of abortions performed for pregnancies at risk appears to be relatively small, though there are no satisfactory systematic data. It seems likely that either decisions regarding childbearing have already been made before the pregnancy, or possibly that in an unplanned pregnancy the risk of HD is considered relatively remote in comparison with a severe congenital malformation.

Presymptomatic and prenatal testing

The application of DNA markers for the HD gene as an option for those at high risk promises to change the situation considerably. The topics of presymptomatic and pregnancy exclusion testing are discussed fully in Chapter 12, but it should be noted here that the wish to have children free from the risk of HD was given as the primary reason for requesting a presymptomatic test in 20% of the series of applicants reported by Morris *et al.* (1989). Likewise the series of pregnancy exclusion tests of Quarrell *et al.* (1987) and Tyler *et al.* (1990) had this as their primary objective and contained couples who had no wish for presymptomatic testing as well as those whose family structure would not permit it. Thus presymptomatic testing, direct prenatal diagnosis and pregnancy-exclusion testing represent three valid and different options by which a person at risk of HD can achieve their objective of having children that are biologically their own yet at low risk for HD. It is clearly essential that those requesting irreversible procedures such as sterilization should be fully informed of these possibilities and their probable future development.

SYSTEMATIC STUDIES OF GENETIC COUNSELLING IN HD

Genetic counselling is not an easy subject to evaluate or study formally, but several workers have attempted to do so. Stern and Eldridge (1975) and Barette and Marsden (1979) both employed postal questionnaires to members of the lay society in the United States and Britain respectively; their results are important but cannot be regarded as representative of HD families generally. The study of Tyler (Tyler, 1982; Tyler and Harper, 1983) analysed data on 92 HD patients and an additional 51 spouses (all those available for interview in a specific geographical area) together with data on 91 randomly selected offspring of HD patients between the ages of 18–49 years and 39 spouses in the same population. All were interviewed personally.

A striking feature of all the studies, confirming the general impressions of most HD surveys, was the high proportion of people who had received little or no genetic information on HD. Where this had been obtained it had come mainly from the lay society in the study of Stern and Eldridge (not surprisingly since respondents were all members) and from the authors' own previous population survey in the study of Tyler and Harper. In this study two-thirds of HD patients and spouses had received either no information at all or information only from relatives before completing their family, while the same was true for one-third of the offspring at risk (Table 11.5). Almost uniformly, the

Table 11.5 First source of genetic information in HD families, prior to completing their families. Based on Tyler and Harper (1983)

	HD patients/ spouses $n = 143$		Offspring at risk $n = 91$	
	No.	%	No.	%
No previous genetic information, or from family members only	95	66	30	33
Family doctor	13	10	6	6
Hospital personnel (including genetics clinic)	25	17	7	8
HD survey team	3	2	41	45
Media/other	7	5	7	8

information given by relatives was inaccurate in one or more important respects.

The picture of widespread unmet need for genetic counselling given by these surveys is unfortunately likely to be a true one in almost all countries and regions. Medical genetics services in Europe and North America almost universally find HD a major reason for referral but are aware that they see only a small minority of families unless they have made a special study of the disorder. It could be argued that most of the genetic counselling needs of families are being met by neurologists, family doctors and other clinicians, but sadly this is most unlikely to be the case at present. The study of Martindale and Yale (1983) of all known HD families in a London health district containing a major teaching hospital is highly disturbing and can only be described as an indictment of the standard of clinical practice in relation to this particular topic. Only one of the 10 families fully studied had received adequate genetic counselling, while the neurologists, psychiatrists and others involved with management of the patients almost universally did not, when interviewed, consider it part of their duty to undertake or to refer family members for genetic counselling. It seems quite possible that any recent change for the better in meeting the genetic counselling needs of families has been the result of greater awareness and positive action by the families themselves in seeking referral, yet involvement of the clinicians directly involved in diagnosis and management is essential if a comprehensive service is to be provided.

Does genetic counselling influence people's pattern of childbearing? Tyler (1982) studied this in two ways, by asking whether individuals had or would have been so influenced, and by analysing the number of children actually born to them. Sixty per cent of those questioned stated that genetic counselling had either modified their plans for having children or would have done if their families had not already been completed, a proportion lower than the 86% found by Stern and Eldridge (1975) but not surprisingly so given the different ascertainment. When the 'influenced' and 'not influenced' groups in Tyler's study were analysed according to numbers of children actually born, the mean family size was 1.19 for the 'influenced' and 1.71 for the 'uninfluenced' group. Further studies that are relevant to this point are those of Carter and Evans (1979) and Carter et al. (1983), who showed a marked reduction in the number of children born to couples at risk for HD who had attended a genetic counselling clinic.

In their 1983 report, Carter et al. also included data on children of affected patients given information at the time of diagnosis in the parent, but not seen in the genetics clinic. For both these groups receiving information before starting their families the number of children born was around half that expected from the general population, while for those children of an affected parent when information had not been received before completing their family the number of children born did not differ from the general population.

The timing of genetic counselling was examined by Tyler and Harper (1983). Overall, out of 181 individuals questioned 80 (44%) would have wished for earlier genetic information, this being particularly the case for the spouses of patients (74%). Twenty-seven per cent overall felt that the timing of the information received had been appropriate; none would have preferred later information, though 15 individuals (8%) stated that they would have preferred no genetic information at all.

GENETIC REGISTERS FOR HD

Any systematic study of HD families in a region will rapidly generate a large quantity of data on affected members and relatives, but this in no way necessarily merits the title of register unless it has special characteristics. The features of a genetic register have been described by Emery *et al.* (1978); the principal ones are that the data should have been systematically collected and that it should be regularly maintained. It is this last property that most HD surveys have lacked, being relatively complete at the time they were performed, but losing this and most of their value, within a surprisingly short period of time.

The design of a genetic register will depend on the aims which it hopes to serve. In the case of HD the principal aim has been to allow up-to-date information on families to be maintained for genetic counselling purposes; an exception to this is the United States national Huntington's disease roster, established by the department of Medical Genetics in Indiana, whose main aim is to provide data for research (Conneally, 1984; Gersting *et al.*, 1984). Here the term 'roster' was deliberately chosen in preference to 'register' as it makes no attempt at completeness of ascertainment but is a collection of data on families that have volunteered this for research use. Its very size has permitted important analyses and conclusions that smaller studies would not have allowed, such as those relating to sex-related effects and age at onset correlations (see Chapter 9).

Most service-orientated genetic registers have been founded on an initial survey and relate to a specific region or small country. The population base to be served is a critical decision to be made at the onset; if it is too small the register will probably not be viable in terms of its service role and will have little epidemiological value also. If it is too big, the large number of families will be more than offset by the inevitable lack of ascertainment and maintaining it will become a bureaucratic exercise rather than representing personal involvement. In general, populations of 1–3 million, often those of a health region, province or small country, are most suitable. These have well-defined boundaries, often corresponding to regions served by a single specialist unit working on HD, and are sufficiently small to allow personal contact between those maintaining the register and the various medical and other staff involved with HD families. A 'national' register is rarely either feasible or desirable except for small countries like Holland or Denmark, or where, like Wales, the country also functions as a single health region. For larger countries, such as England, Germany or the

United States, the size is too great, while there is also the danger of inappropriate governmental involvement in a register perceived as 'national'.

It is possible to make an approximate estimate of the numbers of family members at risk if one knows the numbers that are affected in an area, as pointed out by Conneally (1984). Individuals with HD are symptomatic for around one-third of their lives, meaning that there will be about twice as many asymptomatic gene carriers in the population as affected individuals. Those with the gene, whether symptomatic or not, will in turn have an equivalent number of sibs who are not carrying the gene but who cannot be distinguished, so that the total number of first-degree relatives at risk will be around five times the number of affecteds. Thus in considering a register for a district of 1–2 million population containing 100 living HD patients one might expect around 500 first-degree relatives at high risk. This estimate ignores second-degree relatives and in practice it is preferable to use an age-adjusted risk estimate with a suitable cut-off point, as in the Wales register described below.

The evolution of the register maintained in Wales over the past 18 years serves as a useful example of the way in which a genetic register for HD can be set up, maintained and used, as well as how it has had to evolve and some of the problems encountered. A more general description of HD research in Wales has been given in Chapter 1.

Our initial HD register was confined to the 'study area' of south Wales for which complete ascertainment had been attempted (Walker *et al.*, 1981). This covered a population of 1.7 million, with 131 living HD patients from 101 kindreds and almost 1000 relatives at high (greater than 10% when age-adjusted) risk. Following the later survey of Quarrell (Quarrell *et al.*, 1988; Quarrell, 1989) the register was extended to the whole of Wales (population 2.8 million) and now contains data on around 220 living patients from 270 kindreds, with around 2100 living relatives at over 10% risk, of whom 880 have a risk greater than 30% when age adjusted. We have found that maintaining even simple data on a register of this size is a major task for those involved and have deliberately declined to extend the boundaries of the register, preferring to encourage neighbouring health regions to initiate and maintain their own.

The details of the operation of the Wales HD register and the data recorded are described elsewhere (Harper *et al.*, 1982; Sarfarazi *et al.*, 1987). Here it is worth recording that they have been kept simple (Table 11.6) on the basis that, human nature being what it is, complex or extensive data sheets are less likely to be fully completed. Purely research data, including most of that collected in the original survey, were deliberately excluded when the service register was set up and we would strongly advise others embarking on a HD register to keep theirs correspondingly simple.

Confidentiality is the cornerstone of any genetic register and particularly so for such a sensitive disorder as HD. In our initial register, maintained on a mainframe computer, identifying information was deliberately omitted and a unique code number given, with a linking file kept only in the genetics department. When the register was modified for microcomputer use, this

Table 11.6 The Wales HD register. Summary of principal information recorded

---

Name
Identifying number (kindred, generation, individual)
Address
Clinical status (affected, unaffected, equivocal)
Parent number
Living/dead
Date of birth
Date of and age at death (if deceased)
Sex
Place of birth
Civil state
Year and age of onset (if affected)
Current risk estimate
Number of children
Regular surveillance wished (yes, no, type)
Family doctor
DNA sample taken (date, lab number)

---

restriction was lifted, since data no longer had to be transferred outside the department. Since the beginning of the register, the strict policy has been not to release identifying information on any individual to third parties without specific consent. Under British law the register has to be registered under the 1987 Data Protection Act, like all computerized data bases, and any individual has the right to know what information is recorded about him or herself.

The value of the Wales HD register is threefold. Firstly it allows a systematic recording and updating of the abundant genetic data relating to HD patients and relatives that would otherwise have long ago become unmanageable. It is easy to check whether an apparently new referral in fact relates to a family already known to us or whether a parent or a person at risk is confirmed as being affected; the unique number of each person also avoids the confusion often caused by the small number of surnames in Wales, where it is easy to place people with the same surname in the same kindred erroneously. The second use of our register is to provide regular updating of useful information that might otherwise be overlooked; for example a listing of young people at risk who may need to be offered genetic counselling in their own right rather than just to be regarded as 'children of Mr and Mrs—', or a listing of individuals known to have died, or to have moved out of the area. The final category of use is more epidemiological and consists of the monitoring of birth trends discussed on page 362, which is founded on the register data of new births to at-risk individuals in each year.

The Wales HD register has included two features which have been original as well as valuable. The first is the inclusion of the life-table algorithm of Newcombe (1981) in the structure of the register, thus allowing automatic estimation and easy updating of risk estimates, which would otherwise progressively become outdated as individuals aged. The second is the inclusion of an automatic pedigree-plotting system in the microcomputer version, which allows pedigree data to be displayed and printed on any family or part of family where this is needed. However, an HD register need not be complex or particularly sophisticated in order to be effective. Any system that allows basic information on families to be accurately recorded, easily retrieved and kept securely will be valuable.

POPULATION PREVENTION OF HD

It has been emphasized throughout this chapter that genetic counselling is a subject that relates primarily to individuals and families, not to populations. The discussion has thus far been in terms of how genetic counselling and a general awareness of the genetic aspects of HD affect the attitudes of these individuals and their decisions in relation to childbearing. The question of prevention of HD in the population as a whole is a quite different topic, yet one of great importance, for in the current absence of any significant therapy it is only prevention that is capable of significantly reducing the future burden of the disease.

The different views that workers in the field hold on the relationship between genetic counselling and prevention are well illustrated in the genetic counselling section of the United States Congress Commission report on HD (1977), where one member issued a minority report stating that the concentration on the individual and psychological aspects of counselling had led to the importance of prevention being ignored. The rather frank correspondence published in the report clearly illustrates both the differences between the individual and population goals of the two approaches, and the fact that both are important.

In discussing prevention it is important to be clear from the outset that one is meaning a reduction of numbers born carrying the HD gene, rather than prevention of the disease in those who carry the gene, which is at present no more possible than is the curative therapy of those actually affected. Some epidemiologists may disagree with use of the term 'prevention' for what is essentially avoidance of births at risk. In the context of HD, however, where any individual born with the abnormal gene will in time develop the disorder, prevention seems a perfectly appropriate word to use, provided it is clear to all concerned what it means in this context.

Prevention of HD has until now revolved around two fundamental considerations: firstly, since the proportion of new mutations is extremely low and almost all affected individuals receive the gene from an affected parent, the disorder is in theoretical terms highly preventable. If reduced to the level of new mutation, HD would be an exceptionally rare disease. Secondly, there has been no way of identifying those with the gene except by their belonging to a HD family, which does not distinguish those with the gene from those without it. While this last aspect is now altering with the advent of predictive tests, these have not yet made a significant population impact, Any population approach has thus to be based on a combination of ascertainment, registers and genetic counselling, all of which have been discussed individually, but whose integration and systematic application are essential in any attempt to prevent HD at the population level.

A thorough study giving ascertainment in the population in question is clearly essential; as has already been stated, no estimate of prevalence can be considered accurate or adequate unless a detailed survey has been made over a considerable period of time. Even when this has been done, ascertainment will certainly prove less than complete, as seen by repeat surveys in regions already carefully studied (Wallace and Parker, 1973; Quarrell et al., 1988).

The existence of a genetic register for HD is essential in recording information on those at risk in families; it is these individuals rather than those actually affected who transmit the great majority of HD genes to the next generation and whose pattern of reproduction will determine whether the future prevalence of HD will change. Only an accurate and systematically maintained register designed for this purpose will provide information on this. Furthermore, such a register will allow a prediction of future trends for HD, since a knowledge of the numbers of births at risk will directly determine the future prevalence of the disorder.

The third factor, genetic counselling, is as central as the others to the prevention of HD. The high proportion of individuals in past generations who have received no genetic information has already been noted; it is clear that no reduction in future prevalence is possible if those at risk are unaware of it at the time of having their own families. At the same time, the non-directive and individual nature of genetic counselling has here been emphasized, so that a paradox can be seen if one regards this as part of population prevention.

Leaving aside for the present the question whether any active policy of population prevention for HD can be justified, it seems essential that we should have reliable long-term data on HD at the population level. Since this is something that most surveys have not been able to collect, it is worth describing the ongoing study in South Wales, which was carried out specifically with this in view. After the initial population survey (Walker et al., 1981) was complete, with ascertainment in the study area thought to be almost total (see Chapter 1) a genetic register was established specifically to maintain contact with those at risk and to monitor births, deaths and other relevant data. Since each birth in a particular year could be assigned a specific risk based on age of the parent at risk and whether this parent was a first- or second-degree relative of an HD patient, adding these small portions of risk allowed a prediction to be made for the total likely number of new HD genes born into the population during that year, even though one could not predict which individuals carried the gene.

Using this approach an attempt was made to estimate the births of predicted new cases over the years, and for past years to combine the estimate with the 'real' data from the birth year of known affected patients. Figure 11.5a shows the results (Harper et al., 1979), given as rolling 5-year averages to smoothe the fluctuations of individual years and also compared with the general population birth rate. It can be seen that there is very little variation over a prolonged period, apart from a possible decline close to the time of study. The analysis was repeated to cover births during the years following the study when a systematic attempt had been made to contact all high-risk relatives and offer non-directive counselling. A decline in the number of predicted new cases was observed (Harper et al., 1981) which contrasted with the previous relative constancy. From the study it was not possible to determine what factors might have been responsible; the authors suggested that these might be multiple, with increased awareness and information about HD being relevant in addition to possible effects of counselling itself. The data were examined for a third period by Quarrell et al. (1988) (Figure 11.5b) who showed that data accumulated in the earlier years predicted a less steep, but still significant, decline compared with that estimated previously; they also observed a decline in births at risk in north Wales, an area not included in the previous analysis and where families had not been monitored or given genetic counselling to the same systematic extent as in the South Wales study area.

These studies show that population trends for HD can be measured and predicted with reasonable accuracy, at least in a population that is stable. It is unwise, however, to use studies of this type to draw conclusions on the reasons underlying a trend, which are best analysed by detailed studies for this purpose. It can be concluded, though, that the South Wales results are compatible with the separate data of Carter et al. (1983) and Tyler and Harper (1983) that a considerable proportion of HD families will restrict their intended and actual family size if they are fully aware of their genetic risks.

The studies discussed above have been preventive only in the sense of attempting full ascertainment of those at risk and giving full information including genetic counselling to all those wishing for it. Should a preventive programme for HD be more directive in terms of attempting to reduce births at risk, whether by more direct counselling, encouraging sterilization, or in future more widespread application of predictive testing? Here one enters into dangerous territory, and there is a real risk of crossing the boundary into the final area to be considered in this chapter, the abuse of genetics in relation to HD. There are wider issues to be considered, too, notably those of confidentiality, privacy, freedom of individual decision and avoidance of discrimination, all of which could be breached by any programme which took such a direction. There may, indeed, be a real conflict in goals between the 'public health' aim of reducing and eliminating a serious disorder and the primary goal of those involved in genetic counselling of ensuring that individuals have the full facts on which to base their own decisions. However in our view there need not and should not be such a conflict. Experience with other serious genetic disorders such as the haemoglobinopathies and muscular dystrophies, has shown that once satisfactory options are open to them, most families will act in a way that

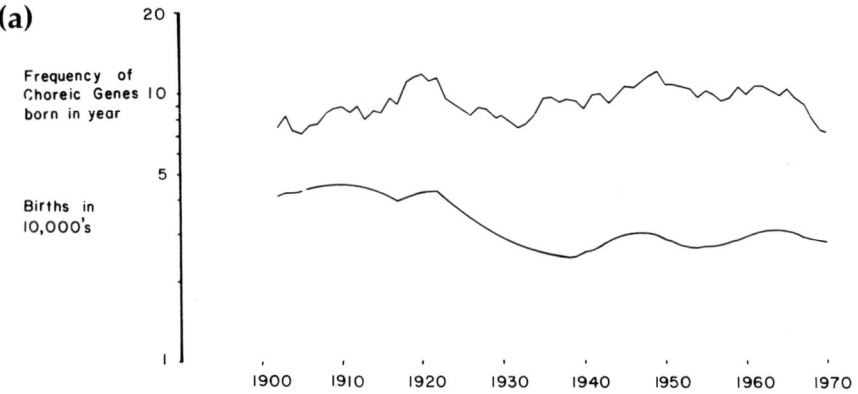

Figure 11.5 Long-term trends in births at risk for HD in the South Wales population (see text for details). (a) Five-year moving averages for frequency of new HD genes born into the population between 1900 and 1970, compared with the general population of the same area (from Harper *et al.*, 1979). (b) Analysis of trends up to year 1984. A: general population annual birth rate, B; estimated number of births heterozygous for HD, C; estimated number of heterozygotes per 10 000 population (from Quarrell *et al.*, 1988).

both satisfies their own individual wishes and at the same time is likely to produce a long-term reduction in prevalence of the disorder. The options open to HD families in having children free from the risks of HD are still far from satisfactory. It seems preferable that we should attempt to improve these, to make them more widely accessible along with genetic counselling and other information, and to continue to monitor the population effects of this, rather than to attempt any measures in the name of prevention that might prejudice the rights of HD patients and family members.

HUNTINGTON'S DISEASE AND THE ABUSE OF GENETICS

In a chapter which has emphasized the importance of genetic counselling in HD, and where the development of genetic registers and active ascertainment of those at risk has been discussed as the basis for effective prevention, it might seem perverse to end with a section on the abuse of these very measures. Yet the line between effective prevention and coercion is a very fine one, which is easy to overstep without realizing. The only safeguard is for medical personnel, scientists and the general public to be fully aware of the work that is going on and is planned for the future, so that it can be regulated as society sees appropriate.

Unfortunately, for genetic disorders in general and HD in particular, this is not enough; abuse has already occurred in the past, and a close examination of this uncomfortable topic is needed, if only to convince others that special care must be taken if it is not to happen again. In the case of HD it is likely that much

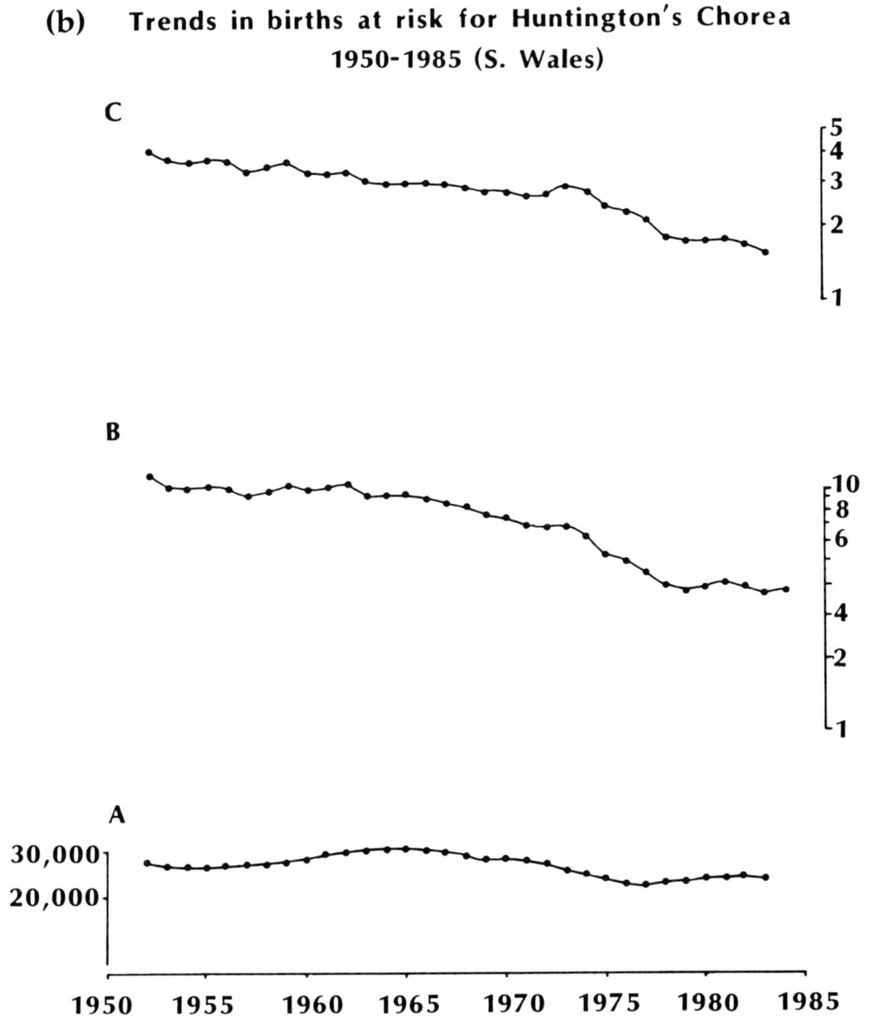

Genetic Counselling in Huntington's Disease 365

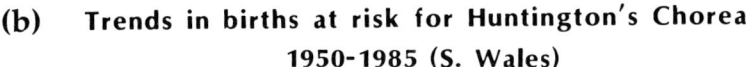

Figure 11.5b

of this abuse is still unrecognized. The information so far obtained is fragmentary and preliminary; the authors would appreciate further information so that the topic can in future be dealt with in the detail that it deserves. The general subject of the abuse of genetics has been dealt with in two books which make vivid and disturbing reading. *In the Name of Eugenics* by Kevles (1985), traces the origins of the eugenics movement in America and Britain and documents particularly well the discriminatory laws in America that resulted in compulsory sterilization. *Murderous Science* by Müller-Hill (1988) brings to light the atrocities committed against the mentally ill and those with genetic disorders in Nazi Germany and documents the complicity of physicians and scientists of that time, many of whom remained in prominent positions after the war.

Professor Müller-Hill's help in tracing aspects relating to HD has been particularly appreciated. A third book by Weindling (1989) also gives considerable detail on the abuses of genetics in Nazi Germany, in the context of a broader political analysis.

It is not surprising that early workers studying families with HD should have been impressed with the need for effective action to prevent its spread. Many of the large families derived from a single immigrant ancestor, prompting the following statement from Davenport regarding the disorder in New England: 'All these evils in our study trace back to some half-dozen individuals including three brothers, who migrated to this country during the 17th century. Had these half-dozen individuals been kept out of this country much of misery might have been saved' (Davenport and Muncey, 1916).

It is in fact likely that individuals at risk for HD may have been prevented from immigration to North America and Australia until recent times, though we have not been able to trace specific examples. It is clear from the spread of the HD gene from many founding immigrant populations that such a policy was ineffective, if it ever formally existed.

If the introduction of the HD gene could not be prevented at source, then legislation to prevent its spread was clearly seen as the only alternative by some of the early eugenicists. Davenport again states his views clearly:

> It would be a work of far-seeing philanthropy to sterilize all those in which chronic chorea has already developed and to secure that such of their offspring as show prematurely its symptoms shall not reproduce. It is for the state to investigate every case of Huntington's chorea that appears and to concern itself with all of the progeny of such. That is the least the state can do to fulfil its duty toward the as yet unborn. A state that knows who are its choreics and knows that half of the children of every one of such will (on the average) become choreic and does not do the obvious thing to prevent the spread of this dire inheritable disease is impotent, stupid and blind and invites disaster. We think only of personal liberty and forget the rights and liberties of the unborn of whom that state is the sole protector. Unfortunate the nation when the state declines to fulfil this duty! (Davenport and Muncey, 1916).

The invoking of the 'duty' of the state presages chillingly the later developments in Germany, something that is not surprising in view of the close links between eugenicists of the two countries (Kevles, 1985). Davenport's Eugenics Record Office specifically supported the German programme of compulsory sterilization (Weindling, 1989).

However it would be wrong to single out America as favouring compulsory sterilization; the need for legislation was also expressed by Spillane and Phillips (1937) in their early study of HD families in Wales: 'Perhaps with repeated advice and education, some would voluntarily abstain from marriage, but the majority would no doubt be prepared to accept the even chance that nature offers them. We are thus left with the conclusion that only legislative measures will eventually succeed in eradicating the disease'.

It was in Nazi Germany, however, that these developments were taken to their logical conclusion. The 'race hygiene' policies of the Third Reich concentrated on two medical groups, the mentally ill and handicapped and those affected by genetic disorders. HD families were thus doubly at risk; the disorder is specifically mentioned in the law enacted 14 July, 1933 (Figure 11.6), though only those actually affected were covered by this law. Family members, especially those with 'anomalies of character' were to be closely supervised, Exactly how many HD patients were forcibly sterilized will probably never be known, but a thesis on the subject at the time mentions nine HD patients among 950 such sterilizations in Leipzig between 1934 and 1940 (Munchow, 1943). Müller-Hill (personal communication) suggests that in view of the total number of 350 000 to 400 000 people compulsorily sterilized, there would have been 3000 to 3500 sterilizations for HD. How many were later killed is another unknown figure; it is clear from the review of Meyer (1988) that HD was specifically listed among those disorders for which extermination was indicated.

Patients to be sterilized with HD and other disorders were brought before a special 'Genetic Health Court' and the façade of legality was maintained, with sterilization of some of the Leipzig patients turned down on grounds of uncertain diagnosis, A comparable court in Düsseldorf also refused some HD

Gesetz zur Verhütung erbkranken Nachwuchses
Vom 14. Juli 1933

(Reichsgesetzblatt I S. 529)

Die Reichsregierung hat das folgende Gesetz beschlossen, das hiermit verkündet wird:

§ 1

(1) Wer erbkrank ist, kann durch chirurgischen Eingriff unfruchtbar gemacht (sterilisiert) werden, wenn nach den Erfahrungen der ärztlichen Wissenschaft mit großer Wahrscheinlichkeit zu erwarten ist, daß seine Nachkommen an schweren körperlichen oder geistigen Erbschäden leiden werden.

(2) Erbkrank im Sinne dieses Gesetzes ist, wer an einer der folgenden Krankheiten leidet:

1. angeborenem Schwachsinn,
2. Schizophrenie,
3. zirkulärem (manisch-depressivem) Irresein,
4. erblicher Fallsucht,
5. erblichem Veitstanz (Huntingtonsche Chorea),
6. erblicher Blindheit,
7. erblicher Taubheit,
8. schwerer erblicher körperlicher Mißbildung.

(3) Ferner kann unfruchtbar gemacht werden, wer an schwerem Alkoholismus leidet.

Figure 11.6 Excerpt from the compulsory sterilization law of 14 July, 1933 in Nazi Germany. HD is specifically listed as one of the indications.

cases on grounds of advanced age and general deterioration in health (Platzek, 1988). Weindling (1989) has documented the procedures used and has emphasized the dominant role of the medical progression in this abuse.

> The sterilisation law shows how professional powers and state authority reinforced one another. Nine 'diseases' were selected: hereditary feeble-mindedness, Schizophrenia, manic-depression, hereditary epilepsy, Huntington's chorea, hereditary blindness, hereditary deafness, hereditary malformations, and (indicating the historical roots of the law) severe alcoholism. A system of hereditary courts was established; each tribunal was composed of a lawyer, a medical officer and a doctor with specialist training in racial hygiene. The medical officer could initiate proceedings as well as adjudicate, and doctors were in the majority. The state established primacy over reproduction, but left the operating of the controls to the medical profession.

The work of Panse (1942), who undertook the first fully systematic survey and analysis of HD in the Rhineland of Germany has already been mentioned, and is frequently quoted in the HD literature. Less noticed, however, has been Panse's role in reporting his cases and their families to the Nazi authorities:

> We proceeded in a manner that we reported all choreic cases, and moreover all suspicious cases and finally all not yet choreic sibs and offspring as being at risk to the health authorities.

> 79 cases located and diagnosed by us were reported to the health administration. They have been passed on to the Genetic Health procedure, if they were of an age to procreate. (Panse, 1942, pp. 232–233.)

Friedrich Panse (1899–1973) was professor at the Psychiatric–Neurological Research Institute in Bonn, whose director was Professor Kurt Pohlisch. Both were Nazi party members and were closely associated with the establishment of the Nazi race-hygiene laws; Panse and Pohlisch were actively involved in the 'euthanasia' mass murder and the genetic health courts described above (Müller-Hill, 1988; Weindling, 1989), but both were acquitted of criminal acts after the war and returned to leading scientific appointments,

After the war work on HD in Germany began again, much as if the wartime abuses had never occurred. It is remarkable that they are not mentioned in the monograph of Wendt and Drohm (1972), which is based on work carried out in the 1950s and 1960s. The only reference to the Nazi period is a note on the difficulty of obtaining complete register data for the period 1933–1945, 'in which it was looked on as a disgrace in Germany to have a genetic disease'. Even the detailed and sensitively written study of 'Ethics and medical genetics in the Federal Republic of Germany' (Schroeder-Kurth and Huebner, 1989), which mentions concerns about testing for HD, makes almost no reference to the Nazi abuses of genetics.

The disturbing saga of the Panse documents is not yet over, for there is currently in Germany considerable debate as to whether they should be used for current research and for tracing families, as described in an article in the German magazine *Stern* (Erb-Sommer and Muller, 1987). Panse's full collection of data on HD families was stored along with many other documents in

Heldburg, but as the bombing of Germany increased they were transported east in three train loads to what subsequently became East Germany. They remained there until 1986, when they were located and moved back to Düsseldorf to form the foundation of a new HD register. Intense controversy ensued as to whether it was right to use the data in this way and the documents are currently under a legal veto.

The sensitivity and even hostility of many people in Germany to issues involving genetics (Schroeder-Kurth and Huebner, 1989) can be readily appreciated in the light of these events. Does this dark picture of abuse now relate solely to the past, or is it possible that it might occur again elsewhere in the future? It would be reassuring to believe that it would not, but there is every reason to believe that we should be extremely reluctant to accept this. Furthermore, scientific advances have increased the potential dangers of such abuse. Computerization of genetic registers and the development of predictive tests are two such advances that would surely have been abused systematically had they been available at the time of the Third Reich. The number of authoritarian political regimes remains very high, despite recent changes in Eastern Europe; some countries outside Europe still have repressive legislation for genetic disorders and could misuse predictive testing.

The most significant reason for extreme wariness, however, comes not from other societies but from our own, as illustrated by the attitude of society to those who carry the HIV virus. There are close parallels between HD and AIDS in the issues involved, not only regarding the testing of symptomless carriers, but also the stigmatization and repression that has emerged for those affected. It requires little imagination to see how readily such discrimination could be transferred to HD if we see the situation purely in terms of public health, rather than as a disorder of individuals and families. It will require great efforts from both professionals and the public to make sure that these problems do not happen.

REFERENCES

Barette J and Marsden CD (1979) Attitudes of families to some aspects of Huntington's chorea. *Psychological Medicine* **9**:327–336.
Caro A (1977) Huntington's chorea: a clinical problem in East Anglia. University of East Anglia: PhD thesis.
Carter CO (1982) In: M Oxtoby (ed), *Genetics in adoption and fostering*. London: British Agencies for Adoption and Fostering.
Carter CO and Evans K (1979) Counselling and Huntington's chorea. *Lancet* **ii**:470–472.
Carter CO, Evans KA and Baraitser M. (1983) Effect of genetic counselling on the prevalence of Huntington's chorea. *British Medical Journal* **286**:281–283.
Conneally PM (1984) Huntington's disease: genetics and epidemiology. *American Journal of Human Genetics* **36**:506–526.
Davenport CB and Muncey EB (1916) Huntington's chorea in relation to heredity and insanity. *American Journal of Insanity* **73**:195–222.
Emery AEH, Brough C, Crawfurd M, Harper PS, Harris R and Oakshott G (1978) A report on genetic registers, *Journal of Medical Genetics* **15**:435–442.
Erb-Sommer M and Muller, LA (1987) *Stern* **49**, 96–98.

Folstein SE (1989) *Huntington's Disease: a Disorder of Families*. Baltimore: Johns Hopkins University Press.
Gersting JM, Conneally PM and Yount EA (1984) Huntington's disease research roster data base support with MEGADATS-3M. *Journal of Medical Systems* **8**:163–171.
Glenndinning N (1975) A study in Huntington's chorea. University of London: MD thesis.
Harper PS (1988) *Practical Genetic Counselling*, 3rd ed, London: Wright.
Harper PS and Tyler A (1985) Huntington's chorea: problems in adoption and fostering. *Adoption and Fostering* **9**:47–51.
Harper PS, Tyler A, Walker DA, Newcombe RG and Davies K (1979) Huntington's chorea: The basis for long-term prevention. *Lancet* **ii**:346–349.
Harper PS, Tyler A, Smith S, Jones P, Newcombe RG and McBroom V (1981) Decline in the predicted incidence of Huntington's chorea associated with systematic genetic counselling and family support. *Lancet* **ii**:411–413.
Harper PS, Tyler A, Smith S, Jones P, Newcombe RG and McBroom V (1982) A genetic register for Huntington's chorea in South Wales. *Journal of Medical Genetics* **19**:241–245.
Hayden MR, Soles JA and Ward RH (1985) Age of onset in siblings of persons with juvenile Huntington's disease. *Clinical Genetics* **28**:100–105.
Kelly TE (1986) *Clinical Genetics and Genetic Counselling*. Chicago: Year Book Medical Publishers.
Kevles DJ (1985) *In the Name of Eugenics*. New York, Knopf.
Martindale B and Yale R (1983) Huntington's chorea: neglected opportunities for preventive medicine. *Lancet* **i**:634–636.
McCormack MK, Leiblum S and Lazzarani A (1983) Attitudes regarding utilization of artificial insemination by donor in Huntington's disease. *American Journal of Medical Genetics* **14**:5–13.
Meyer JE (1988) The fate of the mentally ill in Germany during the Third Reich. *Psychological Medicine* **18**:575–581.
Morris MJ, Tyler A, Lazarou L, Meredith L and Harper PS (1989) Problems in genetic prediction for Huntington's disease. *Lancet* **ii**:601–603.
Müller-Hill B (1988) *Murderous Science*. Oxford: University Press.
Munchow R (1943) 950 Gutachten in Erbgesundheitssachen der Jahre 1934–1940. University of Berlin: Thesis.
Newcombe RG (1981) A life table for onset of Huntington's chorea. *Annals of Human Genetics* **45**:375–385.
Oliver JE (1968) Abortion and Huntington's chorea. *British Medical Journal* **i**:576–577.
Panse F (1942) *Die Erbchorea; eine klinisch-genetische Studie*. Leipzig: Thieme.
Penney JB, Young AB and Shoulson I (1990) Huntington's disease in Venezuela: 7 years of follow-up on symptomatic and asymptomatic individuals. *Movement Disorders* **5**:93–99.
Platzek B (1988) Sterilisations prozesse am beispiel des erbgesundheitgerichts Düsseldorf unter besonderer berucksichtigung der chorea Huntington-Kranken. University of Düsseldorf: Thesis.
Pleydell MJ (1968) Indications for termination of pregnancy. *British Medical Journal* **i**:376.
Quarrell OWJ (1989) Prevention and prediction of Huntington's disease. MD thesis. University of London.
Quarrell OWJ, Tyler A, Meredith AL, Youngman S, Upadhyaya M and Harper PS (1987) Exclusion testing for Huntington's disease in pregnancy with a closely linked DNA marker. *Lancet* **i**:1281–1283.
Quarrell OWJ, Tyler A, Jones MP, Nordin M and Harper PS (1988) Population studies of Huntington's disease in Wales. *Clinical Genetics* **33**:189–195.
Sarfarazi M, Quarrell OWJ, Wolak G and Harper PS (1987) An integrated micro-computer system to maintain a genetic register for Huntington's disease. *American Journal of Medical Genetics* **28**:999–1006.
Schroeder-Kurth TM and Huebner J (1989) Ethics and medical genetics in the Federal Republic of Germany (FRG). In: *Ethics and Human Genetics* (eds T Wertz and J Fletcher). Berlin: Springer.
Shaw MW (1987) Testing for the Huntington's gene: a right to know, a right not to know, or a duty to know? *American Journal of Medical Genetics* **26**:243–246.
Spillane J and Phillips R (1937) Huntington's chorea in South Wales. *Quarterly Journal of Medicine* **6**:403–423.
Stern R and Eldridge R (1975) Attitudes of patients and their relatives to Huntington's disease. *Journal of Medical Genetics* **12**:217–223.
Stevens DL (1973) Heterozygote frequency for Huntington's chorea. *Advances in Neurology* **1**:191–198.

Tyler A (1982) The social, personal and economic burden of Huntington's chorea in South Wales. University of Wales: thesis.

Tyler A and Harper PS (1983) Attitudes of subjects at risk and their relatives towards genetic counselling in Huntington's chorea. *Journal of Medical Genetics* **27**:179–188.

Tyler A, Quarrell OWJ, Lazarou LP, Meredith AL and Harper PS (1990) Exclusion testing in pregnancy for Huntington's disease. *Journal of Medical Genetics* **27**:488–495.

United States Congress Commission (1977) *Report: Commission for the Control of Huntington's Disease and its Consequences*. Washington: US Department of Health Education and Welfare.

Walker DA, Harper PS, Wells CEC, Tyler A, Davies K and Newcombe RG (1981) Huntington's chorea in South Wales: a genetic and epidemiological study. *Clinical Genetics* **19**:213–221.

Wallace DC and Parker N (1973) Huntington's chorea in Queensland: the most recent story. *Advances in Neurology* **1**:223–236.

Weindling P (1989) *Health, Race and German Politics between National Unification and Nazism, 1870–1945*. Cambridge: University Press.

Wendt GG and Drohm D (1972) *Die Huntingtonsche chorea. Eine populations genetische studie*. Stuttgart: Thieme.

Wexler N (1984) Huntington's disease and other late onset genetic disorders. In: *Psychological Aspects of Genetic Counselling* (eds AEH Emery and I Pullen). London: Academic Press.

12

Predictive Tests in Huntington's Disease

INTRODUCTION

The advances in DNA technology and their application to Huntington's Disease (HD), discussed in the previous chapter, have clear implications for the possibility of predicting which individuals at risk for the disorder are carrying the abnormal gene. Even though the HD gene itself has not yet been identified, genetic linkage studies of families can now alter an individual's risk status, often allowing a prediction of considerable accuracy. The implications of this possibility will now be discussed.

EARLY APPROACHES TO PREDICTION

The localization of the HD gene has received considerable publicity, as has the whole field of applying DNA markers to genetic diseases, so that it might be assumed that little thought had been given to the possibilities of prediction in HD before the advent of DNA technology. Nothing could be further from the truth; attempts to detect asymptomatic gene carriers go back half a century, while the need for such a task has been recognized for even longer. Bell (1934) stated the situation clearly:

> The almost continuous anxiety of unaffected members of these families over so long a period must be a great strain and handicap, even if they remain free from disquieting symptoms; it is thus of urgent importance that some means should be sought by which immunity of an individual could be predicted early in life, both from the point of view of relief to those who carry no liability to the disease and as an indication to others that they should abstain from parenthood. No facts in the clinical histories of patients provide definite guidance in this matter prior to the onset of symptoms, but the development of the science of genetics may at some future date enable us to obtain information concerning the inherent characteristics in such cases.

In essence, the early presymptomatic tests tried to demonstrate a difference between HD patients and normal controls and then to ascertain what proportion of those at risk had the abnormality in question. Despite an ingenious

Table 12.1 Early attempts at predictive testing

Clinical studies	
Neurological	
Tremor	Falek, 1969; Myers and Falek 1979
Eye movement	Petit and Milbled, 1973
Psychiatric	
Cognitive tests	Baro, 1973; Lyle and Gottesman, 1979
Personality tests	Palm, 1973
Electrophysiological studies	
Electromyography	Petajan et al., 1979
EEG	Patterson et al., 1948; Chandler 1966, 1969
Biochemical studies	
Prolactin/growth hormone responses	Hayden et al., 1977; Caraceni et al., 1977
Caeruloplasmin	Shokeir, 1975
Pharmacological studies	
l-Dopa provocation test	Klawans et al., 1970, 1980
Radiological studies	
CT scan	Neophytides et al., 1979
Positron emission tomography	(see text)

variety of such tests, most of them had serious methodological shortcomings, such as small sample numbers, the failure of the test to divide the at-risk group into two roughly equal proportions and contradictory results if the tests were combined. The greatest methodological problem has been the lack of prospective follow-up which has only been performed in three studies (Chandler, 1966; Lyle and Gottesman, 1977; Klawans et al., 1980).

In retrospect, it is probably fortunate that none of these tests was sufficiently encouraging to result in widespread use, since little thought seems to have been given to what individuals might have been told. Perry (1981) makes a point which remains highly valid, at least regarding research data:

> While reviewing research proposals dealing with predictive tests for Huntington's chorea I have been struck with the cavalier attitude of some investigators towards the use of the data they hope to generate. I suggest that pending development of an effective form of treatment, scientists who perform preclinical tests on persons at risk should ensure that the results of individual tests are not made available to those tested.

Some of the more relevant early studies on predictive tests are summarized in Table 12.1 and a few are discussed briefly here.

Muscle Tremor

The investigation of fine motor movement in HD has a reasonable basis as a possible presymptomatic test. Muscle tremor assessed by an accelerometer taped to the back of the hand, was examined in a controlled study of nine patients and 15 at-risk individuals (Myers and Falek, 1979). Although clear abnormalities were found in the affected group, only one-third of the at-risk group had altered responses. Longitudinal studies will be needed before deciding whether this approach to prediction is of value.

Eye movements
Petit and Milbled (1973) hypothesized that conjugate ocular abnormalities antedated the onset of other clinical features of HD. They studied a group of 30 at-risk subjects using an opticoelectronic device. Oculomotor abnormalities, such as large saccadic movements, poor fixation and latency of ocular response, were found in 11 individuals. However, these ocular motor findings are non-specific and can occur in a number of other disorders.

Cognitive and personality tests
Neuropsychological investigation of at-risk individuals has been reported in several studies (Baro, 1973; Palm, 1973; Wexler, 1979). These tended to concentrate on intelligence tests although one exception was that by Fedio *et al.* (1979) who suggested that impaired performance on neuropsychological tests of frontal lobe function is a premorbid indicator of HD.

Perhaps the most interesting is the long-term follow-up of 88 subjects tested during the 1950s when they were clinically normal (Lyle and Gottesman, 1977, 1979). Fifteen to twenty years later. 28 of this group had developed HD while the others remained asymptomatic. Marked differences on test scores, including the Wechsler Adult Intelligence Scale (Wechsler), were demonstrated between the two groups. Those who later developed HD scored much closer to the affected group than those who remained symptom-free and there was a gradual decline in intelligence even before the onset of neurological symptoms.

Both self-report procedures and projective techniques have been used in studies of early detection of HD. Palm (1973) administered the Minnesota Multiphasic Personality Inventory (MMPI) (Dahlstrom and Welsh, 1960) to 23 individuals at risk. He found that 13 MMPI profiles were abnormal and that there were three major characteristics in this group: a tendency to be 'rebellious of social and ethical codes', to have a low or relatively high 'activity drive' and procrastination. However, these personality characteristics are not confined to those at risk of HD and Palm concluded that the MMPI should not be used in prediction.

Projective techniques of personality assessment were also used by Palm (1973). He applied the Rorschach test (Kloepfer and Davidson, 1962), which is a series of ink-blots, to the same group of respondents described above and asked them to describe what they saw in each blot. Abnormalities, such as 'lack of emotional control', 'lack of organizational ability' and 'preoccupation with sex', were reported in eight of the 23 individuals and a further five were considered 'neurotic'. However, projective techniques face problems of reliability and validity which reduce their usefulness as a predictive test. Poor correlation was found between the Rorschach and MMPI results.

It is now becoming possible to evaluate the significance of psychological tests by comparing their outcome with that of molecular testing (discussed fully later in the chapter), since most individuals undergoing DNA testing in the early series have had psychological testing as part of the procedure. A very small initial series of 7 with a high risk and 3 with a low risk result (Jason et al., 1988) suggested that those with a high risk result had a higher frequency of abnormalities, but a larger and more carefully controlled study (Strauss and Brandt, 1990) has shown no such difference. It is possible that inclusion of individuals in the prodromal stages of the disorder could have also biased the earlier report; at present the outcome of molecular tests gives no support that any form of psychological testing is of long term predictive significance for the later development of HD.

Electromyography
Electromyography has been studied in relation to the abnormal movements of HD (Petajan et al., 1979). Abnormalities of motor unit activity, including irregular and sudden bursts of activity that manifest in chorea and excessive recruitment of motor units during voluntary movement, have been described in detail (Petajan et al., 1979). In 55% of a group of at-risk individuals, defined electromyographic disturbances including frequent smaller ballistic contractions ('microchorea') were described. Further research using this technique needs to be done on control groups with other movement disorders.

EEG
Patterson et al. (1948) studied 26 at-risk subjects in depth using a variety of methods including psychometric and blood-group tests in an attempt to identify gene carriers. The electroencephalogram (EEG) was thought to be the most useful test in prediction and assessments were made on the degree of EEG abnormality, the presence of paroxysmal features and the presence of slow wave activity. They classified the at-risk group into those 'most likely to develop chorea' (12 cases), those with a 'bare possibility of developing the disease' (seven cases) and those who 'probably would never develop Huntington's chorea' (seven cases). Chandler (1966, 1969) followed up 23 members of this cohort for 18 years and found that 12 had developed HD while the other 11 remained unaffected. However, the EEG in prediction was found to be little better than chance when only five of the 12 were correctly predicted to develop HD. Two other series also reported that the EEG is unhelpful in prediction (Harvald, 1951; Palm, 1973).

CT scan
Neophytides et al. (1979) tested the hypothesis that detectable brain abnormalities preceded the onset of HD by performing CT scans on 18 HD patients and 47 at-risk individuals. Although they found cortical and basal ganglia atrophy in the affected group, they were unable to identify changes in the scans of the at-risk group that would make this investigation useful for prediction.

The use of PET scanning in early and presymptomatic diagnosis is discussed later in this chapter.

The *l*-dopa test
The single likely exception to the negative outcome of these early tests was the use of the drug *l*-dopa as a provocative agent. Originally found to be of benefit in the treatment of Parkinson's disease, choreic movements had been noted as a symptom of excessive dosage, while when given to patients with HD, movements were aggravated. A few patients with the rigid form appeared to benefit, but here too, choreic movements could be induced. These findings stimulated several groups to study the effects of *l*-dopa on the asymptomatic offspring of HD patients. Klawans *et al.* (1970, 1972, 1973, 1980) studied 30 such subjects and found that 10 developed transient choreic movements. None of the 25 control subjects showed abnormalities. They suggested that a positive result indicated a high chance of carrying the HD gene.

Cawein and Turney (1971) observed a similar response in a person at risk who, after a year's continued observation, developed definite HD. It seems likely that this individual was already showing minimal features, but this report raised the natural concern as to whether *l*-dopa might precipitate onset of HD or at least advance it in those at risk. This possibility, together with the obvious and disturbing movements rendering privacy of the information impossible, resulted in this approach being abandoned, though its unpublished use probably continued longer; a further study was performed more recently to assess the effect of *l*-dopa on blink reflex which suggested that abnormalities in those at risk could be induced or enhanced by the drug (Esteban *et al.*, 1981). Disturbingly, this study included a 9-year-old girl, whose decision to participate is stated to have 'been assumed by her guardian'. (Some of the other early attempts at presymptomatic detection also involved the testing of children.) Subsequent follow-up of individuals in the original study has shown both false negatives and false positives. After an 8-year interval Klawans *et al.* (1980) found one of 20 showing a negative test result to have developed HD, while five of the 10 giving a positive result had become affected. Thus there seemed little doubt that the *l*-dopa test does represent a real biological response in a proportion of gene carriers, probably those closest to clinical manifestation where neurochemical changes are already well established.

The *l*-dopa test, not surprisingly, caused considerable controversy, and prompted a series of general discussions on the ethics of predictive testing for a disease with a progressive and fatal course, and with no available treatment (Gaylin, 1972; Hemphill, 1972). Experienced clinicians were against such testing in general (Stevens, 1971; Fahn, 1980; Marsden, 1981), whereas other workers, including those themselves at risk, strongly defended the right of autonomy of choice (Rothstein, 1971; Bates, 1981). The debate continued over the next decade, notably at the regular meetings of the World Federation of Neurology Research Group on Huntington's chorea and in the various lay societies, with considerable thought given to the ways in which a test should be administered, if and when one became available.

ATTITUDES OF FAMILIES TO PREDICTIVE TESTING

The realization that an accurate predictive test would be discovered at some time prompted several systematic studies to find out what families with HD actually felt about this topic. However, many of these studies had methodological defects, the most obvious of which was biased samples. In one study, subjects were recruited 'through public service announcements in local newspapers and on the radio' (Kessler, 1987). Members of lay societies would be likely to show more interest in predictive testing than at-risk individuals who did not belong to such bodies. This hypothesis has much supporting evidence: postal surveys of members of self-help groups have found up to 84% would wish to have a presymptomatic test (Stern and Eldridge, 1975; Barette and Marsden, 1979; Evers-Kiebooms et al., 1987).

Another methodological shortcoming was low response rate (13% in one series, Markel et al., 1987). One of the few population-based surveys of at-risk individuals with a high response rate showed that 56% indicated a desire for presymptomatic testing (Tyler and Harper, 1983). Comparable results were also found by later studies carried out after discovery of genetic linkage had made the possibility of testing more imminent (see Table 12.2).

While the precise proportion favouring the case of a predictive test can be debated, these studies left no doubt that a significant number of families would request it, regardless of the views of their medical advisors. They also highlighted the major service commitment that such testing would involve for preparation to meet this.

GENETIC LINKAGE AND PREDICTIVE TESTING

The implications of genetic linkage for prediction were recognized from the outset, as indicated in the quotation given from the paper of Bell and Haldane (1937) in Chapter 10 (p. 318). A detailed discussion of carrier detection by Neel (1949) shows that HD was high on the list of disorders where a genetic approach to carrier testing was considered important. Despite this, it is likely that none of the early investigators of genetic linkage in HD was expecting to find a result that would allow early application, while the limited number of classical blood-groups and other protein markers meant that even if linkage were to be found, there would be no obvious way to obtain closer markers.

The discovery of polymorphisms in human DNA changed this situation dramatically in two way: their abundance in the human genome meant that a loose initial linkage could be progressively converted into one close enough to

Table 12.2 Summary of surveys on attitudes of individuals at risk for HD towards presymptomatic tests

First author	Year	Country	Original sample	Actual no. sampled	No. of replies	% in favour of tests	Random sampling
Stern	1975	US	2600	1065 (41%)	365 (34%)	77	No
Barette	1979	UK	420	135 (36%)	132 (86%)	80	No
Teltscher	1981	Australia	130	73 (56%)	50 (68%)	84	Yes
Tyler	1983	S. Wales	100	91 (91%)	91 (100%)	56	Yes
Schoenfeld	1984	US	141	55 (39%)	55 (100%)	73	No
Koller	1984	US	75	75 (100%)	75 (100%)	81	Uncertain
Evers-Kiebooms	1987	Belgium	76	76 (100%)	49 (64%)	57	No
Kessler	1987	US	69	69 (100%)	66 (96%)	79	No
Mastromauro	1987	US	360	360 (100%)	131 (36%)	66	No
Meissen	1987	US	106	64 (60%)	56 (87%)	65	No
Markel	1987	US	1200	1200 (100%)	155 (13%)	63	No

allow accurate prediction; of equal importance were the implications of independence of age and tissue – any marker used in prediction could also be used in prenatal diagnosis from a chorionic villus sample or amniotic fluid cells. Thus the unexpectedly rapid and close discovery of the G8 linkage by Gusella *et al.* (1983), already discussed in Chapter 10, produced a situation not just of future but of immediate significance for application in prediction.

It was at this stage that the years of previous discussions on predictive tests probably made their greatest contribution. Pressure for early application from some family members was intense, as was the publicity in the media. It would have been very easy for premature and unwise application of the discovery to have been made at this stage, something that could have caused great harm and could well have discredited the whole topic of DNA-based prediction.

Fortunately, the actual outcome was one of extreme caution and the various groups closely involved concentrated initially on the more immediately necessary but less controversial tasks of excluding genetic heterogeneity, setting up DNA banks to preserve samples from key family members, searching for closer markers and, most importantly, obtaining sufficient resources to undertake planned programmes of prediction. Indeed this interval of around 2 years can be looked on as a self-imposed moratorium, and was of the greatest value in allowing both professional and public opinion to adjust to the implications of the new situation. A renewed debate on both ethical and practical aspects of predictive testing took place (Bird, 1985; Harper, 1986) which also covered the legal and third party aspects (Berg and Fletcher, 1986; Lamport, 1987). The new surveys of attitudes of family members to testing have already been mentioned (Table 12.2).

An issue of general importance that surfaced at this time, and which produced a spirited correspondence in the columns of *Nature*, was whether the originators of a particular DNA probe, such as G8, had the right or duty to restrict its use (Gusella, 1986; Maddox, 1986; Watt *et al.* 1986). While Watt *et al.* considered that any published marker should be made fully available without delay, Gusella stressed the need for further work before releasing it for clinical application and the duty of a research worker to ensure that a discovery is not misapplied. The advent of further markers defused this situation without fully resolving the valid argument involved on both sides. Fortunately the one form of restriction that did not arise, though this was at one time seriously feared, was any commercial restriction or patenting of the diagnostic use of the original or later markers. Indeed the making of such a resource freely available for non-commercial use, something that is now the norm for almost all DNA probes, is an achievement that reflects considerable credit on the molecular scientists and others involved.

The principles of prediction based on genetic linkage

Genetic linkage has already been discussed in Chapter 10 but its application to prediction in a family with HD now needs more detailed consideration. Given the optimal circumstances, it is a topic that is extremely simple; in practice this

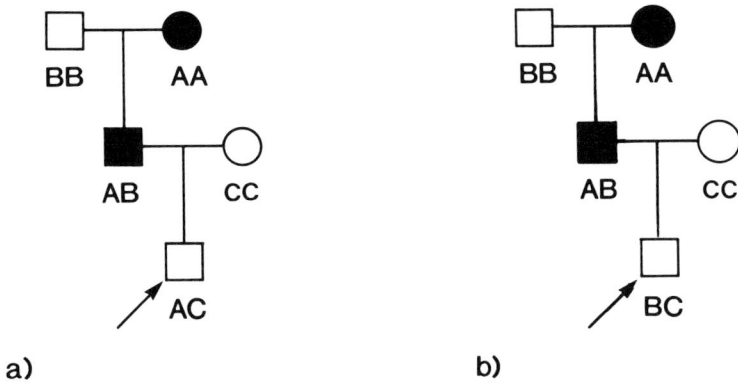

Figure 12.1 Genotype prediction in HD (see text for details). In this family the HD gene has been passed to the affected parent along with marker genotype A. Thus if the person at risk has also inherited genotype A (example (a)) there will be a high risk for developing HD, whereas if genotype B (example (b)) has been inherited, the risk will be low. The precise risks will depend on recombination rate between the marker and HD loci.

is far from being the case and there are major pitfalls for the unwary. Figure 12.1 illustrates the situation at its simplest. The HD gene and the linked marker locus are indicated and it is clear that in family (a) the HD gene has been passed on to the affected parent with the 'A' marker genotype, whereas the 'B' genotype has been received from the healthy grandparent. Thus if the person requesting a predictive test types as A it is likely that he will have also inherited the HD gene, unless recombination has occurred between the marker and disease loci. Likewise if the B type has been transmitted (example (b)), the person should be free from the disorder, again assuming no recombination. Thus if the recombination rate is known to be 5%, a prediction of 95% can be given of having inherited either the HD or alternatively the normal gene.

Even in this simple example some problems can soon be identified. Is there a possibility of more than a single locus for HD? What are the confidence limits for the 5% estimate of recombination? Could the recombination rate vary according to whether the genes are being transmitted through a male or female meiosis? All are relevant points needing to be resolved before embarking on testing.

Frequently the genotypes will not be so clearly distinguishable as in the idealized example given. Figure 12.2 gives an example of the 'partially informative' situation, often encountered in practice. Here both parents of the person at risk are heterozygous for the marker genotype; an informative result will only be possible if the person is homozygous as in examples (a) and (b), i.e. AA (giving a high risk result) or BB (giving a low risk result). In 50% of such cases (example (c)) one would expect an AB result, from which no conclusions can be drawn since the A marker could have come from either parent. Numerous variations on this situation can occur at different points in the pedigree.

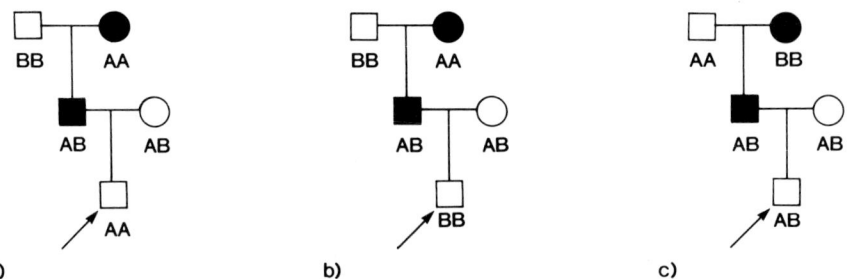

Figure 12.2 Genotype prediction in HD. A 'partially informative' family (see text for details). Since both parents of the person at risk are heterozygous for the marker genotype, a prediction for HD can only be made if the marker genotype is homozygous (examples (a) and (b)), not if heterozygous (example (c)).

So far it has been assumed that both parent and grandparent are living and can be typed, a highly unusual situation, though with the storage of DNA from affected individuals, it is becoming possible to use samples from long-deceased relatives. A more realistic testing scenario is given in Figure 12.3(a), where the genotype of the deceased affected grandparent has been inferred sufficiently to allow the conclusion that he had an 'A' marker gene and transmitted it along with HD. Note also that this further assumes correct paternity. Where neither grandparent is living, as in Figure 12.3(b), no prediction can be made, since although in this case also the A marker has been inherited from the affected parent we do not know whether HD was in this family being transmitted with A or with B.

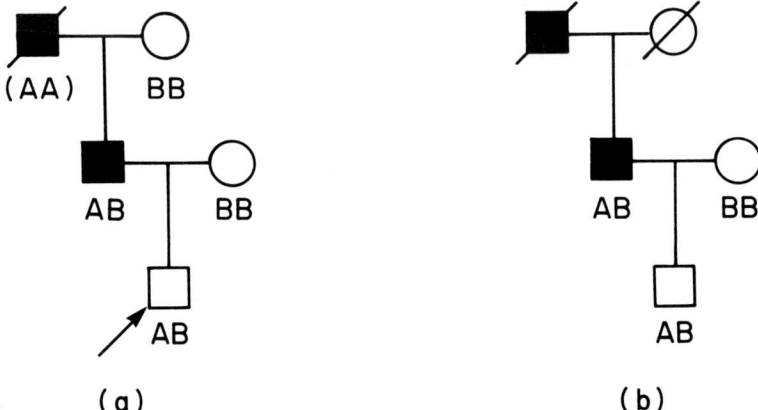

Figure 12.3 Genotype prediction in HD (see text for details). In example (a) the genotype of the deceased grandparent can be inferred, allowing prediction for the individual at risk. In example (b) this is impossible, since both grandparents are dead.

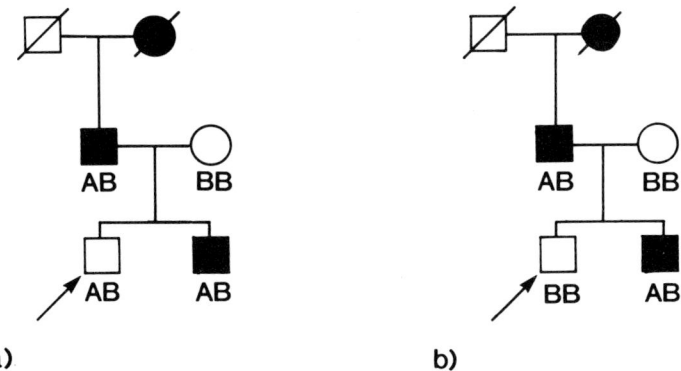

Figure 12.4 Use of an affected sib in prediction for HD (see text for details). In example (a) the individual at risk shares the same marker genotype as the affected sib, increasing the risk of HD. In example (b) the risk is correspondingly reduced since the sibs have inherited different marker genotypes from the affected parent.

To overcome this frequent situation it may be necessary to rely on more distant family members, as shown in Figure 12.4. Here the potential for error rapidly increases, since there is more than one point at which recombination might have occurred. Thus in Figure 12.4(a), the fact that the person at risk has inherited the same marker genotype as their affected sib clearly raises their risk, whereas in Figure 12.4(b) it is correspondingly reduced. However, it is possible that the parent actually received HD along with marker type B and that the affected child is a recombinant, as well as the possibility of recombination in the person requesting prediction. Thus the potential error rate in prediction is approximately doubled. Similar considerations apply when elderly relatives who are considered to be 'escapees' are typed in place of a grandparent (Figure 12.5).

The complexity of risk estimation in even an apparently simple situation means that it is wise to use one of the computerized programmes, along the lines described by Conneally *et al.* (1984) and Sandkuyl and Ott (1989). This type of risk estimation is becoming an integral part of the clinical use of linked markers for a variety of diseases in most medical genetics centres.

In view of these potential and actual difficulties it is relevant to ask what proportion of families have the right structures to allow prediction by genetic linkage to be carried out? Harper and Sarfarazi (1985) examined this, using data from the South Wales HD register (Table 12.3). Of all adults aged 16–45 years whose genetic risk was greater than 10% (the group most likely to request prediction), only 19% had at least one grandparent living (usually the unaffected one) while only 10.5% had affected sibs. Thus less than a quarter of those at risk had a pedigree structure which could be described as satisfactory. Two other series, based on the American Huntington's disease roster (Farrer *et al.*, 1988) and on London families seen for genetic counselling (Misra *et al.*, 1988)

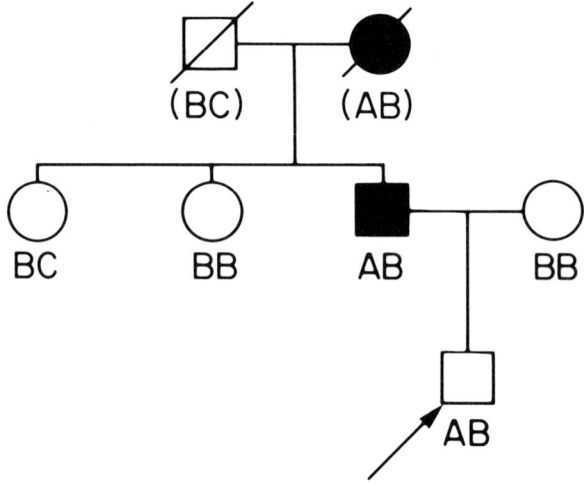

Figure 12.5 Use of 'escapees' in genetic prediction for HD (see text for details). In this family prediction would be impossible without utilizing the genotypes of the elderly unaffected sibs in the middle generation. These individuals type BB and BC, suggesting that the HD gene has been passed to the affected sib along with marker genotype A; thus the risk is raised for the offspring requesting prediction (most likely inferred genotypes of grandparents given in parentheses). Considerable caution is needed in both clinical assessment and risk estimation if such apparently unaffected individuals are to be used in this way.

Table 12.3 Frequency of pedigree structures in HD families on the South Wales register in relation to feasibility of predictive testing. Based on Harper and Sarfarazi (1985)*

	Total number	At least one living grandparent	Affected sib living	At least one living parent
Adults at risk	637	99 (15.5%)	67 (10.5%)	559 (87.8%)
Pregnancies at risk†	100	19 (19.0%)	3 (3.0%)	89 (89.0%)

*Only individuals and pregnancies with risk of HD greater than 10% included.
†Figures refer to parent and grandparent of the fetus.

have shown a somewhat higher proportion (30–40%) of suitably structured families, possibly reflecting a higher than average representation of larger families in these series, but the fact remains that family structure remains the most serious limitation in predictive testing for HD. Computer programmes can again be useful in allowing an estimate to be made of the degree of risk alteration possible given a particular pedigree structure and whether the typing of additional family members will significantly improve the situation (Sandkuyl and Ott, 1989).

The G8 probe and prediction
The unexpectedly strong and close nature of the initial linkage between the G8 probe and HD made it inevitable that this marker would be not only the theoretical model for prediction but the one actually used first in practice. It was to be some time before closer markers were developed and it was thus fortunate that the G8 probe and its corresponding locus, termed D4S10 according to gene-mapping nomenclature, proved to be highly polymorphic and therefore informative for use in prediction.

The initial DNA sequence used proved to show two separate polymorphic sites, each with two alleles (1 and 2), which could be combined to give four distinct haplotypes, termed A,B,C and D. In the original families studied, HD was segregating with the C haplotype in the Venezuela kindred, with the A haplotype in the American kindred; no recombination was found in either at this stage. The band pattern for the four haplotypes has already been shown in Figure 10.3; for practical purposes it has often proved easier to use subclones of the original probe that each contain only one of the polymorphic sites. By using the four haplotypes, over half the population could be made heterozygous and hence informative for prediction, but clearly there was a need for further polymorphisms to make the remainder also informative. Work in a number of laboratories progressively supplied these, while analysis of adjoining DNA sequences produced new polymorphisms that, while distinct from those of the original clone, were still so close that the possibility of recombination between them could be discounted. Table 12.4 summarizes the major polymorphisms at

Table 12.4 Polymorphisms at and adjoining the D4S10 (G8) locus

Probe	Restriction enzyme		Allele frequency (%) Allele 1	Allele 2	Source
PK083	EcoR1		50	50	Gusella et al. (1983)
PK082	Hind III	i)	75	25	Gusella et al. (1983)
		ii)	88	12	
	Nci I		87	13	
	Pst I		88	12	
R7	Bgl I		60	40	Gusella et al. (1983)
	Hind III	i)	75	25	
		ii)	88	12	
F5.52	Msp I		60	40	Bakker et al. (1987)
F5.53	EcoR I		50	50	Bakker et al. (1987)
	Ava II		65	35	

Table 12.5 Closely linked DNA markers to the HD gene listed in probable order (proximal to telomeric)

Locus	(Probe)	Authors
D4S62	(P8)	Hayden et al., 1987
*D4S10	(G8)	Gusella et al., 1983
*D4S95	(BS674)	Wasmuth et al., 1988
*D4S43	(C4H)	Gilliam et al., 1987
*D4S114	(W92)	Whaley et al., 1988
*D4S98	(BS731)	Whaley et al., 1988
*D4S113	(102BN4.0)	Whaley et al., 1988
*D4S96	(BS678)	Smith et al., 1988
D4S141	(2R3)	Snell et al., 1989
D4S90	(D5)	Youngman et al., 1988, 1989

*Indicates markers sufficiently close for use in prediction.

the D4S10 locus and immediately adjoining regions, while Table 12.5 gives comparable information on closer probes isolated subsequently that have proved suitable for clinical use.

Before serious application in prediction could begin, two major questions had to be answered, which required the close collaboration of a number of units. Firstly, an accurate estimation of recombination was essential: this proved to be close to 4% in both sexes with very narrow confidence limits (Conneally et al., 1989). The second problem was that of heterogeneity, since if more than one locus were to be found determining the disorder this would make diagnostic application extremely difficult save in the occasional very large kindred. Fortunately no evidence whatever for more than a single locus has emerged from a collaborative study of over 50 kindreds of both Caucasian and non-Caucasian origin (Conneally et al., 1989). Interestingly a number of distinctive and atypical kindreds have also been shown to be linked (Zweig et al., 1989), while other distinct but possibly allied disorders have been shown not to be allelic to HD, such as benign familial chorea (Quarrell et al., 1988) and dentatorubropallidoluysian atrophy (Kondo et al., 1990). Thus, while diagnostic problems can still cause considerable difficulty in families requesting prediction, as discussed below, the two main questions had been adequately resolved by the end of 1985 and prediction could now begin on a secure foundation.

CLINICAL AND COUNSELLING ASPECTS

The principal fear both of professionals closely concerned with HD and of families involved in the lay societies was that when an effective predictive test was discovered it would be applied without adequate preparation, counselling and subsequent support for those involved. It was the concept of an 'off the shelf' test which had been largely responsible for a number of concerned and experienced clinicians feeling that predictive testing should not be used and that it was likely to cause more harm than benefit. So far this has not happened; indeed the careful and responsible way in which predictive testing has been introduced is likely to become a model for comparable testing in other progressive neurological disorders. Several factors have combined to favour this. The intense debate and public awareness of the issues involved were undoubtedly a major factor; so was the relative difficulty of the test procedure, involving techniques then restricted to a small number of specialist centres and requiring testing of an entire family unit. That these allied activities already involved geneticists, either undertaking or closely associated with established genetic counselling services, meant that predictive testing was seen from the outset more as a test for family members than simply as a 'neurological' procedure.

All the principal centres that initiated predictive testing did so in the framework of a detailed research protocol. There were five such centres originally involved: Boston and Baltimore in the United States, Vancouver in Canada, and Manchester and Cardiff in the UK: further research-based pilot studies soon followed in Holland, Belgium and Scotland. All the original groups used extremely similar approaches and protocols, which was not surprising since all were in close contact, had shared their plans and experience on a regular basis, and had similar attitudes to the subject. Most of the centres have published their detailed methods as well as their early results. Table 12.6 outlines the broad structure of the initial studies and gives the relevant references, while the Cardiff protocol of the authors is given in Appendix 5. Most centres concentrated on presymptomatic testing of those at risk; the allied but distinct topic of prenatal exclusion testing is discussed later.

All programmes had in common the requirement for multiple interviews before a test result was given. The number and complexity varied, partly due to the amount of associated psychological and other evaluation, but the basic structure involved at least the series shown in Table 12.7, based on the authors'

Table 12.6 Early research evaluations of presymptomatic testing in HD

Country	Authors	Source of applicants
United States	Meissen et al. (1988)	Boston area (Initial contact through register)
United States	Brandt et al. (1989)	Maryland and adjacent states (Initial contact through register)
Canada	Hayden et al. (1987)	British Columbia
UK	Brock et al. (1989)	Scotland (Letters sent to family practitioners, announcement in local HD newsletter)
UK	Craufurd et al. (1989)	Northern England (Initial contact through register)
UK	Morris et al. (1989)	Wales and southern England (No prior publicity)

Table 12.7 Cardiff presymptomatic testing programme for HD; summary of protocol

---

INTERVIEW ONE
 Sociodemographic details
 Confirmation of family and clinical data
 Assessment of impact of HD and test results
 Assessment of knowledge of HD and presymptomatic testing
 Reasons for requesting prediction
 Neurological examination

INTERVIEW TWO
 Assessment of psychological, personality and social characteristics using standardised instruments
 Further counselling and discussion of disclosure session
 Nomination of professional supporter
 Signing of consent form
 Final blood sample

INTERVIEW THREE (HELD 4 WEEKS LATER)
 Disclosure of results

FORMAL FOLLOW-UP
 1 week
 3 months
 12 months

---

study. A preliminary interview was found to be most valuable in providing an unstructured general discussion of the implications and limitations of presymptomatic testing, as well as ensuring that the applicant was fully familiar with the genetic and other aspects of HD as a whole. It was frequently found that people had only a partial and sometimes an inaccurate idea of what was involved; some decided at this stage that presymptomatic testing was not the appropriate course for them, while for others it was clear from the outset that family structure would not permit it. In some instances, as where the applicant had already been fully seen for genetic counselling in another regional centre, this introductory interview could be dispensed with, but whenever there was any doubt it was given, even though this might mean considerable extra travelling for the applicant.

The second interview, the first in the Cardiff programme to follow a structured format, provided the detailed foundation for the testing procedure. Detailed pedigree analysis was essential to decide which family members needed to be sampled, whether the pedigree structure was optimal for testing, and if not, what degree of risk modification might reasonably be expected. This last point was checked in the later stages of the programme by computer simulation, as described by Sandkuyl and Ott (1989). Arrangements for sampling relatives needed to be made in detail and the possibility of individual family members refusing had to be considered.

Diagnostic details were documented at this stage and arrangements made to check autopsy results, death certificates and hospital records. The testing process not uncommonly broke down at this point when it became apparent that the diagnosis of HD was far from certain, or when a family member essential for the testing was not confirmed as being affected. In those cases where no neuropathology was available, particular care had to be taken to obtain full clinical details and the possibility of diagnostic error had to be carefully discussed. Motivation for having a presymptomatic tests was also assessed. A full explanation was given of the test procedure, the general aspects being reinforced by a written fact-sheet, while the specific situation was summarized in an individual letter. Again it was emphasized that the applicant could withdraw from testing at any point.

The amount of information requiring to be collected and to be given and absorbed at this interview was considerable, as was the amount of work involved in collecting and analysing samples from the necessary relatives. There was thus a considerable interval between this and the subsequent interview, most of which was taken up with laboratory analyses, often repeated and protracted, to find polymorphisms which would give the maximum risk alteration possible in the light of the pedigree structure. This interval, while frustrating to some applicants, was also valuable in allowing time for reflection as to whether presymptomatic testing was really what was desired in the light of the counselling given at the previous interview.

The final pre-test interview was only scheduled when it was clear that the typing of family members was as complete and informative as could be expected, giving a clear indication as to whether the genetic situation was fully or partially informative (as already shown in Figure 12.2) and the approximate range of risk alteration that was likely. Any diagnostic uncertainties had also to be resolved before this interview. Included in the interview were a further check on full understanding of the implications of testing, while a variety of psychological testing procedures was carried out to assess aspects of personality, depression and other indices which might later prove valuable in any correlations with the long-term outcome. Detailed arrangements were made for local medical and psychological support, especially in the light of a possible adverse outcome, and practical arrangements for the giving of results. Applicants were given a specific date to attend for the final result rather than being phoned as to when a result would be ready. They were also asked to give careful thought concerning to whom they would impart the results. Throughout the testing procedure all applicants were urged to come with a companion, usually spouse or partner, but close friend or relative where appropriate. This was considered essential for the actual disclosure of the results.

It was only at this time that the second and last blood sample was taken from the applicant with a view to keeping the inevitably highly stressful interval between final sampling and disclosure of result to a minimum. It was felt that once this sample was given and a consent form signed, an irrevocable step had been taken. All applicants had been reassured that a sample taken earlier would not be analysed until a late stage. They were again reminded of their right to withdraw at any time. Participants were invited to telephone the testing centre if they wanted any clarification before the disclosure of results but they were also informed that the counsellors themselves would not know the test result until the day before the disclosure interview. At the conclusion of this session, the situation was again summarized in the form of a letter.

The interview at which results were given was inevitably an emotionally charged and at times traumatic one for both applicant and counsellors. Some of the reactions to results, favourable and adverse, are given later in this chapter, but some points, obvious but important, need to be made here. All results were given face to face, never by letter or telephone. They were given by the same team of two counsellors (a clinician and a social worker in the Cardiff project) who had been involved in the previous interviews. The setting was a quiet, informal room designed for genetic counselling, with an additional room where siblings being tested together could be apart with their spouses or companions, or where the counsellors could move to, so that those being tested could be left to express their joy or grief privately. Most applicants in our series had travelled a considerable distance and prior arrangements were always made to avoid the need for driving by the applicant.

Results were always given in terms of risks being lowered or raised to a specific figure, though the impression given was that most people interpreted this as either being 'clear' or having the gene. A subsequent letter reinforced the specific risk estimate. Arrangements for support made previously were put into effect, with a particular effort to speak personally to local medical advisors if a result had been adverse or if there was special concern as to how a person might cope with the result. At the end of the day, an effort was always made by the testing team, including the laboratory staff, to meet together as a group to discuss how the situation had been handled, what could be learned from the particular case and whether any special problems were encountered. There is no doubt that such meetings have a therapeutic effect for those involved. We have found that the laboratory staff (two in number), whose meticulous work and attention to detail has been the foundation of the entire testing procedure, also share in the stresses involved and welcome an informal meeting that involves them as well as the clinical staff.

Regular follow-up visits are an integral part of the testing procedure and include formal psychological assessments to gauge depression and other variables, as well as to ensure that practical support arrangements are in place and that the risk figures given previously have been understood. It is perhaps not surprising that it is the follow-up that has given some difficulty: those receiving a high-risk result may be reluctant to return to the place perceived as responsible for giving the bad news, especially if they are receiving support more locally; those with a low risk may wish to get on with their lives and put the whole episode behind them now they feel freed from the need for medical involvement.

It is clear that the testing procedures outlined above, and the comparable and in some cases more elaborate protocols in other centres, are time-consuming and demand not only the involvement of a centre with existing experience, but also staff additional to those likely to be available in any established service setting, even if the facilities of genetics, psychiatry and neurology services were to be combined. It was recognized from the outset that it would be unrealistic to envisage such detailed procedures being transposed unchanged into a regular service setting. Such a change is taking place currently and is outlined later, but there can equally be no doubt that it was right for HD presymptomatic testing to be evaluated initially as a research procedure. These studies have individually and collectively provided an immense amount of information on the whole subject, not just on which factors are important in providing an optimal service, but on such aspects as the motivation of those seeking testing, the frequency of neurological and psychological abnormality in those tested, as well as the problems, expected and unexpected, that are encountered during testing. The comparability of the different projects means that data on many aspects, notably the long-term outcome of those tested, should be able to be pooled, so that eventually we should have an accurate and detailed picture of presymptomatic testing for HD from which much can be learned regarding such testing in other inherited neurological disorders.

Selection and exclusion
Testing centres have varied in the way in which individuals were selected for testing and in what criteria should be used for inclusion or exclusion. Some centres (e.g. Baltimore and Manchester) wrote to all individuals in their region known to be at risk, offering the procedure, while others relied on spontaneous demand. The limited number of centres initially involved meant that many individuals who had waited years for such a test were prepared to travel long distances, while the close contacts between different genetics centres, and between those involved with HD families generally, facilitated such referrals. Thus in the Cardiff study, 41 of 80 reaching the first formal interview came from outside Wales. Clearly those requesting testing were in no way a random selection of HD family members, and a considerable difference was found in socioeconomic status in the Cardiff series, as shown in Table 12.8.

Most of the original centres only accepted individuals at 50% prior risk, most excluded those already showing definite or equivocal signs of HD, while none accepted minors. In addition, a serious risk of suicide and current significant mental illness were usually exclusion criteria in these studies, though the applicants could be offered the test at a later date if there was a sufficient improvement in mental health.

EARLY STUDIES OF PRESYMPTOMATIC TESTING

The research studies already outlined were designed principally to evaluate the outcome of testing and to define the optimal way in which it could be delivered. While it is still too early to reach definitive conclusions, a considerable amount of preliminary information is already available. Table 12.9

summarizes some basic information from the different series; individual aspects can be considered under several headings.

Why do people request presymptomatic testing? We found that for most applicants there is no single overriding reason but rather a combination of reasons. To estimate the relative importance of various reasons for having the test, we asked applicants to rate them subjectively on a scale from 0 to 10 (Table 12.10). Having the test to enable childbearing decisions to be made was rated highly. If the test result were adverse, many applicants resolved not to have any further children; a few may consider prenatal testing. Others wanted the test so that they would be aware of the risks to existing children. If this was the only reason offered, we encouraged the applicant to postpone having a test until the children would be older and of the age of majority. We asked applicants to discuss testing with their adult offspring, many of whom did not want to know of alteration of their parents' risks, especially if they were considering a career in the armed forces or police.

Table 12.8 Social data on applicants in the Cardiff series attending first interview for presymptomatic testing. (**a**) Age-range of applicants ($n = 80$)

Age	Number
Under 20 years	7
20–29	21
30–39	33
40–49	14
50–59	5
Over 60	0

Table 12.8(b) Civil status ($n = 80$)

Married (includes co-habiting)	56
Single	17
Divorced/separated	7
Widowed	0

Table 12.8(c) Social class ($n = 80$)

Social class	Total	%	Welsh applicants	%	Non-Welsh applicants	%	% General population*
1	11	14	1	1	10	12	6
2	21	26	6	7	15	19	23
3	25	31	17	22	8	10	49
4	4	5	2	3	2	2	17
5	9	11	6	7	3	4	5
6	10	13	7	9	3	4	—

*England and Wales, 1981 census.

Table 12.9 Presymptomatic test results from seven centres

Centre	Result			Total
	Favourable	Adverse	Uninformative	
Vancouver (Hayden et al., 1988; Wiggins, 1989)*	34	21	0	55
Boston (Meissen et al., 1988)	7	4	5	16
Baltimore (Brandt et al., 1989)	30	12	13	55
Manchester (Craufurd et al., 1989)	15	15	3	33
Cardiff (Morris et al., 1989; updated)	23	12	5	40
Scotland (Brock et al., 1989)*	25	19	0	46
Netherlands (Roos et al., 1990)	17	8	2	27

* Laboratory results from several separate counselling centres.

Another major reason offered for having a test was to relieve uncertainty. The feeling that 'I want to know where I stand' was echoed by several applicants. This group in particular was more likely to write or to telephone the testing centre asking about the latest progress report. Occupational or financial reasons for being tested were rated as less important. Having the test to make decisions about marriage did not apply to most of the Cardiff series, 70% of whom were already married (or co-habiting). Evers-Kiebooms et al. (1989) in a questionnaire study of those at both 25% and 50% risk, ascertained in association with the local lay organization, reported not only on reasons for having a test but also on reasons why people declined having a test: among the latter reasons were the absence of treatment and that a 'bad result can be too difficult to live with'.

Table 12.10 Reasons for requesting predictive testing (in order of frequency; Cardiff series)

1. Inform existing children
2. Make childbearing decisions
3. Relieve uncertainty
4. Reduce frequency of HD
5. Inform partner
6. Make financial decisions
7. Make employment decisions
8. Help research
9. Estimate time span of health
10. Make decisions concerning marriage

What type of people request testing? As has been mentioned, they are more likely to have higher levels of income and occupation—a predominance of women has also been a feature of most series. In terms of more detailed psychological assessment the main conclusion is that they are mostly normal people, with low levels of psychiatric morbidity. The preliminary data correlating the results of psychological testing with the outcome of DNA analysis, discussed earlier in this chapter, also on balance support this impression of normality. An exception to this has been the unexpectedly high proportion of individuals found already to have definite onset signs of HD: it is possible that for some of these the sense that all was not well might have prompted the request for testing, or that such a request might have been more acceptable than a direct request for medical help.

What proportion of those making an initial enquiry actually persists with testing? All series have shown a progressive diminution at each stage, some of which is due to unsuitable family structure, uncertain diagnosis or other difficulties. Figure 12.6 illustrates this in the Cardiff series: it is of course likely that some unable to progress to testing because of pedigree or diagnostic

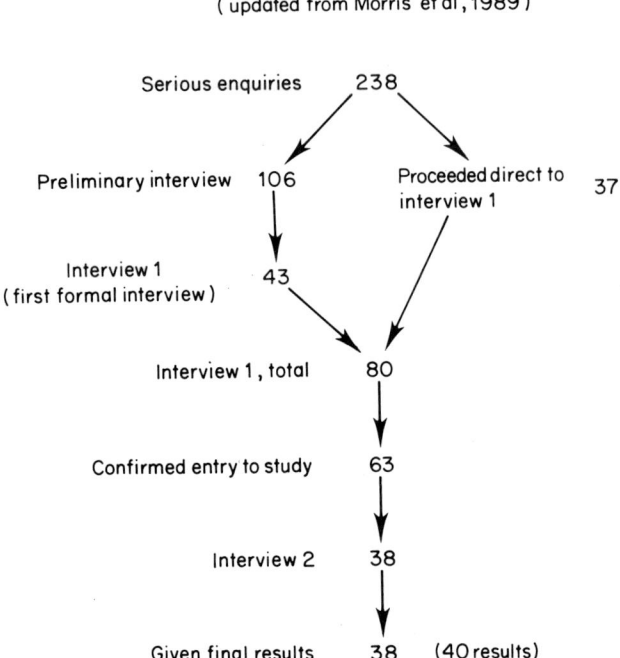

Figure 12.6 Genetic prediction in HD; the Cardiff series. The flow-chart shows the current numbers of individuals reaching successive stages of the protocol (see text for details).

limitation will do so later, as will some others who have at present decided against testing. Those centres basing their testing series principally on those contacted in their regions have found a relatively low response; only 15% of those contacted in the Manchester series took up the offer (Craufurd et al., 1989), while the proportion in the Baltimore study was also low, with only 12% of those contacted requesting to be tested (Quaid et al., 1987). There is clearly a marked difference between the numbers in the various surveys discussed earlier who say they would like to be tested and the number who actually proceed with it. In South Wales the proportion of those known to be at risk who have requested presymptomatic testing is only about 5%.

How many of those wishing to be tested prove to have the right pedigree structure and genotypes to permit this? It is difficult to put an accurate figure on this since a variable degree of preliminary 'filtering off' of applicants with a clearly unsuitable structure will occur, while equally there is considerable variation in the minimum degree of risk alteration that individuals find acceptable and that centres are prepared to accept. What is clear is that inadequate structure and diagnostic limitations are now much more serious limitations than is lack of informativeness of the probes now in use. The rapid development in the number and degree of polymorphism of these markers occurring during the time frame of the studies has greatly reduced the number of such uninformative families.

What proportion of applicants have adverse results? At first sight one might expect this to equal the number of low-risk results, but most series have found a proportion considerably less than this (Table 12.9). This may be partly a function of age, which has in many cases reduced the 50% prior risk considerably. It may also reflect the exclusion of those with definite or equivocal clinical signs of HD.

How accurate can prediction be? While the original G8 probe, giving a 4% recombination rate, was the main marker used in the early stages of the testing programmes, the discovery of closer markers has made a more accurate prediction possible in some cases. Although these new markers are known to be extremely close to the HD locus, an accurate value for recombination is rarely available, so most workers have used a value of 2% even though the true figure may be considerably less. The uncertainty of the location of the HD gene in relation to the telomere (see Chapter 10) has so far prevented the use of 'flanking markers' as a practical means of decreasing error from recombination. Currently most risk estimates are reduced in their definitiveness by ambiguities in genotype resulting from the pedigree structure rather than anything directly related to the markers themselves. Whether one should persist with testing in a situation where only a modest risk alteration is feasible is a difficult decision; our own policy has been to be guided by the wishes of the applicant when such a situation arises during testing, but not to embark on it at all unless there is reasonable expectation that the limits will be above 85% or below 15%.

What has been the immediate reaction of those given test results? At this interview, applicants were given results without delay, although one asked to be left alone in the counselling room and for the result to be put into an envelope. Those who received favourable test results were understandably overjoyed. Some cried with relief that it was 'all over' and that the result was 'worth waiting for' while others hugged their partners. Siblings who had a low-risk result felt sad and guilty when other family members had the opposite result or were already affected. Several asked to be left in private or to be allowed to phone anxious relatives and friends. Although they had persevered to reach this final stage, they then felt impatient after disclosure of results and wanted to return home as quickly as possible. This interview is clearly an inappropriate time to attempt to collect any new clinical details.

The immediate reactions of those who received adverse results depended on their previous attitude. Individuals who felt they were always gene carriers were not as distressed as those who expected the outcome to be favourable. Both groups felt restrained and tried stoically to control their emotions. Some felt that their hopes were dashed whereas the first thought of others was for their children who now had their risks altered upwards. These applicants said it would be an ordeal to tell their children and 1 year later, some still had not told them, as originally planned. Consultands who had adverse results wanted to leave sooner than those who received favourable results and they should not be unnecessarily detained.

We have less data on reactions to the test results 1 year later. Those who had favourable results had a great boost of confidence in the short term but then realized that the test result is not a panacea: 'life goes on and our problems are still there'. Some felt guilty in that if they had a favourable result, 'somebody else might have had a bad result'. However no one in this group regretted taking the test. Applicants who received adverse results needed a lot of time to come to terms with their new status. In some cases they did not do what they said they would do, like informing relatives or taking a holiday. Others have declined follow-up. Siblings who have received opposite results have tended to see less of each other than before the test and have found their relationships strained. Three out of 11 applicants who had adverse results regretted having the test. Two of the five who had uninformative results indicated that they wanted testing to continue and were eventually given an altered risk. To date, there have been no suicides, psychiatric hospital admissions or serious mental disorder on follow-up of our series.

As already indicated, most centres have followed a similar protocol with detailed counselling, psychological assessments and computerized risk estimations. The first reports of presymptomatic testing therefore have much in common, though centres tended to concentrate on different aspects of the test. Meissen et al. (1988), in the first clinical report of predictive testing, studied 47 persons at risk using G8 polymorphisms and disclosed final results to 15 applicants; five results were uninformative. Hayden et al. (1988), who used three linked DNA markers in their series, disclosed results to 20 participants; there was only one uninformative result. The Baltimore group showed that there was no significant increase in psychiatric or social morbidity up to 1 year after the test result (Brandt et al., 1989). This has also been the experience of other centres although Lam et al. (1988) reported self-injurious behaviour (including attempts at self-strangulation using an electric cord) in a woman who was enrolled in a pilot preclinical programme and given a positive clinical diagnosis of HD. The Scottish consortium has reported in detail the laboratory results of 50 predictive tests from several clinical centres (Brock et al., 1989) whereas the Manchester group has concentrated on assessing uptake of the test (Craufurd et al., 1989). In Cardiff, special attention has been given to the documentation of problems arising from testing, as discussed below.

PROBLEMS IN PREDICTIVE TESTING

All the original centres involved in presymptomatic testing were fully aware that a wide variety of problems was to be expected. Despite extensive discussions and preparation of guidelines, it was recognized that there could be no substitute for real experience of testing and that it was likely that many problems would only surface once a programme was underway. This has indeed proved to be the case, and probably one of the principal values of these early evaluations will prove to be not so much their 'success' as such, but the lessons they have provided from problems arising, which can be used to improve subsequent, more service-orientated, programmes. The Cardiff programme in particular has concentrated on the systematic documentation of problems (Morris et al., 1989), and it is worth discussing these here. Table 12.11 summarizes the main groups of problems encountered in a series of 328 individuals referred for testing; 90 of these formed part of the exclusion testing series described later. The problems could be broadly divided into those involving laboratory activities and those involving clinical and counselling aspects, illustrating the need for close liaison between laboratory and clinical staff.

Problems could begin from the point of referral, the most obvious initial problem being referrals which we and most others would regard as inappropriate for presymptomatic testing. The question of testing children is dealt with separately later, as is the special, but related, topic of preadoption testing; we were surprised how frequently such requests were encountered. Three at-risk individuals were referred by their doctors without the individual's permission. Two of these had psychiatric symptoms which were thought to indicate early HD; one patient was detained under the Mental Health Act (1983). We believe that testing without knowledge or permission is unethical; in the case of the detained patient permission could be obtained after recovery from psychosis. Two years after this original referral, we were asked on behalf of a court to test this same individual who was since charged with a criminal offence. Coercion from third parties must be avoided.

Another unforeseen referral was from a prison officer who referred an at-risk prisoner (Figure 12.7). Such an applicant would be unable to attend the clinic for counselling and disclosure of results in the usual way. However, a preliminary interview at the prison was refused because of staff shortages and so we were unable to find out whether the prisoner intended to use an adverse test result in an attempt to reduce his sentence. If that were his purpose, what would be the ethical implications? Shiwach et al. (1990) saw no ethical problem in dealing with a similar referral.

In 15 cases, the family history of HD proved to be insecure or inaccurate; this is quite apart from the larger number where the diagnostic details were inadequate. The high frequency of clinically affected individuals in all series has already been mentioned; we encountered 18 such patients who were excluded from the DNA testing programme. It proved to be a particularly difficult problem to break the news that they not only had the HD gene but that they were already affected. Most of these individuals were unaware of their symptoms; some might well have not proceeded with predictive testing, but the observation of clinical signs effectively forced the issue. It should be

Table 12.11 Problems encountered during predictive testing for HD. Updated from Morris et al. (1989)

Problems before laboratory testing	
Inappropriate use of the test	
Referral without permission of individual	4
Referral of minors by parents	22
doctors	6
social workers	3
Request from adoption societies	7
Lack of clear family history of HD	15
Clinical status of key individuals	
Applicant already clinically affected	18
Applicant showing equivocal clinical signs	3
Affected relatives unknown to applicant	5
Problems during laboratory testing	
Refusal to donate blood sample	
Affected individual	10
Other relative	2
Use of research sample for clinical application	14
Unintentional risk alteration possible	5
Anonymous testing	3
Problems with sample processing	
Inadequate labelling	6
Sample wastage (tube broken in post; DNA degraded)	23
Inappropriate release of sample to other centre	1
Use of pseudonyms by applicant	1
Non-paternity	2
Problems after predictive testing	
Result requested by insurance company	1
Refusal to allow GP access to result	3
Follow-up by testing centre refused	3

H.M Prison and Remand Centre

Dr
Dept. of Medical Genetics
University Hospital of Wales
Heath Park
CARDIFF
CF4 4XY

Your reference

Our reference

Date 7 November

Re:

This   year old man is presently serving a sentence of imprisonment and
informs me that his father suffers from Huntingdon's Chorea. He wishes
to be tested to establish whether he has the potentiality for this
condition and informs me that his brother,   , has
already given blood to University Hospital of Wales for this purpose.

I have explained to him the nature of the transmission of this disease
and its future consequence for child bearing if he is a carrier and
is at risk from this illness. I would be grateful if you could inform
me if this test could be available for him whilst he is serving his
sentence and the procedures, including the application form and containers
that are required as well as the laboratory to which the blood should
be sent.

Yours sincerely

Medical Officer

Figure 12.7 Inappropriate use of predictive testing for HD (see text). A request for testing on a convicted prisoner.

stressed that all those coming for first interview were informed in advance that a neurological examination would be carried out. The issue of how to disclose the diagnosis in such individuals has been discussed in Chapter 3, but our general policy was not to give an immediate opinion that the person was affected, but to advise further neurological and other investigations, usually as an inpatient, as a means of giving time for both patient and clinicians to adjust to the situation before the diagnosis. A related and equally difficult problem was the finding of equivocal features of HD, an exclusion criterion in our own study but only encountered three times.

Diagnostic problems were not confined to the applicants but could also involve key relatives. In addition to those already mentioned where it was uncertain whether HD was indeed the diagnosis, uncertainty over the status of other individuals could cause difficulty, either where a relative was thought to be affected but had never been investigated medically, or where the testing team knew that the person was affected but the applicant did not.

Refusal to donate a blood sample was encountered on 12 occasions, usually by uncooperative affected individuals or by their spouses wishing to protect them. Less often, unaffected relatives refused for reasons ranging from revenge to concern about the children's welfare. One healthy divorced parent refused to donate a sample because he 'wished to punish my children' who had requested prediction. In this instance, the ethical aspects were compounded by a well-meaning family doctor who, knowing of the difficulty, obtained a blood sample when seeing this individual for a separate complaint and posted it to the testing centre without informing him! Another reason offered by a parent for refusing a blood sample was that 'my son is not mature enough for the test'.

The question of who should obtain the necessary blood samples from relatives is not an easy one. We know of numerous instances from outside our own series where alarm and distress have been caused by an unexpected and unexplained request for a blood sample, often through the mail. It may naturally be concluded by such relatives that this will produce a result for themselves that they do not wish to have, and unless handled very carefully, this can constitute a serious invasion of privacy. Our policy is to ask the applicant to approach the individuals involved and then we would post the necessary blood container and mailing instructions to the relative or family doctor, explaining that the relative's status would remain unaltered and offering counselling, if requested. We prefer to use the term 'genetic studies' rather than 'predictive testing' when corresponding with family members. In fact, it should only rarely be necessary to sample individuals at risk and to do so can cause serious problems from the testing centre having too much information.

The inappropriate use of research samples is a matter of serious concern. Many HD research units hold large numbers of blood samples on patients and relatives at risk which could be used for DNA typing. It is often far from clear what individuals donating them have been told about their use and storage, an issue already brought up in Chapter 11. In our view, it is unethical to use such samples as part of predictive testing unless specific permission has been given for this. Since such samples have additional problems of degradation and possible misidentification, we have always insisted on new samples being obtained unless the person is dead. We have also tried to maintain a clear distinction between the use of service and research samples and results.

The number of wasted samples due to breakage or degradation in our series appears high, but reflects the high proportion sent by mail from relatives of applicants living at a distance. Non-paternity was encountered twice; DNA fingerprinting is now routinely included in our, and most other, programmes but non-paternity can be suspected from unexpected typing results. A final,

```
12th August

Medical Genetics
University Hospital of Wales
Heath Park
Cardiff

Dear

We have received an application from your patient,
            , requesting private medical insurance.  In order that we
may proceed with the application, we would be grateful if you
could provide us with the results of the blood tests taken
recently regarding the hereditary disease of Huntingdon's Chorea.

We have been given permission by             for us to request
this information and he is aware that any fee which may be
charged is not recoverable from

We look forward to hearing from you.

Yours sincerely
```

```
    POLICY_ADMINISTRATION_MANAGER
```

Figure 12.8 Inappropriate use of predictive testing for HD (see text). A request for results of testing from an insurance company.

quite unanticipated, laboratory problem arose from an applicant who insisted on using a pseudonym, but whose relatives were given by their true names on some samples, but with a pseudonym on others!

Problems do not end with the testing procedure. Refusal of follow-up has already been mentioned, but refusal to allow the family doctor to be given the result can create difficulty, the reasons given being perceived lack of confidentiality in the doctor's premises; that the doctor was a personal friend; or that the result might be disclosed to third parties (Ashbury, 1986). In a number of cases the applicants stated that the family doctor was not aware of the person's risk status. We only encountered one direct example of a request from an insurance company for a test result (Figure 12.8). As already discussed there is a serious need for an agreed code of conduct, and possibly legislation, regarding the use of such results. In the UK, the problem of confidentiality has been discussed at a conference where representatives of the major testing centres produced recommendations on good clinical practice in predictive testing (Tyler and Morris, 1990). Consensus was reached on many other issues such as criteria for testing, pre- and post-test counselling, and collection and storage of DNA.

After reading this account of the problems encountered in predictive testing, some workers may be deterred from embarking on it at all. This is probably reasonable if it ensures that only those with adequate training, experience, sensitivity and resources embark on it. As testing moves from the research phase to becoming an established service it is becoming increasingly important that all those involved with HD families are fully aware of the problems involved if they are to provide a service of the highest standard.

TESTING CHILDREN FOR THE HD GENE

It has already been mentioned that a surprising number of requests were received to test children at risk. Such testing is specifically excluded both in our research protocol and international guidelines (World Federation of Neurology, 1989, 1990), but could such testing ever be justified, particularly once programmes became established in a service setting? We also realized that we and others had little information on why such requests were being made.

Most requests came from parents, motivated partly by a hope that being tested might spare their children the life-long worry they themselves had gone through, but also from a wish to ensure that their children were known to be free from the gene before having their own families in future. A few parents hoped that the result would reduce their own feelings of guilt and anxiety. Barette and Marsden (1979) found that 68% of those questioned stated that they would like their family members to be tested before the age of 18. Most parents accepted after discussion that for a late-onset disorder such as HD these were decisions that the child must make for himself or herself when grown up, and were reassured by the knowledge that minor features they had noticed were unlikely to be related to HD. There were no instances of suspected juvenile HD. Occasionally parents could be remarkably persistent in their demand for testing. The mother of a 2-year-old boy, married to a man at 50% risk for HD, was adamant that she could not accept her child unless she knew whether he carried the gene; when testing was declined she sought opinions elsewhere, including America (all without success). The marriage broke up shortly afterwards. It is perhaps relevant that this woman had not been told of the family history of HD by her husband or family before her marriage; her family doctor first told her the relevant medical details early in pregnancy and she felt that 'nothing was right after this' (Tyler et al., 1990).

The principal reason against presymptomatic testing in children is that it removes their later possibility of choice; it also raises the possibility of stigmatization within the family and outside, and could have serious educational and career implications. There is no clear medical justification for testing; HD in childhood is rare and preclinical detection would not alter management. The question of the diagnostic use of DNA testing in juvenile HD is a different issue that is considered below. There seems no doubt that in the absence of therapy, childhood presymptomatic testing for HD is unethical. Whether a clinician or laboratory is justified in performing or refusing it in the face of insistence by the parents may prove to be a difficult legal question, but our view is that childhood testing should not be performed. Bloch and Hayden (1990), who recently reviewed the subject, came to a similar conclusion.

The related question of pre-adoption testing of infants at risk for HD has also arisen and because of the very real likelihood that it might be undertaken, we have given this considerable attention (Morris et al., 1988; Tyler, 1988). The main reason for requesting testing in this situation has been the right of adoptive parents to know any potential medical disorder that might arise in their child and the possibility that without such information the child might not be able to be adopted. Both are valid points but they conflict with the primary responsibility to the child itself. We believe that preadoption testing of at-risk children should not be undertaken.

The topic of childhood testing for HD has made it clear that the general subject of testing children for late-onset genetic disorders has been largely ignored. Testing of children is frequently being performed without clear indications and without sufficient thought concerning possible adverse consequences (Harper and Clarke, 1990). General guidelines should be developed that will be of help in the increasing number of neurological and other disorders that can be detected presymptomatically.

MOLECULAR TECHNIQUES IN THE DIAGNOSIS OF HD

The new techniques of molecular genetics now play a major role in the diagnosis of a number of important neurological disorders. Thus deletion of the dystrophin gene is frequent in Duchenne muscular dystrophy and can also be useful in distinguishing the X-linked late-onset Becker dystrophy from the clinically similar autosomal recessive limb girdle form (Norman et al., 1989). Specific mutations can be identified in disorders where the gene has been isolated, such as forms of amyloid neuropathy, Leber's optic atrophy and prion dementia. It would clearly be useful if we were able to do the same for HD, especially for the significant number of puzzling or atypical cases. Such a situation will undoubtedly materialize once the HD gene itself has been isolated, but the present linked markers have a very limited diagnostic role. This is principally because the isolated case, which produces the greatest diagnostic difficulties, cannot be evaluated by linked markers, while it would require a family of considerable size to be certain whether or not it was linked to the same markers as HD. The only present useful application is for the established family with HD containing a member that appears to have a different neurological disorder. Here the demonstration that the individual has either a very high or very low chance of having inherited the HD gene will be likely to influence the most likely diagnosis, though it does not on its own prove or disprove it. One such application in a child thought to have juvenile HD has been reported (Hammer et al., 1987; Schomig-Spingler et al., 1989), as mentioned in Chapter 2.

If such testing is to be used, great care must be taken to be sure that one is not in fact undertaking a predictive test. Thus testing a patient with psychiatric symptoms alone would seem unjustified without the use of the guidelines described for predictive testing. The precise borderline between diagnosis and prediction in such a situation is always going to be difficult to define and demands careful thought in each case. Craufurd et al. (1990) have pointed out some of the dangers.

IMAGING TECHNIQUES IN PRESYMPTOMATIC DETECTION

The use of various techniques as aids to the early diagnosis of HD has already been discussed in Chapter 2. Positron emission tomography (PET) has in particular been shown to provide early evidence of caudate hypometabolism (Hayden *et al.*, 1986) though it also shows abnormality in other disorders of the basal ganglia (Hosokawa *et al.*, 1987). Studies of relatives at risk have been conflicting; Mazziota *et al.* (1987) found abnormality in 18 (31%) of 58 individuals at risk, whereas Young *et al.* (1987) found no abnormalities in a series of 29 individuals. Hayden *et al.* (1987a) studied 13 at-risk family members with both DNA markers and PET scans and showed that of eight individuals predicted to have a high risk by DNA typing, three had abnormal caudate glucose metabolism; however one of the four predicted to have a low DNA risk also showed an abnormal PET result.

The most likely interpretation of these results is that the PET scan is able to show abnormality in those individuals who already have established structural and functional changes in the caudate nucleus but do not yet show symptoms. This would fit with the observation of a clinically normal person dying accidentally who at autopsy showed marked caudate atrophy (Carrasco and Mukherji, 1986). Thus an abnormal PET scan is likely to indicate onset of HD in the relatively near future, though precise time relationship and accuracy remain to be determined. It will also be important to know whether a normal PET scan gives a high probability of the individual shown to have a high risk for the HD gene remaining clinically normal for some years. This information could be of real value to the growing number of people who have had a high-risk result from DNA-based presymptomatic testing.

PRENATAL DIAGNOSIS

Pregnancies at high risk for HD are frequently terminated and the surveys of attitudes to genetic aspects of HD already discussed have shown a high proportion of relatives who would support termination of a pregnancy at risk in a family member. A major advance of DNA testing for other genetic disorders, such as Duchenne muscular dystrophy and cystic fibrosis, is that it has greatly reduced the number of normal pregnancies in those at risk that are terminated on grounds of genetic risk alone.

Prenatal diagnosis has so far played a very small part in relation to DNA-based prediction in HD. This is not just because it is a late-onset disorder, allowing many years of healthy life before onset; the attitudes of those seeking presymptomatic testing have clearly shown that planning reproductive decisions is a major reason for requesting this. Few pregnancies occur to those who are already affected, so that undertaking a specific prenatal diagnosis inevitably implies a presymptomatic test for the healthy person at risk. Few couples will wish to undergo such a double ordeal; it is much more likely that most of those with an adverse presymptomatic test result will refrain from

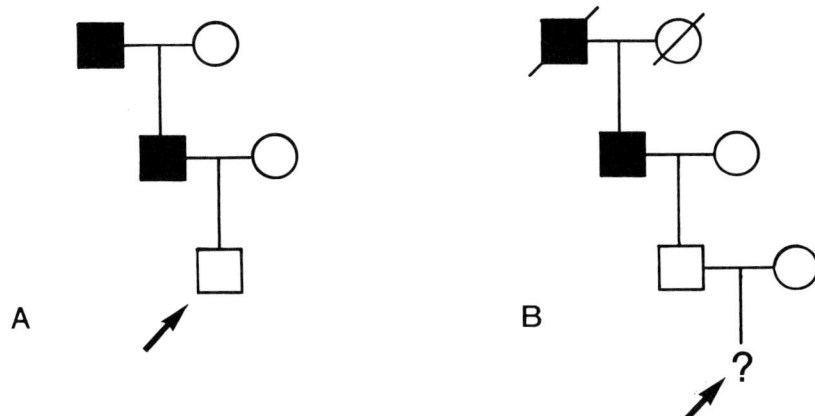

Figure 12.9 Pregnancy exclusion testing in HD and pedigree structure. (A) Optimal pedigree structure for presymptomatic testing. It is relatively uncommon for grandparents to be living when the individual requesting prediction (arrowed) is adult. (B) When exclusion testing in a pregnancy (arrowed) is requested, typing of the top generation is not required; the appropriate pedigree structure is thus present in the majority of families. See Table 12.3 and text for details.

childbearing entirely, while those with a favourable result will go ahead without the need for prenatal tests.

A somewhat different approach to prenatal diagnosis is what has been termed 'prenatal exclusion testing' (Quarrell et al., 1987; Hayden et al., 1987b; Tyler et al., 1990) which requires some explanation. It was noted earlier that only a quarter of individuals at risk for HD in South Wales had a pedigree structure appropriate for presymptomatic testing (Harper and Sarfarazi, 1985). This study also examined whether testing would be feasible for a pregnancy and found by contrast that for this group it would be possible in almost 90%, the same figure being the case for a series of pregnancies that had actually occurred in those on the HD register (Table 12.3). The reason for this difference is best illustrated in Figure 12.9. The testing procedure is essentially shifted down a generation, requiring only the parents of the individual intending a pregnancy to be living, rather than at least one grandparent.

There is however, a major difference between this approach and full prenatal diagnosis, giving rise to the term 'exclusion test'. If the marker coming from its non-HD grandparent is the one transmitted to the fetus, then its risk of HD will clearly be reduced greatly to around half the recombination rate (1–2%). However, if the marker inherited by the fetus is that from its HD grandparent, this does not mean that it will be affected by HD but simply that it will have the same risk as its parent, i.e. an increase from 25% to around 50%. This produces two problems: the first is a decision to terminate taken on the basis of a 50% risk. Secondly, the risk to the pregnancy is linked to that of the at-risk parent; should the pregnancy continue and the parent later develop HD, then that

Table 12.12 Exclusion testing in and prior to pregnancy. Cardiff series (Tyler et al., 1990)

Couples referred	90
Withdrew before testing complete	23
Fully informative for pregnancy testing	61
Pregnancies tested (15 couples)	24
Risk increased	10
Risk decreased	14

same high risk will apply to the child and an 'involuntary' presymptomatic test will have been done.

Regardless of the outcome of the prenatal testing, the situation of the parent at risk will not be altered. Thus the test enables individuals in this situation who do not wish to know their own status nevertheless to be able to have children at low genetic risk. Some younger at-risk individuals who have a considerable reproductive span ahead of them have requested this form of predictive test. In an unselected series of 50 couples where one partner was at 50% risk Tyler (1987) found that 20 (40%) wished to consider prenatal testing though six of these had doubts about whether their attitude would remain the same when pregnant. In the United States, surveys of attitudes towards prenatal testing in HD found that 32–65% of individuals at risk would use it (Kessler, 1987; Markel et al., 1987).

The Cardiff centre (Quarrell et al., 1987) embarked on an evaluation of exclusion testing before starting presymptomatic testing; in part, this was to give experience of the organizational aspects of the latter through a testing programme that was free of some of the unique problems of presymptomatic testing. A series of 90 couples has since been referred, and 74 tested with a view to future exclusion testing in pregnancy (Tyler et al., 1990). Twenty-four pregnancies from 15 couples have been tested, 10 giving an increased risk and 14 an 'exclusion'. All the increased risk pregnancies were terminated, while the low-risk ones have come to normal delivery apart from one miscarriage. Five women have undergone testing in more than one pregnancy, all achieving a low-risk result in at least one of the pregnancies (Table 12.12).

The majority of couples in this series were tested on behalf of genetic centres elsewhere, but a detailed analysis of social factors and attitudes was possible in 37, including reasons for withdrawing from the programme after prepregnancy testing. Many ethical and counselling problems arose including lack of understanding of the test. It can be concluded from this experience that prenatal exclusion testing is very different from presymptomatic testing but needs equally thorough counselling.

PREDICTIVE TESTING AS A SERVICE

The work described so far has all been research-based, with the aims of evaluating the consequences of predictive testing and the best ways in which to undertake it. The long-term follow-up of those who have been tested will need to continue for many years, but already it has become clear that testing is feasible and that there is a considerable demand for it, though this currently amounts to only a small proportion of those at risk. We are now beginning to see the evolution of service-based testing programmes; it is important to examine how they should be set up and audited, as well as what developments are likely to occur.

The laboratory aspects of testing have improved rapidly, with availability of a considerable number of closely linked and highly polymorphic probes (see Table 12.5) increasing the numbers of families that can be helped and decreasing the workload, as well as allowing a more definite risk estimate. In Britain the pattern that has developed is of regionally-based molecular genetics services closely linked to genetic counselling centres and undertaking DNA analysis for HD together with other disorders (Meredith et al., 1988). In Scotland a consortium arrangement has resulted in a single laboratory undertaking HD analysis for several clinical centres (Brock et al., 1989; Brock, 1990). Probably the best developed arrangement of this type have been reached by Canada, where 14 different genetics centres undertaking presymptomatic testing have adopted a common protocol and send samples to a single laboratory (Dayton, 1988; Wiggins, 1989). In Britain also, a common service protocol has been drawn up by those centres involved in testing; this is considerably simpler than the research-based protocols and should be usable by existing service staff, yet provide a satisfactory framework for undertaking essential clinical evaluation and counselling.

The introduction of these common service protocols is of particular importance in allowing long-term analysis of a large body of pooled data, especially the relatively small number of individuals predicted as being at high-risk. Over 150 completed presymptomatic tests were known to have been completed by January 1990 in the UK, a body of data greatly exceeding any individual research series. The existence of such protocols and their use by all those involved will also provide strong peer pressure against the future involvement of others who are unable or unwilling to provide the high standard of counselling and assessment involved. Whether or not it will deter commercial laboratories from accepting samples direct without families being adequately counselled remains to be seen, but it seems likely that the establishment of norms recognized to represent good clinical practice, could well act as a powerful legal deterrent. This will be of particular importance when the discovery of specific mutations makes prediction possible on a sample from a single individual and no longer requires analysis of the whole family.

FUTURE DEVELOPMENTS

The HD gene will undoubtedly be isolated within the coming few years; this will revolutionize our understanding of the disorder and should eventually

Table 12.13 Predictive testing based on detection of specific mutations

Advantages
　Specificity. No error from recombination
　Privacy. No need to test multiple family members unless requested
　Simplicity. PCR-based testing cheaper and quicker when mutation defined
　Only single test required when mutation already defined in family
　Population studies may show limited number of mutations in region

Disadvantages
　Ease and specificity of test could result in inadequate or absent counselling and support
　Potential abuse from testing of research and other samples taken for different purposes

---

allow rational approaches to its treatment, but there will also be an immediate impact on predictive testing. We can learn much from the experience of comparable work on such disorders as Duchenne muscular dystrophy and cystic fibrosis, where the gene and gene product were also isolated by the 'reverse genetics' approach as described in Chapter 10.

Much will depend on whether HD proves to be due to a large number of mutations of approximately equal frequency, as in Duchenne muscular dystrophy, or whether a single mutation is common or predominant, as for cystic fibrosis where a single 3-base deletion accounts for 70% of cases, at least in northern Europe (Rommens *et al.*, 1989). The evidence already discussed suggests that mutations for HD are few and that a particular population could well contain a predominance of a single mutation. If this proves to be so, it should be relatively simple to develop a series of molecular tests to detect the specific mutations. Techniques, such as the polymerase chain reaction (PCR) and other methods, should make the laboratory analyses much easier and more rapid than they are currently, while the work load will be greatly reduced by being able to dispense with the analysis of relatives and with much of the complex risk estimation that is currently required.

Regardless of the precise number of HD mutations, material from at least one affected member of a family will remain essential, since it is the identification of a specific mutation in the particular affected person that will indicate what should be looked for in the relatives at risk requesting genetic prediction. Genetic registers will have an added importance in the light of this information; it is likely that many currently separate families will prove to have common descent when they are shown to carry the same mutation, especially those in the same region, or families that have emigrated from a known place. The discovery of linkage disequilibrium (Snell *et al.*, 1989; Theilmann *et al.*, 1989) and development of specific DNA haplotypes associated with HD are steps in this direction (see Chapter 10).

Predictive testing based on a specific mutation will have both advantages and difficulties, some of which are summarized in Table 12.13. Privacy will be easier without the need for typing other family members, but should the knowledge of what type of mutation is present be made available to the entire kindred or be the private concern of the individual? Samples taken for other purposes could easily be typed without knowledge of the donor, and specific conclusions drawn. Even population screening, as currently being evaluated for cystic fibrosis carriers, might be feasible, though it is most unlikely to be considered desirable or ethical. Thus while predictive testing for HD will undoubtedly become easier in a technical sense when the gene is isolated, there will be even greater need for it to be carried out carefully and with the same degree of clinical and counselling involvement as at present.

Perhaps the greatest change in relation to presymptomatic detection will occur when we have a therapy which gives significant improvement in the course of the disease. There will then be a clear and direct benefit for the individual to know that he or she has the gene rather than, as at present, benefit mainly from the knowledge of not having it. Such a development would be likely to cause a major and probably rapid increase in demand, which should be borne in mind when services are being planned. Nothing would give greater pleasure to those involved in the demanding and challenging field of predictive testing for HD than to know that the detection of the gene could be followed by an effective means of delaying or preventing the onset of the disease itself.

REFERENCES

Asbury JFP (1986) Confidentiality and third party interests (letter). *Lancet* **ii**:1388.
Bakker E, Skraastad MI, Fisser-Groen YM *et al*. (1987) Two additional RFLPs at the D4S10 locus, useful for Huntington's disease (HD) family studies. *Nucleic Acids Research* **15**:9100.
Barette J and Marsden CD (1979) Attitudes of families to some aspects of Huntington's chorea. *Psychological Medicine* **9**:327–336.
Baro F (1973) A neuropsychological approach to early detection of Huntington's chorea. *Advances in Neurology* **1**:329–338.
Bates M (1981) Ethics of provocative test for Huntington's disease (letter). *New England Journal of Medicine* **304**:175–176.
Bell J (1934) Huntington's chorea. In: *The Treasury of Human Inheritance* (ed RA Fisher), pp. 1–77. Cambridge: University Press.
Bell J and Haldane JBS (1937) The linkage between the genes for colour-blindness and haemophilia in man. *Proceedings of the Royal Society* B **123**:119–150.
Berg K and Fletcher J (1986) Ethical and legal aspects of predictive testing (letter). *Lancet* **i**:1043.
Bird SJ (1985) Presymptomatic testing for Huntington's disease. *Journal of the American Medical Association* **253**:3286–3291.
Bloch M and Hayden MR (1990) Predictive testing for Huntington's disease in childhood: challenges and implications. *American Journal of Human Genetics* **46**:1–4.
Brandt J, Quaid K, Folstein SE *et al*. (1989) Presymptomatic diagnosis of delayed-onset disease with linked DNA markers: the experience in Huntington's disease. *Journal of the American Medical Association* **261**:3108–3114.
Brock DJH (1990) The Scottish Molecular Genetic Consortium. *Scottish Medical Journal* **34**:483–484.
Brock DH, Mennie M, Curtis A *et al*. (1989) Predictive testing for Huntington's disease with linked DNA markers. *Lancet* **ii**:463–466.
Caraceni TA, Panerai AE, Parati EA *et al*. (1977) Altered growth hormone and prolactin responses to dopaminergic stimulation in Huntington's chorea. *Journal of Clinical Endocrinology and Metabolism* **44**:870–875.

Carrasco LH and Mukherji CS (1986) Atrophy of corpus striatum in normal male at risk of Huntington's chorea. *Lancet* **i**, 1388–1389.
Cawein M and Turney F (1971) Test for incipient Huntington's chorea (letter). *New England Journal of Medicine* **284**:504.
Chandler JH (1966) EEG in prediction of Huntington's chorea: an eighteen year follow-up. *Electroencephalography and Clinical Neurology* **21**:79–80.
Chandler JH (1969) EEG in the prediction of Huntington's chorea. *Progress in Neurogenetics* **1**:564–585.
Conneally PM, Wallace MR, Gusella JF and Wexler NF (1984) Huntington's disease: estimation of heterozygote status using linked genetic markers. *Genetic Epidemiology* **1**:81–88.
Conneally PM, Haines JL, Tanzi RE et al. (1989) Huntington disease: no evidence for locus heterogeneity. *Genomics* **5**:304–308.
Craufurd D, Dodge A, Kerzin-Storrar L et al. (1989) Uptake of presymptomatic predictive testing for Huntington's disease. *Lancet* **ii**:603–605.
Craufurd D, Donnai D, Kerzin-Storrar L and Osborn M (1990) Testing of children for 'adult' genetic diseases (letter). *Lancet* **i**:1406.
Dahlstrom WG and Welsh GS (1960) *An MMPI Handbook: a Guide to Use in Clinical Practice and Research*. Minnesota: University Press.
Dayton S (1988) Canada pioneers national screening for Huntington's disease. *New Scientist* **119**:(22 Sept) 26.
Esteban A, Mateo D and Gimenez-Roldan S (1981) Early detection of Huntington's disease: blink reflex and levodopa lead in presymptomatic and incipient subjects. *Journal of Neurology, Neurosurgery and Psychiatry* **44**:43–48.
Evers-Kiebooms G, Cassiman JJ and van den Berghe H (1987) Attitudes towards predictive testing in Huntington's disease: a recent survey in Belgium. *Journal of Medical Genetics* **24**:275–279.
Evers-Kiebooms G, Swerts A, Cassiman JJ and van den Berghe H (1989) The motivation of at-risk individuals and their partners in deciding for or against predictive testing for Huntington's disease. *Clinical Genetics* **35**:29–40.
Fahn S (1980) Levodopa provocative test for Huntington's disease (letter). *New England Journal of Medicine* **303**:884.
Falek A (1969) Preclinical detection of Huntington's chorea. Preliminary report. In: *Progress in Neurogenetics* (eds A Barbeau and JR Brunette), pp. 529–533. Amsterdam: Excerpta Medica Foundation.
Farrer LA, Myers RH, Cupples LA and Conneally PM (1988) Considerations in using linkage analysis as a presymptomatic test for Huntington's disease. *Journal of Medical Genetics* **25**:577–588.
Fedio P, Cox CS, Neophytides A, Canal-Frederick G and Chase TN (1979) Neuropsychological profile of Huntington's disease: patients and those at risk. *Advances in Neurology* **23**:239–255.
Gaylin W (1972) Genetic engineering: the ethics of knowing. *New England Journal of Medicine* **286**:1361–1362.
Gilliam TC, Bucan M, MacDonald ME et al. (1987) A DNA segment encoding two genes very tightly linked to Huntington's disease. *Science* **238**:950–952.
Gusella JF (1986) Probes in Huntington's chorea (letter). *Nature* **320**:21–22.
Gusella JF, Wexler NS, Conneally PM et al. (1983) A polymorphic DNA marker genetically linked to Huntington's disease. *Nature* **306**:234–238.
Hammer J, Machler M, Schmid W and Schomig-Spingler M (1987) Linked DNA markers in clinical diagnosis of juvenile Huntington's disease (letter). *Lancet* **ii**:1088.
Harper PS (1986) The prevention of Huntington's chorea. The Milroy lecture. *Journal of the Royal College of Physicians of London* **20**:7–14.
Harper, PS and Clarke, A (1990) Should we test children for 'adult' genetic disease? *Lancet* **i**:1205–1206.
Harper PS and Morris M (1989) Predictive testing for Huntington's disease: progress and problems [Editorial]. *British Medical Journal* **298**:404–405.
Harper PS and Sarfarazi M (1985) Genetic prediction and family structure in Huntington's chorea. *British Medical Journal* **290**:129–131.
Harvald B (1951) Prediction of Huntington's chorea by EEG. *American Journal of Psychiatry* **108**:295–297.
Hayden MR, Vinik AI, Paul M and Beighton P (1977) Impaired prolactin release in Huntington's chorea: evidence for dopaminergic excess. *Lancet* **ii**:423–426.
Hayden MR, Hewitt J, Stoessl AJ, Clark C, Ammann W and Martin WRW (1987a) The combined use of positron emission tomography and DNA polymorphisms for preclinical detection of Huntington's disease. *Neurology* **37**:1441–1447.

Hayden MR, Hewitt J, Kastelein JJP et al. (1987b) First trimester prenatal diagnosis for Huntington's disease with DNA probes. *Lancet* **i**:1284–1285.
Hayden MR, Robbins C, Allard D et al. (1988) Improved predictive testing for Huntington disease by using three linked DNA markers. *American Journal of Human Genetics* **43**:689–694.
Hemphill, M (1972) Tests for presymptomatic Huntington's chorea. *New England Journal of Medicine* **287**:823–824.
Hosokawa S, Ichiya Y, Kuwabara Y et al. (1987) Positron emission tomography in cases of chorea with different underlying diseases. *Journal of Neurology, Neurosurgery and Psychiatry* **50**:1284–1287.
Jason GW, Pajurkoua EM, Sachowersky O et al. (1988) Presymptomatic neuropsychological impairment in HD. *Archives of Neurology* **45**:769–773.
Kessler S (1987) Psychiatric implications of presymptomatic testing for Huntington's disease. *American Journal of Orthopsychiatry* **87**:212–219.
Klawans HL, Paulson GW and Barbeau A (1970) Predictive test for Huntington's chorea. *Lancet* **ii**:1185–1186.
Klawans HL, Paulson GW, Ringel SP and Barbeau A (1972) *l*-Dopa in the detection of presymptomatic Huntington's chorea. *New England Journal of Medicine* **286**:1332–1334.
Klawans HL, Paulson GW, Ringel SP and Barbeau A (1973) The use of *l*-Dopa in the presymptomatic detection of Huntington's chorea. *Advances in Neurology* **1**:295–300.
Klawans HL, Goetz CG, Paulson GW and Barbeau A (1980) Levodopa and presymptomatic detection of Huntington's disease: eight-year follow up. *New England Journal of Medicine* **302**:511–512.
Kloepfer B and Davidson HH (1962) *The Rorschach Technique: an Introductory Manual*. Philadelphia: Harcourt Brace Jovanovich.
Koller WC and Davenport S (1984) Genetic testing in Huntington's disease. *Annals of Neurology* **16**:511–512.
Kondo L, Ohta H, Yazaki M, Ikeda JE, Gusella JFV and Kanazawa I (1990) Exclusion mapping of the hereditary dentatorubropallidoluysian atrophy gene from the Huntington's disease locus. *Journal of Medical Genetics* **27**:105–106.
Lam RW, Bloch M, Jones BD et al. (1988) Psychiatric morbidity associated with early clinical diagnosis of Huntington disease in a predictive testing program. *Journal of Clinical Psychiatry* **49**:444–447.
Lamport AT (1987) Presymptomatic testing for Huntington's chorea: ethical and legal issues. *American Journal of Medical Genetics* **26**:307–314.
Lyle OE and Gottesman II (1977) Premorbid psychometric indicators of the gene for Huntington's disease. *Journal of Consulting and Clinical Psychology*. **45**:1011–1022.
Lyle OE and Gottesman II (1979) Subtle cognitive deficits as 15 to 20-year precursors of Huntington's disease. *Advances in Neurology* **23**:227–238.
Maddox J (1986) Proprietary rights to research. *Nature* **320**:11.
Markel DS, Young AB and Penney JB (1987) At-risk persons' attitudes toward predictive and prenatal testing of Huntington's disease in Michigan. *American Journal of Medical Genetics* **26**:295–305.
Marsden CD (1981) Prediction of Huntington's disease (letter). *Annals of Neurology* **10**:202–203.
Mastromauro C, Myers RH and Berkman B (1987) Attitudes towards predictive testing in Huntington disease. *American Journal of Medical Genetics* **26**:271–282.
Mazziotta J, Phelps M, Pahl JJ et al. (1987) Reduced cerebral glucose metabolism in asymptomatic subjects at risk for Huntington's disease. *New England Journal of Medicine* **316**:357–362.
Meissen GJ and Berchek RL (1988) Intentions to use predictive testing by those at risk for Huntington's disease: implications for prevention. *American Journal of Community Psychology* **16**:261–277.
Meissen GJ, Myers RH, Mastromauro CA et al. (1988) Predictive testing for Huntington's disease with use of a linked DNA marker. *New England Journal of Medicine* **318**:535–542.
Meredith AL, Upadhyaya M, Lazarou LP et al. (1988) Molecular genetics in clinical practice: the evolution of a DNA diagnostic service. *British Medical Journal* **297**:843–846.
Misra VP, Baraitser M, and Harding AE (1988) Genetic prediction in Huntington's disease: what are the limitations imposed by pedigree structure? *Movement Disorders* **3**:233–236.
Morris M, Tyler A and Harper PS (1988) Adoption and genetic prediction for Huntington's disease. *Lancet* **ii**:1069–1070.
Morris M, Tyler A, Lazarou L, Meredith L and Harper PS (1989) Problems in genetic prediction for Huntington's disease. *Lancet* **ii**:601–603.

Myers RH and Falek A (1979) Quantifications of muscle tremor of Huntington's disease patients and their offspring in an early detection study. *Biological Psychiatry* **14**:777–789.

Neel JV (1949) The detection of the genetic carriers of hereditary disease. *American Journal of Human Genetics* **1**:19–36.

Neophytides AN, Di Chiro G, Barron SA and Chase TN (1979) Computed axial tomography in Huntington's disease and persons at-risk for Huntington's disease. *Advances in Neurology* **23**:185–191.

Norman A, Thomas N, Coakley J and Harper PS (1989) Distinction of Becker from limb-girdle muscular dystrophy by means of dystrophin cDNA probes. *Lancet*, **466**–468.

Palm JD (1973) Longitudinal study of a preclinical test program for Huntington's chorea. *Advances in Neurology* **1**:311–324.

Patterson RM, Bagghi BK and Test A (1948) The prediction of Huntington's chorea: an electroencephalographic and genetic study. *American Journal of Psychiatry* **104**:786–797.

Perry TL (1981) Some ethical problems in Huntington's chorea. *Canadian Medical Association Journal* **125**:1098–1100.

Petit H and Milbled G (1973) Anomalies of conjugate ocular movements in Huntington's chorea: application to early detection. *Advances in Neurology* **1**:287–294.

Quaid K, Brandt J and Folstein SE (1987) The decision to be tested for Huntington's disease (letter). *Journal of the American Medical Association* **257**, 3362.

Quarrell OWJ, Meredith AL, Tyler A, Youngman S, Upadhyaya M and Harper PS (1987) Exclusion testing for Huntington's disease in pregnancy with a closely linked DNA marker. *Lancet* **i**:1281–1283.

Quarrell OWJ, Youngman S, Sarfarazi M and Harper PS (1988) Absence of close linkage between benign hereditary chorea and the locus D4S10 (probe G8). *Journal of Medical Genetics* **25**:191–194.

Rommens JM, Iannuzzi MC, Kerem B-S et al. (1989) Identification of the cystic fibrosis gene: chromosome walking and jumping. *Science* **245**:1059–1065.

Roos RA, Vegter van der Vlis M, Tibben A, Skraastad MI and Pearson PL (1990) Procedure en eerste resultaten van presymptomatisch DNA-onderzoek bij de chorea van Huntington. *Nederlands Tijdschrift voor Geneeskunde* **134**:704–707.

Rothstein E (1971) Huntington's chorea: optimistic view. *New England Journal of Medicine* **285**:751.

Sandkuyl L and Ott J (1989) Determining informativity of marker typing for genetic counselling in a pedigree. *Human Genetics* **82**, 159–162.

Schoenfeld M, Myers RH, Berkman B et al. (1984) Potential impact of a predictive test on the gene frequency of Huntington disease. *American Journal of Human Genetics* **18**:423–429.

Schomig-Spingler M, Hammer J and Kruse K (1989) DNA analysis in juvenile Huntington disease. *European Journal of Pediatrics* **148**:447–449.

Shokeir, MHK (1975) Investigations on Huntington's disease in the Canadian prairies. *Clinical Genetics* **7**:345–348.

Smith B, Skarecky D, Bengtsson U, Magenis RE, Carpenter N and Wasmuth JJ (1988) Isolation of DNA marker in the direction of the Huntington disease gene from the G8 locus. *American Journal of Human Genetics* **42**:335–344.

Snell RG, Lazarou L, Youngman S et al (1989) Linkage disequilibrium in Huntington's disease: an improved localisation for the gene. *Journal of Medical Genetics* **26**:673–675.

Spillane J and Phillips R (1937) Huntington's chorea in South Wales. *Quarterly Journal of Medicine* **6**:403–423.

Stern R and Eldridge R (1975) Attitudes of patients and their relatives to Huntington's disease, *Journal of Medical Genetics* **12**:217–223.

Stevens DL (1971) Tests for Huntington's chorea. *New England Journal of Medicine* **285**:413–414.

Strauss ME and Brandt J (1990) Are there neuropsychologic manifestations of the gene for Huntington's disease in asymptomatic, at-risk individuals? *Archives of Neurology* **47**:905–908.

Teltscher B and Polgar S (1981) Objective knowledge about Huntington's disease and attitudes towards predictive tests of persons at risk. *Journal of Medical Genetics* **18**:31–39.

Theilmann J, Kanani S, Shiang R et al. (1989) Non-random association between the alleles detected at D4S95 and D4S98 and the Huntington's disease gene. *Journal of Medical Genetics* **26**:676–681.

Tyler A and Harper PS (1983) Attitudes of subjects at-risk and their relatives towards genetic counselling in Huntington's chorea. *Journal of Medical Genetics* **20**:179–188.

Tyler A (1987) Genetic counselling in Huntington's chorea. In: *Genetic Risk, Risk Perception, and Decision Making* (eds G Evers-Kiebooms, J -J Cassiman, H Van den Berghe and G d'Ydewalle), pp. 85–97. New York: Liss.

Tyler A, Quarrell OWJ, Lazarou LP, Meredith AL and Harper PS (1990) Exclusion testing in pregnancy for Huntington's disease. *Journal of Medical Genetics* **27**:488–495.

Wasmuth JJ, Hewitt J, Smith B et al. (1988) A highly polymorphic locus very tightly linked to the Huntington's disease gene. *Nature* **332**:734–736.

Watt DC, Lindenbaum RH, Jonasson JA and Edwards JH (1986) Probes in Huntington's chorea (letter). *Nature* **320**:21.

Wexler NS (1979) Perceptual-motor, cognitive and emotional characteristics of persons at risk for Huntington's disease. *Advances in Neurology* **23**:257–271.

Whaley WL, Michiels F, MacDonald ME et al. (1988) Mapping of D4S98/S114/S113 confines the Huntington's defect to a reduced physical region at the telomere of chromosome 4. *Nucleic Acids Research* **16**:11769–11780.

Wiggins S (1989) Early follow-up of persons participating in the Canadian National Collaborative Study of Predictive Testing. *American Journal of Human Genetics* **45 (suppl.)**:A282.

World Federation of Neurology Research Group on Huntington's Disease (1989) Ethical issues policy statement on Huntington's disease molecular genetics predictive test. *Journal of the Neurological Sciences* **94**:327–332.

World Federation of Neurology Research Group on Huntington's Disease (1990) Ethical issues policy statement on Huntington's disease molecular genetics predictive test. *Journal of Medical Genetics* **27**:34–38.

Youngman S, Sarfarazi M, Quarrell OWJ et al. (1986) Studies of a DNA marker (G8) genetically linked to Huntington disease in British families. *Human Genetics* **73**:333–339.

Youngman S, Shaw DJ, Gusella JF et al. (1988) A DNA probe, D5 (D4S90) mapping to human chromosome 4p16.3. *Nucleic Acids Research* **16**: 1648.

Youngman S, Sarfarazi M, Bucan M et al. (1989) A new DNA marker (D4S90) is terminally located on the short arm of chromosome 4, close to the Huntington's disease gene. *Genomics* **5**:807–809.

Zweig RM, Koven SJ, Hedreen JC, Maestri NE, Kazazian HH and Folstein SE (1989) Linkage to the Huntington's disease locus in a family with unusual clinical and pathological features, *Annals of Neurology* **26**:78–84.

Appendix 1

HUNTINGTON'S DISEASE RESEARCH IN WALES.
A CHRONOLOGICAL LIST OF PUBLICATIONS

Spillane J and Phillips R (1937) Huntington's Chorea in South Wales. *Quarterly Journal of Medicine* 6:405–425.
Harper PS (1976) Genetic variation in Wales. *Journal of the Royal College of Physicians of London* 10:321–332.
Harper PS (1978) Benign hereditary chorea: clinical and genetic aspects. *Clinical Genetics* 13:85–95.
Harper PS, Walker DA, Tyler A, Newcombe RG and Davies K (1979) Huntington's chorea: the basis for long-term prevention *Lancet* ii:346–349.
Walker D (1979) Huntington's chorea in South Wales. University of Liverpool:MD thesis.
Tyler A (1980) Marriage, sex and counselling in Huntington's disease. *Journal of Sexuality and Disability* 3:159–160.
Volkers WS, Went LN, Vegter van der Vlis M, Harper PS and Caro A (1980) Genetic linkage studies in Huntington's chorea. *Annals of Human Genetics* 44:75–79.
Newcombe RG (1981) A life table for onset of Huntington's chorea. *Annals of Human Genetics* 45:375–385.
Newcombe RG, Walker DA, Harper PS (1981) Factors influencing age at onset and duration of survival in Huntington's chorea. *Annals of Human Genetics* 45:387–396.
Harper PS, Tyler A, Smith S, Jones P, Newcombe RG and McBroom V (1981) Decline in the predicted incidence of Huntington's chorea associated with systematic genetic counselling and family support. *Lancet* ii:411–413.
Walker DA, Harper PS, Wells CGC, Tyler A, Davies K and Newcombe RG (1981) Huntington's chorea in South Wales: a genetic and epidemiological study. *Clinical Genetics* 19:213–221.
Harper PS, Tyler A, Smith S, Jones P, Newcombe RG and McBroom V (1982) A genetic register for Huntington's chorea in South Wales. *Journal of Medical Genetics* 19:241–245.
Tyler A (1982) Sharing the diagnosis of Huntington's chorea. Community care, September.
Tyler A (1982) The social, personal and economic burden of Huntington's chorea in South Wales. University of Wales: MSc thesis.
Tyler A, Harper PS, Walker DA, Davies K and Newcombe RG (1982) The socioeconomic burden of Huntington's chorea in South Wales. *Journal of Biosocial Science* 14:379–389.
Harper PS (1983) A genetic marker for Huntington's disease. *British Medical Journal* 287:1567–1568.
Harper PS (1983) Genetics of neurological disorders. In: Swash M (ed.) *Scientific Basis of Clinical Neurology*.
Tyler A and Harper PS (1983) Attitudes of subjects at-risk and their relatives towards genetic counselling in Huntington's chorea. *Journal of Medical Genetics* 20:179–188.
Tyler A, Harper PS, Davies K and Newcombe RG (1983) Family breakdown and stress in Huntington's chorea. *Journal of Biosocial Science* 15:127–138.
Walker DA, Harper PS, Newcombe RG and Davies K (1983) Huntington's chorea in South Wales: mutation, fertility, and genetic fitness. *Journal of Medical Genetics* 20:12–17.
Harper PS (1984) Localisation of the gene for Huntington's chorea. *Trends in Neurosciences* 7:1–2.
Harper PS (1985) Huntington's disease. *Medicine International*: 601–603.

Harper PS and Tyler A (1985) Huntington's chorea: problems in adoption and fostering. *Adoption and Fostering* **9**:47–51.

Harper PS and Sarfarazi M (1985) Genetic prediction and family structure in Huntington's chorea. *British Medical Journal* **290**:129–131.

Harper PS, Youngman S, Anderson MA et al. (1985) Genetic linkage between Huntington's disease and the DNA polymorphism G8 in south Wales families. *Journal of Medical Genetics* **22**:447–450.

Upadhyaya M, Reynolds GP and Harper PS (1985) Recombinant DNA studies on stored necropsy brain samples from patients with Huntington's chorea. *Journal of Clinical Pathology* **38**:1093–1095.

Harper PS (1986) The prevention of Huntington's chorea: the Milroy lecture, *Journal of the Royal College of Physicians of London* **20**:7–14.

Quarrell OWJ and Harper PS (1986) Huntington's chorea without dementia. *British Journal of Psychiatry* **148**: 612–613.

Quarrell OWJ, Tyler A, Cole G and Harper PS (1986) The problem of isolated cases of Huntington's disease in South Wales 1974–1984. *Clinical Genetics* **30**:433–439.

Sarfarazi M (1986) Report on genetic linkage analysis between Huntington's disease and the G8 DNA polymorphism. *Genetic Epidemiology* (**suppl. 1**):259–264.

Youngman S, Sarfarazi M, Quarrell OWJ et al. (1986) Studies of a DNA marker (G8) genetically linked to Huntington disease in British families. *Human Genetics* **73**:333–339.

Quarrell OWJ and Harper PS (1987) Is Huntington's chorea predictable and preventable? In: *More Dilemmas in the Management of the Neurological Patient* (eds C Warlow and J Garfield), pp. 36–44. Edinburgh: Churchill Livingstone.

Quarrell OWJ, Meredith AL, Tyler A, Youngman S, Upadhyaya M and Harper, PS (1987) Exclusion testing for Huntington's disease in pregnancy with a closely linked DNA marker. *Lancet* **i**:1281–1283.

Sarfarazi M, Quarrell OWJ, Wolak G and Harper PS (1987) An integrated microcomputer system to maintain a genetic register for Huntington disease. *American Journal of Medical Genetics* **28**:999–1006.

Tyler A (1987) Genetic counselling in Huntington's chorea. In: *Genetic Risk, Risk Perception, and Decision Making* (eds G Evers-Kiebooms, J-J Cassiman, H Van den Berghe, and G d'Ydewalle), pp. 85–97. New York: Liss.

Harper PS, Quarrell OWJ and Youngman S (1988) Huntington's disease: prediction and prevention. *Philosophical Transactions of the Royal Society of London* **319**:285–298.

Meredith AL, Upadhyaya M, Lazarou LP et al. (1988) Molecular genetics in clinical practice: the evolution of a DNA diagnostic service. *British Medical Journal* **297**:843–846.

Morris M, Tyler A and Harper PS (1988) Adoption and genetic prediction for Huntington's disease. *Lancet* **ii**:1069–1070.

Quarrell OWJ, Tyler A, Jones MP, Nordin M and Harper PS (1988) Population studies of Huntington's disease in Wales. *Clinical Genetics* **33**:189–195.

Quarrell OWJ, Youngman S, Sarfarazi M and Harper PS (1988) Absence of close linkage between benign hereditary chorea and the locus D4S10 (probe G8). *Journal of Medical Genetics* **25**:191–194.

Tyler A (1988) Adoption policy in relation to presymptomatic testing for Huntington's disease. *Adoption and Fostering* **12**:52.

Youngman S, Shaw DJ, Gusella JF et al. (1988) A DNA probe, D5 (D4S90) mapping to human chromosome 4p16.3. *Nucleic Acids Research* **16**:1648.

Harper PS and Morris M (1989) Predictive testing for Huntington's disease: progress and problems. *British Medical Journal* **298**:404–405.

Morris M (1989) Predictive testing in Huntington's chorea. In: *Presymptomatic Diagnosis* (eds JJP van de Kamp, RAC Roos and HGM Rooijmans), pp. 37–44. Leiden: University Press.

Morris M and Harper PS (1989) Recent advances in Huntington's disease. *Royal Society of Medicine Current Medical Literature (Neurology)* **5**:67–70.

Morris M and Harper PS (1989) Genetic counselling for Huntington's disease. *Irish Medical Journal* **82**:99–100.

Morris M, Tyler A, Lazarou J, Meredith L and Harper PS (1989) Problems in genetic prediction for Huntington's disease. *Lancet* **ii**:601–603.

Quarrell, OWJ (1989) Prevention and prediction of Huntington's disease. University of London: MD thesis.

Robbins C, Theilmann J, Youngman S et al. (1989) Evidence from family studies that the gene causing Huntington disease is telomeric to D4S95 and D4S90. *American Journal of Human Genetics* **44**:422–425.

Snell RG, Lazarou L, Youngman S et al. (1989) Linkage disequilibrium in Huntington's disease: an improved localisation for the gene. *Journal of Medical Genetics* **26**:673–675.

Youngman S (1989) Studies on Huntington's disease using recombinant DNA techniques. University of Wales College of Medicine: PhD thesis.

Youngman S, Sarfarazi M, Bucan M et al. (1989) A new DNA marker (D4S90) is terminally located on the short arm of chromosome 4, close to the Huntington's disease gene. *Genomics* **5**:802–809.

Bucan M, Zimmer M, Whaley WL et al. (1990) Physical maps of 4p16.3, the area expected to contain the Huntington disease mutation. *Genomics* **6**:1–15.

Harper PS and Clarke A (1990) Should we test children for 'adult' genetic diseases? *Lancet* **i**:1205–1206.

Harper PS, Morris M and Tyler A (1990) Genetic testing for Huntington's disease. *British Medical Journal* **300**:1089–1090.

Morris M (1990) Huntington's disease: presymptomatic testing. *Current Opinion in Neurology and Neurosurgery* **3**:337–341.

Morris M and Harper PS (1990) Prevention and prediction in Huntington's disease. In: *Advances in Psychiatric Genetics* (eds P McGuffin, and R Murray). London: Heinemann.

Morris M and Tyler A (1990) World Federation of Neurology Research Group in Huntington's disease. *Journal of Medical Genetics* **27**:211–212.

Tyler A and Morris M (1990) National symposium on predictive testing. *Journal of Medical Ethics* **16**:41–42.

Tyler A, Quarrell OWJ, Lazarou L, Meredith AL and Harper PS (1990) Exclusion testing in pregnancy for Huntington's disease. *Journal of Medical Genetics* **27**:488–495.

Harper PS and Morris MJ (1991) *Predictive and Presymptomatic Testing for Genetic Disorders: Lessons from Huntington's Disease*. London: Galton Institute.

Appendix 2

CRITERIA FOR QUANTIFIED STAGING OF FUNCTIONAL CAPACITIES.
Reproduced from Shoulson (1986) by permission of the author and publishers

A. Engagement in occupation
   3. *Usual level* – full-time salaried employment, actual or potential (e.g. job offer or qualified), with normal work expectations and satisfactory performance.
   2. *Lower level* – full- or part-time salaried employment, actual or potential, with a lower than usual work expectation (relative to patient's training and education) but with satisfactory performance.
   1. *Marginal level* – part-time voluntary or salaried employment, actual or potential, with lower expectation and less than satisfactory work performance.
   0. *Unable* – totally unable to engage in voluntary or salaried employment.

B. Capacity to handle financial affairs
   3. *Full* – normal capacity to handle personal and family finances (income tax, balancing checkbook, paying bills, budgeting, shopping).
   2. *Requires slight assistance* – mildly impaired ability to handle financial affairs, such that accustomed routine responsibilities require some organization and assistance from family member or financial advisor.
   1. *Requires major assistance* – moderately impaired ability to handle financial affairs, such that patient comprehends the nature and purpose of routine financial procedures and is competent to handle funds but requires major assistance in the performance of these tasks.
   0. *Unable* – patient is unable to comprehend the financial process and is totally unable to perform tasks related to routine financial procedures.

C. Capacity to manage domestic responsibilities
   2. *Full* – no impairment in performance of routine domestic tasks (cleaning, laundering, dishwashing, table setting, recipes, lawn care, answering mail, civic responsibilities).
   1. *Impaired* – moderate impairment in performance of routine domestic tasks, such that patient requires some assistance in carrying out these tasks.
   0. *Unable* – marked impairment in function and marginal performance; requires major assistance.

D. Capacity to perform activities of daily living
   3. *Full* – complete independence in eating, dressing, and bathing.
   2. *Mildly impaired* – somewhat laboured performance:
      in eating (avoids certain foods that cause chewing and swallowing problems)
      in dressing (difficulty in fine tasks only, e.g. buttoning or tying shoes)
      in bathing (difficulty in fine performance only, e.g. brushing teeth); requires only slight assistance.

Appendix 2

    1. *Moderately impaired* – substantial difficulty
        in eating (swallows only liquid or soft foods and requires considerable assistance)
        in dressing (performs only gross dressing activities and requires assistance with everything else)
        in bathing (performs only gross bathing tasks, otherwise requires assistance).
    0. *Severely impaired* – requires total care in activities of daily living.
E. Care can be provided at:
    2. *Home* – patient living at home, and family readily able to meet care needs.
    1. *Home or extended care facility* – patient may be living at home, but care needs would be better provided at an extended care facility,
    0. *Total care facility only* – patient requires full-time, skilled nursing care.

Appendix 3

VOLUNTARY SOCIETIES CONCERNED WITH HUNTINGTON'S DISEASE

Australia
Australian Huntington's Disease Association
PO Box 247
Lidcombe
New South Wales 2141
Australia

Austria
Oestreichische Huntington Hilfe
Belvedeweg 17/22
1040 Vienna
Austria

Belgium
Huntington Liga,
Neervelden 12
B2130 Brasschaat
Belgium

Canada
Huntington Society of Canada
13 Water Street North
PO Box 33
Cambridge
Ontario N1R 5T8
Canada

Czechoslovakia
Dr Eva Seemanova
Geneticke Oddeleni UVVD
V Uvalu 84
150 06 Praha 5 – Motol FDL UK
Czechoslovakia

Denmark
Landsforeningen mod Huntington's Chorea
Blegdamsvej 3
DK-2200 Copenhagen N.
Denmark

Finland
Ms Marjatta Sipponen
Dept of Medical Genetics
Vaistoliitto
Kalerankath 16
SF-00100 Helsinki
Finland

France
Association Huntington de France
Residence Manin
119 rue Manin
75019 Paris
France

Germany
Huntington Gesellschaft, e.V.
Oberstadtstrasse 23
7452 Haigerloch
Germany

Huntington Gruppen in der Familienholfe, e.V.
Bahnhofstr. 7A
3550 Marburg
Germany

Dr O. Riess
Humbolt Universität
Institut für Medizinische Genetik
Postfach 140
Berlin
Germany

India
Mrs Mano Singh
F-30 Geetanjali Enclave
New Delhi 49
India

Appendix 3

Ireland
Huntington's Disease Association of Ireland
279 Sutton Park
Dublin 13
Republic of Ireland

Israel
Mira Dangoor
Kibbutz Ein Hashojet
19232 Israel

Italy
Associazione per Combattere la Corea di Huntington
Instituto Neurologica
'C Besta'
Via Celoria 11
Milano 20133
Italy

Mexico
Dr Ma Elisa Alonson Vilatela
Dept de Genetica
Insituto de Neurologia y Neurocirugia
Insurgentes sur 3877 Col lo Fama
Del Tlalpan
DIP 14410 Mexico DF

The Netherlands
Vereniging van Huntington
Postbus 30470
2500 GL s'Gravenhage
Netherlands

New Zealand
New Zealand Huntington's Disease Association
GPO 25-088 Christchurch
New Zealand

Norway
Landsforeningen for Huntingtons Sykdom
Department of Social Medicine
Kattegat 1
4500 Mandal
Norway

South Africa
Huntington's Disease Society of South Africa
PO Box 70
Steenberg Cape
South Africa 7945

Spain
Ms Ascuncion Martinez-Descals
c/Puerto De Santa Maria 122 30
28043 Madrid
Spain

Sweden
Mr Patrick Strand
Skallbergsgaten 11c
S-772 21 Vasteras
Sweden

Switzerland
Mrs Regula Bischof-Brunold
Wartensee
CH 9400 Rorschacherberg
Switzerland

United Kingdom
Huntington's Disease Association
108 Battersea High Street,
London SW11 3HP
United Kingdom

The Huntington's Society
PO Box 26
Selby
North Yorkshire
YO8 0GZ
United Kingdom

United States
Huntington's Disease Society of America, Inc.
140 W. 22nd Street
6th Floor
New York
NY 10011
United States

The Hereditary Disease Foundation
606 Wiltshire Boulevard
Suite 504 Santa Monica
California 90401
United States

Appendix 4

DRUGS USED TO TREAT CHOREA

Author (Full reference listed in Chapter 7)		No. of patients	Double-blind	Placebo	Rating	Effective
DRUGS AFFECTING DOPAMINERGIC TRANSMISSION						
Neuroleptics						
1. *Phenothiazines*						
(a) Aliphatic side chain						
chlorpromazine						
Azima and Ogle	1954	1	N	N	N	Y
Vaughan et al.	1955	3	N	N	N	N
(b) Piperazine side chain						
trifluoperazine						
Hawks and Silverstone	1962	1	N	N	Y	N
Cohen	1962	7	N	N	Y	Y
perphenazine						
Larecchia and Baseri	1958	1	N	N	N	Y
Pakenham-Walsh	1960	2	N	N	N	Y
Merskey et al.	1961	4	N	N	N	N
Straub and Melina	1962	16	N	N	?	Y
Fahn	1973	8	Y	N	Y	Y
thiopropazate						
Mathews	1958	10	N	N	N	Y
Fulghum et al.	1960	5	N	N	Y	Y
Lyon	1962	5	N	N	N	Y
Crawford	1969	2	N	N	N	Y
(c) Piperidine sidechain						
mesoridizine						
Lehnoff	1973	6	N	N	N	Y
2. *Butyrophenones*						
haloperidol						
Kent	1965	6	N	N	N	Y
Barr et al.	1988	20	N	N	Y	Y
trifluoperidol						
Tarighati and A'Brook	1968	4	N	N	N	Y

Appendix 4

Drugs used to treat chorea (*continued*)

Author (Full reference listed in Chapter 7)		No. of patients	Double-blind	Placebo	Rating	Effective
buronil						
Mattsson and Boman	1974	7	N	Y	Y	Y
3. *Benzamides*						
sulpiride						
Quinn and Marsden	1984	11	Y	Y	Y	Y
tiapride						
Girotti *et al.*	1984	12	Y	Y		
4. *Diphenylbutylpiperidines*						
pimozide						
Bobon	1968	1	N	N	N	N
Foy and Pakkenberg	1970	12	N	N	N	Y
5. *Benzoquinolizines*						
tetrabenazine						
Sattes	1960	6	N	N	N	Y
Brandrup	1960	2	N	N	N	Y
Sattes and Haze	1964	13	N	N	N	Y
Pakkenberg	1968	11	N	N	N	Y
Dalby	1969	8	N	N	Y	Y
Soutar	1970	2	N	N	N	Y
Swash *et al.*	1972	2	N	N	Y	Y
McLellan *et al.*	1974	9	Y	Y	Y	Y
Snaith and de B. Warren	1974	1	N	N	N	N
6. *Dibenzodiazepines*						
clozapine						
Caine *et al.*	1979	2	Y	Y	Y	Y
7. *Reserpine*						
Chandler	1955					
Morgan	1956	3				
Forrest	1957	4	Y	Y	Y	Y
Chhuttani and Singh	1959	3				
Brandrup	1960	7				
Kempinsky *et al.*	1960	10	N	N	Y	Y
Markham *et al.*	1963	10				
Mackiewicz and Reid	1965	4	N	N	N	N
DRUGS AFFECTING GABAergic TRANSMISSION						
GABA						
Fisher *et al.*	1974	7	N	N	N*	
aminooxyacetic acid						
Perry *et al.*	1980	7	Y	Y	Y	N

(*continued*)

Drugs used to treat chorea (*continued*)

Author (Full reference listed in Chapter 7)		No. of patients	Double-blind	Placebo	Rating	Effective
muscimol						
Shoulson et al.	1978	10	Y	Y	Y	N
γ-acetylenic GABA						
Tell et al.	1981	14	N	Y	Y	N
baclofen						
Shoulson et al.	1989	60	Y	Y	Y	N
isoniazid						
Perry	1977	6	N	N	N	Y
McLean et al.	1982	8	Y	Y	Y	N
Benzodiazepines						
diazepam						
Nilsen	1964	1	N	N	N	Y
clonazepam						
Peiris et al.	1976	3	N	N	Y	Y
Stewart	1988	1	N	N	Y	Y
DRUGS AFFECTING CHOLINERGIC TRANSMISSION						
choline						
Aquilonius and Eckernas	1977	5	N	N	Y	N
2-dimethylaminoethanol						
Walker et al.	1973	7	N	N	N	Y
Laterre	1975	7	N	N	N	N
Caraceni et al.	1978	9	Y	Y	Y	N
physostigmine						
Tarsy et al.	1973	6	N	N	Y	N
MISCELLANEOUS DRUGS						
biperiden						
Crawford	1969	1				
cysteamine						
Shults et al.	1986	5	Y	Y	Y	N
dextromethorphan						
Walker and Hunt	1989	11	N	N	Y	N
diltiazem						
Walker et al.	1985	11	Y	Y	Y	N†
dihydrochloride						
Souder	1959	3	N	N	N	Y
dipropylacetic acid						
Bachman et al.	1977	2	N	N	Y	N
hydrolazine						
Hoerster et al.	1961	5	N	N	N	Y‡
estrogen						
Koller et al.	1982	21	Y	Y	Y	N

Drugs used to treat chorea (continued)

Author (Full reference listed in Chapter 7)		No. of patients	Double-blind	Placebo	Rating	Effective
lithium						
Mattsson	1973					
Aminoff et al.	1974	9	Y	N	Y	N
Carmen et al.	1974	6	Y	N	Y	N
Dalen	1974	6	N	N	N	Y
Bleiweiss	1989	9	N	N	N	Y
methysergide						
Klawans et al.	1972	7	N	N	Y	N
penicillamine						
Haslam	1967	2	N	N	Y	N
procaine amide						
Cohen	1956	1	N	N	N	Y
Bruyn	1958	1	N	N	Y	Y
Merskey	1958	8	N	N	Y	N
Goldman	1952	9	N	N	N	Y

N, No; Y, Yes.
*Two patients improved.
†Improvement not statistically significant.
‡'Slight improvement'.

Appendix 5

CARDIFF PRESYMPTOMATIC TESTING PROTOCOL FOR HD

Inclusion criteria
   (1) Confirmed family history of HD
   (2) 50% *a priori* risk
   (3) Potentially informative family structure
   (4) Aged 18 or over
   (5) Consent freely given

Exclusion/postponement criteria
   (1) Clinically affected with HD
   (2) Significant risk of suicide
   (3) Recent history of significant mental illness
   (4) Recent history of significant drug misuse
   (5) Pregnancy (unless requesting prenatal diagnosis)

All interviews conducted by two counsellors
   Interview one (applicant asked to be accompanied by partner if appropriate)
      (1) Collection of sociodemographic details
      (2) Confirmation of family and clinical data
      (3) Assessment of impact (e.g. personal, financial, social) of HD on applicant and if appropriate on partner. Includes history of learning about HD
      (4) Assessment of knowledge of HD and presymptomatic testing
      (5) Reasons for requesting prediction and potential impact of result on applicant's life
      (6) Detailed explanation of how the test works, its possible outcomes and its limitations. Includes, if appropriate, consideration of insurance issues and problems of non-paternity
      (7) Discussion on how and from whom blood samples from key relatives will be obtained. Two separate blood samples required from key relatives. First sample from applicant may be taken
      (8) Neurological examination
      (9) Information sheet explaining the test and research protocol given to applicant
      (10) Applicants asked to indicate in writing if they wish to proceed with testing

   Interview two (arranged after DNA typing becomes informative, usually several months after interview one)
      (1) Explanation of the possible results (calculated by the programme MLINK–also includes age-modified risks)
      (2) Further counselling concerning the test and its potential impact. Includes to whom applicant plans to disclose the results
      (3) Review of informal social supports
      (4) Nomination of professional supporter who would be willing to offer support if and when necessary
      (5) Rehearsal of disclosure session and travel arrangements
      (6) Assessment of psychological, personality and social characteristics using standardized instruments (see below)
      (7) Signing of consent form. Copy given to applicant

(8) Final blood sample
(9) Specific appointment made for interview three (i.e. applicant not phoned when result is available) and applicant is reminded that withdrawal from testing can take place at any time

Instruments used in interview two
   A. Assessment of present and previous psychiatric morbidity
      (1) Past history schedule of the present state examination
      (2) Present state examination
      (3) General health questionnaire-60 (also given to partner)
      (4) Irritability, anxiety, depression scale
      (5) Beck depression inventory
      (6) Alcohol questionnaire
      (7) Illness behaviour questionnaire
   B. Assessment of personality
      (1) Eysenck personality questionnaire
      (2) Ways of coping scale
   C. Assessment of social adjustment and life events
      (1) Standardized interview to assess social maladjustment
      (2) Life events schedule

Interview three (held 4 weeks later)
(Applicant asked to be accompanied by partner, if appropriate, and another person who can provide support especially in the event of an adverse result)
Disclosure of results

Formal follow-up
   1 week (by telephone)
   3 months (clinic)
   12 months (clinic)

Index

Abortion, counselling, 355
Abstraction, Stroop Test, 115
Accommodation
  adapting family home, 225
  nursing homes, 226
  residential care, 225–226, 228
  specialized centres, 227
  voluntary and charitable homes, 227
Acetylcholine, levels in HD, 158–160
Addresses (*Appendix 3*), voluntary societies, 420–421
Adolescents
  counselling, 351–353
  testing for HD gene, 401–402
Adoption and fostering
  at-risk children, 195–196
  as options for high-risk family members, 354–355
Adrenal autografts, 215–216
Aetiology theories, 165–169
Affective (mood) disorders
  familial association, 94–95
  nature of association with HD, 95–96
  racial variation, 95
  see also Depression
Africa, epidemiology of Huntington's disease, 271–273
Age
  at death, 134–135
  at diagnosis, 133–134
  at onset, 127–132, 134
    age-at-onset curves, prediction of risks, 343, 345
    parent-child differences, 304
  risk curves, second-degree relatives, 346
  risk table, 347
  and survival of HD patients, 50–52
Aggressive behaviour
  legal aspects, 194–195
  management, 217
  psychiatric aspects, 101–102

Akinesia, in clinical picture, 44
Alcohol, misuse in HD, 105
Allowances and benefits, 229–231
Alzheimer's disease, 116–117
Animal models of Huntington's disease, 169–170
Aphasia, Boston Diagnostic Aphasia Examination, 113
Aphonia, genetic heterogeneity study, 302
Artificial insemination by donor, 355
Asia, various countries, epidemiology of Huntington's disease, 267–268, 269
Athetosis, paroxysmal choreoathetosis, 71–73
Attitudes, to predictive test, 379
Attorney, power of attorney, 239
Australia
  epidemiology of Huntington's disease, 268–270
  genetic heterogeneity, 301, 302
  self-help groups, 234
Autosomal dominant inheritance, criteria applied to HD, 282

Ballismus
  defined, 38
  and hemiballismus, 163
Basal ganglia, neural network, 164, 165
Bathing and toileting aids, 224
Benefits and allowances, 229–231
Benign familial chorea, differential diagnosis, 70–71
Blink reflexes, 172–173
Boston Diagnostic Aphasia Examination, 113
Bradykinesia
  in clinical picture, 44
  ratio of large:small neurones in lentiform nucleus, 148
Brain
  coronal sections, 145–148
  lateral view, 144

Brain—*cont.*
  midbrain pathology, and saccadic movements, 152–154
  pathology *see* Caudate nucleus; Lentiform nucleus; Striatum; Substantia nigra
Brain damage, psychometric tests, 108–110
Brain size, 143–144
Brain stem, microscopic changes in HD, 150
Brain stem auditory evoked potentials, 172
Bulimia, 105

Cachexia, 47–48
Canada
  epidemiology of Huntington's disease, 260
  self-help groups, 234
Carers
  attendants, 219
  care attendant schemes, 236
  day care services, 219–220
  effects of HD, 191–192
  'good neighbour schemes', 236
  Griffith Report on Community Care, 235
  holiday breaks, 236
  laundry services, 219
  management, 234–237
  National Institute of Social Work, 235–236
  non-contributory benefits, 231
  patterns of care, 186–190
  respite, 236
Caudate nucleus
  atrophy, pathology, 143–154
  'C'-shaped development, 142
  hypometabolism, 65
  microscopic change, 149
  neuropathological grading, 154–156
  nucleus accumbens, nuclear membrane pathology, 157–158
Cell implantation
  adrenal autografts, 215–216
  fetal brain cells, 216
Centennial Symposium on Huntington's disease, 20–21
Cerebellum
  abnormalities, 49–50
  loss in Purkinje cells, 151–152
  microscopic changes in HD, 150
Cerebral cortex, microscopic changes in HD, 150
Charcot-Marie-Tooth disease, and HD, 300
Children
  adoption and fostering, 195–196
  counselling, 351–353
  and predictive tests, 402–403

Choline acetyltransferase, levels in HD, 160
Cholinergic therapies, 211–212
Chorea
  antichoreic agents, 207
  benign familial chorea, 70–71
  chorea lasciva, 8
  chorea scale *see* Neurological examination, quantified
  chorea-acanthocytosis, 72–73
  pathology, 153
  clinical neurology, 38–39
  early concepts, 7–9
  paroxysmal choreoathetosis, 71–73
  'senile' chorea, 52
  Sydenham's chorea, 69
  tardive dyskinesia, 69–70
  various causes, 67–68
'Choreopathy', historical notes, 103
Chromosomes
  abnormalities, 298–300
  flow-sorting, 324
  linking and hopping, 328
  see also Genetic aspects of Huntington's disease
Chronically Sick and Disabled Persons Act (1970), 238
Clinical assessment
  cachexia, 47–48
  consent to treatment, 237–238
  functional assessment scale (*Appendix 2*), 418–419
  historical background, 15–18
  pattern of progression of HD, 44
  quantitative scales, 59–61
  suspected HD patient, 57–59
Clinical investigation, suspected HD patient
  DNA analysis, 62–63
  EEG, 62
  functional imaging, 63–65
  neuropathology, 65–67
  structural imaging, 63
Clinical neurology
  cachexia, 47–48
  cerebellar abnormalities, 49–50
  chorea, 38–39
  dysarthria, 46–47
  dysphagia, 47
  elderly patients, 50–52
  epilepsy, 49
  eye movements, 45–46
  gait, 45
  incontinence, 48–49
  juvenile Huntington's disease, 52–57
  motor abnormalities, 39–45

Index

sleep, 48
tendon reflexes and pyramidal tract
      abnormalities, 50
Clinical staging
   and neuropathological grading, 154–157
   and rate of decline, 136–137
Cognitive tests, as predictive test, 375–376
Committee to Combat Huntington's disease,
      21–22, 23
Community psychiatric nurses, 221–222
Consent to treatment
   admission for assessment, 237–238
   legal aspects, 237–238
Corticospinal pathway, integrity, testing, 173
Cost of Huntington's disease, 190
Counselling
   social aspects of HD, 200–201
   see also Genetic counselling
Crime and antisocial behaviour, 194–195
CT scan
   FH/CC ratio, 63
   as predictive test, 376
   suspected HD patient, 63

D4S10 locus see G8 marker
Dancing Mania, 8–9
Davenport, Charles, contribution to HD, 19
Day care, 219–220
Death
   age at death, 134–135
   causes, 137–138
   death rates
      England and Wales, 255, 256
      various countries, 255
Delinquency, crime and antisocial behaviour,
      194–195
Delusional disorders
   induced, 92
   persistent, 89–92
Dementia
   Alzheimer's disease, 116–117
   atypical motor neurone disease, 118
   causes, 116
   cortical and subcortical, differential
      diagnosis, 112
   definition and criteria, 107–108
   differential diagnosis, 116
   general paresis, 118
   insight, 116
   language, 113–114
   memory, 111–113
   mental set, 115–116
   Pick's disease, 117–118

presenile, main features, 116
prevalence, 108–110
psychometric tests, 108–110
subcortical dementia, 110–111
visuospatial praxis, 114–115
Dentato-rubro-pallido-luysian atrophy
   clinical features, 74
   decreased GABA, 164
   genetic heterogeneity study, 303
Depression
   classification, 93
   drug therapies, 213–214
   familial association, 94–95
   nature of association with HD, 92–96
      coincidental hypothesis, 95–96
      organic hypothesis, 96
      reactivity hypothesis, 95–96
   prevalence, 94
   racial variation, 95
   and suicide, 96–99
Differential diagnosis, 67–75
DNA analysis
   suspected HD patient, 62–63
   see also Genetic aspects; Gene mapping
DNA markers, closely linked to HD gene, 386
Dominant inheritance, criteria applied to HD,
      282
l-Dopa test, as predictive test, 377
Dopamine
   antidopaminergic therapies, 209
   levels in HD, 158–160
   postsynaptic dopamine blockade, 209
   presynaptic depletion, 209
Driving, medical fitness, 240–241
DRPLA see Dentato-rubro-pallido-luysian
      atrophy
Drugs
   antichoreic agents, 206–209
   antidopaminergics, 209–210
   antipsychotics, 214, 248–249
      WHO classification, 207
   benzodiazepines, 211, 213, 424
   choice, 208
   cholinergic therapies, 211–212, 424
   cysteamine, 212, 424
   extrapyramidal rigidity as a side-effect, 213
   GABAergics, 210–211, 424
   haloperidol, 207, 209
   historical notes, 206
   lists (*Appendix 4*), 422–425
   lithium salts, 212, 425
   neuroleptics, 207–208, 422
   neurotoxins, 212
   phenothiazines, 207, 422

Drugs—*cont.*
    replacement therapies, 208–209
    reserpine, 207, 209, 423
    steroids, 212
    sulpiride, 209
    tetrabenazine, 209
Dystonia *see* Myoclonic dystonia; Torsion dystonia
Dystonic movements, defined, 43

Ekbom's syndrome, 91–92
Electromyography, as predictive test, 376
Electrophysiology
    brain stem auditory evoked potentials, 172
    cognitive assessment, 173
    EEG as predictive test, 376
    experimental observations, 171–173
    suspected HD patient, 62
    synaptic evoked potentials, 172
    visual evoked potentials, 172
Epidemiology of Huntington's disease
    analysis of death rate, 254–256
    family studies, 252–253
    migration and spread of the HD gene, 273–276
    other inherited disorders, 253–254
    various countries, 256–273
Epilepsy, 49
Eugenics, abuse of genetic knowledge of HD, 364–369
Europe, various countries, epidemiology of Huntington's disease, 266–267
Evoked potentials *see* Electrophysiology
Excitotoxic theory of HD, 165–168
    neurotoxins, 212
Extrapyramidal rigidity, as a side-effect of drug therapy, 213
Eye movements
    as predictive test, 375
    scale *see* Neurological examination, quantified

Family, effects of HD, 192–194
Feeding
    aids, 222
    dietitians, 222
Fertility and genetic fitness, 294–298
Fetal brain cells, cell implantation, 216
Flanking marker DNA probe, 325, 395
Fleas and other insects, Ekbom's syndrome, 91
Fluorodeoxyglucose, in PET scan, 65

Functional assessment scale (*Appendix 2*), 418–419

G8 probe
    closely linked DNA markers, 386
    and prediction, 298, 302, 321–327, 384–386
GABA
    GABAergic therapies, 210
    levels in HD, 158–160
    neural network in basal ganglia, 164
    in spiny neurons, 158–159
GABAergics, 210–211, 423–424
GAD *see* Glutamic acid decarboxylase
Gene mapping
    fine mapping of HD locus, 324–327
    first localization of HD gene (on chromosome 4), 321–327
    candidate genes and the nature of HD mutation, 331–334
        conflicting results, 328–329
        current research efforts, 330–331
        linkage disequilibrium, 329–330
        new technology, 327–328
    Huntington's disease linkage, 320–324
    linkage, 317–319
    lod score, 318
    nature of HD gene
        'gain of function' mutation, 332
        lack of direct test, 333
        and parental origin effect, 332–333
        'position-effect variegation', 332
        true dominance of HD gene, 332
    restriction fragment length polymorphism, 319–320
General paresis, dementia, 118
Genetic aspects of Huntington's disease
    chromosomal abnormalities, 298–300
    closely linked DNA markers, 386
    fertility and genetic fitness, 294–298
    genetic heterogeneity, 300–303
    heterozygote and gene frequency, 292–294
    homozygosity for HD gene, 289–292
    Mendelian laws, 282–285
    modifying genes, 303–305
    new mutation cases, 286–289
    sex-related effects, 305–311
        paternal inheritance of juvenile HD, 306–311
    twin studies, 285–286
    see also Gene mapping; Genetic counselling
Genetic counselling
    abuse of genetic knowledge, 364–369
    adolescents, 351–353

Index

communication and counselling, 345–348
defined, 337
diagnostic difficulties, 338–341
extended family, 349–361
genetic registers, 358–361
options for high-risk family members
  abortion, 355
  adoption and fostering, 354–355
  AI (donor), 355
  presymptomatic and prenatal testing, 355–356
  sterilization, 353
pedigree analysis and risk estimation, 341–345
population prevention of HD, 361–364
presence of HD in applicant, 340
  Quantitative Neurological Examination, 341
principles, 337–338
support following counselling, 348–349
systematic studies, 356–358
Genetic imprinting, 308
Genetic linkage *see* Predictive tests
Germany
  epidemiology of Huntington's disease, 265
  historical notes, abuse of genetic knowledge, 364–369
Glial cells
  abnormalities, 157
  glial index, 149
Globus pallidus
  microscopic changes in HD, 150
  neurone and volume reduction, 150
  neuropathological grading, 154–156
Glucose intolerance, HD patients, 171
Glutamate
  excitotoxic theory of HD, 165–168
  neurotoxicity, 167
Glutamic acid decarboxylase
  activity in rat model, 169
  loss in HD, 159
Griffith Report on Community Care, 235
Guardianship, powers of guardians, 238–239
Guthrie, Woody, 21–22

Halstead-Reitan Battery, 110
Health care
  bathing and toileting aids, 224
  community psychiatric nurses, 221–222
  dietitians, 222
  district nurses, 221
  family doctors, 221
  health visitors, 221

incontinence aids, 224
occupational therapists, 223
physiotherapists, 223
rehabilitation aids, 223–224
speech therapists, 222–223
wheelchairs, 223–224
Hemiballismus, lesions of subthalamic nucleus, 163
Hereditary ataxias, differential diagnosis, 75
Hereditary Disease Foundation, Venezuela Project, 23
Historical background
  clinical picture, 15–18
  descriptions of HD (before 1872), 9–12
  development of research, 20
  early concepts of chorea, 7–9
  George Huntington
    hereditary chorea, 3–5
    life and background, 6–7
  inheritance, 18–19
  juvenile Huntington's disease, 17
  landmarks in study of HD, 20
  neuropathology, 18
  psychiatric syndromes, 16
  rigid form of HD, 17
  spread of knowledge on HD, 13–14
    New England, 14–15
    various countries, 13
  treatment, 18
  Venezuela Project, 23–25
  Wales project, 25–31
  William Osler and HD, 12–13
Holland, genetic heterogeneity, 301
Home helps, and 'meals on wheels', 218–219
Huntington, George life and background, 6–7
Huntington's disease
  research *see* Wales, (*Appendix 1*)
  self-help groups, 232–234
3-Hydroxyanthranilate oxygenase, increase in HD brains, 167
Hypothalamus, lateral tuberal nucleus, microscopic changes in HD, 150

Imaging techniques, predictive tests, 404
Incontinence, 48–49
Incontinence aids, 224
Induced delusional disorder, 92
Inheritance
  George Huntington *quoted*, 18
  historical background, 18–19
Insight, retention, 115
Insurance, 241–242

Intelligence, Wechsler Adult Intelligence
    Scale, 109
International Huntington Association, 233
Irritability, 101–102

Japan
  dentato-rubro-pallido-luysian atrophy, 74
  epidemiology of Huntington's disease, 270
  fertility and genetic fitness, 296
Juvenile Huntington's disease
  age at diagnosis, 133–134
  age at onset, 129–132
  clinical neurology, 52–57
  course, 56
  distinguishing features, 55
  genetic aspects, 56
  historical background, 17
  paternal inheritance of juvenile HD, 306–311
  prediction of risk, age-at-onset curves, 345

Kainic acid, excitotoxic theory of HD, 165–168
Kynurenine pathway, tryptophan metabolism, 166, 168

Laboratory investigations, 61–67
  see also Clinical investigations
Landmarks in study of HD, 20
Language, 113–114
Late onset see Age of HD patients
Laundry services, 219
Legal aspects, management, 237–242
Lentiform nucleus
  atrophy, pathology, 143–154
  development, 142
  large:small neurones, ratio, 148
Lipofuscin pigments, accumulation, 157
Lod score, gene mapping, 318
Lund, Johan, *quoted*, 10–11
Lyon, Irving, *quoted*, 11–12

Machado Joseph disease, differential diagnosis, 75
Madelung deformity, 300
Magnetic resonance imaging, suspected HD patient, 63
Management
  accommodation, 225–228
  carers, 234–237
  legal aspects, 237–242
  social welfare entitlements, 228–231

special factors, 205
specific approaches to treatment
  memory aids, 218
  neurosurgery, 215–216
  pharmacotherapy, 206–214
  psychological treatment, 216–218
support services
  health services, 221–224
  social services, 218–221
voluntary agencies, 231–234
Mania, 99–100
Mapother, Edward, *quoted*, 16
Marital breakdown
  ignorance of hereditary implications, 186
  South Wales sample, 185–186
Maternal factor hypothesis, 307
Medical fitness to drive, 240–241
Memory, 111–113
Memory aids, 218
Mental Health Act (1983), 237–238
Mental set in dementia, 115–116
Met-enkephalin fibres
  early loss, 159, 163
  neural network in basal ganglia, 164, 165
Metabolism, defects outside brain, research, 174
N-Methyl-d-aspartate-receptor
  drug therapies, 212
  excitotoxic theory of HD, 166, 167–168
  and quinolinic acid agonist, 166
Microscopic changes, Huntington's disease, summary, 150
Mini-Mental State Examination, 109
Minnesota Multiphasic Personality Inventory, as predictive test, 375
Misdiagnosis, psychiatric assessment, 84
Mitochondrial factor hypothesis, 307
Motor abnormalities other than chorea, 39–45
Motor impairment scale see Neurological examination, quantitative
Motor neurone disease, atypical in dementia, 118
Muscle tremor, as predictive test, 374–375
Mutations
  estimates of mutation rate, 288–289
  new, criteria applied to HD, 287
Myoclonic dystonia, differential diagnosis, 38, 75

NADPH-diaphorase
  levels in HD, 160

neurone resistance to quinolinic acid
   toxicity, 167
National Assistance Act (1948, Amendment
   1951), 238
National Institute of Social Work, 235–236
Natural history of Huntington's disease
   age at death, 134–135
   age at diagnosis, 133–134
   age at onset, 127–132, 134, 304, 343, 345
   causes of death, 137–138
   clinical staging and rate of decline, 136–137
   survival, duration, 135–136
Neuro-acanthocytosis, 73–74
Neuroendocrine abnormalities, 170–171
Neurofibromatosis and HD, 300
Neurological disorders, estimates of mutation
   rate, 289
Neurological examination
   points of particular diagnostic significance,
      58
   quantitative, 59–61
Neurology *see* Clinical
   neurology
Neuropathology
   aetiology theories, 165–169
   animal models, 169–170
   basal ganglia, 142–143
   changes outside the brain, 173–174
   excitotoxic theory of HD, 165–168
   experimental electrophysiological
      observations, 171–173
   grading, 154–157
      and clinical staging, 154–157
   historical background, 18
   indications and importance, 66
   loss of spiny neurones in HD, 158–159
   microscopic changes, summary, 150
   neuroendocrine abnormalities, 170–171
   neuropathological grading, 154–157
   pathological changes, 143–154
   quinolinic acid model, 168
   selective neuronal loss in the striatum, 158–
      165
   suspected HD patient, 65–67
   ultrastructural changes, 157–158
Neuropeptide Y, levels in HD, 160
Neurosurgery, 215–216
Neurotransmitters
   drug replacement therapies, 208–209
   striatum, 158–161
NMDA *see* N-Methyl-*d*-aspartate
Nucleus accumbens, caudate nucleus, nuclear
      membrane pathology, 157–158
Nursing homes, 226

Occupational therapists, 223
Osler, William
   and Huntington's disease, 12–13
   quoted, 5, 12–13
Othello syndrome, 90–91

Pacific Islands, epidemiology of Huntington's
      disease, 270
Paracelsus, on chorea, 8
Paranoia, 89–90
Parkinson's disease, differential diagnosis, 75
Paroxysmal choreoathetosis, 71–73
Patient's property and affairs
   appointee, 240
   civil law relating to patient's property, 239
   Court of Protection, 239
   power of attorney, 239
   testamentary capacity, 240
Pedigree analysis, and risk estimation, 361–
      364
Persistent delusional disorders
   Ekbom's syndrome, 91–92
   Othello syndrome, 90–91
   paranoia, 89–90
Personality disorder
   antisocial (psychopathic) personality, 104
   at-risk individuals, 104–105
   historical notes, 102–103
   recent studies, 103–104
   standardized tests, 100–101
Personality tests, as predictive test, 375–376
PET scan, suspected HD patient, 63–65
Pharmacotherapy, 206–214
Photon emission tomography, caudate
      nucleus hypometabolism, 65
Physiotherapy, 223
Pick's disease, 117–118
Polymorphisms, at G8 locus, 385
Population prevention of HD, 361–364
Power of attorney, 239
Predictive tests
   accuracy, 395
   attitudes of families, 378
   children, 401–402
   clinical and counselling aspects, 386–391
   early studies of presymptomatic testing,
      391–396
   future developments, 407–409
   and genetic linkage, 378–386
   historical notes, 373–378
   imaging techniques, 403–404
   molecular techniques, 403

Predictive tests—*cont.*
  prenatal diagnosis, 355–356, 404–406
  problems, 397–401
  reasons for requesting, 393
  results, seven centres, 393
  selection and exclusion, 391
  as a service, 406–407
Presymptomatic and prenatal testing, 355–356, 404–406
Presymptomatic testing protocol, Wales, (*Appendix 5*), 426–427
Prevalence estimates for Huntington's disease, UK, 263, 264
Primates *see* Animal models
Property
  civil law relating to patient's property, 239
  Court of Protection, 239
Psychiatric assessment
  cognitive assessment, 82
  initial manifestations of HD, 82–84
  misdiagnosis, 84
  taking a history, 81–82
Psychiatric disorder
  affective (mood) disorders, 92–96
  classification, 85–86
  deliberate self harm, 99–100
  dementia, 107–118
  irritability, 101–102
  mania, 100–101
  persistent delusional disorders, 89–92
  personality disorder, 102–105
  prevalence in HD, 84–85
  psychodynamics, 106
  schizophrenia, 86–89
  sexual disorders and deviations, 106
  and suicide, 96–99
Psychological treatment, 216–218
  drug therapies in psychosis, 207, 214
Pulsed-field gel electrophoresis, 327–328
Purkinje cells, loss in cerebellum, 49–50, 151–152
Putamen *see* Lentiform nucleus

Quinolinic acid agonist, and *N*-methyl-*d*-aspartate, 166
Quinolinic acid model, problems, 168

Racial origin, and HD, 271, 272–273
Rare-cutter enzymes, 327–328
Reality Orientation Therapy, 218
Residential care, 225–226, 228
  see also Accommodation

Restriction fragment length polymorphism, 319–320
Rigid form of HD
  clinical neurology, 52–53
  historical background, 17
  paternal transmission, 306
  prevalence, 43
Rigidity, extrapyramidal rigidity as a side-effect of drug therapy, 213
Risk estimation
  and pedigree analysis, 361–364
  see also Genetic counselling

Saccadic movements, 46
  and midbrain pathology, 152–154
St Vitus' dance, origins of term, 9
Schizophrenia
  historical notes, 86
  misdiagnosis, 87
  precursor of HD, 89
  prevalence in HD, 87–89
Self harm, deliberate, 99–100
Sexual disorders and deviations, 105
Sleep function, 48
Social aspects of HD
  adoption and fostering of at-risk children, 195–196
  counselling, 200–201
  delinquency, crime and antisocial behaviour, 194–195
  economic cost, 190
  effects on carer, 191–192
  effects on family, 192–194
  effects on patient, 190–191
  experience of being at risk, 196–198
  patterns of care, 186–190
  social management, 198–200
  South Wales sample, 180–186
Social welfare
  contributory benefits, 229
    Invalidity Benefit, 229
    Sickness Benefit, 229
    Statutory Sick Pay, 229
  entitlements, 228–231
  non-contributory benefits
    Attendance Allowance, 229–230
    Dependent Relatives' Allowance, 231
    home responsibilities protection, 231
    hospital travel costs, 230–231
    Invalid Care Allowance, 231
    Mobility Allowance, 230

Severe Disablement Allowance, 231
  tax allowances, 231
  social services, 218–221
Somatostatin, levels in HD, 160
Somatostatin-neuropeptide Y-NADPH-
    diaphorase
  aspiny interneurones, 159, 160, 162–163
  selective sparing, lack of reproducibility, 168
South Africa, epidemiology of Huntington's
    disease, 271–273
South America, epidemiology of Huntington's
    disease, 260–262
Spastic paraplegia, and HD, 300
SPECT *see* Photon emission tomography
Speech
  prosody, 114
  speech profile ratings, 113
  speech therapists, 222–223
Spinal cord, microscopic changes in HD, 150
Stereotactic surgery, 215
Sterilization, 353, 356
Striatum
  aspiny interneurones, 159, 160, 162–163
  classification of neurones, 158
  dopaminergic nigrostriatal pathway, 160
  excitotoxic theory of HD, 165–168
  glial cells, 149
  ligand binding to dopaminergic receptors,
    161
  loss of dopaminergic receptors, 161–162
  microscopic changes in HD, 150
  morphological studies, 158
  neuropathological grading, 155
  neurotransmitter studies, 158–161
  patch-matrix compartments, 161–162
  pathology, 143–154
  selective neuronal loss, 158–165
Stroop Test, abstraction, 115
Substance P, striatum, 163
Substantia nigra
  microscopic changes in HD, 150
  neuron and volume reduction, 150–151
Suicide
  defined, 96
  and parasuicide, 99–100
  rates, various countries, 97–99
Surgery
  cell implantation, 215–216
  procedures and results, 217
  stereotactic surgery, 215
Survival, from diagnosis, duration, 135–136
Sydenham, Thomas, and St Vitus' dance, 9
Sydenham's chorea, differential diagnosis, 69
Symposium, centennial, 20–21

Tardive dyskinesia, differential diagnosis, 69–
    70
Testamentary capacity, 240
Thalamus, neural network, 165
Thalamus and subthalamus
  atrophy, 150
  microscopic changes in HD, 150
Tics, defined, 38
Torsion dystonia, differential diagnosis, 75
Trail Making Test, 115
Treatment
  historical background, 18
  see also Management
Tryptophan metabolism, kynurenine
    pathway, 166, 168
Twin studies, 285–286
Tyrosine hydroxylase
  and dopamine, 161
  excitotoxic theory of HD, 166–168

United States
  counselling, and predictive tests, 387
  epidemiology of Huntington's disease, 257–
    259
  fertility and genetic fitness, 295–296, 297
  genetic heterogeneity study, 302
  HD in New England, 14–15
  paternal transmission study, 307
  self-help groups, 233–234

Venezuela, epidemiology of Huntington's
    disease, 260–262
Venezuela Project
  Hereditary Disease Foundation, 23
  historical background, 22–24
Visuospatial praxis in dementia, 114–115
VNTRs *see* Restriction fragment length
    polymorphisms
Voluntary agencies, 232–234
Voluntary societies, addresses (*Appendix 3*),
    420–421

Wales study
  genetic heterogeneity, 301
  genetic registers, 358–360
  HD research, 25–31
  HD research D chronological list,
    (*Appendix 1*), 415–416
  long-term trends in births at risk, 364–365
  migration and spread of the HD gene,
    274–276

Wales study—*cont.*
  presymptomatic testing protocol, 388–391
    (*Appendix 5*), 426–427
    flow-chart, with numbers at each stage, 394
    selection and exclusion, 391
    social data on applicants, 392
  prevalence estimates, 263
  South Wales sample
    age and sex, 181
    education, 182–183
    employment, 183–185
    marital breakdown, 185–186
    social class, 182
Waters, Charles, *quoted*, 10
Wechsler Adult Intelligence Scale
  HD patients, 109
  as predictive test, 375
'Westphal' variant *see* Rigid form of HD
Wheelchairs, 223–224
Will, testamentary capacity, 240
Wilson's disease, differential diagnosis, 75
Wisconsin Card Sorting Test, 115
Wolf–Hirschhorn syndrome
  chromosome 4 deletion, 323, 326
  clinical features, 298–299
Writing disturbances, 114

Yeast artificial chromosome system, cloning human DNA, 329–331